*Principles of Ideal-Fluid
Aerodynamics*

Principles of Ideal-Fluid Aerodynamics

Krishnamurty Karamcheti

Professor of Aeronautics and Astronautics
Stanford University

ROBERT E. KRIEGER PUBLISHING COMPANY
MALABAR, FLORIDA

Original Edition 1966
Reprint Edition 1980 (with corrections)

Printed and Published by
ROBERT E. KRIEGER PUBLISHING COMPANY, INC.
KRIEGER DRIVE
MALABAR, FLORIDA 32950

Library of Congress Cataloging in Publication Data

Krishnamurty Karamcheti.
 Principles of ideal-fluid aerodynamics.

 Reprint of the edition published by Wiley, New York.
 Bibliography: p.
 Includes index.
 1. Aerodynamics.   I. Title.
[QA930.K74 1980]   533'.62'0151   79-26876
ISBN 0-89874-113-0

TO
MY GURU
HANS WOLFGANG LIEPMANN

Preface

The aim of this book is to explain the basic principles and analytical methods underlying the theory of the motion of an ideal fluid (an inviscid incompressible fluid) and the role of the theory in describing and predicting the flows associated with the motion of certain bodies of aerodynamic interest such as wings and bodies of revolution. I have attempted to describe ideal-fluid aerodynamics, although restricted to certain problems, as a branch of theoretical physics.

The subject is developed from basic principles showing clearly the complementary features of physical understanding and the mathematical handling of the theory. The intention is to show the role of physical understanding in mathematical formulation, to bring out the motivation for the mathematical language and methods employed and the necessity for applying a certain amount of mathematical rigor in arriving at physically appealing solutions.

The book is written to serve as a self-contained text at the senior undergraduate or first-year graduate level. The idea is not to give inadequately explained solutions to many special problems, but rather to present, for a few selected practical problems, a unified treatment leading from basic principles to practically meaningful results. A large part of the book deals with general concepts and mathematical methods, always related, however, to the solution of problems. In this way I hope that the book will perform the valuable function of teaching subject matter related to a broader methodology that will lead logically to more advanced topics and methods in fluid mechanics; it should be of interest to students in various disciplines, such as applied mathematics, physics, and engineering.

This book has grown out of lectures on aerodynamic theory which I have offered for the last decade and which have been received with considerable enthusiasm. It is because of the students' encouragement that I venture to publish them.

I am greatly indebted to Professor Irmgard Flügge Lotz for reviewing the manuscript and for many valuable suggestions and discussions. My special thanks are due to Dr. Maurice L. Rasmussen who read the manuscript and offered valuable criticism. Many students have helped me enthusiastically with the preparation of the book, and my deep appreciation goes to all of them.

I am very grateful to Professor O. G. Tietjens for furnishing me with original prints of many of the flow photographs.

The original source of the photographs for the plates 8 and 9 is the National Physical Laboratory, England, and I am greatly obliged to its Director for permission to reproduce the photographs which are Crown copyright. The original photographs for plates 3, 4, 6a, and 7 are all from prewar German publications, and I wish to record my indebtedness to their respective sources. Plates 3 and 4 are after F. Homann, *Forschung auf dem Gebiete des Ingenieurwesens*, 7 (1936). Plate 6a is after L. Prandtl, *Handbuch der Experimentalphysik*, 4, Part 1 (Leipzig, 1931). Plate 7 is after Prandtl, *The Physics of Solids and Fluids* (London, 1930).

The typing was capably handled by Mrs. Katherine Bradley, Miss Gail Lemmond, and Mrs. Elaine Morris. My sincere thanks to them.

Finally I wish to express my appreciation to John Wiley and Sons for the understanding, patience, and encouragement they have extended me over the years.

Krishnamurty Karamcheti

Stanford, California
August 1966

Contents

1. INTRODUCTION 1

1.1 Fluid as a Continuous Medium 2
1.2 Properties of a Fluid at Rest: Thermodynamic properties; Compressibility; Incompressible fluid; Heat conduction and the coefficient of thermal conductivity 2
1.3 Properties of a Fluid in Motion: Friction or viscosity; Coefficient of viscosity; Compressibility; Heat transfer. 4
1.4 Laminar and turbulent motions 10
1.5 Some Relevant Parameters: Relative magnitude of the forces, Froude number, Reynolds number, and Mach number; Parameters characterizing compressibility; Prandtl number; Parameters on which force and heat transfer depend 13
1.6 Range of Some Parameters 23
1.7 Conditions for Neglecting Compressibility Effects; Case of liquids; Case of gases 23
1.8 Conditions for Neglecting Gravity Effects 25
1.9 Nature of the Problem when Compressibility Effects are Negligible 25
1.10 Variation of Flow Patterns with Reynolds Number; Flow past bluff bodies; Flow past streamlined bodies 26
1.11 Variation of Flow Pattern with Mach number 35
1.12 Effects of Viscosity at High Reynolds Numbers: The Boundary layer: Boundary layer concept; Some characteristics of the laminar boundary layer; Turbulent boundary layer; Separation; Wakes 40
1.13 Consequences of the Boundary-Layer Concept 52
1.14 Ideal Fluid Theory 55

2. ELEMENTS OF VECTOR ALGEBRA AND CALCULUS . . 56

2.1 Representation of a Vector 57
2.2 Addition and Subtraction 58
2.3 Definition of a Vector 60
2.4 Multiplication by a Number 60
2.5 Unit Vector 61
2.6 Zero Vector 61
2.7 Scalar Product of Two Vectors 61

2.8 Vector Product of Two Vectors 62
2.9 Plane Area as a Vector 65
2.10 Velocity of a Point of a Rotating Rigid Body 66
2.11 Polar and Axial Vectors 69
2.12 Multiple Products: Scalar triple product; Vector triple
     product 69
2.13 Components of a Vector 72
2.14 Specification of a Vector 74
2.15 Cartesian Coordinates and the **i, j, k** System of Unit Vectors 75
2.16 Notion of Curvilinear Coordinates 77
2.17 Orthogonal Curvilinear Coordinates: Examples—Cylindrical
     and spherical coordinates 79
2.18 Products of Vectors in Terms of Their Components 83
2.19 Functions Involving Vectors and Scalars 84
2.20 Scalar and Vector Fields 87
2.21 Differentiation of a Vector Function of a Scalar Variable . . 89
2.22 Changes in the Unit Vectors of Cylindrical and Spherical
     Coordinates 94
2.23 Frames of Reference 96
2.24 Differentiation of a Scalar Function of a Vector: Concept of
     a gradient 98
2.25 Differentiation of a Vector Function of a Vector: Concept of
     tensor gradient of a vector; Relation of divergence, strain,
     rotation, and curl to the tensor gradient 103
2.26 Del, the Vector Differential Operator 108
2.27 Integration of a Vector Function of a Scalar 111
2.28 Line Integrals: Circulation 111
2.29 Surface Integrals 113
2.30 Volume Integrals 115
2.31 Integral Definition of the Gradient 116
2.32 Divergence of a Vector Field 119
2.33 Curl of a Vector Field 120
2.34 Components of a Curl as Circulation 125
2.35 Some Related Remarks 128
2.36 Relations Between Surface and Volume Integrals: Gradient
     theorem; Divergence theorem 130
2.37 Theorem of Stokes 131
2.38 Further Operations 133
2.39 Laplace Operator 133
2.40 Green's Theorem 134
2.41 Irrotational Field: Scalar potential 136
2.42 Solenoidal Field: Vector potential 137
2.43 Laplace's Equation 138
2.44 Poisson's Equation 139
2.45 Expressions in General Orthogonal Curvilinear Coordinates:
     Unit vectors; Infinitesimal distance between two neighboring

points; Differential volume and surface elements; Gradient;
Divergence; Curl; Laplacian 140
2.46 Some Useful Relations 146

3. STRESS IN A FLUID 148

3.1 Surface Forces and Body Forces 148
3.2 Concept of Stress and the Specification of Stress at a Point . 149
3.3 Stress in a Fluid at Rest: Hydrostatic pressure 153
3.4 Stress in a Fluid in Motion 154
3.5 Stress in a Non-Viscous Fluid in Motion: Pressure 155
3.6 Pressure Distribution in a Fluid at Rest 155
3.7 Concluding Remarks 157

4. DESCRIPTION OF FLUID MOTION 158

4.1 Lagrangian Method 158
4.2 Eulerian Method 159
4.3 Connection between the Lagrangian and Eulerian Descriptions 160
4.4 Steady and Unsteady Motions 162
4.5 Path Line 162
4.6 Streamlines 162
4.7 Stream Surfaces and Stream Tubes 164
4.8 Reference Frame and Streamline Pattern 165
4.9 Stream Functions 165
4.10 Stream Function For Two-dimensional Flow 168
4.11 Stream Function For Axisymmetric Motion 170
4.12 Stagnation Points 172

**5. EULERIAN EQUATIONS
FOR THE MOTION OF AN IDEAL FLUID** 175

5.1 Local, Convective and Material Derivatives 175
5.2 Euler's Equation 178
5.3 Equation of Conservation of Mass: Change of volume of a
    fluid element 181
5.4 Equation of Energy 184
5.5 Equation of State 186
5.6 Equations for an Inviscid Compressible Fluid 187
5.7 Condition of Incompressibility 187
5.8 Consequences of Incompressibility 188
5.9 Equations for an Ideal Fluid 190
5.10 Initial Conditions 190
5.11 Boundary Conditions for an Ideal Fluid: Condition at a solid
    fluid boundary; Condition at a free surface 190
5.12 Conditions at Infinity 194

5.13 Stream Functions for Incompressible Flow 194
5.14 Vector Potential for Incompressible Flow and Its Relation in
the Stream Functions 196
5.15 Elimination of the Body Forces from the Equation of Motion
for a Certain Incompressible Flow Problem 197

6. ALTERNATE FORMS OF THE EQUATIONS 198

6.1 Equation of Change 198
6.2 Conservation of Mass 199
6.3 Conservation of Momentum 199
6.4 Conservation of Energy 201
6.5 Integral Form of the Equations From the Point of View of a
Fixed Region of Space 202
6.6 Integral Form of the Equations From the Point of View of a
Finite Fluid Region 203
6.7 Rate of Change of a Quantity Following a Fluid Region . . 205
6.8 Equations of Change of an Ideal Fluid 207
6.9 Rate of Change of a Quantity Following a Moving Region of
Space 207

7. EQUATIONS OF DISCONTINUOUS MOTION 210

7.1 A Stationary Discontinuity in a Steady Flow 210
7.2 A Moving Discontinuity in the Unsteady Flow of an Ideal
Fluid 213
7.3 Discontinuity in the Flow of an Inhomogeneous Incompressible
Fluid 217
7.4 Remarks 219

8. INTEGRATION OF EULER'S EQUATION
IN SPECIAL CASES 221

8.1 Mathematical Character of the Equations 221
8.2 Integration of Euler's Equation in Steady Rotational Motion 223
8.3 Spatial Variation of H_S 224
8.4 Integration of Euler's Equation in Irrotational Motion . . . 226
8.5 Remarks on an Irrotational Force Field 227
8.6 Remarks on Bernoulli's Equation 229

9. IRROTATIONAL MOTION 231

9.1 Most General Motion of a Fluid Element 231
9.2 Rotation and Vorticity 233
9.3 Circulation and Vorticity 235
9.4 Rate of Change of Vorticity: Helmholtz's Theorem 236

9.5   Rate of Change of Circulation: Kelvin's Theorem   239
9.6   Irrotational Motion   244
9.7   Velocity Potential   244
9.8   Equations for Irrotational Motion of an Ideal Fluid 245
9.9   Irrotational Motion as an Impulsively Generated Motion:
      Velocity Potential as the Potential of an Impulse 246
9.10 Boundary Conditions: Condition at a solid-fluid boundary;
      Conditions at a free surface   249
9.11 Problems of Concern   250
9.12 Some Topological Notions: Connectivity; Reconcilable and
      irreconcilable paths; Reducible and irreducible circuits;
      Reconcilable and irreconcilable circuits; Simply connected
      region; Doubly connected region; Multiply connected
      region; Barriers   252
9.13 Irrotational Motion in a Simply Connected Region   258
9.14 Irrotational Motion in a Doubly Connected Region   259
9.15 Summary   263
9.16 Conditions at Infinity   263
9.17 Velocity Components at Infinity 268
9.18 Some Further Properties of Irrotational Motion: Simply
      connected region; Doubly connected region   269
9.19 Stream Functions and the Velocity Potential: Two-dimensional
      motion; Axisymmetric motion   276

10. UNSTEADY ACYCLIC MOTION   278

10.1 Mathematical Problem   278
10.2 Expanding Sphere   279
10.3 Problem for a Translating Body in Terms of Body Fixed
      Reference Frame   281
10.4 Translating Sphere   284
10.5 Force on a Translating Body of Arbitrary Shape   291
10.6 Impulse   297
10.7 The Apparent Mass Tensor   297
10.8 Kinetic Energy and Impulse   301
10.9 Moment on a Translating Body   304
10.10 Uniform Transiation   309
10.11 Permanent Translation   311
10.12 Remarks   311

11. STEADY ACYCLIC MOTION   312

11.1 Statement of the Problem   312
11.2 Simple Polynomial Solutions   313
11.3 The Source Potential   317
11.4 Source in a Uniform Flow (Axisymmetric Flow over a Semi-
      infinite Body of Revolution)   322

11.5   Source and Sink in a Uniform Flow (Axisymmetric Flow over
       a Closed Body of Revolution)   327
11.6   Line Distribution of Sources and Sinks in a Uniform Flow:
       Axisymmetric Flow over Slender Bodies of Revolution . . . 331
11.7   The Doublet Potential   333
11.8   Doublet in a Uniform Stream; Flow over a Sphere   339
11.9   Line Distribution of Doublets in a Uniform Stream: Lateral
       and Axisymmetric Flow Past a Body of Revolution   342
11.10 Flow Past Arbitrary Bodies of Revolution 343
11.11 Flow Past an Arbitrary Body   344
11.12 Pressures   348
11.13 Discussion 350
11.14 Force on an Arbitrary Body: d'Alembert's Paradox 354
11.15 Circulation as the Agency of Force 355

12. STEADY TWO-DIMENSIONAL ACYCLIC MOTION 359

12.1   Recapitulation 359
12.2   Further Considerations Relating to the Stream Function . . 361
12.3   Problem in Terms of the Stream Function   363
12.4   Uniform Stream 364
12.5   Source Flow   365
12.6   Combination of a Source and a Sink of Equal Strength . . . 367
12.7   Doublet   369
12.8   Source and Sink of Equal Strength in a Uniform Stream . . 371
12.9   Doublet in Uniform Stream: Flow over a Circular Cylinder   372
12.10 Flow Past an Arbitrary Cylinder   375

13. CIRCULATION AND LIFT
     FOR AN INFINITE WING IN STEADY FLOW 376

13.1   Circulatory Flow with Constant Vorticity 376
13.2   Circulatory Flow without Rotation: Vortex Flow   378
13.3   Circulation as the Strength of a Vortex Flow   381
13.4   Stream and Potential Functions for a Vortex Flow 382
13.5   Uniform Flow Past a Circular Cylinder with Circulation . . 382
13.6   Flow with Circulation Past an Arbitrary Cylinder 389
13.7   Kutta-Joukowski Theorem and the Problem of the Circulation
       Theory of Lift 389
13.8   Airfoils, Circulation, and the Kutta Condition 390
13.9   The Generation of Circulation   393
13.10 Mathematical Problem 395

14. ELEMENTS OF THE THEORY OF FUNCTIONS
     OF A COMPLEX VARIABLE 402

14.1   General Solution of Laplace's Equation in Two Dimensions:
       Introduction of the Complex Variable   402

14.2 Nomenclature and Algebra of Complex Numbers 404
14.3 Geometrical Interpretation 405
14.4 Polar and Exponential Forms of a Complex Number 407
14.5 Function of a Complex Number 409
14.6 Analytic Function 410
14.7 Cauchy-Riemann Conditions 411
14.8 Some Consequences of Cauchy-Riemann Equations 413
14.9 Remarks 414
14.10 Some Analytic Functions 415
14.11 Geometrical Significance of a Complex Function: Notion of
      Mapping 416
14.12 Some Simple Transformations 417
14.13 Conformal Transformation: Transformation by Analytic
      Functions 420
14.14 Critical Point of a Transformation 422
14.15 Complex Integrals 425
14.16 The Cauchy Integral Theorem 426
14.17 Integration in Multiply Connected Regions 428
14.18 Some Simple Integrals 431
14.19 The Cauchy Integral Formula 435
14.20 Unlimited Differentiability of an Analytic Function 437
14.21 Taylor Series 438
14.22 Laurent Series 440
14.23 Integration of a Function with Singularities: The Residue
      Theorem 441

**15. TWO-DIMENSIONAL MOTION
     AND THE COMPLEX VARIABLE** 444

15.1 Complex Potential and Complex Velocity 444
15.2 Flows Represented by Some Simple Functions 446
15.3 Circulation and Source Strength 450
15.4 Flow Past an Arbitrary Cylinder 451
15.5 Flow Past a Circular Cylinder 452
15.6 Complex Representation of Forces and Moments Acting on
     an Arbitrary Body: Blasius' Relations 453
15.7 Force and Moment on an Arbitrary Cylinder: Kutta-
     Joukowski Theorem 454
15.8 Mapping of Flows 456
15.9 Transformation of Circulation and Source Strength 458
15.10 Transformation of Flow Past an Arbitrary Cylinder into that
      Past a Circular Cylinder: Transformation; Complex potential
      and complex velocity for the flow past the circle; Velocity field
      in the plane of the arbitrary cylinder; The pressure field;
      Force and moment on the arbitrary cylinder; Remarks . . 458

16. PROBLEM OF THE AIRFOIL 466

16.1 Nomenclature 466
16.2 Mapping of the Trailing Edge 467
16.3 Kutta Condition and the Value of Circulation 468
16.4 Lift on the Airfoil 470
16.5 Moment on the Airfoil: Moment at no-lift; Aerodynamic
       center and the moment about the aerodynamic center 471
16.6 Velocity and Pressure Distributions on the Airfoil Surface . . 474
16.7 Transformation of a Circle into an Airfoil 475
16.8 The Joukowski Transformation 477
16.9 The Joukowski Airfoils 479
16.10 Properties of Joukowski Airfoils 484
16.11 Other Airfoils: Karman-Trefftz airfoils 487
16.12 Theodorsen's Method for the Arbitrary Airfoil 490

17. ELEMENTS OF THIN AIRFOIL THEORY 492

17.1 Formulation of the Problem in Terms of the Perturbation Field 492
17.2 Simplification of the Boundary Condition 494
17.3 Transfer of the Boundary Condition 495
17.4 Frame Work of the Theory of the Thin Airfoil 496
17.5 Pressure Relation in the Simple Theory 499
17.6 Symmetrical Airfoil at Zero Angle of Attack: Solution by
       Source Distribution 500
17.7 Cambered Airfoil of Zero Thickness at Zero Angle of Attack:
       Solution by Vortex Distribution 506
17.8 Flat Plate Airfoil at an Angle of Attack: Solution by Vortex
       Distribution 514
17.9 Aerodynamic Characteristics of a Thin Airfoil 516

18. SOME FEATURES OF FLOW WITH VORTICITY 518

18.1 Recapitulation 518
18.2 Vortex Line, Surface, Tube and Filament 519
18.3 Vorticity Field is a Divergenceless Field 520
18.4 Spatial Conservation of Vorticity: Strength of a Vortex Tube 520
18.5 Consequences of the Theorems of Helmholtz and Kelvin . . 523
18.6 Velocity Field Due to a Vortex Distribution in an Incom-
       pressible Fluid 524
18.7 Velocity Field of a Vortex Filament: Biot-Savart Law . . . 526
18.8 Simple Applications 528
18.9 Vortex Sheet 530
18.10 Solution for the Vector Potential 532

Contents

19. ELEMENTS OF FINITE WING THEORY 535

19.1 Prandtl's Theory 538
19.2 Problems of Interest 548
19.3 Elliptic Lift Distribution: Elliptic Wing 548
19.4 Solution for the Arbitrary Wing: Trefftz's Method 554
19.5 Forces and Moments on an Arbitrary Wing 562
19.6 Question of the Smallest Drag 564
19.7 Determination of the Coefficients A_n: Methods of Glauret
       and Irmgard Lotz 564

**20. ELEMENTS OF THE THEORY FOR THE FLOW
       PAST A SLENDER BODY OF REVOLUTION** 568

20.1 Formulation of the Problem in Terms of the Perturbation
       Field 568
20.2 Boundary Condition for a Slender Body of Revolution:
       Axisymmetric Flow Past a Slender Body of Revolution;
       Cross Flow or Lateral Flow Past a Slender Body of Revolution 572
20.3 Axisymmetric Flow Past a Slender Body of Revolution:
       Solution by Source Distribution 574
20.4 Cross Flow Past a Slender Body of Revolution: Solution by
       Doublet Distribution 578
20.5 Pressure Distribution; Axial Flow; Lateral Flow; Flow at an
       Angle of Yaw 581
20.6 Forces on the Body of Revolution 584
20.7 Moment on the Body of Revolution 587

Selected Problems 589
References 601
Some Books 605

Appendix A: Theorems of Linear and Angular Momentum 607
Appendix B: Characteristics of the Flow Fields of Two-Dimensional
                 Source and Vortex Distributions 612
Appendix C: Poisson's Integral Formulas 617
Appendix D: Conjugate Fourier Series 621
Appendix E: Some Integrals 624

Index 629

Chapter 1

Introduction

Aerodynamics is one branch of fluid mechanics, the science of the motion of liquids and gases, both of which together are known as fluids. Fluid mechanics is a very extensive subject encompassing widely diverse topics such as the motion of airplanes and missiles through the atmosphere, satellites through the outer atmosphere, submarines and ships through water, the swimming of microscopic organisms, the flow of liquids and gases through ducts, the transfer of heat and mass by fluid motion, propagation of sound through gases and liquids, the study of ocean waves and tides, the study of air masses in the atmosphere, and many astro-physical, geophysical, and meteorological problems. *Aerodynamics* deals primarily with the motion of air, or, more generally, of any gas.* The science dealing with the motion of water, or, more generally, of any liquid is called *hydrodynamics*. A great deal of aerodynamics is concerned with the phenomena on which flight (such as that of an airplane) depends, and aerodynamics is usually thought of as the *science of flight*.

The possibility of flight rests on the nature of the force experienced by a body moving through air. This force on the body is usually resolved into two components: one called *lift*, in a direction normal to the flight direction and a fixed direction in the body, and the other called *drag*, in a direction opposed to that of motion. In the practical problem of flying lift is desirable to sustain or lift the body against its own weight, whereas drag is undesirable because it hinders the motion of the body, forcing us to expend energy (by means of a propulsive device) to compensate for it. The principal requirement of mechanical flight is to have a body that experiences a large lift and a low drag. Fortunately, there are certain bodies, such as the wings of an airplane, which are capable of producing more lift than drag. The study of lift and drag is an important part of aerodynamics. The scope of this book is limited to certain aspects of the study of lift and drag. In particular, we shall be concerned mainly with the study of flows related to the motion of a wing or an airshiplike body moving with a constant velocity through otherwise undisturbed air, the

---

* It is then referred to as gasdynamics.

magnitude of the velocity being sufficiently small compared to the speed of sound in the undisturbed air. As is known from observation, essential differences exist between the phenomena underlying the motion of a body through air at a sufficiently low speed and those underlying the motion of the body at a high speed as compared to the speed of sound in the undisturbed air.

We now describe briefly some of the properties of a fluid that are called into play in determining flows pertaining to our objective and outline the approximate basis on which we shall analyse these flows.

1.1 Fluid As a Continuous Medium

Fluids, like all matter, are made up of molecules. Thus the properties of fluid motion such as are observed may be studied on the basis of the mechanics of the molecules composing the fluid. Although such a procedure may appear feasible in principle, it will indeed be a formidable task to achieve solutions of practical problems. Apart from this consideration, we are generally not interested in the details of the mechanics of the molecules. What we wish to do is to establish relations between various macroscopically observable quantities pertaining to a fluid at rest or in motion. Such observable properties are called macroscopic or bulk properties. They are mean values in space and time obtained by taking the average over a sufficiently large volume containing a considerable number of molecules and a sufficiently long time compared to a certain time related to the mechanics of the molecules. From the macroscopic point of view this will mean, in many practical flow problems, such extremely small volumes and short time intervals that the variations in the bulk properties of the fluid, whether at rest or in motion, could hardly be observed within those volumes and time intervals. For instance, at normal temperature and pressure, a volume of 10^{-12} cc (a cube of width 1/1000 mm) will contain about 2.7×10^7 molecules, a large number. This being the case, it is a reasonable approximation to regard a fluid, whether at rest or in motion, as a continuous distribution of matter. We then speak of the fluid as a *continuous medium* or as a *continuum*. With such a picture of the fluid, we call an infinitely small *fluid element* a *fluid particle*.

We now consider *some of the bulk properties* of a fluid. In doing so we restrict ourselves to a fluid in which chemical reactions, electromagnetic processes, radiation, and the like are absent.

1.2 Properties of a Fluid at Rest

Thermodynamic Properties. The properties *density*, *pressure*, and *temperature* at a point in a liquid or a gas in static equilibrium are well known. To fix our ideas about the nature of a fluid, let us recall the notion of pressure. At any point it is the magnitude of the normal stress acting on an elemental plane area passing through that point (see Chapter 3 for more details on the concept of stress, pressure, and so forth). It is known from experience that in a fluid at rest only normal stresses occur and that, in general, they are "compressive" in nature. When only normal stresses occur, it can be shown easily that at any point they should be equal in all directions. It is also a matter of experience that the force exerted by a fluid in static equilibrium on a solid body submerged in it is due to only normal stresses acting on the surface of the body.

The fact that only normal stresses occur for a fluid in static equilibrium is in contrast with the state of affairs for a solid in static equilibrium (both being under the action of external forces). It is known that for a solid in static equilibrium both normal and tangential stresses occur in general. *The absence of tangential stresses in a state of static equilibrium is what distinguishes a fluid from a solid and may be considered as the property that defines a fluid.*

For a fluid in static equilibrium, besides pressure, density, and temperature, the properties such as internal energy, enthalpy, and entropy are also familiar concepts. These various thermodynamic properties and others like them are not all independent of each other. As shown by observations, functional relations exist between them. Such relations are known as *characteristic equations* or *equations of state*. For example, $p = \rho RT$, where R is a constant, is an equation of state for a perfect gas. Equations of state for liquids and other gases are not of such simple form.

Compressibility. All fluids undergo changes in volume under changes of pressure and temperature. For fluids in static equilibrium the changes in pressure result from the applied external forces and the changes in temperature result from nonuniform heating of the fluid. *The ability for changes in volume of a mass of fluid is known as compressibility.* It is well known that gases are more easily compressed than liquids. When a fixed mass of a fluid undergoes changes in volume, its density also changes. Thus the capacity for changes in the density of a mass element of a fluid is also known as compressibility.

Under normal circumstances, in liquids, the changes in density due to pressure changes are practically unobservable. Density changes due to temperature differences are, however, *not* negligible in general. If the temperature differences are sufficiently small, the density changes in a liquid are almost nil and the liquid may then be regarded as an *incompressible fluid. An incompressible fluid is one whose elements undergo no changes in volume or density.*

Compressibility is easily noticeable in gases. In certain circumstances, however, the changes in volume or density of an element of a gas may be negligibly small. In such a case, as a reasonable approximation, the gas may be regarded as an incompressible fluid.

Heat Conduction. When a fluid in static equilibrium is heated non-uniformly, heat may be transferred (without causing motion of the fluid) from points at which the temperature is high to those at which it is low by what is called *thermal conduction*. Consider a surface element situated at some point in the fluid. Observations show that under usual circumstances the *heat flux* (which is the amount of heat transferred across the surface element per unit time per unit area) in the direction of the normal to the element is proportional to the spatial rate of change of the temperature at that point in the direction of the normal. The heat flow occurs in the direction of decreasing temperature. Thus, if q_n denotes the heat flux and $\partial T/\partial n$ denotes the rate of increase of temperature with distance in the direction of the normal, we have

$$q_n = -k \frac{\partial T}{\partial n} \tag{1.1}$$

where k is a proportionality factor known as the *coefficient of thermal conductivity* or simply as the thermal conductivity. It is a material property of the fluid, and thus its value differs from fluid to fluid. The thermal conductivity of a fluid is always positive and, in general, a function of temperature and pressure. In the context of macroscopic considerations k has to be known from experimental observations. The temperature variation of k at atmospheric pressure for water and air is shown in Fig. 1.1.

It is found that the static equilibrium of a fluid in which the temperature is not constant is unstable unless certain conditions are fulfilled. This instability leads to the appearance of *convection currents* in the fluid, which tend to mix the fluid in such a way that the temperature is equalized.

1.3 Properties of a Fluid in Motion

The density of an element of a fluid in motion is defined in the same way as a fluid at rest. The concepts of pressure, temperature, and other properties as known in thermodynamics are assumed to apply equally well to a fluid in motion. It is further assumed that the equations of state obtained for a fluid in thermodynamic equilibrium are equally valid for fluids in motion. The reasonableness of these assumptions rests on the fact that for many flow problems they lead to results that are in satisfactory agreement with observations.

Friction or Viscosity. It is a matter of experience that even smoothly shaped bodies* moving with a constant velocity through an otherwise undisturbed fluid encounter a resistance to their motion. Similarly, a fluid flowing through a pipe offers resistance. These observations suggest that for a fluid in motion tangential stresses in addition to normal stresses

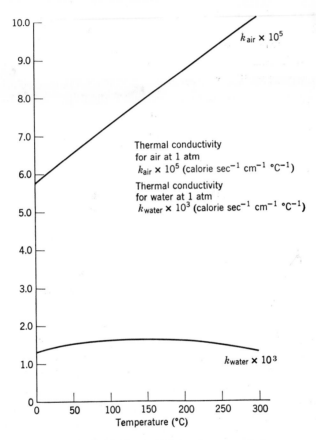

Thermal conductivity
for air at 1 atm
$k_{air} \times 10^5$ (calorie sec^{-1} cm^{-1} $°C^{-1}$)
Thermal conductivity
for water at 1 atm
$k_{water} \times 10^3$ (calorie sec^{-1} cm^{-1} $°C^{-1}$)

$k_{air} \times 10^5$

$k_{water} \times 10^3$

Temperature (°C)

Fig. 1.1 Thermal conductivity of water and air.

occur on any elemental plane passing through a point in the fluid. The appearance of these tangential stresses only when the fluid is in motion constitutes the phenomenon of *internal friction* or *viscosity* in a fluid. Such stresses give rise to a resistance to nonuniform motion of a fluid.

* For example, a body of revolution moving in the direction of its axis or a thin flat plate moving in its plane.

Nonuniform motion refers to the situation in which the velocity differs from point to point in the fluid.

In general, viscosity gives rise not only to resistive tangential stresses but also to resistive normal stresses. Such normal and tangential stresses are called *frictional* or *viscous stresses*. They appear only when the fluid is in nonuniform motion and disappear when the nonuniformities disappear. They do not occur at all when the fluid is in static equilibrium. For a fluid in motion it is assumed that the viscous stresses occur over and above the normal stresses associated with pressure.

For a wide range of flow situations observations tell us that, because of the phenomenon of viscosity, there can be no relative velocity at all between a moving fluid and a solid body at their surface of contact.

Coefficient of Viscosity. Consider the following experiment.* A fluid is contained between two parallel plates of indefinite extent that are placed at a small distance h (see Fig. 1.2). One of the plates is at rest

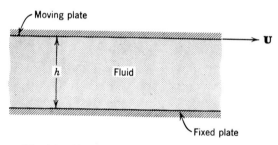

Fig. 1.2   Illustrating definition of viscosity.

while the other is moving with a constant velocity U parallel to itself. Because of viscosity, the fluid will also be in motion, its velocity at the moving plate being U and that at the stationary plate being zero. Resistive tangential stresses occur in the fluid and at the plates. Experiment shows that over a wide range of conditions the tangential stress τ acting on either of the plates is proportional to the relative velocity between the plates and inversely proportional to the distance h. Thus we have

$$\tau = \mu \frac{U}{h} \tag{1.2}$$

where μ is a factor of proportionality that is independent of U and h and depends only on the nature of the fluid. This factor is a measure of the

---

* Such an experiment is perhaps not easily realizable, but it affords the simplest example for our purposes. A realizable experiment analogous to the one under consideration is that of the motion of a fluid contained in the narrow annular space formed by two concentric cylinders, one of which rotates at a constant speed and the other is stationary.

viscosity of the fluid and is called the *coefficient of shear viscosity* or *simply the coefficient of viscosity.*

Equation (1.2) states that the viscous stress is proportional to the average spatial rate of change of the velocity in the fluid over the distance h. Now consider a similar situation in which the fluid is moving in parallel plane layers with the same direction everywhere. Let y denote distance from a fixed point measured perpendicular to the layers and let u denote the velocity of a fluid layer at a distance y. With u as a function of y, the fluid is said to be in simple shearing motion. There is a shearing or tangential stress between adjacent layers of the fluid. In light of the interpretation of Eq. (1.2), we assert that the shear stress τ at any point of the fluid for the motion under consideration is given by

$$\tau = \mu \frac{du}{dy} \tag{1.3}$$

where μ, as before, is the coefficient of viscosity for the fluid in question.

In generalizing (1.3) for a fluid in a more general motion, it is assumed that in a wide range of flow conditions the viscous stresses are linearly related to the rates of strain in the fluid. These rates of strain are given by certain combinations of the spatial derivatives of the velocity components. In these relations there appear two factors of proportionality, both of which are referred to as coefficients of viscosity. One of them is the shear-viscosity coefficient μ already discussed, and the other is expressible in terms of μ and the *bulk modulus* (also known as the *coefficient of bulk viscosity*) of the fluid. In many investigations of fluid flow it is assumed that the bulk-viscosity coefficient is zero. For motions in which the fluid may be regarded as incompressible it turns out that because there are no volume changes the viscous stresses do not contain any terms involving the bulk viscosity coefficient. Thus in many fluid-flow problems only one viscosity coefficient occurs, and this is μ. Fluids characterized by a relation of the form (1.3) are called *Newtonian fluids.*

The coefficient of viscosity, which henceforth shall mean only the shear viscosity μ, has different values for different fluids and for a particular fluid is, in general, a function of both temperature and pressure. In the context of macroscopic considerations, viscosity, just like thermal conductivity, has to be known from experimental observations. Within a given range of temperatures and pressures the dependence of μ on temperature is markedly noticeable, whereas that on pressure is hardly seen. For gases μ is an increasing function of temperature; for liquids it is a decreasing function of temperature. For both liquids and gases μ increases slightly with pressure, the changes in μ being very small for pressure changes that are likely to occur in many circumstances.

Table 1.1 gives the values of μ for some selected fluids, at 15°C and atmospheric pressure. For air and water, which are of particular interest to us the variation of μ with temperature at atmospheric pressure is shown in Figs. 1.3 and 1.4. As we shall see later, the effect of viscosity on the motion of a fluid is determined by the ratio of μ to the density ρ rather than by μ alone. This ratio μ/ρ is known as the *kinematic viscosity* and is usually denoted by ν. Table 1.1 and Figs 1.3 and 1.4 include the corresponding values of ν.

Table 1.1.   *Viscosity μ and Kinematic Viscosity ν, in CGS Units, for Gases and Liquids at 15°C and Atmospheric Pressure*

	$\mu \times 10^3$	$\nu \times 10$
Air	0.18	1.5
Nitrogen N_2	0.17	1.5
Oxygen O_2	0.20	1.5
Hydrogen H_2	0.09	10.5
Helium He	0.20	1.2
Neon Ne	0.31	3.7
Argon A	0.22	1.3
Carbon dioxide CO_2	0.14	0.77
Water H_2O	11.4	0.114
Mercury Hg	16.0	0.012
Paraffin oil	200	2.5
Olive oil	1000	10
Glycerine	13,000	100
Castor oil	15,000	150
Pitch	$\simeq 10^{13}$	$\simeq 10^{11}$

Compressibility. Density and volume changes for an element of a fluid in motion arise, as in a fluid at rest, from temperature and pressure changes. When a fluid is in motion, these changes occur because of various factors in addition to external forces and nonuniform heating of the fluid. These factors are those that affect the motion of a fluid element. Pressure changes would result from changes of momentum of the element and the action of viscous stresses, in addition to the action of the external forces. Temperature changes would result from exchanges between kinetic and internal energies, which are due in turn to work done by external forces, pressure forces, and viscous forces, in addition to non-uniform heating. The situation in general is rather complicated. Later we shall examine the parameters that characterize the compressibility of a

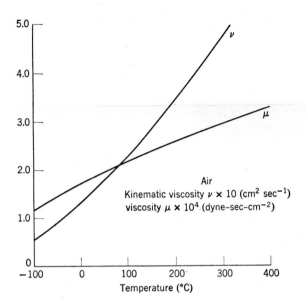

Fig. 1.3   Viscosity and kinematic viscosity of air.

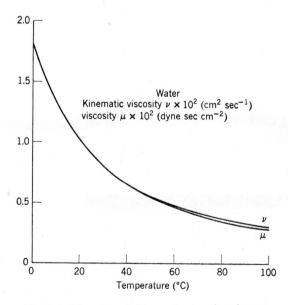

Fig. 1.4   Viscosity and kinematic viscosity of water.

fluid element in motion. At that time we shall also see under what conditions the compressibility of the element may be neglected.

Heat Transfer. It is a matter of common experience that a heated body immersed in a moving fluid cools considerably more rapidly than one in a fluid at rest, in which heat transfer is accomplished only by conduction; in a moving fluid heat transfer is accomplished by the combined action of conduction and *convection*. By convection we mean the motion of a nonuniformly heated fluid. If the motion is caused by the action of gravity alone, it is known as *free convection*. If the motion is caused by agencies other than gravity, it is known as *forced convection*. The velocities occurring in free convection are small. Generally, forced convection is the governing process for heat transfer in a flowing fluid.

When heat transfer occurs between a flowing fluid and a body immersed in it, convection is absent at the surface of the body itself. The transfer of heat between the fluid and the body takes place primarily by heat conduction through a thin layer close to the surface. Under normal circumstances it is found that at the surface of the body the temperature of the fluid is equal to the temperature of the body.

1.4 Laminar and Turbulent Motion

It is a matter of observation that there are two fundamentally different types of fluid flow, which we call *laminar* and *turbulent* flows. The difference between the two flows is readily illustrated by the famous experiment of Reynolds (1883). He observed the nature of the flow of water through a long glass tube connected to a reservoir. The observation was made by introducing a dye at the entrance of the tube. At small velocities the dye forms a thin, straight thread parallel to the axis of the tube, indicating that the flow is steady and orderly in nature. This type of flow is referred to as laminar flow. As the velocity is gradually increased at a certain velocity (depending on the dimensions of the tube), the flow suddenly changes in character; the dye thread becomes violently agitated, and the dye soon spreads over the whole tube. The flow becomes one of an irregular character. We call such a flow turbulent flow. These observations are illustrated in Plate 1.

Regular laminar motion is exceptional in nature; turbulent flow is much more common. This is true for the flow of water in a river, for a moving stream of gas, as for the atmosphere which, as a whole, may be at rest.

Turbulent motion, unlike laminar motion, is characterized by random fluctuations at a fixed point in the fluid quantities such as velocity and pressure. Such fluctuations result in pronounced *mixing*. Fluctuations in fluid velocity may be detected by what is called the *hot-wire anemometer*

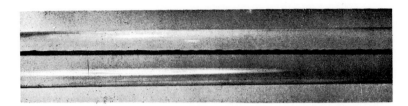

Reynolds No. 1500

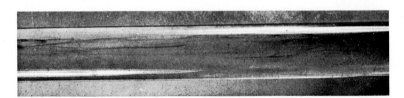

Reynolds No. 2000

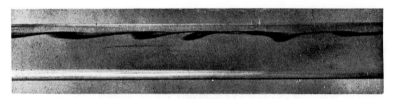

Reynolds No. 3000

Plate 1 Illustrating Reynolds' observations of laminar and turbulent flows of water through a glass tube. Courtesy of Professor F. J. Bayley and Wiley-Interscience. Figure 6.2 of F. J. Bayley: An Introduction to Fluid Dynamics, 1958.

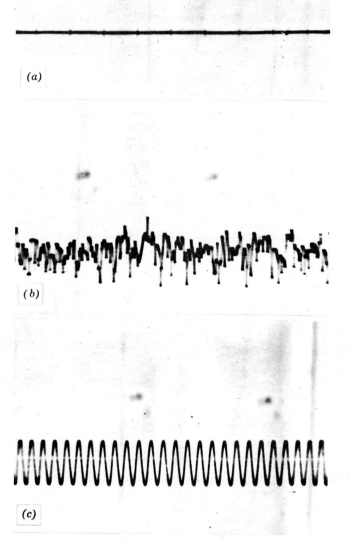

Plate 2 Nature of velocity fluctuations: (*a*) in the laminar region of an air jet; (*b*) in the turbulent region of an air jet; (*c*) in the jet region of the flow field of an edge tone of discrete frequency. Time along the horizontal axis and speed along the vertical axis.

(see, for instance, Dryden and Kuethe, 1929). When fluctuations exist, a mean velocity may be defined by taking an average over a time sufficiently large compared to the time scale of the fluctuations. By velocity in a turbulent flow we mean such a mean velocity. The turbulent fluctuations are generally composed of a wide range of frequencies. The magnitude of the fluctuations can range from extremely small, almost immeasurable fractions of the mean speed to as high as 30 to 50 per cent of the mean speed. The random character of turbulent flow is illustrated in Plate 2, which shows records of the velocity in two different parts, laminar and turbulent, of a thin rectangular air jet issuing from a slit. When such a jet impinges on a wedge placed a short distance from the slit, a sound tone, known as the edge tone, of discrete frequency is generated. In such a situation the fluctuations in the fluid velocity are of discrete frequency. A record of these fluctuations is also included in the Plate 2.

The analysis of turbulent flow requires special methods. In our studies here we will have no occasion to enter into the details.

1.5 Some Relevant Parameters

We now seek the parameters that characterize the effects of external forces, viscosity, compressibility, and heat transfer in a fluid in motion. For this purpose and to fix our ideas we consider specifically the problem of a rigid body moving with a constant velocity through an infinitely extending fluid which is initially in a state of static equilibrium under the action of gravity. Gravity forces are the only external forces acting on the fluid. We assume that in the undisturbed state the temperature and the internal energy (measured per unit mass) are constant throughout the fluid. We denote this temperature and internal energy by T and e, respectively. The pressure and density in the undisturbed state would vary from point to point of the fluid. Let p and ρ denote some reference values of the pressure and density in the undisturbed state.

The geometric shape of the surface of the body is fixed. The surface is then specified by a length l characteristic of the body. Let V denote the magnitude of the constant velocity with which the body is moving.

When a body is set into motion with a constant velocity through an otherwise undisturbed fluid, after a sufficient length of time the surface of the body would attain a temperature different from that of the undisturbed fluid. This is so even if the body and the fluid have the same temperature before the body is set into motion. For the following considerations we assume that when the body is moving with constant velocity the temperature of its surface is maintained at a constant temperature T_w different from the temperature T of the undisturbed fluid.

Relative Magnitudes of the Forces. Froude Number, Reynolds Number, and Mach Number. We now make an estimate of the relative magnitudes of the various forces acting on a fluid element in motion. According to Newton's second law of motion, we have

the time rate of change of momentum of a fluid element

= the gravity force + the pressure force

+ the viscous force acting on the element   (1.4)

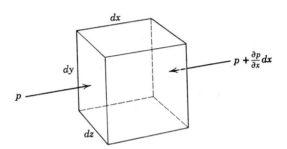

Fig. 1.5   Pressure force on an element.

The time rate of change of momentum of the element is the negative of the *inertial* force acting on the element. The equation of motion thus expresses simply the fact that the result of all the forces acting on the element is zero. We obtain an estimate of the relative magnitudes of these forces as follows.*

Consider a fluid element of volume $\delta\tau$. The speed of the element is proportional to V, the speed of the body. The change in momentum of the element is proportional to $\rho V\delta\tau$, where ρ is the density in the undisturbed state. We may say that this change would occur in a time interval that is proportional to the time l/V taken by the body to move a distance equal to the characteristic length l. Thus the rate of change of momentum of the element or the *inertial force* acting on it *is proportional to*

$$\frac{\rho V\delta\tau}{l/V} = \frac{\rho V^2}{l}\delta\tau$$

Let g denote the force due to gravity, measured per unit mass. Then the *gravity force acting on the fluid element is equal to* $\rho g\delta\tau$.

To estimate the pressure force, we examine the order of magnitude of one of the components of that force. For this purpose we consider

---

* In the following discussion we are concerned only with the magnitudes of the various quantities under consideration.

an element as shown in Fig. 1.5 and the component of the pressure force in the x-direction. The magnitude of this component is equal to

$$\frac{\partial p}{\partial x}\, dx\, dy\, dz = \frac{\partial p}{\partial x}\, \delta \tau$$

Assuming that all components of the pressure force are of about equal magnitude, we say that *the pressure force on the element is proportional to*

$$\frac{\delta p}{l}\, \delta \tau$$

where δp is a representative change in pressure.

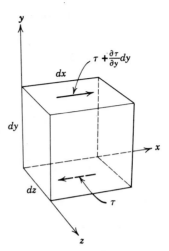

Fig. 1.6 Viscous force on an element.

To estimate the magnitude of the viscous force, we proceed in a similar way. As shown in Fig. 1.6, we consider only the contribution to the x-component of the viscous force due to the appropriate shear stresses acting on the y-faces (the faces normal to y) of the element. Thus the x-component of the viscous force is given by

$$\frac{\partial \tau}{\partial y}\, dx\, dy\, dz = \frac{\partial \tau}{\partial y}\, \delta \tau$$

Now, we may say that the spatial rate of change $\partial \tau / \partial y$ is proportional to τ / l, where τ is the shear stress at the point the element is situated. Furthermore, we say that

$$\tau = \mu \frac{\partial u}{\partial y}$$

where u is the velocity in the x-direction, and that

$$\frac{\partial u}{\partial y} \sim \frac{V}{l}$$

We then have

$$\frac{\partial \tau}{\partial y}\, \delta \tau \sim \mu \frac{V}{l^2}\, \delta \tau$$

We assume that all components of the viscous force on the element are of about the same magnitude, and *the viscous force on the element may be said to be proportional to* $(\mu V/l^2)\, \delta \tau$.

The relative magnitudes of the various forces on the element are then as follows:

inertial force : gravity force : pressure force : viscous force =

$$\frac{\rho V^2}{l} : \rho g : \frac{\delta p}{l} : \mu \frac{V}{l^2} \tag{1.5}$$

The ratio of any one of the forces to another is a nondimensional number. The various nondimensional numbers obtained by taking the ratio of the inertial force to each of the others play an important part in fluid mechanics. They are known by special names after the scientists who have initially pointed out their importance.

The ratio of the inertial force to the gravity force is given by the parameter V^2/gl, whose square root is usually known as the Froude number and denoted by Fr. Thus we have

$$\text{Fr} = \frac{V}{\sqrt{gl}} \tag{1.6}$$

and

$$\text{Fr}^2 \sim \frac{\text{inertial force}}{\text{gravity force}} \tag{1.7}$$

The ratio of the inertial force to the viscous force is given by the parameter $\rho Vl/\mu$. This is known as the Reynolds number and usually denoted by R. We thus have

$$\text{R} \equiv \frac{\rho Vl}{\mu} = \frac{Vl}{\nu} \sim \frac{\text{inertial force}}{\text{viscous force}} \tag{1.8}$$

where ν is the kinematic viscosity.

The ratio of the inertial to the pressure forces is given by

$$\frac{V^2}{\delta p/\rho} = \frac{V^2}{p/\rho} \frac{1}{\delta p/p}$$

It can be shown that the speed of propagation of small (soundlike) disturbances through a fluid that is initially at rest and in a uniform state* is proportional to $(p/\rho)^{1/2}$, where p and ρ are the pressure and density in the undisturbed state. *Such speed of propagation is known as the speed of sound.* Denoting this speed by a, we have

$$\frac{p}{\rho} \sim a^2$$

* This implies the absence of any external forces.

and

$$\frac{V^2}{p/\rho} \sim \frac{V^2}{a^2}$$

The nondimensional parameter V/a, where a is a characteristic sound speed in the undisturbed state (p, ρ), is known as the Mach number and denoted by M. We thus have

$$M \equiv \frac{V}{a} \tag{1.9}$$

and

$$M^2 \frac{1}{\delta p/p} \sim \frac{V^2}{\delta p/\rho} \sim \frac{\text{inertial force}}{\text{pressure force}}$$

For a perfect gas (for which $p = \rho RT$) with constant specific heats, it is found that

$$a^2 = \gamma \frac{p}{\rho}$$

and that therefore

$$\frac{V^2}{p/\rho} = \gamma M^2$$

where γ is the ratio of the specific heat at constant pressure to that at constant volume.

Parameters Characterizing Compressibility. The changes in volume or density of a fluid element, as already pointed out, arise out of pressure and temperature changes brought about by the motion. Thus the factors that characterize compressibility are those on which pressure and temperature changes depend. We shall now examine these factors.

On the basis of the equation of motion (1.4) and the consideration that the pressure force per unit volume of the element is proportional to the ratio of the change δp in the pressure to a distance, we may say that $\delta p/l$ is proportional to each of the forces, inertia, gravity, and viscous, acting on the element. Thus, we have

$$\delta p \sim \rho V^2, \quad \rho g l, \quad \frac{\mu V}{l}$$

For the relative change in pressure, $\delta p/p$, we therefore have

$$\frac{\delta p}{p} \sim \frac{V^2}{p/\rho}, \quad \frac{gl}{p/\rho}, \quad \frac{\mu V}{lp} \tag{1.10}$$

Introducing the nondimensional parameters Fr, M, and R, we may state that

$$\frac{\delta p}{p} \sim M^2, \qquad \frac{M^2}{Fr^2}, \qquad \frac{M^2}{R} \tag{1.11}$$

To seek the factors influencing temperature changes we note that such changes are proportional to the changes in internal energy of the element. The factors influencing the changes in internal energy are exhibited by the equation of energy, which states

> the rate of change of internal energy of an element + the rate of change of its kinetic energy = the rate at which work is done on the element by the gravity force + the rate at which work is done on the element by the viscous forces + the rate at which work is done on the element by the pressure forces + the rate at which heat is added to the element by heat conduction
> (1.12)

Denoting by Δe the change in internal energy per unit mass of an element of volume $\delta\tau$ and assuming that the various changes take place in a time interval proportional to l/V as before, we may say that

the rate of change of internal energy $\sim \Delta e \, \dfrac{\rho \, \delta\tau}{l/V}$

the rate of change of kinetic energy $\sim \dfrac{V^2}{2} \dfrac{\rho \, \delta\tau}{l/V}$

the rate at which work is done by the gravity force $\sim gV\rho \, \delta\tau$

the rate at which work is done by the pressure force $\sim \dfrac{p}{l} V \, \delta\tau$

the rate at which work is done by the viscous force $\dfrac{\mu V^2}{l^2} \, \delta\tau$

To estimate the rate of heat addition by conduction we consider the element shown in Fig. 1.7 and the heat flow through its x-faces. Denoting by q_x the flux through the face situated at x, the amount of heat added to the element per unit time is given by

$$\frac{\partial q_x}{\partial x} \, dx \, dy \, dz = \frac{\partial q_x}{\partial x} \, \delta\tau$$

We may say that $\partial q_x/\partial x$ is proportional to q_x/l. According to the equation of heat conduction (1.1), we have

$$q_x = -k \frac{\partial T}{\partial x}$$

where k is the thermal conductivity and T is the temperature. With $(T - T_w)$ being the temperature difference between the undisturbed fluid and the body, we may say that $\partial T/\partial x$ is proportional to $(T - T_w)/l$. This being the case, the heat flow through the x-faces and thus *the rate of heat addition to the element by heat conduction may be said to be proportional to*

$$\frac{k(T - T_w)}{l^2}\, \delta\tau$$

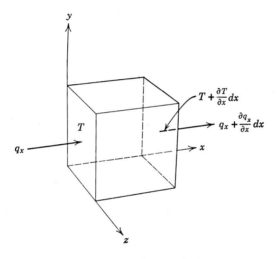

Fig. 1.7   Heat flow into an element.

For the relative changes of the internal energy or, equivalently, of the temperature of the element we therefore have

$$\frac{\Delta T}{T} \text{ or } \frac{\Delta e}{e} \sim \frac{V^2}{2e}, \quad \frac{gl}{e}, \quad \frac{p}{\rho e}, \quad \frac{\mu}{\rho Vl}\frac{V^2}{e}, \quad \frac{k(T - T_w)}{\rho Vle} \qquad (1.13)$$

Introducing the speed of sound a and the nondimensional parameters Fr, M, and R, we may state that

$$\frac{\Delta T}{T} \text{ or } \frac{\Delta e}{e} \sim \frac{\mathrm{M}^2}{2}\frac{a^2}{e}, \quad \frac{\mathrm{M}^2}{\mathrm{Fr}^2}\frac{a^2}{e}, \quad \frac{a^2}{e}, \quad \frac{\mathrm{M}^2}{\mathrm{R}}\frac{a^2}{e}, \quad \frac{k(T - T_w)}{\rho Vle}$$
$$(1.14)$$

Since the relative changes, $\Delta\rho/\rho$, in the density of the element are proportional to the relative changes in pressure and temperature, we obtain by using (1.11) and (1.14)

$$\frac{\Delta\rho}{\rho} \sim M^2, \quad \frac{M^2}{Fr^2}, \quad \frac{M^2}{R}, \quad \frac{M^2 a^2}{2e}, \quad \frac{M^2 a^2}{Fr^2 e}, \quad \frac{M^2 a^2}{Re}, \quad \frac{k(T - T_w)}{\rho Vle}$$

$$(1.15)$$

These, then, are the parameters that characterize the compressibility of a fluid element. Among them are only five independent parameters—M^2, M^2/Fr^2, M^2/R, a^2/e, $k(T - T_w)/\rho Vle$.

For a perfect gas with constant specific heats we have

$$a^2 = \gamma(\gamma - 1)e$$

and

$$e = \frac{C_p T}{\gamma}$$

where C_p is the specific heat at constant pressure. We may then write

$$\frac{a^2}{e} = \gamma(\gamma - 1)$$

and

$$\frac{k(T - T_w)}{\rho Vle} = \gamma \frac{k}{\mu C_p} \frac{\mu}{\rho Vl} \frac{T - T_w}{T}$$

$$= \gamma \frac{k}{\mu C_p} \frac{1}{R} \frac{T - T_w}{T}$$

The ratio $\mu C_p/k$ is nondimensional and is known as the *Prandtl number*, usually denoted by σ. Introducing σ, for a perfect gas with constant specific heats we have

$$\frac{k(T - T_w)}{\rho Vle} = \gamma \frac{1}{\sigma R} \frac{T - T_w}{T}$$

Thus for a perfect gas with constant specific heats we may state that

$$\frac{\Delta\rho}{\rho} \sim M^2, \quad \frac{M^2}{Fr^2}, \quad \frac{M^2}{R}, \quad \frac{\gamma(\gamma - 1)}{2} M^2, \quad \gamma(\gamma - 1)\frac{M^2}{Fr^2},$$
$$\gamma(\gamma - 1)\frac{M^2}{R}, \quad \frac{\gamma}{\sigma R} \frac{T - T_w}{T}$$

$$(1.16)$$

For such a gas the five independent parameters characterizing compressibility are

$$M^2, \quad \frac{M^2}{Fr^2}, \quad \frac{M^2}{R}, \quad \gamma(\gamma - 1), \quad \frac{\gamma}{\sigma R} \frac{T - T_w}{T}$$

Prandtl Number.   For any fluid, introducing the Prandtl number σ, we may write

$$\frac{k(T - T_w)}{\rho V l e} = \frac{1}{\sigma R} \frac{C_p T}{e} \frac{T - T_w}{T}$$

The Prandtl number, like the viscosity and the thermal conductivity, is a material property and it thus varies from fluid to fluid. For any particular fluid it is in general a function of temperature and pressure. For gases it is of the order of unity; for liquids σ varies more widely. For very viscous liquids it may be very large. Some representative values of σ at 20°C and 1 atm pressure for various substances are as follows:

Air	0.733
Water	6.75
Mercury	0.044
Glycerine	7250

For air at atmospheric pressure σ varies from 0.719 at $-50°F$ to 0.722 at 600°F. For water at atmospheric pressure σ varies from 6.88 at 68°F to 1.7 at 212°F.

Parameters on Which Force and Heat Transfer Depend.   Let us denote by F the magnitude of the force acting on the body and by Q the magnitude of the heat transferred to the body per unit time. We may regard F as meaning the total force or a component of the total force in some chosen direction. We expect the force and the heat transfer to depend on the properties $g, p, \rho, T, e, \mu, k, V, l, T_w$, the shape and orientation of the body. We regard the geometric shape of the boundary of the body as fixed. The shape is then completely specified by the characteristic length l. The orientation of the body is specified by two angles, for example, α and β, which the velocity vector forms with the axes of a coordinate system fixed in the body. The force and heat transfer may then be expressed in the following functional form:

$$F = F_1(\alpha, \beta, l, V, T_w, g, p, \rho, T, e, \mu, k)$$

and   (1.17)

$$Q = F_2(\alpha, \beta, l, V, T_w, g, p, \rho, T, e, \mu, k)$$

By using the methods of *dimensional analysis** we may further express these functional relations in nondimensional form. To describe the force and the heat transfer and the 12 quantities on which they depend we employ the four fundamental dimensions: mass, length, time, and

---

* For the methods of dimensional analysis see the cited references.

temperature. This means that according to dimensional analysis the force and heat transfer, suitably nondimensionalized, may each be expressed as a function of eight independent nondimensional parameters formed from the 12 variables α, β, l, V, T_w, g, p, ρ, T, e, μ, and k. Since the combination $\rho V^2 l^2$ has the dimensions of a force, we introduce the parameter $F/\rho V^2 l^2$ in place of F. Similarly, because $lk(T - T_w)$ has the dimensions of energy per unit time, the same as those of Q, we introduce the parameter $Q/lk(T - T_w)$ in place of Q. The eight independent parameters on which $F/\rho V^2 l^2$ and $Q/lk(T - T_w)$ depend may be given in different forms. We choose those forms that would lead us to the parameters which have already appeared in our considerations. The results may be expressed as follows:

$$\frac{F}{\rho V^2 l^2} = f_1\left(\alpha, \beta, \frac{V}{\sqrt{gl}}, \frac{\rho V l}{\mu}, \frac{V}{\sqrt{p/\rho}}, \frac{p}{\rho e}, \frac{kT}{\mu e}, \frac{T_w}{T}\right)$$

$$\frac{Q}{lk(T - T_w)} = f_2\left(\alpha, \beta, \frac{V}{\sqrt{gl}}, \frac{\rho V l}{\mu}, \frac{V}{\sqrt{p/\rho}}, \frac{p}{\rho e}, \frac{kT}{\mu e}, \frac{T_w}{T}\right)$$

(1.18)

Introducing Fr, R, M, σ, a and C_p, we may write (1.18) as

$$\frac{F}{\rho V^2 l^2} = f_1\left(\alpha, \beta, \text{Fr}, \text{R}, \text{M}, \frac{a^2}{e}, \frac{1}{\sigma}\frac{C_p T}{e}, \frac{T_w}{T}\right)$$

$$\frac{Q}{lk(T - T_w)} = f_2\left(\alpha, \beta, \text{Fr}, \text{R}, \text{M}, \frac{a^2}{e}, \frac{1}{\sigma}\frac{C_p T}{e}, \frac{T_w}{T}\right)$$

(1.19)

The determination of the functions f_1 and f_2 is the central problem of many theoretical and experimental studies in aerodynamics and hydrodynamics.
For a perfect gas with constant specific heats we have

$$\frac{a^2}{e} = \gamma(\gamma - 1)$$

$$\frac{C_p T}{e} = \gamma$$

In such a case (1.19) may be expressed in the following form:

$$\frac{F}{\rho V^2 l^2} = f_1\left(\alpha, \beta, \text{Fr}, \text{R}, \text{M}, \gamma, \sigma, \frac{T_w}{T}\right)$$

$$\frac{Q}{lk(T - T_w)} = f_2\left(\alpha, \beta, \text{Fr}, \text{R}, \text{M}, \gamma, \sigma, \frac{T_w}{T}\right)$$

(1.20)

1.6 Range of Some Parameters

The problem we have been considering, namely that of the motion of a body with a constant speed through a fluid, covers a wide range of length-scales, speeds, and fluids. The length-scales may vary from about 10^{-3} mm to about 300 meters. The speeds may range from a few centimeters per second (or even less) to thousands. Fluids with widely differing values of kinematic viscosity may be of concern. This being the case, the range of Reynolds numbers is indeed very wide. Reynolds numbers from about 10^{-3} to about 10^7 to 10^8 occur in practice*.

The motion of aircraft through the atmosphere involves length-scales of about 1 to 50 meters, speeds of about 30 meters/sec to about 1000 meters/sec, and Reynolds numbers of about 10^5 or 10^6 to 10^7 or 10^8. The motion of submarines and torpedoes through water involves length-scales of about 5 to 100 meters, speeds of about 10 meters/sec, and Reynolds numbers of about 10^9. Our concern in this book is with problems of this type and thus with motions involving high Reynolds numbers, $R \geq 10^4$.

With speeds and length-scales varying over a wide range, the range of Froude number, $Fr = V/\sqrt{gl}$, occurring in practice is also wide. For fluid motions involving aircraft the Froude numbers may have values of 1 to 100 or more. For submarine and torpedo motions Fr may have values in the range of 0.3 to 1.5.

We now consider the range of Mach numbers. Under normal conditions the speed of sound in gases is about 300 to 400 meters/sec. In liquids the speed of sound is much higher. For instance, at normal temperature, the speed of sound in air is 370 meters/sec, whereas in water it is about 1600 meters/sec.

Speeds of bodies moving through liquids are usually limited to small values. Possible maximum speeds are in the range of 20 to 30 meters/sec. This being the case, the Mach numbers in liquid flows are extremely small—the order of 0.01.

Speeds of bodies moving through gases, on the other hand, may range from considerably small values to large values compared to the speed of sound. This means that the Mach numbers for gas flows may vary from values much less than unity to values much greater than unity.

1.7 Conditions for Neglecting Compressibility Effects

We have seen that the changes in density of a fluid element come about from pressure and temperature changes arising out of the motion of the fluid. Density changes caused by pressure changes would be negligible if

* See Lighthill (1963).

the following conditions were satisfied (refer to relation 1.10):

$$\frac{V}{\sqrt{p/\rho}} \sim \frac{V}{a} = \mathrm{M} \ll 1$$

$$\frac{gl}{p/\rho} \sim \frac{gl}{a^2} = \frac{\mathrm{M}^2}{\mathrm{Fr}^2} \ll 1 \tag{1.21}$$

$$\frac{\mu V}{lp} \sim \frac{\mathrm{M}^2}{\mathrm{R}} \ll 1$$

Density changes caused by temperature changes would be negligible if the following conditions were fulfilled (refer to relation 1.13):

$$\frac{V^2}{e} = \mathrm{M}^2 \frac{a^2}{e} \ll 1$$

$$\frac{gl}{e} = \frac{\mathrm{M}^2}{\mathrm{Fr}^2} \frac{a^2}{e} \ll 1$$

$$\frac{p}{\rho e} \sim \frac{a^2}{e} \ll 1$$

$$\frac{\mu V}{\rho le} = \frac{\mathrm{M}^2}{\mathrm{R}} \frac{a^2}{e} \ll 1$$

$$\frac{k(T - T_w)}{\rho V le} = \frac{1}{\sigma \mathrm{R}} \frac{C_p T}{e} \frac{T - T_w}{T} \ll 1$$

By putting these conditions together we may state that compressibility effects would be negligible under the following conditions:

$$\mathrm{M} \ll 1 \qquad \frac{\mathrm{M}^2}{\mathrm{Fr}^2} \ll 1 \qquad \frac{\mathrm{M}^2}{\mathrm{R}} \ll 1$$

$$\frac{a^2}{e} \ll 1 \qquad \frac{1}{\sigma \mathrm{R}} \frac{C_p T}{e} \frac{T - T_w}{T} \ll 1 \tag{1.22}$$

It may be noted that the parameter $C_p T/e$ is of the order of unity.

Case of Liquids. In the motion of a liquid the Mach number is very small, $\mathrm{M} \simeq 0.01$. The parameters gl/a^2 and $\mu V/lp$ are also extremely small. Consequently, in the motion of a liquid the density changes caused by pressure changes are indeed negligible. This, however, does not mean that there are no density changes, for they may result from temperature changes. Temperature changes occur primarily in heat transfer between the fluid and the body. In the absence of heat transfer we may state that the motion of a liquid occurs without any noticeable changes in the density of a fluid element. The motion then is essentially one of an incompressible fluid.

Case of Gases. In the motion of gases the Mach number may range from values much less than unity to values much greater than unity, and, in general, density changes arise from both pressure and temperature changes. However, when the Mach number is much less than unity, that is, when the speed is small compared to the speed of sound, and there is no heat transfer between the fluid and the body, the changes in the density of an element would be negligible. Under such circumstances the motion of a gas may also be regarded as that of an incompressible fluid.

1.8   Conditions for Neglecting Gravity Effects

The force due to gravity on a fluid element may be neglected in comparison with the inertial force, the pressure force, and the viscous force when the following conditions are fulfilled:

$$\text{Fr}^2 = \frac{V^2}{gl} \sim \frac{\text{inertial force}}{\text{gravity force}} \gg 1$$

$$\frac{\delta p}{p}\frac{\text{Fr}^2}{\text{M}^2} = \frac{a^2}{gl}\frac{\delta p}{p} \sim \frac{\text{pressure force}}{\text{gravity force}} \gg 1$$

$$\frac{\text{Fr}^2}{\text{R}} = \frac{vV}{gl^2} \sim \frac{\text{viscous force}}{\text{gravity force}} \gg 1$$

The first two conditions are satisfied whenever the length-scale and the speed are such that both V and a are much greater than $\sqrt{gl}$. Since g is about 9.8 meters/sec^2, the speeds V and a should be greater than say $3.2\sqrt{l}$ meters/sec.

The condition that $a \gg \sqrt{gl}$ is one of the conditions for neglecting the compressibility effects due to gravity, the other being that $e \gg gl$.

1.9   Nature of the Problem when Compressibility Effects are Negligible

We have seen that when the Mach number is much less than unity, Fr and R are greater than unity, and that when there is no heat transfer compressibility effects are negligible. Under such circumstances the problem of a body moving with a constant speed through a fluid may be regarded as that of a body moving through an incompressible fluid. The main objective of the problem, then, is that of determining the force on the body. The functional dependence of the force is represented as follows:

$$\frac{F}{\rho V^2 l^2} = f_1(\alpha, \beta, \text{Fr}, \text{R}) \tag{1.23}$$

When the motion can be regarded as incompressible, it is possible (as we shall see later) to separate the effect of gravity from the dynamical problem of the motion of the fluid. Gravity effects may be accounted for by considering the corresponding static situation. This being the case, the Froude number may be dropped from consideration, and the problem reduces to that of determining the following functional relation for the force:

$$\frac{F}{\rho V^2 l^2} = f(\alpha, \beta, R) \tag{1.24}$$

Our concern is with flows at high Reynolds numbers, that is, for $R \geq 10^4$. For such flows and for certain types of body that are of practical interest, and for certain orientations of these bodies, the effects of viscosity are generally confined to a very thin layer surrounding the body. In the major part of the flow the viscous effects are negligible. In such a situation it is possible to analyze certain important aspects of the problem by assuming that the viscosity is zero or equivalently that the fluid is *frictionless or inviscid*. The Reynolds number then does not enter the problem and the functional dependence for the force is simply

$$\frac{F}{\rho V^2 l^2} = f(\alpha, \beta) \tag{1.25}$$

The central problem in this book is the determination of the function f for certain types of body such as wings and airshiplike bodies.

Although under certain circumstances the role of viscosity and that of the Reynolds number could be dropped from direct considerations, it is necessary always to bear in mind the important role played by viscosity or, more significantly, the Reynolds number (for as we have seen, what matters is not viscosity per se but the nondimensional parameter Vl/ν), in determining the nature of the flow and the type of a meaningful approximate treatment that may be attempted for the problem. To gain some appreciation in this direction we describe briefly in the next section some qualitative features of certain flows at different Reynolds numbers.

1.10 Variation of Flow Patterns with Reynolds Number

To describe the flow patterns qualitatively we use illustrations of the *streamlines* of the flow. A streamline is a line with the property that at each point of the line the direction of the line is the same as the direction of the velocity vector of the flow at that point. Suitable experimental methods permit visualization and photography of the streamline patterns of fluid flow.*

For a rigid body translating with a constant velocity through an

* See references cited at the end of the book.

unbounded fluid the streamline pattern appears the same at all times if observed from a frame fixed with respect to the body. In fact, it is equivalent to that due to steady flow past a body that is kept stationary, and the velocity far ahead of the body is equal to the negative of the velocity of the body in the initial problem (see Fig. 1.8). By steady flow we mean that

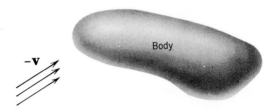

Fig. 1.8 Steady flow past a fixed body.

the velocity at every point is constant in time. In the following we refer to our problem as that of steady flow past a stationary body. In steady flow streamlines and paths described by fluid elements are the same.

In describing the variation of flow patterns with Reynolds number we concern ourselves with situations in which compressibility effects are negligible, for this is our main concern. Later we describe some qualitative features of the effects of Mach number on the flow pattern at high Reynolds number.

Flow Past Bluff Bodies. We consider two-dimensional flow past a circular cylinder (Fig. 1.9). By this we mean that the flow pattern is

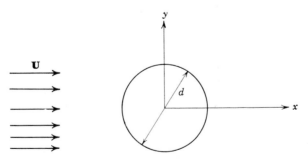

Fig. 1.9 Flow past a circular cylinder.

identical in all planes normal to the axis of the cylinder. Such a flow may be realized approximately over the central region of a long cylinder. Plates 3 and 4 show the flow patterns for various Reynolds numbers. (In these examples, $R = Ud/v$, where d is the diameter of the cylinder.)

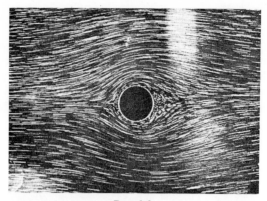

R = 3·9

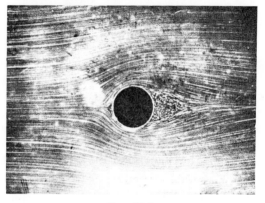

R = 18·6

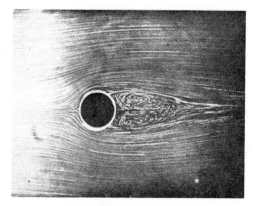

R = 33·5

Plate 3   Flow past a circular cylinder at small Reynolds numbers, after Homann (1936). Courtesy of Oxford University Press.   Plate 31 of Goldstein (1938).

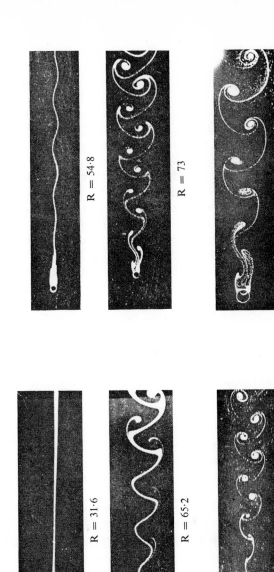

R = 54·8

R = 73

R = 161

R = 31·6

R = 65·2

R = 101·5

Plate 4 Flow past a circular cylinder at high Reynolds numbers, after Homann (1936). Courtesy of Oxford University Press. Plate 32 of Goldstein (1938).

29

For very small Reynolds numbers (for R less than about 4) the stream-lines near the surface are closely parallel to it. The flow pattern is almost symmetrical about the diameters of the cylinder parallel and normal to the undisturbed stream. As R increases, the streamlines at the rear of the cylinder widen out more and more to form a closed region behind it. This region is known as a *separation region* or *separation bubble*. It is followed by a layer known as a *wake* (see also Plate 4). The separation bubble is separated from the main flow by a pair of *separation lines* originating from some point on each side of the rear of the cylinder. The points are referred to as *separation points*. The flow in the separation bubble consists of two regions in which the fluid is in circulatory motion, and each of these regions is known as a *vortex*. Thus the separation bubble consists of a vortex pair.

As R increases, the separation bubble becomes more and more elongated in the direction of the main stream. This happens until a certain value of R (about 40) is reached and instability sets in. In the range below this R, the separation point moves forward from the rear as R increases.

As R rises above this critical value, instabilities set in and the vortices become asymmetrical shortly after the beginning of the motion, leave the cylinder, and move downstream (see Plates 4 and 5). Vortices form, grow, and leave the cylinder periodically. As they move downstream, they form a regular pattern consisting of a double row of alternating vortices known as *Karman's vortex street* (see Plate 5). A distinct vortex street, as shown in Plate 5, occurs for Reynolds numbers between 40 and about 200. For larger Reynolds numbers the processes become more and more irregular and complicated as R increases.

The variation with R shown by the flow pattern around a circular cylinder is characteristic also of two-dimensional steady flow past cylinders of some other shapes. Bodies that show such flow characteristics are known as *bluff bodies*. We may describe a bluff body as one for which separation of the flow from the surface takes place well ahead of the rear part leading to a large wake. It should be noted that whether or not a body is bluff depends not only on the shape of the body but also on its orientation. This is illustrated by Plate 6, which shows the flow past an elliptic cylinder. When the cylinder is oriented with its major axis normal to the stream, it behaves like a bluff body. When the cylinder is oriented with its major axis parallel to the stream, it behaves unlike a bluff body.

The variation with Reynolds number of the flow patterns around three-dimensional bluff bodies, such as a sphere, is in some ways similar to that around two-dimensional bodies. There are, however, certain differences, but there is less information on three-dimensional bluff bodies.

Flow Past Streamlined Bodies.   A streamlined body may be described as one for which separation of flow, if it occurs, does so near the rear of the

Plate 5 Kármán vortex street $R = 250$: (*a*) as viewed with respect to the cylinder; (*b*) as viewed with respect to the undisturbed fluid. Courtesy of Professor O. G. Tietjens: Plate 24 of Prandtl and Tietjens (1934). Also, O. G. Tietjens: Strömungslehre Vol. I, Springer-Verlag, 1960.

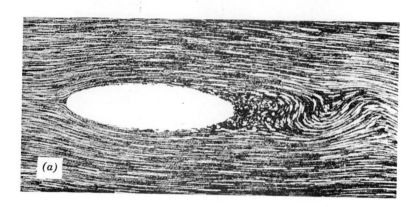

Plate 6   Flow past an elliptic cylinder: (a) with the major axis in the direction of the undisturbed stream;   (b) with the major axis normal to the stream.   (a) Courtesy of Oxford University Press.   Plate II of Goldstein (1938).   (b) Courtesy of Professor O. G. Tietjens.   Figure 66, Plate 26 of Prandtl and Tietjens (1934).

body and the consequent wake is very narrow. An example, shown in Plate 6, is the elliptic cylinder with its major axis in the direction of the stream. Experiments reveal that to achieve a streamlined body the body must be well rounded and slender and elongated in the direction of motion and that the surface of the body should come to a point or an edge at the rear with a gentle curvature. A slender body of revolution is another example of a streamlined body. So also is a cylinder of the "airfoil" shape, which is employed for lifting wings. At very low Reynolds numbers the flow pattern due to steady two-dimensional flow past a streamlined cylinder, such as the slender elliptical cylinder shown in Plate 6 is no different from that for flow past a bluff body.

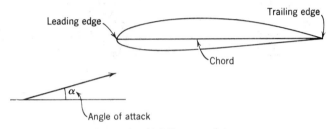

Fig. 1.10   Airfoil nomenclature.

Significant differences, however, are observed at high Reynolds numbers for which the flow pattern depends on the orientation of the body with respect to the main stream. We consider the case of flow past an airfoil (see Plate 7). The orientation of the airfoil with respect to main stream is known as the angle of attack, denoted by α. It is measured between the direction of the undisturbed main stream and a reference line, known as the chord, drawn from the leading edge to the trailing edge (see Fig. 1.10). At small values of α the streamlines near the body follow the airfoil surface closely, right to the trailing edge (see *a* of Plate 7). There is a very narrow wake.

As the angle α increases, changes in the flow pattern occur, primarily on the upper surface of the airfoil. Separation begins at the rear on the upper side and moves forward as α increases. Correspondingly, a wake is generated which grows as α increases. At a certain value of α the flow separates near the leading edge, giving rise to a large wake, just as in the flow past a bluff body (see *b* of Plate 7). In practical applications we are interested in situations at small α.

The flow pattern for finite wings large in a direction normal to that of the main stream is in many ways similar, at least over a major portion of the wing, to that for an airfoil. Complications develop for short wings.

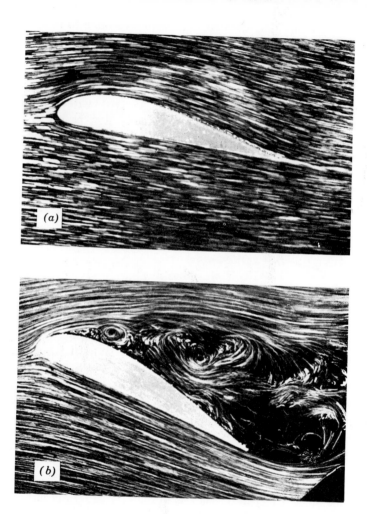

Plate 7 Flow past an airfoil: (a) at a gentle angle of attack; (b) at a large angle of attack. Note the separation on the top surface. Courtesy of Oxford University Press. Plate 12 of Goldstein (1938).

For steady flow past a slender body of revolution with its axis in the direction of the main stream we find that separation occurs close to the rear and that the wake is very thin.

Our concern would be with steady flow at high Reynolds numbers past airfoils and large finite wings at small angles of attack and past bodies of revolution with their axes parallel to the stream. Furthermore, we are interested only in speeds for which the compressibility effects are negligible. To have some appreciation of the effects of compressibility we give in the next section a brief description of the effect of Mach number on the flow pattern, at high Reynolds numbers around an airfoil and a body of revolution.

1.11   Variation of Flow Pattern with Mach Number

We may consider steady flow at high Reynolds number past an airfoil at a gentle angle of attack. Plate 8 shows the variation of the flow pattern with increasing Mach number $M = U/a_\infty$, where U is the speed of the undisturbed stream far ahead of the airfoil and a_∞ is the speed of sound in the undisturbed stream. The photographs of flow were obtained by what is called the shadowgraph method, which responds to variations of density (strictly to the second spatial derivative) in the flow field.*

These photographs show that the flow pattern changes considerably as the Mach number is increased. The flow fields for M below a certain value are completely different from those for M above that value. Above this value, which may be referred to as the *critical Mach number*, the flow patterns are characterized by the appearance of narrow regions through which considerable spatial variations of density occur. Such regions are known as *shocks*.† As we proceed in the direction of the main stream, the density increases, almost abruptly, through the shocks. As M increases from the critical value but still remains less than unity, shocks first appear on the forward part of the airfoil and move back with increasing strength towards the trailing edge.

As M exceeds unity but is still close to it, a detached shock wave appears in front of the leading edge and the other shocks appearing on the surface of the airfoil occur at or near the trailing edge. As M continues to increase, the leading edge shock moves closer to that edge. If the nose of the airfoil is sharp, as is common for "supersonic airfoils," the leading edge shock is attached to that edge when M exceeds a certain value greater than unity.

On the basis of the flow patterns exhibited at various Mach numbers, it is usual to divide the whole range of possible mach numbers into several parts, each part being associated with a different type of flow. The range of

* See cited references for a description of this and other methods.
† For a detailed description of the origin of shocks consult cited references.

←
Direction of Flow

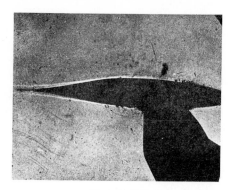

$M_1 = 0.628$

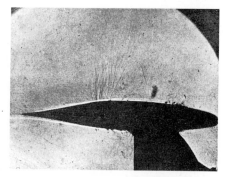

$M_1 = 0.732$

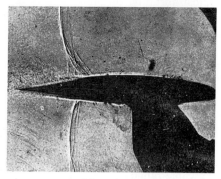

$M_1 = 0.835$

<-----------
Direction of Flow

$M_1 = 0.683$

$M_1 = 0.784$

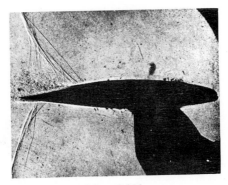

$M_1 = 0.876$

Plate 8 Shadowgraphs of subsonic flow past an airfoil at a gentle angle of attack (critical Mach number is 0.695). Courtesy of Oxford University Press. Plate 16 of *Modern Developments in Fluid Dynamics: High Speed Flow*, L. Howarth, editor, 1953.

M = 0·86

M = 1·01

M = 1·26

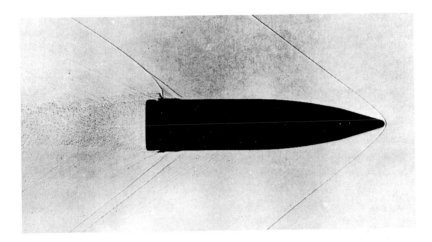

M = 1·76

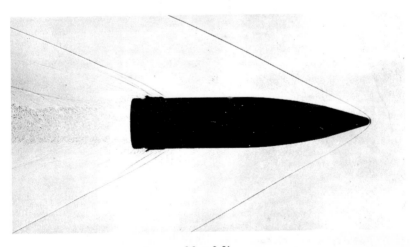

M = 2·51

Plate 9   Shadowgraphs at various Mach numbers of a projectile in flight.   Courtesy of Oxford University Press.   Plates 21 and 22 of *Modern Developments in Fluid Dynamics: High Speed Flow*, L. Howarth, editor, 1953.

M extending from zero up to the critical Mach number is known as the *subsonic range*. In this range the flow is characterized by the absence of any shocks. It exhibits all the features of incompressible flow, although gradual density variations take place as M increases. In this range the local Mach number (the ratio of the local speed to local speed of sound) is subsonic at every point of the flow field. At subsonic Mach numbers compressible flow past a slender airfoil may be related to an equivalent incompressible flow problem.

The range of M extending from the critical value to some value above unity but close to unity is usually referred to as the *transonic range*. In this range the flow field is characterized by the appearance of shocks on the surface of the body, and a detached leading edge shock when M is close to unity. In such a range the rather complicated flow field consists of partly (locally) subsonic and partly (locally) supersonic regions.

The range in which M is greater than the value for which there are only leading edge and trailing edge shocks with no others on the airfoil surface (except perhaps near the trailing edge) is known as the *supersonic range*. In this range the flow almost everywhere is supersonic. A small locally subsonic flow region may, however, exist near the leading edge, particularly if the airfoil has a round nose. The range of Mach numbers exceeding a certain supersonic Mach number (which depends on the situation at hand) is referred to as the *hypersonic range*.

Steady flow past a body of revolution exhibits similar features at different Mach numbers. Plate 9 shows shadowgraph pictures of a projectile in free flight.

1.12 Effects of Viscosity at High Reynolds Number: The Boundary Layer

We now return to the consideration of the problem of steady flow at high Reynolds number past a fixed rigid streamlined body. Although we are primarily concerned with flows with negligible compressibility effects, the following considerations are equally applicable in their essentials when such effects are present. When compressibility effects exist, account must be taken of energy exchanges and temperature differences. In transonic and supersonic flows complications develop from the appearance of shocks.

Boundary Layer Concept. In 1904 Prandtl introduced a far-reaching idea, now known as the *boundary layer concept*, and showed the way to treat satisfactorily the flow past a streamlined body at high Reynolds numbers (Prandtl, 1904). He showed that *at high Reynolds numbers the effects of viscosity are confined to a very thin layer close to the body and a thin wake extending from the body*. When a fluid flows past a fixed body, because of the effect of viscosity, no matter how small, the layer of fluid immediately adjacent to the surface is at rest. Away from the wall the fluid is in motion with a certain velocity. This means that as the solid surface is approached fluid layers are retarded. The retardation arises out of the action of the viscous forces. Depending on the viscosity of the fluid, the retarding effect may extend only to short distances or to large distances from the body. It is observed that for fluids with small viscosity, such as water and air (strictly speaking, in flows at large Reynolds numbers), the retardation effects are confined only to a very thin region close to the body.

In such a region the velocity rises rapidly from zero at the wall to its value in the main stream. In that region the spatial rate of change of velocity (velocity gradient) in a direction normal to the body is large, and consequently the viscous forces would not be negligible even if the viscosity were small. Outside that region the velocity gradients are small and the viscous forces there would be negligible. The thin region close to the body in which the viscosity effects are confined is called the retardation layer or, more popularly, *the boundary layer*. Outside the boundary layer the fluid may be regarded to a high degree of accuracy as an inviscid fluid. Both theory and experiment have supported the correctness of Prandtl's boundary-layer concept.

Some Characteristics of a Laminar Boundary Layer. The flow inside a boundary layer may be laminar or turbulent or partly laminar and partly turbulent. We now consider the characteristics of a laminar boundary layer.

For the sake of clarity we consider steady two-dimensional flow past a thin flat plate (Fig. 1.11). We introduce the coordinates x and y as shown. According to the concept of the boundary layer, we expect the following: just ahead of the leading edge the fluid approaches the velocity U, the velocity of the undisturbed stream, at all distances normal to the plate; at any section x lying on the plate the velocity will be zero at the plate, $y = 0$, and rise to the value U a short distance, say $y = \delta$, from the plate. This situation is illustrated in Fig. 1.11, in which the distribution of u, the velocity component parallel to the plate, with distance y from the plate is shown. The scale for y is greatly exaggerated.

We call δ the *boundary layer thickness*. It varies along the plate, being zero at the leading edge; thus

$$\delta = \delta(x) \qquad (1.26)$$

We now wish to estimate $\delta(x)$. We do this on the basis of the consideration that *within the boundary layer the viscous forces are of the same order of magnitude as the inertial forces*. In estimating these forces for a fluid element in the boundary layer, we have to bear in mind the fact that within the layer the changes in the y-direction occur in a much smaller distance than in the x-direction, so that the flow situation in the boundary layer at any section x does not depend sensibly on what happens behind that section but mostly on what happens ahead. This means that within the boundary layer *the characteristic length for changes in the x-direction is the distance x itself of the section under consideration from the leading edge of the plate. The characteristic length for change in the y-direction is the boundary-layer thickness δ(x) at the section under consideration.*

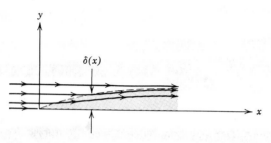

Fig. 1.11 Schematic representation of the boundary layer on a flat plate.

Now consider an element situated at the section x. Using considerations similar to those given in Section 1.5 and noting that x is the characteristic length for changes in the x-direction, we find that the x-component of the inertial force acting on the element per unit volume is proportional to $\rho U^2/x$. Similarly, we find that the x-component of the viscous force on the element per unit volume is proportional to τ_0/δ or equivalently to $\mu U/\delta^2$, where τ_0 is the shear stress on the plate at the section x; $\tau_0 = \tau_0(x)$. Since the viscous and inertial forces are of the same magnitude, we have

$$\frac{uU}{\delta^2(x)} \quad \text{or} \quad \frac{\tau_0}{\delta(x)} \sim \frac{\rho U^2}{x}$$

Using this relation and introducing a Reynolds number R_x defined by $R_x = Ux/\nu$, we obtain

$$\delta(x) \sim \sqrt{\frac{\nu x}{U}} \tag{1.27}$$

$$\frac{\delta(x)}{x} \sim \sqrt{\frac{1}{R_x}} \tag{1.28}$$

$$\frac{\tau_0(x)}{\rho U^2} \sim \sqrt{\frac{1}{R_x}} \tag{1.29}$$

We note that the boundary layer grows as the square root of the distance from the leading edge. The boundary-layer thickness is proportional to the square root of the kinematic viscosity ν. Thus the thickness δ decreases with ν, with $\delta \to 0$ as $\nu \to 0$. Since $\delta(x)/x$ is proportional to $\sqrt{1/R_x}$, then

$$\frac{\delta(x)}{x} \ll 1 \quad \text{if} \quad R_x \gg 1$$

Thus, if R_x becomes very large, δ/x becomes very small, and the flow over the plate with its very thin boundary layer becomes nearly that of a fluid without viscosity. However, no matter how small δ/x is, there is always the boundary layer with a decisive effect on the flow at the plate.

Consider now the behavior of the viscous stress τ_0 at the plate. We observe that $\tau_0/\rho U^2$ varies as $\sqrt{1/R_x}$; thus the stress becomes smaller as R_x becomes larger. Even at large values of R_x there is a finite, although small, viscous stress on the plate. The stress τ_0 is proportional to $\sqrt{\nu}$ and is smaller, the smaller the kinematic viscosity of the fluid. However, no matter how small ν is, there is a finite viscous stress and consequently a viscous drag on the plate.

We now consider the variation of the velocity and pressure within the boundary layer. Let u and v denote the x- and y-components of the velocity. In the boundary layer, at any x, u is a function of y, with $u = 0$ at $y = 0$ and $u = U$ at $y = \delta$. Theoretically, u approaches U asymptotically and therefore the definition of δ cannot be made precise. We may, however, define δ as the distance within which u reaches a certain percentage of U, say for instance 99 per cent of U. Now, since δ varies with x, u must also vary with x; thus $u = u(x, y)$. Since τ_0 is equal to $\mu(\partial u/\partial y)$ at the plate and τ_0 decreases with x, the slope $\partial u/\partial y$ at the plate decreases with x. Using this result and the fact that δ increases with x, we may represent the variation of u schematically (see Fig. 1.12). We note that at a given distance from the plate u decreases with the distance from the leading edge.

Since u is a function of x and y, there will be a v that will also be a function of x, y. This is seen in the following manner. Consider the rectangular box shown in Fig. 1.13; the thickness of the box is unity in the direction normal to the xy-plane. Two sides of the box are normal to the x-direction and situated at an infinitesimal distance dx. The other two sides are normal to the y-direction, one of them being the surface of the plate and the other located at $y = y$. Fluid flows into and out of the box through its sides. However, there is no accumulation of mass in the box. We therefore require the rate of flow of mass into the box to balance the rate of flow of mass out of the box. Fluid flows into the box through the x-face (i.e., normal to x-direction) situated at x and flows out through the x-face

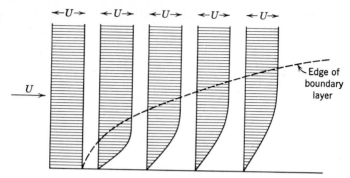

Fig. 1.12 Schematic representation of the u-velocity distribution in the boundary layer over a flat plate.

situated at $x + dx$. The rate of inflow through the face at x is equal to

$$\rho \int_0^y u(x, y)\, dy$$

whereas the rate of outflow through face at $x + dx$ is equal to

$$\rho \int_0^y u(x + dx, y)\, dy$$

Since $u(x + dx, y)$ is less than $u(x, y)$, the outflow does not balance the inflow. This means that because there is no accumulation of mass in the box, there must be a flow of fluid through the y-faces of the box to balance the net inflow through its x-faces. There must therefore be a v velocity. However, there is no flow through the y-face at $y = 0$, which is the plate itself. If $v(x, y)$ is the y-component of the velocity at the point x, y, the

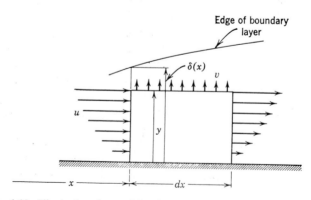

Fig. 1.13 Illustrating the need for the v-velocity in the boundary layer.

rate of outflow of mass through the y-face at y is equal to

$$\rho v \, (x, \, vy) \, dx$$

Equating this rate of outflow to the net inflow through the x-faces, we have

$$v(x, \, y) \, dx = \int_0^y [u(x, \, y) - u(x + dx, \, y)] \, dy$$

We may set, neglecting higher order terms,

$$u(x + dx, \, y) = u(x, \, y) + \left(\frac{\partial u}{\partial x}\right)_{x,y} dx$$

We then have

$$v(x, \, y) = -\int_0^y \frac{\partial u}{\partial x} \, dy$$

Note that $\partial u/\partial x$ is negative.

At the plate v is zero. At the edge of the boundary layer $y = \delta(x)$ the v-component does not vanish but has a value given by

$$v(x, \, \delta(x)) = -\int_0^{\delta(x)} \frac{\partial u}{\partial x} \, dy \qquad (1.30)$$

On the basis of this relation we may state that

$$v(x, \, \delta) \sim \frac{U \, \delta(x)}{x}$$

or

$$\frac{v(x, \, \delta)}{U} \sim \frac{\delta}{x} \sim \sqrt{\frac{1}{R_x}} \qquad (1.31)$$

We thus conclude that the v-velocity is small, the ratio v/U being of the same order of magnitude as δ/x.

We next consider the variation of pressure across the boundary layer. The pressure force in the y-direction, $\partial p/\partial y$ per unit volume, on a fluid element in the boundary layer is of the same order of magnitude as that of the inertial force, $\rho v^2/\delta$ per unit volume, on the element in that direction.* We therefore have

$$\frac{\partial p}{\partial y} \sim \rho \frac{v^2}{\delta} \sim \rho U^2 \frac{\delta}{x^2} \qquad (1.32)$$

The magnitude of the pressure change Δp across the boundary layer is therefore given by

$$\Delta p \sim \rho v^2 \sim \rho U^2 \left[\frac{\delta(x)}{x}\right]^2 \sim \frac{\rho U^2}{R_x} \qquad (1.33)$$

We conclude that in the boundary layer spatial variation of pressure in the

* The viscous force in that direction is of the same order of magnitude.

y-direction and the actual change in pressure across the layer are negligibly small. As we shall see later, this conclusion has far-reaching implications.

The solution for the problem of steady incompressible flow in a laminar boundary layer along a flat plate, as formulated by Prandtl, was obtained by Blasius in 1908. His solution shows that δ (defined as the value of y when u is within 0.6 per cent of U) is given by

$$\frac{\delta(x)}{x} = \frac{5.2}{\sqrt{R_x}} \qquad (1.34)$$

The local skin friction coefficient is given by

$$C_f \equiv \frac{\tau_0(x)}{\frac{1}{2}\rho U^2} = \frac{0.664}{\sqrt{R_x}} \qquad (1.35)$$

The distribution of the velocity across the boundary layer is similar at different x-stations along the plate. This being the case, variation of u with x and y may be represented by a single curve. Such a curve, as obtained by Blasius, is shown in Fig. 1.14, where u/U is plotted against the parameter $\eta = y\sqrt{U/vx}$. Also shown in the figure are experimental results obtained by Nikuradse (1942). It is seen that the theory and the experiment are in excellent agreement. This agreement is also found for the local skin friction coefficient, shown in Fig. 1.15. The experimental results are obtained by Liepmann and Dhawan (1951, 1953) by direct measurement of the skin friction coefficients.

So far we have considered the features of a laminar boundary layer along a plane surface. Similar results also hold for the boundary layer along a curved surface if certain conditions are met. It is necessary that the radius of curvature of the surface is everywhere large compared to the boundary-layer thickness and that there is none of the variation in curvature that would occur near sharp edges. Under such conditions the flat-plate results may be applied to the problem of boundary layer along a curved surface if we now regard the coordinates x, y as defining a suitable system of curvilinear coordinates given by curves parallel to the given surface and straight lines normal to the surface. The surface itself is given by $y = 0$ (see Fig. 1.16).

For flow past a curved surface the velocity component parallel to the wall at the edge of the boundary layer is not a constant, as it is in the case of the flat plate, but a function of the distance x. Similarly the pressure at the edge of the boundary layer is also a function of x. We therefore introduce the following notation:

$$\begin{aligned} U_e &= U_e(x) = u(x, \delta(x)) \\ p_e &= p_e(x) = p(x, \delta(x)) \end{aligned} \qquad (1.36)$$

The subscript e signifies that the quantity refers to the edge of the boundary layer.

Turbulent Boundary Layer. Observations show that at high Reynolds numbers the flow in a boundary layer does not remain laminar all along the surface of a solid body but usually becomes turbulent at some distance

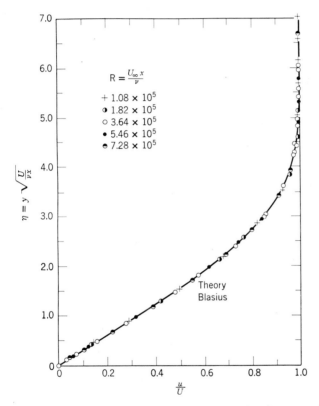

Fig. 1.14 Velocity distribution in the laminar boundary layer on a flat plate. Experimental results from Nikuradse (1942).

from the leading edge. The transition from laminar to turbulent motion originates from the instability (in some form or other) of the laminar motion and depends on various factors.* The transition takes place over a region and not at a point. The value of the Reynolds number R_x at which transition to turbulent flow begins is known as the critical Reynolds number denoted by $R_{x,\text{crit}}$. Under normal circumstances $R_{x,\text{crit}}$ for the boundary

* For some details on this subject (which is still not fully understood) see, for example, Kuethe and Schetzer (1959).

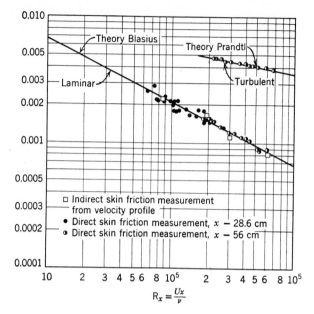

Fig. 1.15 Local skin friction on a flat plate. Direct measurements by Liepmann and Dhawan (1951, 1953).

layer along a flat plate is in the range of about 10^5 to 10^6. This means that for the problems that will concern us later the boundary layer is likely to be turbulent over most parts of the body.

Since turbulent motion results in pronounced mixing of the fluid, the retarding effect of a solid surface on the fluid spreads farther in the direction normal to the surface in turbulent motion that in laminar motion. In general, however, the retardation effect still takes place prominently in a layer that is very thin in comparison to the characteristic distance along the body, and the concept of a boundary layer is still applicable. Even when the boundary layer is turbulent, since the turbulent fluctuations should

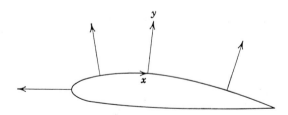

Fig. 1.16 Boundary layer coordinates for flow over a curved surface.

die out as the solid surface is approached, there is always a layer of fluid
next to the surface that is laminarlike. In such a layer, which is known as
the *laminar sublayer*, the velocity rises rapidly from zero at the wall to a
certain value. In the sublayer the velocity gradient normal to the wall is
severer than it would be if the boundary layer were wholly laminar. This
being the case, the local shear stress on the solid surface is much greater
than it would be if the boundary layer were laminar. This is illustrated in
Fig. 1.15, in which measured friction coefficients for a turbulent boundary
layer along a plate are also shown. The velocity distribution across a

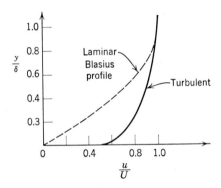

Fig. 1.17   Comparison of laminar and turbulent boundary layer profiles.

turbulent boundary layer is shown schematically in Fig. 1.17, in which the
Blasius distribution for laminar layer is also included for comparison.

   The mechanism of the flow in a turbulent boundary layer is complicated
and different from that of the flow in a laminar boundary layer. Consequently, the order of magnitude estimates given for the laminar layer do
not apply directly to the turbulent layer. There is still no detailed theory
for the flow in a turbulent boundary layer. There are, however, semi-
empirical solutions for the turbulent boundary-layer thickness and local
friction coefficients that are useful. It is found that for Reynolds numbers
in the range 10^5 to 10^6 the boundary-layer thickness and the local shear
stress τ_0 show the following dependence on the Reynolds number R_x:

$$\frac{\delta(x)}{x} \sim \frac{\tau_0(x)}{\rho U^2} \sim \frac{1}{R_x^{0.2}} \tag{1.37}$$

Recall that in the laminar case δ/x and $\tau_0(x)/\rho U^2$ vary as $R_x^{-0.5}$.

   Separation.   An important characteristic of the boundary layer is that
under certain circumstances it leads to a reverse flow over some region
close to the wall (that is the solid surface). By reverse flow we mean flow in

a direction opposite to that of the main stream outside the layer. When reverse flow appears, it usually leads to separation of the boundary layer and consequently of the main stream from the body. It produces a flow pattern that is completely different from the pattern that would exist if there were no separation. Illustrations of separated flows have already been given. We shall not enter here into a description of the time development of the reverse flow and the consequent separation; for such a description reference may be made to Prandtl (1935) or Prandtl and Tietjens (1934) or Schlichting (1955).

In Fig. 1.18 we show schematically the steady flow pattern in the vicinity of separation after separation is completed. Also included in the figure

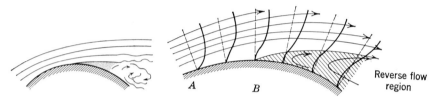

Reverse flow region

A   B

Fig. 1.18   Flow near separation.

are distributions of u, the velocity component parallel to the wall, with y the distance normal to the wall, the y-scale being exaggerated. The figure shows that the streamline branching from B divides the flow coming from the left and right. Such a streamline is known as the *dividing streamline* or as mentioned before as the separation streamline. The point B is the separation point. We see that *at the separation point* $\partial u/\partial y = 0$. At the wall ahead of the separation, $\partial u/\partial y$ is positive; downstream of the separation it is negative.

In the separated region, close to the wall, the flow is in the reverse direction, and consequently the pressure there must increase in the direction of the outside main stream. Thus, to produce the state of affairs shown in Fig. 1.18 initially the flow in the boundary layer close to the wall is subjected to an increasing pressure in the forward direction. Now, according to the boundary-layer concept, when the boundary layer is not separated from the wall, the pressure at the wall is approximately equal to $p_e(x)$, the pressure at the edge of the layer. We therefore conclude that the onset of reverse flow at the wall and the subsequent separation of the main flow should have resulted from an increasing pressure in the forward direction, that is, from a situation in which the gradient $\partial p_e/\partial x$ is positive. Such a gradient is referred to, for obvious reasons, as an *adverse pressure gradient*. We may thus state that *separation of the boundary layer results from an adverse pressure gradient*.

The separation point, *B* in Fig. 1.18, is not the point of minimum pressure for which $\partial p_e/\partial x = 0$. The minimum pressure occurs at some point *A* ahead of *B*. The greater the adverse pressure gradient the shorter the distance *AB*.

In the presence of an adverse pressure gradient a laminar boundary layer may either separate or first become turbulent and then separate. A

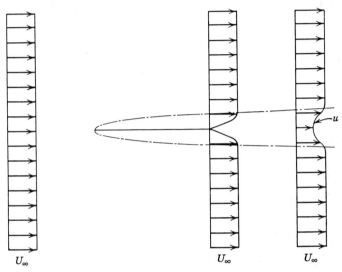

Fig. 1.19   Illustrating the wake behind a flat plate.

turbulent boundary layer, because of the intense mixing due to turbulence, is better able to resist separation than a laminar layer.

Boundary-layer considerations do not apply at and downstream of separation. In particular, the assumption that the pressure gradient $\partial p/\partial y$ normal to the wall is small does not hold any longer. Additional complications also arise, for the flow in the extended separated region is usually unsteady and turbulent.

Wakes. Consider steady flow, at high Reynolds number, past a stationary flat plate of finite extent at zero angle of attack (Fig. 1.19). Behind the trailing edge of the plate, the boundary layers along the two sides of the plate leave the plate and coalesce into one region behind the plate. Such a region is known as a *wake*, a region where viscous effects are still dominant. With increasing distance downstream the width of the wake increases while its mean velocity decreases.

A similar wake appears in the case of steady flow at high Reynolds numbers past a stationary streamlined body at a gentle angle of attack

(See picture a of Plate 7). It is found that the flow in the wake is usually turbulent even though it might have originated from laminar boundary layers along the body. For streamlined bodies at gentle angle of attack, in high Reynolds number flows, the wakes are still thin compared to the characteristic distance along the mainstream direction.

Thin wakes appear only in two instances; either the boundary layer is not separated from the body or, if separated, remaining very close to the rear end of the body. When the boundary layer separates, as it does for bluff bodies and streamlined bodies at "adverse" angles of attack, an extended or a thick wake involving vortices and turbulence appears in the flow (See Plates 4, 5, 6, and 7). Thin-wake flows, as contrasted with thick-wake flows, are not only practically important but also are amenable to approximate calculations. We see in the next section how this may be done.

1.13 Consequences of the Boundary-Layer Concept

Consider steady high Reynolds number flow past a stationary streamlined body which is at a gentle angle of attack (Fig. 1.20). On the basis of

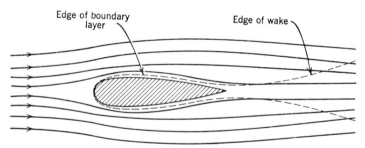

Edge of boundary layer   Edge of wake

Fig. 1.20 Flow past a streamlined body at high Reynolds number.

the boundary layer concept and the characteristics of the flow within such a layer, we may summarise the main features of the flow as follows:

The viscous effects are confined to the boundary layer and the wake behind the body. In the flow outside the boundary layer the viscous effects are negligible.

In the boundary layer the velocity component parallel to the wall rises rapidly from zero at the wall to the mainstream value $U_e(x)$ at the edge of the layer.

The velocity component normal to the wall is very small throughout the layer. At the edge of the layer it is finite but still very small.

The change in pressure across the layer is negligibly small. The pressure at the wall is, therefore, approximately the pressure at the edge of the boundary layer:

$$p(x, 0) = p[x, \delta(x)] = p_e(x)$$

We say that the outer flow "impresses" its pressure on the boundary layer.

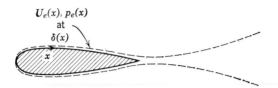

Fig. 1.21   Inner flow corresponding to Fig. 1.20.

In view of these features, the flow field can be divided, for the purposes of mathematical analysis into two regions: (1) An inner region consisting of the boundary layer flow along the body and the wake behind it, and (2) an outer region consisting of the main flow. The inner flow is said to occur under the impressed pressure $p_e(x)$, with a velocity $U_e(x)$ at the edge $\delta(x)$ (Figs. 1.21 and 1.22). In constructing the solution for the inner flow, the fluid is to be treated as viscous, taking into account the simplifications that result from the features of the boundary-layer flow. The solution should satisfy the so-called *no-slip condition* $u = 0$, $v = 0$ at the wall.

The inner boundary of the outer flow is the edge of the boundary layer and the wake. In constructing the solution for the outer flow the fluid may be treated, to a high degree of accuracy, as nonviscous. The solution should satisfy the boundary condition $u = U_e(x)$ and $p = p_e(x)$ at the inner boundary.

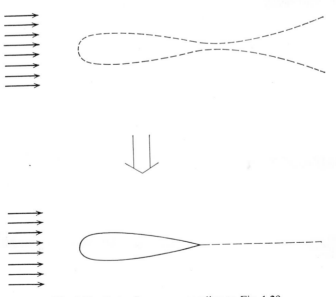

Fig. 1.22   Outer flow corresponding to Fig. 1.20.

We note that *a priori* neither the edge of the boundary layer, nor $U_e(x)$, nor $p_e(x)$ are known. However, on the basis of the characteristics of the boundary-layer flow, the two problems outlined above may be solved approximately as follows:

First we solve the problem of the outer flow. To do so we assume that the inner boundary of the outer flow is coincident with the surface of the body and a surface behind the body, such a surface being obtained by letting the thickness of the wake go to zero while retaining the effect of the wake on the main flow. The problem then is that of the steady flow of an inviscid fluid past the given body with the effects of the wake properly taken into account. The boundary condition then is that the flow be tangential to the wall, the so-called *slip condition*. Some appropriate condition should also be satisfied on the wake surface.

Having determined the outer flow, we attack the problem of the inner flow. We assume that the conditions at the edge of the inner flow, namely $U_e(x)$ and $p_e(x)$, are approximately equal to the corresponding values at the wall and the wake surface given by the inviscid solution for the outer problem.

The procedure may be repeated if so desired. Under normal circumstances one evaluation of the outer and inner solutions is satisfactory.

The inviscid solution gives the pressure distribution over the body and the force and moment resulting from the pressure distribution. The solution for the inner problem gives the boundary-layer thickness, the wake, and most importantly the frictional force, the so-called *skin-friction drag*, on the body. Naturally, the inviscid solution cannot furnish any answers for the viscous drag.

Our concern in this book is with the analysis of the outer problem. We have seen that this is the first step in analysing high Reynolds number flow past a streamlined body, a problem of considerable practical interest.

Prandtl's boundary-layer concept is a landmark in the science of fluid mechanics. Its role in fluid mechanics can be summed up as follows.

The concept furnishes a method for relating the motion of viscous fluid past a body with the corresponding motion of a nonviscous fluid past that body. It shows that for flow at high Reynolds numbers past streamlined bodies, viscosity does not influence the pressure field. The flow pattern around such bodies, the pressure fields, and the pressure forces on the bodies can, therefore, be computed to a high degree of satisfaction on the basis of inviscid fluid flow. It shows that the first step in the analysis of the motion of a viscous fluid past a body is that of solving the corresponding inviscid motion past the body. In formulating the inviscid problem, proper account must be taken of the effects of the wake on the main flow. In this way the boundary-layer concept has brought out the important role played by inviscid flow theory and has lent credence to the extensive use of such a theory in hydrodynamics and aerodynamics. The concept leads to a viscous flow theory known as the "boundary-layer theory" valid for flows at large Reynolds numbers. Such a theory furnishes a method for computing skin friction and also heat transfer when compressibility effects occur.

The concept explains the origin of separation and the flow patterns associated with bluff bodies. It shows that under an adverse pressure gradient a boundary layer would separate.

In this way it emphasizes that solutions and flow patterns furnished by inviscid flow theory are unrealistic when they involve severe adverse pressure gradients along solid surfaces present in the fluid.

On the other hand, on the basis of the inviscid solution, boundary-layer theory would indicate the possible location of separation.

Before the introduction of the boundary-layer concept, a large part of hydrodynamics and aerodynamics was a disturbingly mysterious subject. Theoretical studies were entirely based on inviscid motion and had either no connection with practical problems or no appreciation for such connection where it existed. On the other hand there was a large body of empirically collected knowledge, under the name of hydraulics, which enabled individual treatment of various practical problems. Hydraulics lacked an unified theory. With the appearance of the boundary-layer concept and the boundary-layer theory, the two subjects, theoretical hydro- and aerodynamics, and hydraulics, have been brought together under an unified theory.

1.14   Ideal Fluid Theory

In this book we are concerned with inviscid flow where compressibility effects are practically negligible. This means that *we may regard the fluid as incompressible besides being inviscid. An incompressible inviscid fluid is known as an ideal fluid.*

Naturally, no real fluid is an ideal fluid. However, on the basis of all our previous considerations, we may state that under many practically interesting situations, the motion of a real fluid may be analysed to a high degree of accuracy by regarding the fluid as an ideal fluid. The study of *ideal fluid theory*, apart from its practical utility, is a suitable introduction to the principles and methods underlying the general science of fluid mechanics which encompasses many diverse and complex phenomena We now proceed to the development of the theory of ideal fluid flow pas wings and bodies of revolution.

Chapter 2

Elements of Vector Algebra and Calculus

In the analysis of fluid motion, as in the analysis of many physical phenomena, we are concerned with quantities that may be classified according to the information needed to specify them completely. Quantities such as mass, density, and temperature need (after a choice of units) specification of their magnitudes only, that is, a single number is all that is necessary to specify each of them. Such quantities are called *scalar quantities* or simply *scalars*. A quantity such as force or velocity requires the specification of a magnitude and a direction, that is, of a *directed magnitude*. Quantities of this type are called *vector quantities*. Vector quantities that obey certain rules (such as the parallelogram law of addition) are defined as *vectors*. As we shall see later, not all vector quantities are vectors. Quantities that require specification of more information than needed for vectors also occur in physical problems. For example, to describe a quantity such as stress we need to give a force (i.e., a directed magnitude) and a surface (i.e., the orientation or direction of the surface) on which the force acts. Such quantities are known as *tensors*. There are various kinds of tensors, and, generally speaking, vectors and scalars are degenerate tensors.

Operations of algebra and calculus, such as are known for scalar quantities, are also developed for vectors and tensors. Algebra, applied to vectors, is known as *vector algebra*, while calculus applied to vectors is known as *vector calculus* or *vector analysis*. Similarly, we have tensor algebra and tensor calculus.

In analyzing a physical phenomenon, we set up interrelations between the various quantities that characterize the phenomenon by using laws of nature (such as the laws of Newton, the law of energy conservation). To write a natural law, one introduces a coordinate system in a chosen frame of reference and expresses the various physical quantities involved by means of measurements made with respect to that system. When we choose such a procedure, the expression for the law contains terms that are dependent on the chosen coordinate system and consequently

56

appears differently in different systems. But the laws of nature are independent of the artificial choice of a coordinate system. Therefore we may seek to express the natural laws in a form not related to a particular coordinate system. A way of doing this is provided by vector or tensor analysis. Vector notation exhibits quantities such as displacements, velocities, forces, accelerations, moments, and angular velocities in their natural color, that is, as quantities that possess directions besides magnitudes. When vector notation is used, a coordinate system need not be introduced. Thus use of vector notation in formulating physical laws leaves them in *invariant form* (i.e., in a form independent of coordinate description). Studying a physical phenomenon by means of equations written in invariant form often leads to a deeper understanding of the phenomenon. Besides, the use of vector notation brings considerable simplicity into the analysis of problems.

It is our desire to develop the equations of fluid motion and obtain many of the basic results of our studies in a form not related to any particular coordinate system. To this end we shall employ vector notation, vector algebra, and vector analysis. We shall, therefore, begin our studies by acquainting ourselves in this chapter with the elements of vector algebra and calculus. Vector methods are adequate to treat the aspects of fluid motion presented in this book. Hence we shall not concern ourselves with tensor algebra or calculus.* The concepts of vector analysis are closely associated with the concepts of fluid mechanics.

2.1   Representation of a Vector

If P and Q are any two points in space, the directed straight-line segment from P to Q locates the position of point Q with respect to the point P. Such a directed line segment is called a *position vector*. It is the simplest example of a vector quantity. Graphically we represent the position vector from P to Q by a straight arrow running from P to Q as shown in Fig. 2.1. The length of the arrow gives the magnitude of the distance from P to Q, while the sense of the arrow indicates the direction from P to Q. Following the example of the position vector we represent any vector quantity (e.g., velocity, force) by an arrow pointing in the same direction as the vector. The length of the arrow is made proportional to the magnitude of the quantity.

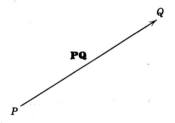

Fig. 2.1   Representation of a vector.

* In discussing the mechanics of a viscous fluid, tensor analysis becomes useful.

A suitable convention is adopted to denote a vector quantity. In printed work it is usually symbolized by a boldface letter. For example, the letter **r** may be used for the position vector, the letter **V** for the velocity vector and so on. In writing it is customary to place an arrow or a bar over the letter that denotes a vector quantity. Thus for a position vector we write $\vec{r}$, for a velocity vector $\vec{V}$, and so on. If we wish to show that a directed line segment from a point P to a point Q represents a certain vector quantity, we sometimes use the notation $\overrightarrow{PQ}$ to denote the vector.

The magnitude of a vector **A** is denoted by $|\mathbf{A}|$ or simply by the letter A.

Two vectors **A** and **B** are *equal* if the magnitude of **A** is equal to the magnitude of **B** and if the direction of **A** is the same as the direction of **B**. Thus a vector is not changed if it is moved parallel to itself. This means that generally the position of a vector in space may be chosen arbitrarily. In certain applications, however (as in the calculation of the moment of a force) the actual point of location of a vector may be important. A vector when associated with a particular point is known as a *localized* or *bound vector*; otherwise it is known as a *free vector*.

When two or more vectors are parallel to the same line, they are said to be *collinear*. When two or more vectors are parallel to the same plane, they are said to be *coplanar*.

2.2   Addition and Subtraction

Let P and Q be two points in space and let $\overrightarrow{OP}$ and $\overrightarrow{OQ}$ be the respective position vectors from a reference point O (Fig. 2.2). $\overrightarrow{PQ}$ denotes the vector from P to Q. From O the point Q may be reached along the

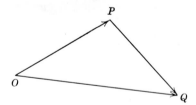

Fig. 2.2   Addition of position vectors.

vector $\overrightarrow{OQ}$ or, alternatively, along the vector $\overrightarrow{OP}$ to P and then along vector $\overrightarrow{PQ}$ to Q. We define that $\overrightarrow{OQ}$ is the sum of the vectors $\overrightarrow{OP}$ and $\overrightarrow{PQ}$. Accordingly we write that

$$\overrightarrow{OQ} = \overrightarrow{OP} + \overrightarrow{PQ}$$

This notion of addition is readily applicable to vector quantities other than position vectors. If **A** and **B** are any two vectors, we can represent them by arrows drawn such that the initial point of **B** coincides with the terminal point of **A** (Fig. 2.3*a*). Then the vector sum **A** + **B** is given by the vector **C**, which extends from the initial point of **A** to the terminal

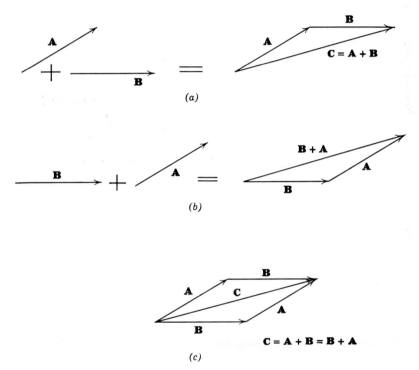

Fig. 2.3   Addition of two vectors.

point of **B**. In the same manner **A** may be added to **B** and the sum **B** + **A** obtained as shown in Fig. 2.3*b*. Putting together Figs. 2.3*a* and 2.3*b* (which are equal triangles), we obtain a parallelogram as shown in Fig. 2.3*c*. The vectors **A** and **B** drawn from a common origin form the sides of the parallelogram. The diagonal **C** drawn from the common origin represents the sum **A** + **B** or **B** + **A**. Thus we say that vectors are added according to the "*parallelogram law*" of addition. Repeated application of the parallelogram law determines the sum of any number of vectors. Since **A** + **B** = **B** + **A**, vectors may be added in any order whatsoever. We, therefore, say vector addition is *commutative*.

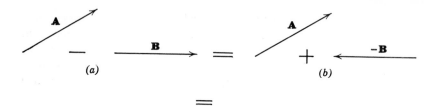

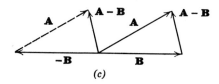

Fig. 2.4   Subtraction of two vectors.

Subtraction of vectors is carried out on the same lines as their addition. To form the vector difference **A** − **B** we write

$$\mathbf{A} - \mathbf{B} = \mathbf{A} + (-\mathbf{B})$$

and reduce the operation of subtraction to one of addition (Fig. 2.4). The negative vector −**B** has the same magnitude as **B** but points in a direction opposite to that of **B**.

2.3   Definition of a Vector

We now *define a vector as a quantity that possesses both a magnitude and a direction and obeys the parallelogram law of addition.* Obeying this law is important, for there are quantities that have both magnitude and direction but do not add according to the parallelogram law. A finite rotation of a rigid body, although possessing magnitude and direction, is not a vector, for such rotations do not obey the parallelogram law. On the other hand, an infinitesimal rotation of a rigid body is a vector. The reader should verify these statements.

2.4   Multiplication by a Number

If a vector **A** is multiplied by a number m, we obtain another vector the magnitude of which is m times the magnitude of **A**, and the direction of which is the same as that of **A**.

$$|m\mathbf{A}| = m\,|\mathbf{A}| \equiv mA$$

2.5 Unit Vector

A vector of unit length (i.e., of unit magnitude) is called a *unit vector.* Considering any vector **A**, form the product

$$\frac{1}{A}\,\mathbf{A}$$

The result is simply a unit vector in the direction of **A**. Denoting this unit vector by $\mathbf{e}_A$, we write

$$\mathbf{e}_A = \frac{\mathbf{A}}{A}$$

or

$$\mathbf{A} = A\mathbf{e}_A$$

That is, *any vector may be represented as the product of its magnitude and a unit vector.*
A unit vector is used to designate a direction.

2.6 Zero Vector

A vector of zero magnitude is called a *zero vector.* It has no definite direction associated with it.

2.7 Scalar Product of Two Vectors

Besides addition, subtraction, and multiplication by a number, two further algebraic operations, known as the *scalar product* and the *vector product,* can be defined for vector quantities. To introduce the scalar product, we recall the concept of work. When a force **F** acts on a mass point, and if under its action the mass experiences an infinitesimal displacement **s**, we define the work done by the force as equal to the orthogonal projection of the force along the direction of the displacement times the magnitude of the displacement (Fig. 2.5).

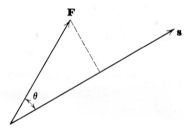

Fig. 2.5   Concept of work.

If θ is the angle between **F** and **s**, we express the work done by

$$W = |\mathbf{F}|\cos\theta\,|\mathbf{s}|$$
$$= Fs\cos\theta$$

where F and s denote, respectively, the magnitudes of **F** and **s**. The work done, W, is a scalar quantity that is obtained by a certain kind of product operation between two vectors, namely **F** and **s**. Such an operation may be given a name and defined for any two vectors. Since the result of the product is a scalar, it is called the *scalar product* of two vectors. It is *defined as the scalar quantity equal to the product of the magnitudes of the vectors times the cosine of the angle between their directions.* If **A** and **B** are any two vectors, their scalar product is denoted by **A · B** and read as **A** dot **B**. Thus we write

$$\mathbf{A} \cdot \mathbf{B} \equiv |\mathbf{A}|\,|\mathbf{B}|\cos\theta$$
$$\equiv AB\cos\theta \tag{2.1}$$

where θ is the angle between the vectors. The scalar product is also known as the *dot product* or as the *inner product*.

By using the notation of the scalar product the work done by a force **F** during an infinitesimal displacement **s** may be represented by

$$W = \mathbf{F} \cdot \mathbf{s}$$

A few simple results follow immediately from the definition (2.1) of the scalar product.

Since $\mathbf{A} \cdot \mathbf{B} = \mathbf{B} \cdot \mathbf{A}$, the scalar product is commutative.

If the vectors **A** and **B** are perpendicular to each other, their scalar product is zero, since $\theta = \pi/2$ and $\cos\theta = 0$. Conversely, if $\mathbf{A} \cdot \mathbf{B} = 0$, it follows that either the vectors are mutually perpendicular or at least one of them is zero.

If two vectors **A** and **B** are parallel to each other, their scalar product becomes simply equal to the product of their magnitudes (i.e., AB), since $\theta = 0$ and $\cos\theta = 1$.

The scalar product of a vector by itself is equal to the square of its magnitude. Thus we have

$$\mathbf{A} \cdot \mathbf{A} = AA = A^2$$

The product $\mathbf{A} \cdot \mathbf{A}$ is sometimes denoted by $\mathbf{A}^2$.

The orthogonal projection of a vector **A** in any direction **e** is given by the product $\mathbf{A} \cdot \mathbf{e}$.

2.8 Vector Product of Two Vectors

To introduce this product we consider the concept of the *moment* due to a force. Suppose we wish to describe the moment about a point O of a force **F** acting at the point P (Fig. 2.6). To describe the moment completely we must give a magnitude and a direction, that is, we must specify a vector quantity. Let us, therefore, denote the moment by **M**. By definition the magnitude of the moment is equal to the product of the magnitude of the force and the shortest distance from the reference point to the line

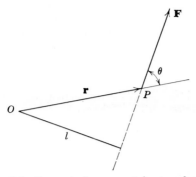

Fig. 2.6 Concept of a moment due to a force.

of action of the force (i.e., the lever arm). Denoting these quantities, respectively, by M, F, and l we have

$$M = Fl$$

If $\mathbf{r}$ denotes the vector $\overrightarrow{OP}$ and θ the angle measured from $\mathbf{r}$ to $\mathbf{F}$ such that $O \leq \theta \leq \pi$ (see Fig. 2.6), the magnitude of the moment becomes

$$M = Fr \sin \theta$$

The direction of the moment is that of a rotation about O in the plane formed by the vectors $\mathbf{r}$ and $\mathbf{F}$. Drawing the vectors $\mathbf{r}$ and $\mathbf{F}$ from the common origin O, we observe that the direction of rotation due to the moment tends to bring the vector $\mathbf{r}$ into the vector $\mathbf{F}$ (Fig. 2.7). To express these ideas symbolically we first set up at O an axis of rotation such that it is perpendicular to the plane $\mathbf{r}$ and $\mathbf{F}$ and points in the direction in which a right-handed screw would advance when turned in the direction of rotation due to the moment (i.e., from $\mathbf{r}$ to $\mathbf{F}$). Along this axis of rotation we then draw a unit vector $\mathbf{e}_m$ and agree that it represents the direction of the moment vector $\mathbf{M}$ (Fig. 2.8). Thus we write

$$\mathbf{M} = Fr \sin \theta \mathbf{e}_m \qquad (2.2)$$

and represent it as shown in Fig. 2.8.

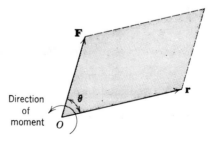

Fig. 2.7 Direction of the moment.

According to (2.2) the moment vector may be looked on as resulting from a certain type of product operation between two other vectors. Thus Eq. (2.2) may be made the basis for defining such a product between any two vectors. Since the result of such a product is a vector, it may be called the *vector product* and defined as follows.

The vector product of the vectors **A** *and* **B** *is a vector* **C** *whose magnitude is equal to the product of the magnitudes of* **A** *and* **B** *times the sine of the*

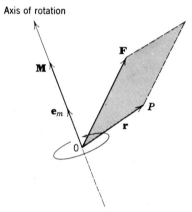

Fig. 2.8   Representation of a moment vector.

angle θ measured from **A** *to* **B** *such that* $0 \leq \theta \leq \pi$, *and whose direction is specified by the condition that* **C** *is perpendicular to the plane of the vectors* **A** *and* **B** *and points in the direction in which a right-handed screw advances when turned so as to bring* **A** *toward* **B**.

The vector product is usually denoted by writing the vectors with a cross between them as

$$\mathbf{A} \times \mathbf{B}$$

and read **A** *cross* **B**. For this reason it is also called the *cross product*.* If A and B are the respective magnitudes of **A** and **B** and if **e** denotes the direction of **A** × **B**, we write

$$\mathbf{A} \times \mathbf{B} = \mathbf{e}AB \sin \theta \qquad (2.3)$$

and represent it geometrically as shown in Fig. 2.9.

By using the notation of the vector product (2.2) can now be abbreviated to the form

$$\mathbf{M} = \mathbf{r} \times \mathbf{F}$$

---

* Skew product and outer product are also used.

We shall now state a few simple results that follow readily from the definition of the vector product.

The products $\mathbf{A} \times \mathbf{B}$ and $\mathbf{B} \times \mathbf{A}$ are not equal. In fact, we have

$$\mathbf{A} \times \mathbf{B} = -\mathbf{B} \times \mathbf{A}$$

This means that the vector product is *not commutative*. It is, therefore necessary to preserve the order of the vectors when dealing with operations that involve vector products.

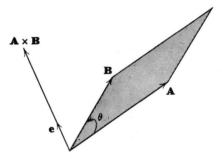

Fig. 2.9   Vector product.

If two vectors $\mathbf{A}$ and $\mathbf{B}$ are *parallel* to each other, their cross product is zero. Conversely, if $\mathbf{A} \times \mathbf{B} = 0$, the vectors $\mathbf{A}$ and $\mathbf{B}$ are either parallel or at least one of them is zero. It follows that the vector product of a vector into itself is zero.

2.9   Plane Area as a Vector

The magnitude of the vector $\mathbf{A} \times \mathbf{B}$ is equal to the area of the parallelogram formed by the vectors $\mathbf{A}$ and $\mathbf{B}$. In fact, the vector $\mathbf{A} \times \mathbf{B}$ may be considered to represent both in magnitude and direction the area of the parallelogram whose sides are $\mathbf{A}$ and $\mathbf{B}$, if a plane area can be represented as a vector. Now, any plane area may be regarded as possessing a direction besides a magnitude, the directional character arising out of the need to specify the orientation in space of the plane area. It is customary to denote the direction of a plane area by means of a unit vector drawn in the direction of the *normal* to that plane. To fix the direction of the normal we first assign a certain *sense* of *travel* along the contour that forms the boundary of the plane area in question. Then the direction of the normal is taken (by convention) as that in which a right-handed screw advances as it is rotated according to the sense of travel along the boundary curve or contour. Thus if $\mathscr{C}$ is a curve enclosing an area S on the plane P and the direction along $\mathscr{C}$ is assigned as shown in Fig. 2.10, the direction of the area is given by the unit vector $\mathbf{n}$ normal to the plane

and directed as shown in the figure.* The area itself can now be represented by a vector **S** whose magnitude is S and whose direction is **n**.
Thus, symbolically, we write

$$\mathbf{S} = S\mathbf{n}$$

According to these ideas, the vector product **A** × **B** represents both in

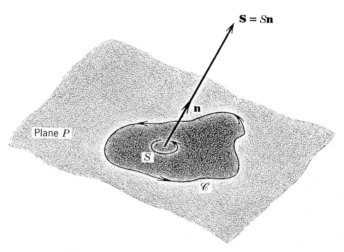

Fig. 2.10   Plane area as a vector quantity.

magnitude and direction the area of the parallelogram whose sides are
A and **B**.

2.10   Velocity of a Point of a Rotating Rigid Body

The vector product of two vectors has many geometrical and physical
applications. The description of the moment of a force about a point
is one important example. The description of the velocity of a point of a
rotating rigid body is another, and we shall consider it here. Suppose
a rigid body is rotating with an angular speed ω about a certain axis.
We wish to describe the velocity of a point P of the body. Let the vector **V**
denote the velocity of the point. Each point of the body describes a
circle that lies in a plane perpendicular to the axis and with its center on
the axis. The radius of the circle is the perpendicular distance from the
point of interest to the axis (Fig. 2.11). The magnitude of the velocity
of the point is simply equal to the product of the angular velocity ω and
the radius, say a, of the circle. The velocity **V**, directed as shown in the

---

* We have used the common notation **n** instead of $\mathbf{e}_n$.

figure, is perpendicular to the radius and to the axis of rotation. Denoting
the direction of **V** by the unit vector **e**, we have

$$\mathbf{V} = \omega a \mathbf{e} \tag{2.4a}$$

Let O be any point on the axis of rotation and let **r** denote the position
vector $\overrightarrow{OP}$ (Fig. 2.11). If θ is the angle between **r** and the axis (measured

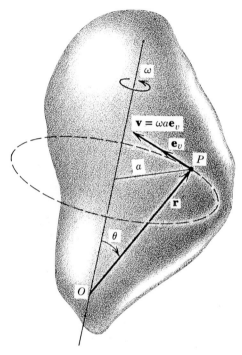

Fig. 2.11   Velocity of a point of a rotating rigid body.

such that $O \le \theta \le \pi$), the radius a is equal to $r \sin \theta$, r being the magnitude of **r**. Then from (2.4a) we obtain

$$\mathbf{V} = \omega r \mathbf{e} \sin \theta \tag{2.4b}$$

Now, angular velocity, like the moment due to a force, is a vector
quantity possessing a direction and a magnitude and, as can be verified,
obeys the parallelogram law of addition. It is thus a vector and we denote
it by **ω**. The direction of **ω**, as in the case of the moment, is a sense of
rotation about a certain axis. To represent it by means of a unit vector
we adopt the convention (or the rule) of the right-handed screw. According to this rule, the direction of the angular velocity is given by a

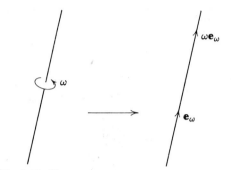

Fig. 2.12   Vector representation of angular velocity.

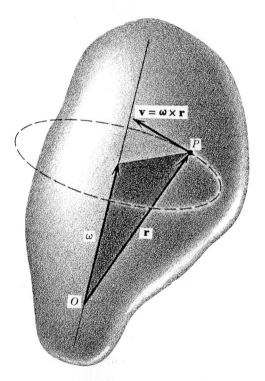

Fig. 2.13   Velocity of a point of a rotating rigid body as a vector product.

unit vector drawn along the axis of rotation and pointing in that direction in which a right-handed screw would advance when it is turned in the direction of rotation about the axis. Denoting this unit vector by e_ω, we write

$$\boldsymbol{\omega} = \omega e_\omega$$

and represent it as shown in Fig. 2.12.

Now, if e_r is a unit vector in the direction of r we observe that

$$e_\omega \times e_r = e \sin \theta$$

With this relation, Eq. (2.4b) may be written as

$$\mathbf{V} = \omega r e_\omega \times e_r = \omega e_\omega \times r e_r$$

$$= \boldsymbol{\omega} \times \mathbf{r} \tag{2.5}$$

This states that the velocity of a point of a rigid body rotating about an axis is given by the vector product of the angular velocity and the position vector drawn from any point on the axis of rotation to the point under consideration (see Fig. 2.13).

2.11   Polar and Axial Vectors

It might have been noticed during the preceding considerations that there is a certain difference between vectors such as angular velocity and the moment of a force and vectors such as velocity, force, and displacement. The difference between the two types of vectors lies in the way they are represented by directed line segments (i.e., arrows). In the case of vectors such as force and velocity the direction of the arrow is the true direction of the vector it represents. Vectors that can be represented in this way are called *polar vectors*. In quantities such as angular velocity and moment the direction of the arrow is not the actual direction of the quantity it represents, for the actual direction in this case is that of a rotation about an axis; and what we have done is to choose to represent this direction of rotation by means of a directed segment along the axis of rotation. To specify the direction of that segment we have adopted, arbitrarily of course, the right-hand rule (i.e., the convention of the right-handed screw). Vectors that are represented this way are called *axial vectors.*

2.12   Multiple Products

We return to the products of vectors. Products between three vectors are called *triple products*. If $\mathbf{A}, \mathbf{B}, \mathbf{C}$ are any three vectors, triple products of the form

$$\mathbf{A}(\mathbf{B} \cdot \mathbf{C}); \quad \mathbf{A} \cdot \mathbf{B} \times \mathbf{C}; \quad \mathbf{A} \times (\mathbf{B} \times \mathbf{C})$$

are easily defined.

The product $A(\mathbf{B} \cdot \mathbf{C})$ is simply a multiplication of the vector $\mathbf{A}$ by a scalar that is equal to $\mathbf{B} \cdot \mathbf{C}$.

Scalar Triple Product.   The result of the product $\mathbf{A} \cdot \mathbf{B} \times \mathbf{C}$ is a scalar. Therefore such a product is called a *scalar triple product*. Drawing the vectors $\mathbf{A}, \mathbf{B}, \mathbf{C}$ from a common origin, we readily see that the product $\mathbf{A} \cdot \mathbf{B} \times \mathbf{C}$ is equal to the volume of the parallelepiped formed by the vectors $\mathbf{A}, \mathbf{B}, \mathbf{C}$ (Fig. 2.14). The following simple results* are important in applications.

In a triple scalar product, the dot and cross can be interchanged without changing the value of the result. Symbolically, we have

$$\mathbf{A} \cdot \mathbf{B} \times \mathbf{C} = \mathbf{A} \times \mathbf{B} \cdot \mathbf{C} \tag{2.6}$$

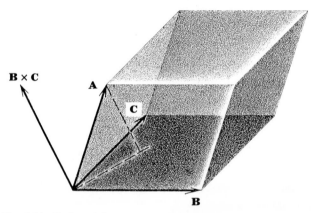

Fig. 2.14   Scalar triple product as the volume of a parallelepiped.

A cyclical permutation of the order of the vectors in a scalar triple product leaves the product unchanged. This is expressed by

$$\mathbf{A} \cdot \mathbf{B} \times \mathbf{C} = \mathbf{B} \cdot \mathbf{C} \times \mathbf{A} = \mathbf{C} \cdot \mathbf{A} \times \mathbf{B} \tag{2.7}$$

It further follows that

$$\mathbf{A} \cdot \mathbf{B} \times \mathbf{C} = -\mathbf{A} \cdot \mathbf{C} \times \mathbf{B} = -\mathbf{C} \cdot \mathbf{B} \times \mathbf{A} = -\mathbf{B} \cdot \mathbf{A} \times \mathbf{C} \tag{2.8}$$

Vector Triple Product.   The result of the product $\mathbf{A} \times (\mathbf{B} \times \mathbf{C})$ is a vector. Therefore such a product is called a *vector triple product*. The vector $\mathbf{A} \times (\mathbf{B} \times \mathbf{C})$ is normal to the plane formed by $\mathbf{A}$ and $(\mathbf{B} \times \mathbf{C})$. The vector $\mathbf{B} \times \mathbf{C}$ is, however, perpendicular to the plane formed by $\mathbf{B}$ and $\mathbf{C}$. This means that the vector $\mathbf{A} \times (\mathbf{B} \times \mathbf{C})$ lies in the plane formed

* Proof of these results is left as an exercise.

by **B** and **C** and is perpendicular to the vector **A** (Fig. 2.15). In such a case **A** × (**B** × **C**) can be expressed as a linear combination of the vectors **B** and **C** (see 2.13). Thus we write

$$\mathbf{A} \times (\mathbf{B} \times \mathbf{C}) = m\mathbf{B} + n\mathbf{C} \tag{2.9a}$$

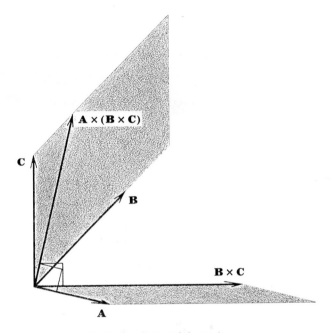

Fig. 2.15   Vector triple product.

where m and n are numbers. It can be shown that

$$m = \mathbf{A} \cdot \mathbf{C}$$

$$n = -\mathbf{A} \cdot \mathbf{B}$$

We thus have

$$\mathbf{A} \times (\mathbf{B} \times \mathbf{C}) = (\mathbf{A} \cdot \mathbf{C})\mathbf{B} - (\mathbf{A} \cdot \mathbf{B})\mathbf{C} \tag{2.9b}$$

Since the vector product of two vectors changes sign when the order of the vectors is changed, it follows that

$$\mathbf{A} \times (\mathbf{B} \times \mathbf{C}) = -\mathbf{A} \times (\mathbf{C} \times \mathbf{B}) = (\mathbf{C} \times \mathbf{B}) \times \mathbf{A} = -(\mathbf{B} \times \mathbf{C}) \times \mathbf{A}$$

Products involving more than three vectors can be readily evaluated in terms of triple products and others we have already considered.

2.13 Components of a Vector

The study, so far, of vector algebra has proceeded on the basis of the geometrical description of a vector as a directed line segment. We now proceed toward an analytical description of a vector and of the operations on it. Such a description yields a connection between vectors and ordinary numbers and relates the operations (of algebra and calculus) on vectors with those on numbers. The analytical description of a vector is based on the notion of *components* of a vector.

From the idea of addition it follows that any vector may be represented as the sum of a number of arbitrarily chosen *noncoplanar* vectors. When this is done we say that a certain vector is *resolved* into a number of *component vectors*.

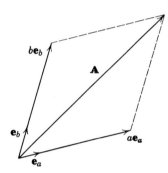

Let us consider, for simplicity, a vector **A** in a plane. The minimum number of *nonparallel* vectors into which **A** may be resolved are two. Thus let us designate, as shown in Fig. 2.16, two nonparallel directions by the unit vectors $\mathbf{e}_a$ and $\mathbf{e}_b$. Choosing suitably two numbers a and b we can draw vectors $a\mathbf{e}_a$ and $b\mathbf{e}_b$ such that their sum is represented by the vector **A** (see figure). Thus $a\mathbf{e}_a$ and $b\mathbf{e}_b$ become the *vector components* of **A** in the directions $\mathbf{e}_a$ and $\mathbf{e}_b$.

Fig. 2.16 Decomposition of a vector in two dimensions.

tions $\mathbf{e}_a$ and $\mathbf{e}_b$. To exhibit analytically the component form of **A** we write

$$\mathbf{A} = a\mathbf{e}_a + b\mathbf{e}_b$$

If it is so desired we could decompose **A** into a number of nonparallel vectors. But, selecting any two of these vector components, we can express each of the rest in terms of the selected two. Thus the number of independent components that are necessary and sufficient to decompose a vector in a plane is two.

In a similar manner, any vector in space (i.e., in *three-dimensional space*) can be resolved into vector components that are noncoplanar. Now, the number of independent components that are necessary and sufficient to decompose a vector in space is *three*. Thus if we designate, as shown in Fig. 2.17, three noncoplanar directions by the unit vectors $\mathbf{e}_a, \mathbf{e}_b, \mathbf{e}_c$, any vector **A** may be represented as made up of the component vectors $a\mathbf{e}_a, b\mathbf{e}_b, c\mathbf{e}_c$, where a, b, c are suitably chosen numbers. Thus the *component form* of a three-dimensional vector **A** is expressed by

$$\mathbf{A} = a\mathbf{e}_a + b\mathbf{e}_b + c\mathbf{e}_c \tag{2.10}$$

The numbers a, b, and c, which may be positive or negative, are called the *scalar components** of **A** or the *measure numbers* of the respective vector components of **A**. It is usual to refer to the numbers a, b, c simply as the *components* of **A**; it being understood that they are the scalar components of **A** in the respective directions e_a, e_b, and e_c.

To show the utility of expressing vectors in component form, let us first set up a *basic system* of three noncoplanar vectors. Then, with respect

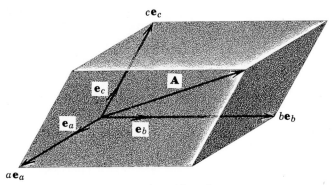

Fig. 2.17   Decomposition of a vector.

to this basic system, let all vector quantities be expressed in their component forms. Once this is done, all operations involving the vector quantities will reduce to operations involving their scalar components. For example, consider equality of two vectors A_1 and A_2.

If e_a, e_b, and e_c denote a basic system of unit vectors, and if a_1, b_1, c_1 are the respective components of A_1 and a_2, b_2, c_2 are the respective components of A_2, we have for

$$A_1 = A_2 \qquad (2.11a)$$

the component relation

$$a_1 e_a + b_1 e_b + c_1 e_c = a_2 e_a + b_2 e_b + c_2 e_c \qquad (2.11b)$$

This further reduces to three scalar equations

$$a_1 = a_2 \qquad (2.11c)$$
$$b_1 = b_2 \qquad (2.11d)$$
$$c_1 = c_2 \qquad (2.11e)$$

As another example consider the addition of two vectors A_1 and A_2.

* Note that unless e_a, e_b, and e_c are mutually perpendicular, $a \neq A \cdot e_a$, $b \neq A \cdot e_b$, and $c \neq A \cdot e_c$.

If the sum $A_1 + A_2$ is denoted by the vector A_3 and its components by a_3, b_3, c_3, we have for

$$A_3 = A_1 + A_2 \tag{2.12a}$$

the form

$$a_3 e_a + b_3 e_b + c_3 e_c = a_1 e_a + b_1 e_b + c_1 e_c + a_2 e_a + b_2 e_b + c_2 e_c \tag{2.12b}$$

This can be rewritten as three scalar equations

$$a_3 = a_1 + a_2 \tag{2.12c}$$

$$b_3 = b_1 + b_2 \tag{2.12d}$$

$$c_3 = c_1 + c_2 \tag{2.12e}$$

Incidentally, the foregoing examples show how the use of vector notation leads to simplicity of expression. A single vector relation such as (2.11a) is equivalent to three scalar relations such as (2.11c) to (2.11e). Similarly, relation (2.12a) is represented by three equations, (2.12c) to (2.12e).

2.14   Specification of a Vector

In the preceeding section we have seen that if two vectors are equal, their components with respect to a chosen reference system of three noncoplanar unit vectors are also equal. Conversely, if the corresponding components of two vectors are equal, the vectors must themselves be equal in magnitude and direction. This means a vector is uniquely determined by a set of three numbers that form the components of the vector.

To specify a vector, quantities other than its components can also be used. To illustrate this consider a vector A. Let a, b, c denote its components with respect to the unit vectors e_a, e_b, e_c, and let α, β, γ denote the angles A makes with e_a, e_b, e_c. Then each of the sets of numbers (a, b, c), (a, β, γ), (b, γ, α), (c, α, β), (a, b, γ), etc. determines completely the direction and magnitude of A.

We may thus describe a vector analytically as a *set of three numbers* that, in some fashion, are related to a chosen reference system of unit vectors, say e_1, e_2, e_3. The set of numbers is *ordered* in the sense that the first number in the set corresponds to the direction e_1, the second to e_2, and the third to e_3. Furthermore, the three numbers representing a vector must obey certain specific rules, for not every ordered set of three numbers expresses a vector. To see how such rules may be set up, consider any vector and describe it in two different reference systems. As we pass from one reference system to another, the vector itself remains unchanged, while the sets of numbers that describe it will be different in the two systems. However, since both the sets describe the same vector, we can set up a rule to express one set of numbers in terms of the other. Then we may require that the three numbers representing a vector obey such a transformation rule.

According to these ideas, we define a vector, analytically, as an ordered set of three numbers that obey certain specific rules. If the set of numbers q_1, q_2, q_3 represent a vector **A**, we express this symbolically by writing

$$\mathbf{A} = (q_1, q_2, q_3)$$

2.15   Cartesian Coordinates and i, j, k System of Unit Vectors

The three unit vectors that constitute a reference system for the various vector quantities under consideration are usually taken along the directions of the axes of a coordinate system used to define a point in space. A

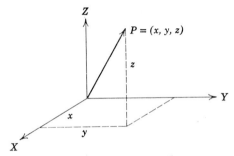

Fig. 2.18   Cartesian coordinates.

familiar example of a coordinate system is the rectangular *Cartesian coordinate system.* In this system a point in space is specified by three *coordinates* measured with respect to three mutually perpendicular axes. We denote the axes by X, Y, Z and the coordinates by x, y, z (Fig. 2.18). The coordinate axes may be designated X, Y, Z arbitrarily, but it is necessary to adopt a consistent convention. As shown in Fig. 2.19 there are two distinct ways of disposing the axes. The method shown in Fig.

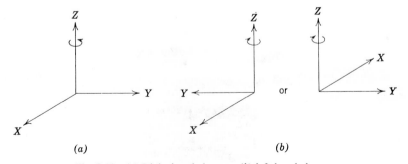

Fig. 2.19   (*a*) Right-handed axes; (*b*) left-handed axes.

2.19a may be characterized by the convention that a *right-handed* rotation about the positive direction of the Z-axis through 90° brings the positive X-axis into the positive Y-axis. A system of axes oriented according to this *rule* of *right-handed rotation* is known as a *right-handed system.* On the same lines the method of orienting the axes as shown in Fig. 2.19b may be characterized by a *left-handed rotation* about the axis Z. In our work here we shall use right-handed systems of axes exclusively.

Let P be a point in space designated by the coordinates x, y, z referred to

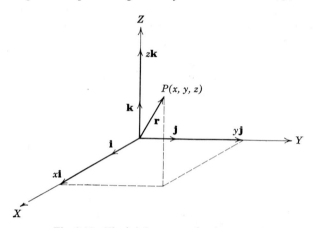

Fig. 2.20   The i, j, k system of unit vectors.

a Cartesian coordinate system XYZ. Let O be the origin of coordinates and let **r** denote the position vector $\overrightarrow{OP}$. To describe the component form of **r** we choose three unit vectors along the positive directions of the axes X, Y, Z. We denote these unit vectors by **i**, **j**, **k**, respectively* (Fig. 2.20). They *form a right-handed system of orthogonal* (i.e, mutually perpendicular) *vectors.* The components of **r** with respect to the **i**, **j**, **k** system are simply x, y, z. Thus we write

$$\mathbf{r} = \overrightarrow{OP} = (x, y, z) = x\mathbf{i} + y\mathbf{j} + z\mathbf{k} \tag{2.13}$$

Since **i**, **j**, **k** are mutually perpendicular, the magnitude of **r** is given by

$$r \equiv |\mathbf{r}| = \sqrt{x^2 + y^2 + z^2} \tag{2.14}$$

If α, β, γ are the angles **r** makes, respectively, with **i**, **j**, **k**, the direction cosines of **r** are given according to the following relations:

$$x = \mathbf{r} \cdot \mathbf{i} = r \cos \alpha$$
$$y = \mathbf{r} \cdot \mathbf{j} = r \cos \beta \tag{2.15}$$
$$z = \mathbf{r} \cdot \mathbf{k} = r \cos \gamma$$

* It is customary to use **i**, **j**, **k** instead of $\mathbf{e}_x$, $\mathbf{e}_y$, $\mathbf{e}_z$.

Now, the same system of **i**, **j**, **k** unit vectors may be set up at any point P and used to describe the component form of any vector associated with the point. Thus if **A** is such a vector quantity and if a_x, a_y, a_z denote the components of **A** with respect to **i**, **j**, **k**, we write

$$\mathbf{A} = (a_x, a_y, a_z) = a_x\mathbf{i} + a_y\mathbf{j} + a_z\mathbf{k} \tag{2.16}$$

The magnitude of **A** is given by

$$\mathbf{A} \equiv |\mathbf{A}| = \sqrt{a_x{}^2 + a_y{}^2 + a_z{}^2} \tag{2.17}$$

2.16   Notion of Curvilinear Coordinates

In analyzing many physical problems it is often advantageous to use coordinates of greater generality than the Cartesians. We shall now see how such general coordinates may be introduced and characterized.

In the Cartesian system various points in space are defined by assigning different values to the coordinates x, y, z. In such an XYZ space, consider a system of three independent functions expressed by

$$q_1 = q_1(x, y, z)$$
$$q_2 = q_2(x, y, z) \tag{2.18}$$
$$q_3 = q_3(x, y, z)$$

such that there is a unique correspondence between (x, y, z) and (q_1, q_2, q_3). By means of these functions we can determine for any point P with the coordinates x, y, z a set of three new numbers q_1, q_2, q_3. Conversely, if q_1, q_2, q_3 are chosen, a point with the coordinates x, y, z can be determined. This means that the position of a point P in the XYZ space may be specified either by the set of numbers (x, y, z) or by (q_1, q_2, q_3). Thus to each point $P(x, y, z)$ we can assign the corresponding values (q_1, q_2, q_3) as a set of new coordinates. In this sense the system of functions expressed by (2.18) may be interpreted as defining a *transformation of coordinates.** The coordinates q_1, q_2, q_3 are known as the *general coordinates of a point*. Note that q_1, q_2, q_3 are coordinates and that they need not necessarily possess the dimensions of length. In other words, they are not necessarily the components of the position vector describing the point P.

Let us now recall the geometrical significance of a function of the form $f(x, y, z) = $ const. Such a function represents a family of surfaces, with each surface of the family corresponding to a different value of the constant. We concern ourselves with cases where only one surface of the family will pass through a chosen point. Consider now the system of

---

* Another interpretation of these equations is that they define a mapping of the XYZ space on to the space of q_1, q_2, q_3 coordinates. For a detailed discussion on "systems of functions, transformations, and mappings," see Courant (1934).

equations expressed by

$$q_1 = q_1(x, y, z) = \text{const.}$$
$$q_2 = q_2(x, y, z) = \text{const.} \qquad (2.19)$$
$$q_3 = q_3(x, y, z) = \text{const.}$$

They represent three independent families of surfaces such that, in general, one surface of each family passes through a chosen point. Then any point in space may be located as the point of intersection of three independent surfaces represented by a system of equations such as (2.19). The values of

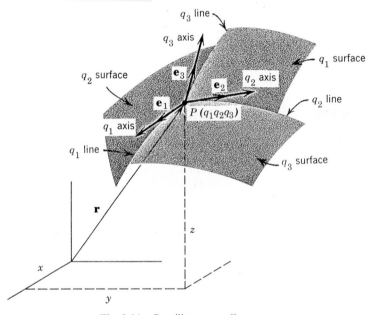

Fig. 2.21   Curvilinear coordinates.

q_1, q_2, q_3, which belong to the three surfaces passing through a point, are then assigned to the point as coordinates. These are nothing but the general coordinates previously expressed by the functions (2.18). The surfaces (curved in general) described by the equations (2.19) are called *curvilinear coordinate surfaces*. The coordinates q_1, q_2, q_3 are therefore also known as *curvilinear coordinates*.

The coordinate surfaces passing through any point $P(q_1, q_2, q_3)$ intersect in pairs and give rise to three space curves passing through that point. These curves of intersection are called the *coordinate lines*. The surfaces $q_2 = \text{const.}$ and $q_3 = \text{const.}$ intersect in a curve along which the coordinate q_1 alone varies. Thus this curve is called the q_1-line or the q_1-curve. Similarly, we have a q_2-line and a q_3-line (see Fig. 2.21).

At the point P we draw a tangent to each of the coordinate lines. These tangents are taken as the coordinate axes at the point P (see Fig. 2.21). The axes are chosen positive in the direction in which q_1, q_2, q_3 increase from the point P. Along the coordinate axes thus formed we mark out from the point P three unit vectors e_1, e_2, e_3 (see Fig. 2.21).* This system of unit vectors at the point P can then be used as a reference system for all vector quantities associated with that point.

It should be readily noted that in a system of curvilinear coordinates, the axes and the reference system of unit vectors are not, in general, of fixed directions in space. Their directions change from point to point in space. We should bear in mind this particular aspect of curvilinear coordinates.

2.17 Orthogonal, Curvilinear Coordinates

In many problems, when curvilinear coordinates are used, one chooses the coordinates in such a way that the coordinate surfaces intersect at right angles at each point in space. Such coordinates are called *orthogonal curvilinear* coordinates. In our studies here we shall be concerned only with orthogonal systems. As examples of such systems we shall consider the following two special cases.†

Cylindrical Coordinates. In this system a point in space is located by the coordinates

$$q_1 = r$$
$$q_2 = \theta$$
$$q_3 = z$$

shown in Fig. 2.22. (Note that here r is not the magnitude of the position vector $\overrightarrow{OP}$.) The transformation between Cartesian coordinates and cylindrical coordinates is expressed by the equations

$$r = q_1(x, y, z) = \sqrt{x^2 + y^2}$$
$$\theta = q_2(x, y, z) = \arctan \frac{y}{x}$$
$$z = q_3(x, y, z) = z$$

or, inversely, by

$$x = r \cos \theta$$
$$y = r \sin \theta$$
$$z = z$$

* In Fig. 2.21 the naming of the coordinate surfaces is initially so chosen as to make e_1, e_2, e_3 a right-handed system.
† For further examples and for a discussion on nonorthogonal systems see Margenau and Murphy (1956).

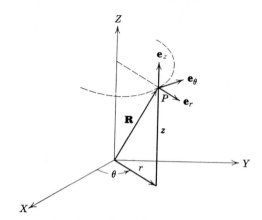

Fig. 2.22 Cylindrical coordinates.

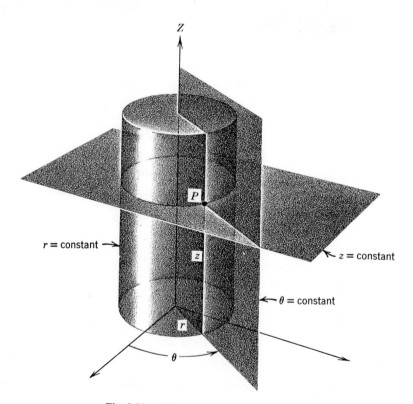

Fig. 2.23 Cylindrical-coordinate surfaces.

The coordinate surfaces given by (1) $r = $ const. are cylinders coaxial with the Z-axis, (2) $\theta = $ const. are half planes through the Z-axis, (3) $z = $ const. are planes perpendicular to the Z-axis (Fig. 2.23).

At any point $P(r, \theta, z)$ the vectors e_r, e_θ, e_z denote the reference system of unit vectors drawn respectively in the directions of increasing r, θ, and z (Fig. 2.22). The unit vectors are orthogonal to one another and form, in the order e_r, e_θ, e_z, a right-handed system. These unit vectors except for e_z are generally of different directions at different points in space.

Let **R** denote the position vector $\overrightarrow{OP}$ from the origin O to the point $P(r, \theta, z)$. The component form of **R** is then expressed by

$$\mathbf{R} = r\mathbf{e}_r + z\mathbf{e}_z$$

If **A** is any vector associated with the point $P(r, \theta, z)$, and if A_r, A_θ, A_z are the components of **A** with respect to the unit vectors e_r, e_θ, e_z at P, we write

$$\mathbf{A} = (A_r, A_\theta, A_z) = A_r\mathbf{e}_r + A_\theta\mathbf{e}_\theta + A_z\mathbf{e}_z$$

Spherical Coordinates. In this system a point in space is located by the coordinates

$$q_1 = r$$
$$q_2 = \theta$$
$$q_3 = \varphi$$

as shown in Fig. 2.24. The transformation between Cartesians and spherical coordinates is expressed by the equations

$$r = \sqrt{x^2 + y^2 + z^2}$$

$$\theta = \arccos \frac{z}{\sqrt{x^2 + y^2 + z^2}} = \arcsin \frac{\sqrt{x^2 + y^2}}{\sqrt{x^2 + y^2 + z^2}}$$

$$\varphi = \arccos \frac{x}{\sqrt{x^2 + y^2}} = \arcsin \frac{y}{\sqrt{x^2 + y^2}}$$

or, inversely, by

$$x = r \sin \theta \cos \varphi$$
$$y = r \sin \theta \sin \varphi$$
$$z = r \cos \theta$$

The coordinate surfaces given by (1) $r = $ const. are concentric spheres about the origin, (2) $\theta = $ const. are circular cones with vertex at the origin and axis along the Z-axis, (3) $\varphi = $ const. are half planes through the Z-axis (Fig. 2.25).

At any point $P(r, \theta, \varphi)$ the vectors e_r, e_θ, e_φ denote the reference system of unit vectors drawn, respectively, in the directions of increasing r, θ, and φ

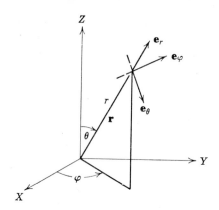

Fig. 2.24 Spherical coordinates.

(Fig. 2.24). The unit vectors are orthogonal to one another and form, in the order $\mathbf{e}_r$, $\mathbf{e}_\theta$, $\mathbf{e}_\varphi$, a right-handed system. These vectors are, in general, of different directions at differential points in space.

The component form of the position vector $\mathbf{r} = \overrightarrow{OP}$ is expressed by

$$\mathbf{r} = r\mathbf{e}_r$$

If $\mathbf{A}$ is any vector associated with the point $P(r, \theta, \varphi)$ and if A_r, A_θ, A_φ

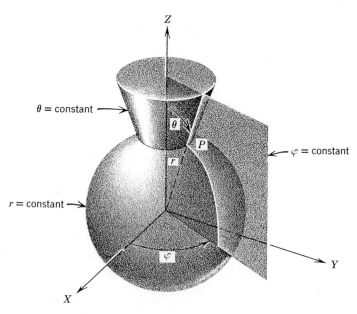

Fig. 2.25 Spherical-coordinate surfaces.

are the components of **A** with respect to the unit vectors $\mathbf{e}_r$, $\mathbf{e}_\theta$, $\mathbf{e}_\varphi$ set up at P, we write

$$\mathbf{A} = (A_r, A_\theta, A_\varphi) = A_r\mathbf{e}_r + A_\theta\mathbf{e}_\theta + A_\varphi\mathbf{e}_\varphi$$

2.18 Products of Vectors in Terms of Their Components

We shall now express in component form the various products we have previously considered. For this purpose we choose an *orthogonal right-handed* system of unit vectors. We denote the unit vectors by $\mathbf{e}_1$, $\mathbf{e}_2$, $\mathbf{e}_3$ and the components of any vector **A** with respect to these by A_1, A_2, A_3.*

First we consider the scalar and vector products between the unit vectors.

A scalar product of the form $\mathbf{e}_1 \cdot \mathbf{e}_1$ is equal to unity. A scalar product of the form $\mathbf{e}_1 \cdot \mathbf{e}_2$ is zero. We thus have

$$\mathbf{e}_1 \cdot \mathbf{e}_1 = \mathbf{e}_2 \cdot \mathbf{e}_2 = \mathbf{e}_3 \cdot \mathbf{e}_3 = 1 \qquad (2.20a)$$

and

$$\mathbf{e}_1 \cdot \mathbf{e}_2 = \mathbf{e}_2 \cdot \mathbf{e}_3 = \mathbf{e}_3 \cdot \mathbf{e}_1 = 0 \qquad (2.20b)$$

A vector product of the form $\mathbf{e}_1 \times \mathbf{e}_1$ is zero. A vector product of the form $\mathbf{e}_1 \times \mathbf{e}_2$ is equal to $\mathbf{e}_3$, and of the form $\mathbf{e}_2 \times \mathbf{e}_1$ is $-\mathbf{e}_3$. We thus have

$$\mathbf{e}_1 \times \mathbf{e}_1 = \mathbf{e}_2 \times \mathbf{e}_2 = \mathbf{e}_3 \times \mathbf{e}_3 = 0 \qquad (2.21a)$$

and

$$\mathbf{e}_1 \times \mathbf{e}_2 = -\mathbf{e}_2 \times \mathbf{e}_1 = \mathbf{e}_3 \qquad (2.21b)$$

$$\mathbf{e}_2 \times \mathbf{e}_3 = -\mathbf{e}_3 \times \mathbf{e}_2 = \mathbf{e}_1 \qquad (2.21c)$$

$$\mathbf{e}_3 \times \mathbf{e}_1 = -\mathbf{e}_1 \times \mathbf{e}_3 = \mathbf{e}_2 \qquad (2.21d)$$

We now consider the products between two vectors **A** and **B**. For the scalar product $\mathbf{A} \cdot \mathbf{B}$ we have

$$\mathbf{A} \cdot \mathbf{B} = (A_1\mathbf{e}_1 + A_2\mathbf{e}_2 + A_3\mathbf{e}_3) \cdot (B_1\mathbf{e}_1 + B_2\mathbf{e}_2 + B_3\mathbf{e}_3)$$

Using the relations (2.20a) and (2.21b), this becomes

$$\mathbf{A} \cdot \mathbf{B} = A_1B_1 + A_2B_2 + A_3B_3 \qquad (2.22)$$

which states that the scalar product of two vectors is equal to the sum of the products of their corresponding components.

For the vector product $\mathbf{A} \times \mathbf{B}$ we have

$$\mathbf{A} \times \mathbf{B} = (A_1\mathbf{e}_1 + A_2\mathbf{e}_2 + A_3\mathbf{e}_3) \times (B_1\mathbf{e}_1 + B_2\mathbf{e}_2 + B_3\mathbf{e}_3)$$

By using the relations (2.21a) through (2.21d), this becomes

$$\mathbf{A} \times \mathbf{B} = (A_2B_3 - A_3B_2)\mathbf{e}_1 + (A_3B_1 - A_1B_3)\mathbf{e}_2 + (A_1B_2 - A_2B_1)\mathbf{e}_3$$

* $\mathbf{e}_1$, $\mathbf{e}_2$, $\mathbf{e}_3$ may represent the system of unit vectors set up at a point described by any orthogonal, curvilinear, coordinate system.

This may be written in the determinant form

$$\mathbf{A} \times \mathbf{B} = \begin{vmatrix} \mathbf{e}_1 & \mathbf{e}_2 & \mathbf{e}_3 \\ A_1 & A_2 & A_3 \\ B_1 & B_2 & B_3 \end{vmatrix} \tag{2.23}$$

which is more convenient to remember.

We now consider triple products between the vectors $\mathbf{A}, \mathbf{B}, \mathbf{C}$. The result of the triple scalar product $\mathbf{A} \cdot \mathbf{B} \times \mathbf{C}$ may be written in the determinant form

$$\mathbf{A} \cdot \mathbf{B} \times \mathbf{C} = \begin{vmatrix} A_1 & A_2 & A_3 \\ B_1 & B_2 & B_3 \\ C_1 & C_2 & C_3 \end{vmatrix} \tag{2.24}$$

which is easy to remember.*

The result of the triple vector product $\mathbf{A} \times (\mathbf{B} \times \mathbf{C})$ may be written in the determinant form*

$$\mathbf{A} \times (\mathbf{B} \times \mathbf{C}) = \begin{vmatrix} \mathbf{e}_1 & \mathbf{e}_2 & \mathbf{e}_3 \\ A_1 & A_2 & A_3 \\ (B_2 C_3 - B_3 C_2) & (B_3 C_1 - B_1 C_3) & (B_1 C_2 - B_2 C_1) \end{vmatrix}$$
$$\tag{2.25}$$

This concludes the essentials of vector algebra. We now pass on to the elements of vector calculus.

2.19 Functions involving Vectors and Scalars

We are familiar with the concepts of calculus as applied to scalars that are functions of other scalars. Extension of such concepts to functions involving vectors and scalars forms vector analysis. We shall first look at the various types of functions that are likely to arise in the vector description of physical problems.

If we wish to describe the motion of a mass point in space, we will do so by specifying at different instances of time its position with respect to some point fixed in a chosen frame of reference. That is, we describe the position vector $\mathbf{r}$ as a function of time t, a scalar variable. Such a functional relation is symbolically expressed by

$$\mathbf{r} = \mathbf{r}(t)$$

This is an example of a vector as a function of a scalar. Similarly, if for each value of a scalar variable t there corresponds a certain value of a

* Proof is left as an exercise.

vector $\mathbf{A}$, we say that the vector $\mathbf{A}$ is a function of the scalar t and write

$$\mathbf{A} = \mathbf{A}(t)$$

Suppose we want to describe the temperature at every point of a heated body. To do this we specify each point of the body by means of a position vector drawn from an arbitrarily chosen reference point and say that the temperature T is a function of the position vector $\mathbf{r}$. Symbolically, we write

$$T = T(\mathbf{r})$$

Here we have an example of a scalar as a function of a vector. Similarly, if for each value of a vector $\mathbf{r}$ there corresponds a certain value of a scalar ϕ, we say that the scalar ϕ is a function of the vector $\mathbf{r}$ and write

$$\phi = \phi(\mathbf{r})$$

When $\mathbf{r}$ denotes the position vector, we say that ϕ is a *scalar function of position*. The distributions of pressure, density, and temperature in the atmosphere are examples of scalar functions of position.

Let us consider next a rigid body rotating with a constant angular velocity $\boldsymbol{\omega}$. As seen in Section 2.10, the velocity of a point of the body is given by

$$\mathbf{V} = \boldsymbol{\omega} \times \mathbf{r}$$

where $\mathbf{r}$ is the position vector from a reference point taken on the axis of rotation. Different values of $\mathbf{r}$ yield the velocities of the different points of the body. We say that the velocity is a function of the position vector and write symbolically

$$\mathbf{V} = \mathbf{V}(\mathbf{r})$$

This is an example of a vector as a function of another vector. If for each value of a vector $\mathbf{r}$ there corresponds a certain value of another vector $\mathbf{A}$, we say that $\mathbf{A}$ is a function of $\mathbf{r}$ and write

$$\mathbf{A} = \mathbf{A}(\mathbf{r})$$

When $\mathbf{r}$ signifies the position vector, we say that $\mathbf{A}$ is a *vector function of position*. The gravitational force experienced by one body in the presence of another is an example of a vector function of position. Similarly, the Coulomb force acting on one electrically charged body in the presence of another charged body is a vector function of position.

The functional relations we have introduced above are only special forms of the more general relations expressed by

$$\phi = \phi(\mathbf{r}, t)$$

(a scalar as a function of another scalar and a vector) and

$$\mathbf{A} = \mathbf{A}(\mathbf{r}, t)$$

(a vector as a function of another vector and a scalar). When $\mathbf{r}$ and t signify position and time, respectively, we say that ϕ is a *scalar function of position and time* and that $\mathbf{A}$ is a *vector function of position and time*. If in the case of a heated body the temperature at any point of the body varies with time, we say that the temperature T is a scalar function of position and time and write $T = T(\mathbf{r}, t)$. Similarly, if in the case of a rotating rigid body the angular velocity changes with time, the velocity at any point of the body varies with time, and we say that the velocity $\mathbf{V}$ is a vector function of position and time and write $\mathbf{V} = \mathbf{V}(\mathbf{r}, t)$.

Using the principle of decomposition of a vector into scalar components, we can interpret the preceding functions involving vectors in terms of scalar functions of scalar variables. Such an interpretation sets up a correspondence between the operations of vector calculus with those of scalar (or ordinary) calculus.

Consider first the function $\mathbf{A} = \mathbf{A}(t)$. Let A_1, A_2, A_3 be the components of $\mathbf{A}$ with respect to a fixed system of three unit vectors denoted by $\mathbf{e}_1, \mathbf{e}_2, \mathbf{e}_3$. We thus write

$$\mathbf{A} = A_1\mathbf{e}_1 + A_2\mathbf{e}_2 + A_3\mathbf{e}_3 = (A_1, A_2, A_3)$$

With this representation we can interpret the function $\mathbf{A}(t)$ as equivalent to the three functions

$$A_1 = A_1(t), A_2 = A_2(t), A_3 = A_3(t)$$

Thus a vector function of a scalar variable is equivalent to a system of three independent scalar functions of the same scalar variable.

Consider next the function $\phi = \phi(\mathbf{r})$. If q_1, q_2, q_3 are the components of $\mathbf{r}$ with respect to a system of unit vectors $\mathbf{e}_1, \mathbf{e}_2, \mathbf{e}_3$ we can interpret $\phi(\mathbf{r})$ as equivalent to the function

$$\phi = \phi(q_1, q_2, q_3)$$

This means that a scalar function of a vector is equivalent to a scalar function of three independent scalar variables.

Consider now the function $\mathbf{A} = \mathbf{A}(\mathbf{r})$. If, as before, A_1, A_2, A_3 are the components of $\mathbf{A}$ and q_1, q_2, q_3 are the components of $\mathbf{r}$, we can write $\mathbf{A} = \mathbf{A}(\mathbf{r})$ as equivalent to the system of functions expressed by

$$A_1 = A_1(\mathbf{r}) = A_1(q_1, q_2, q_3)$$
$$A_2 = A_2(\mathbf{r}) = A_2(q_1, q_2, q_3)$$
$$A_3 = A_3(\mathbf{r}) = A_3(q_1, q_2, q_3)$$

Thus a vector function of a vector is equivalent to a system of three independent scalar functions of three scalar variables.

In a similar manner, the functions $\phi = \phi(\mathbf{r}, t)$ and $\mathbf{A} = \mathbf{A}(\mathbf{r}, t)$ can be expressed in terms of scalar functions of scalar variables.

2.20   Scalar and Vector Fields

A scalar or a vector function of position assigns to each point of a portion of space a definite value of a scalar or a vector quantity. The various points of the given region together with the corresponding values of the quantity, scalar or vector, form what is called a *field*. If the quantity concerned is a scalar, the field is called a *scalar field*; if the quantity is a vector, the field is called a *vector field*. If we are dealing with a scalar or vector function of position and time, the values of the scalar or vector quantity at the various points of the region change from instant to instant and the field becomes an *unsteady* or *nonstationary* field. If we are dealing with a scalar or vector function of position only, the field retains the same structure for all times and we say that it is a *steady* or a *stationary* field. The concept of a field helps us to show what is happening simultaneously at all points of a region of space.

An arbitrary point in a field is usually called a *field point*.

Consider a scalar field represented by a single-valued function $\phi = \phi(\mathbf{r})$. In such a field we can draw a family of surfaces such that each surface passes through all those points that have the same value of the scalar quantity ϕ. The surfaces are thus surfaces of constant ϕ and are represented by

$$\phi = \phi(\mathbf{r}) = \text{const.}$$

with the constant taking a different value for each surface. Such surfaces are commonly referred to as *level surfaces*. Surfaces of constant density or of constant temperature or of constant pressure in the atmosphere are all examples of level surfaces. If a scalar field is steady, its level surfaces remain constant with time. If the scalar field is unsteady, the level surfaces change from instant to instant.

One may picture a vector field by imagining arrows placed at various points of the region of space, each arrow pointing in the direction of the vector quantity associated with the point and having a length proportional to the magnitude of the quantity. As an example, the velocity field of a rigid body rotating with a constant angular velocity is shown in Fig. 2.26. In a vector field one can draw a system of curves such that each curve is tangent at each point on it to the direction of the vector quantity associated with that point. Such curves are called *field lines*. If the vector field is a force field, the field lines are known as the *lines of force*; if the field is the velocity field of a flowing fluid, they are known as *streamlines*.

A familiar example of field lines is the picture of curves formed by iron filings in the presence of a magnet. The field lines of the velocity field of a rigid body rotating with a constant angular velocity are shown in Fig. 2.27. If the vector field we are dealing with is stationary, the picture of its field

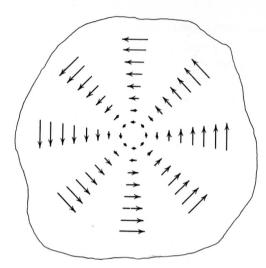

Fig. 2.26   Velocity field of a rotating rigid body as seen in a plane normal to the axis of rotation.

lines remains unchanged with time; otherwise the picture changes from instant to instant. To construct analytically the field lines of a vector field $A = A(r, t)$, we proceed as follows. Consider the field line passing through a point r at some instant of time. Let ds denote an element of the line through r. By definition ds has the same direction as that of the vector A associated with the point r at the instant considered. That is, ds and A

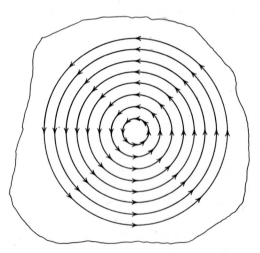

Fig. 2.27   Field lines of the velocity field of Fig. 2.26.

are parallel vectors. Recalling that the vector product of two parallel vectors is zero, we write

$$d\mathbf{s} \times \mathbf{A} = 0 \tag{2.26}$$

This, then, is the (differential) equation that determines (at any instant) the field lines of the vector field $\mathbf{A} = \mathbf{A}(\mathbf{r}, t)$. In determining the streamlines of the velocity field of a fluid in motion we shall have occasion to study the integration of an equation of the type (2.26). Using the principle of decomposition of a vector into components, Eq. (2.26) can be readily specialized to any chosen coordinate system.

Since the function $\mathbf{A} = \mathbf{A}(\mathbf{r}, t)$ is equivalent to three scalar functions of position and time, any vector field may be regarded as equivalent to three scalar fields.

2.21 Differentiation of a Vector Function of a Scalar Variable

If a vector $\mathbf{A}$ changes from a value $\mathbf{A}_1$ to a value $\mathbf{A}_2$, the increment in $\mathbf{A}$ denoted by $\Delta\mathbf{A}$ is simply the vector difference between $\mathbf{A}_2$ and $\mathbf{A}_1$. That is,

$$\Delta\mathbf{A} = \mathbf{A}_2 - \mathbf{A}_1$$

A change in a vector may be brought about by a change in its magnitude or by a change in its direction or by a change in both magnitude and direction.

If a vector $\mathbf{A}$ is a function of a scalar variable t, the increment $\Delta\mathbf{A}$ in $\mathbf{A}$ corresponding to an increment Δt in t from t to $t + \Delta t$ is given by

$$\Delta\mathbf{A} = \mathbf{A}(t + \Delta t) - \mathbf{A}(t)$$

If the ratio $\Delta\mathbf{A}/\Delta t$ (i.e., the average variation of $\mathbf{A}$ with respect to t in the interval Δt) tends to a limit when Δt tends to zero, that limit is called the derivative of $\mathbf{A}$ with respect to t (compare the definition of the derivative of a scalar function of a scalar variable). Following the usual convention of differential calculus, we denote this derivative by $d\mathbf{A}/dt$ and write

$$\frac{d\mathbf{A}(t)}{dt} = \lim_{\Delta t\to 0}\frac{\Delta\mathbf{A}}{\Delta t} = \lim_{\Delta t\to 0}\frac{\mathbf{A}(t + \Delta t) - \mathbf{A}(t)}{\Delta t} \tag{2.27}$$

Let us now look at the geometrical interpretation of this derivative. If we represent the various values of the continuously varying vector $\mathbf{A}$ by means of arrows drawn from a common origin, denoted by O, the terminus of the vector will describe a curve, denoted by $\mathscr{C}$, in space (see Fig. 2.28a). Let $\overrightarrow{OP}$ represent $\mathbf{A}$ at time t and $\overrightarrow{OQ}$ represent it at time $t + \Delta t$ (Fig. 2.28b). Then the increment $\Delta\mathbf{A}$ is represented by the vector chord $\overrightarrow{PQ}$ of the curve $\mathscr{C}$. Thus we have

$$\frac{\Delta\mathbf{A}}{\Delta t} = \frac{\text{chord } \overrightarrow{PQ}}{\Delta t}$$

and

$$\frac{d\mathbf{A}}{dt} = \lim_{\Delta t \to 0} \frac{\Delta \mathbf{A}}{\Delta t} = \lim_{\Delta t \to 0} \frac{\overrightarrow{PQ}}{\Delta t} \tag{2.28}$$

To interpret the limit we proceed as follows. A point such as P or Q on the curve $\mathscr{C}$ can be specified by giving either the vector $\mathbf{A}$ or the distance s measured along the curve from some initial point taken as reference (see figure). As t varies, s will change just as $\mathbf{A}$ does; so $s = s(t)$, and $\mathbf{A}$ may be

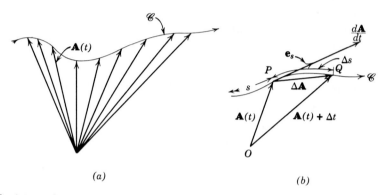

(a)   (b)

Fig. 2.28   (a) Space curve traced by $\mathbf{A}(t)$;   (b) illustrating the differentiation of $\mathbf{A}(t)$.

considered as depending on s. Let Δs denote the increment in s from P to Q. Therefore

$$\Delta s = \text{length of the arc } PQ$$

Introducing Δs, we rewrite Eq. (2.28) as

$$\frac{d\mathbf{A}}{dt} = \lim_{\Delta t \to 0} \frac{\overrightarrow{PQ}}{\Delta s} \frac{\Delta s}{\Delta t} = \left(\lim_{\Delta s \to 0} \frac{\overrightarrow{PQ}}{\Delta s} \right) \left(\lim_{\Delta t \to 0} \frac{\Delta s}{\Delta t} \right)$$

$$= \frac{ds}{dt} \lim_{\Delta s \to 0} \frac{\overrightarrow{PQ}}{\Delta s} \tag{2.28a}$$

Now, $\overrightarrow{PQ}/\Delta s$ is a vector along $\overrightarrow{PQ}$ with a magnitude equal to

$$\frac{\text{length of the chord } \overrightarrow{PQ}}{\text{length of the arc } PQ}$$

As $\Delta s \to 0$,

$$\left| \frac{\overrightarrow{PQ}}{\Delta s} \right| \to 1$$

and the direction of $\overrightarrow{PQ}$ becomes that of the tangent to the curve $\mathscr{C}$ at the point P. Denoting by $\mathbf{e}_s$ a unit tangent vector* at P (see figure), we write

$$\lim_{\Delta s \to 0} \frac{\overrightarrow{PQ}}{\Delta s} = \mathbf{e}_s$$

With this, Eq. (2.28) becomes

$$\frac{d\mathbf{A}}{dt} = \frac{ds}{dt} \mathbf{e}_s \qquad\qquad (2.29)$$

expressing the derivative as the product of a magnitude and a direction.

As an example of the above considerations, let us consider the motion of a mass particle. At any instant let its position be denoted by $\mathbf{r} = \mathbf{r}(t)$ measured from a point fixed in a chosen frame of reference. The path (or the trajectory) of the particle is given by the curve $\mathscr{C}$ traced by the vector $\mathbf{r}$ as t varies. The velocity $\mathbf{V}$ of the particle at any instant (i.e., at the position $\mathbf{r}$) is given by the derivative $d\mathbf{r}/dt$. Thus we have

$$\mathbf{V} = \frac{d\mathbf{r}(t)}{dt} = \frac{ds}{dt} \mathbf{e}_s = V\mathbf{e}_s$$

stating that the velocity is tangential to the trajectory at the instant considered and that the magnitude V of the velocity is equal to the rate of change of distance along the trajectory (i.e., to the speed).

As a simple result following Eq. (2.29) we note that *the direction of the derivative $d\mathbf{A}/dt$ when $\mathbf{A}$ is of constant length but of changing direction is perpendicular to the vector $\mathbf{A}$.*

We consider next the differentiation of the sums and products of vector functions all of which depend on the same scalar variable. In all such cases the formal methods of differentiation as employed in scalar calculus are equally applicable except that in cases involving vector products, the order of the vectors must be preserved. This is, of course, a natural consequence of the fact that vector products are not commutative. Accordingly, we have the following results.

The higher derivatives of the function $\mathbf{A} = \mathbf{A}(t)$ are constructed by successive differentiation just as in scalar calculus.

If $\mathbf{U} = \mathbf{U}(t)$ is the sum of two functions such that

$$\mathbf{U}(t) = \mathbf{A}(t) + \mathbf{B}(t)$$

we have

$$\frac{d\mathbf{U}}{dt} = \frac{d\mathbf{A}}{dt} + \frac{d\mathbf{B}}{dt}$$

* The direction of the tangent vector is taken in the direction of increasing s.

If $u = u(t)$ and $\mathbf{A} = \mathbf{A}(t)$, we have

$$\frac{d}{dt}(u\mathbf{A}) = u\,\frac{d\mathbf{A}}{dt} + \frac{du}{dt}\,\mathbf{A}$$

Here the order of the factors involved need not be preserved. If $\mathbf{A} = \mathbf{A}(t)$ and $\mathbf{B} = \mathbf{B}(t)$, we obtain

$$\frac{d}{dt}(\mathbf{A} \cdot \mathbf{B}) = \mathbf{A} \cdot \frac{d\mathbf{B}}{dt} + \frac{d\mathbf{A}}{dt} \cdot \mathbf{B}$$

Since the scalar product is commutative, the order of the vectors in this differentiation need not be preserved.

The derivative of the cross product $\mathbf{A}(t) \times \mathbf{B}(t)$ is given by

$$\frac{d}{dt}(\mathbf{A} \times \mathbf{B}) = \mathbf{A} \times \frac{d\mathbf{B}}{dt} + \frac{d\mathbf{A}}{dt} \times \mathbf{B}$$

and is not equal to

$$\mathbf{A} \times \frac{d\mathbf{B}}{dt} + \mathbf{B} \times \frac{d\mathbf{A}}{dt}, \text{ etc.}$$

Since the vector product is not commutative, here the order of the vectors should be preserved.

Considering triple products, we have

$$\frac{d}{dt}(\mathbf{A} \cdot \mathbf{B} \times \mathbf{C}) = \frac{d\mathbf{A}}{dt} \cdot \mathbf{B} \times \mathbf{C} + \mathbf{A} \cdot \frac{d\mathbf{B}}{dt} \times \mathbf{C} + \mathbf{A} \cdot \mathbf{B} \times \frac{d\mathbf{C}}{dt}$$

and

$$\frac{d}{dt}[\mathbf{A} \times (\mathbf{B} \times \mathbf{C})] = \frac{d\mathbf{A}}{dt} \times (\mathbf{B} \times \mathbf{C}) + \mathbf{A} \times \left(\frac{d\mathbf{B}}{dt} \times \mathbf{C}\right) + \mathbf{A} \times \left(\mathbf{B} \times \frac{d\mathbf{C}}{dt}\right).$$

Here again the order of the vectors has to be preserved.

In concluding this section we relate the derivative of the vector $\mathbf{A}(t)$ with the derivatives of its components. To do this we choose a system of unit vectors $\mathbf{e}_1$, $\mathbf{e}_2$, $\mathbf{e}_3$ and express

$$\mathbf{A}(t) = A_1\mathbf{e}_1 + A_2\mathbf{e}_2 + A_3\mathbf{e}_3$$

We therefore obtain

$$\frac{d\mathbf{A}}{dt} = \frac{d}{dt}(A_1\mathbf{e}_1) + \frac{d}{dt}(A_2\mathbf{e}_2) + \frac{d}{dt}(A_3\mathbf{e}_3) \qquad (2.30a)$$

If the *unit vectors* are *constant*,* this reduces to

$$\frac{d\mathbf{A}}{dt} = \frac{dA_1}{dt}\mathbf{e}_1 + \frac{dA_2}{dt}\mathbf{e}_2 + \frac{dA_3}{dt}\mathbf{e}_3 \qquad (2.30b)$$

* A change in a unit vector can be brought about by only a change in its direction, for by definition its magnitude is always unity.

If, however, the unit vectors are also changing with the scalar t, we have

$$\frac{d\mathbf{A}}{dt} = \frac{dA_1}{dt}\mathbf{e}_1 + A_1\frac{d\mathbf{e}_1}{dt} + \frac{dA_2}{dt}\mathbf{e}_2 + A_2\frac{d\mathbf{e}_2}{dt} + \frac{dA_3}{dt}\mathbf{e}_3 + A_3\frac{d\mathbf{e}_3}{dt} \quad (2.30c)$$

To illustrate the case in which the reference unit vectors are also changing let us consider the description in cylindrical coordinates of the motion of a mass particle. Accordingly, we denote at any instant the position of the particle by r, θ, z and by $\mathbf{e}_r$, $\mathbf{e}_\theta$, $\mathbf{e}_z$ the corresponding unit vectors. We ask for the velocity, $\mathbf{V}$, of the particle at the instant considered. By definition, the velocity of the particle is equal to the rate of change of its position. Thus if $\mathbf{R} = \mathbf{R}(t)$ gives the position of the particle with respect to a fixed point, we have

$$\mathbf{V} = \mathbf{V}(t) = \frac{d\mathbf{R}}{dt}$$

In cylindrical coordinates

$$\mathbf{R} = r\mathbf{e}_r + z\mathbf{e}_z$$

Therefore we obtain

$$\mathbf{V} = \frac{d}{dt}(r\mathbf{e}_r + z\mathbf{e}_z)$$

Since the direction of the unit vector $\mathbf{e}_r$ changes with change of location of the mass particle, this equation expands to

$$\mathbf{V} = \frac{dr}{dt}\mathbf{e}_r + r\frac{d\mathbf{e}_r}{dt} + \frac{dz}{dt}\mathbf{e}_z$$

To evaluate the rate of change of the unit vector $\mathbf{e}_r$ we proceed in the same way as in deriving Eq. (2.29) and obtain

$$\frac{d\mathbf{e}_r}{dt} = \frac{d\theta}{dt}\mathbf{e}_\theta$$

With this relation, the velocity expressed in cylindrical coordinates becomes

$$\mathbf{V} = \frac{dr}{dt}\mathbf{e}_r + r\frac{d\theta}{dt}\mathbf{e}_\theta + \frac{dz}{dt}\mathbf{e}_z \quad (2.31)$$

If we did not recognize that $\mathbf{e}_r$ is changing, we would have arrived at the *incorrect* result that the velocity of the particle is equal to

$$\frac{dr}{dt}\mathbf{e}_r + \frac{dz}{dt}\mathbf{e}_z$$

2.22 Changes in the Unit Vectors of Cylindrical and Spherical Coordinates

When we move from one point to another in cylindrical or spherical coordinates, changes occur in the directions of the reference unit vectors. We shall now determine these changes.

Cylindrical Coordinates. We move from a point $\mathbf{R} = (r, \theta, z)$ over an infinitesimal distance

$$d\mathbf{s} = dr\,\mathbf{e}_r + r\,d\theta\,\mathbf{e}_\theta + dz\,\mathbf{e}_z$$

in some direction. Here $\mathbf{e}_r$, $\mathbf{e}_\theta$, $\mathbf{e}_z$ are the unit vectors associated with the point $\mathbf{R}$. Let $\mathbf{e}_r'$, $\mathbf{e}_\theta'$, $\mathbf{e}_z'$ be the unit vectors associated with the point $\mathbf{R} + d\mathbf{s}$. As shown in Fig. 2.29a, the system $\mathbf{e}_r'$, $\mathbf{e}_\theta'$, $\mathbf{e}_z'$ results from an

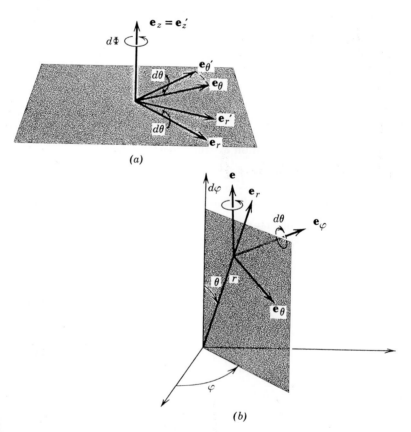

Fig. 2.29 Changes in unit vectors: (a) cylindrical coordinates; (b) spherical coordinates.

infinitesimal rotation

$$d\mathbf{\Phi} = d\theta\, \mathbf{e}_z \tag{2.32}$$

of the system $\mathbf{e}_r$, $\mathbf{e}_\theta$, $\mathbf{e}_z$. Then the change in any of the unit vectors is given by

$$d\mathbf{e}_i \equiv \mathbf{e}_i' - \mathbf{e}_i = d\mathbf{\Phi} \times \mathbf{e}_i \tag{2.33}$$

where the subscript i may be r, θ, or z. Using relations (2.32) and (2.33) we obtain the changes in the unit vectors as

$$\begin{aligned}
d\mathbf{e}_r &= d\theta\, \mathbf{e}_z \times \mathbf{e}_r = d\theta\, \mathbf{e}_\theta \\
d\mathbf{e}_\theta &= d\theta\, \mathbf{e}_z \times \mathbf{e}_\theta = -d\theta\, \mathbf{e}_r \\
d\mathbf{e}_z &= d\theta\, \mathbf{e}_z \times \mathbf{e}_z = 0
\end{aligned} \tag{2.34}$$

Spherical Coordinates.   Now we denote a point in space by $\mathbf{r} = (r, \theta, \varphi)$. We move in some direction, over an infinitesimal distance

$$d\mathbf{s} = dr\, \mathbf{e}_r + r\, d\theta\, \mathbf{e}_\theta + r \sin\theta\, d\varphi\, \mathbf{e}_\varphi$$

where $\mathbf{e}_r$, $\mathbf{e}_\theta$, $\mathbf{e}_\varphi$ are the unit vectors associated with the point $\mathbf{r}$. Denote by $\mathbf{e}_r'$, $\mathbf{e}_\theta'$, $\mathbf{e}_\varphi'$ the unit vectors associated with the point $\mathbf{r} + d\mathbf{s}$. We observe, as indicated in Fig. 2.29b, that the system $\mathbf{e}_r'$, $\mathbf{e}_\theta'$, $\mathbf{e}_\varphi'$ results from an infinitesimal rotation

$$d\mathbf{\Phi} = d\varphi\, \mathbf{e} + d\theta\, \mathbf{e}_\varphi \tag{2.35}$$

of the system $\mathbf{e}_r$, $\mathbf{e}_\theta$, $\mathbf{e}_z$, where $\mathbf{e}$ is a unit vector in the direction of the axis from which θ is measured. Expressing $\mathbf{e}$ in terms of $\mathbf{e}_r$ and $\mathbf{e}_\theta$ by the relation

$$\mathbf{e} = \cos\theta \mathbf{e}_r - \sin\theta \mathbf{e}_\theta$$

we write Eq. (2.35) as

$$d\mathbf{\Phi} = d\varphi \cos\theta \mathbf{e}_r - d\varphi \sin\theta \mathbf{e}_\theta + d\theta \mathbf{e}_\varphi \tag{2.36}$$

By using the relation for $d\mathbf{\Phi}$, the changes in the unit vectors may be determined from the equation

$$d\mathbf{e}_i \equiv \mathbf{e}_i' - \mathbf{e}_i = d\mathbf{\Phi} \times \mathbf{e}_i$$

where the subscript i may be r, θ, or φ. We thus obtain

$$\begin{aligned}
d\mathbf{e}_r &= d\theta\, \mathbf{e}_\theta + d\varphi \sin\theta \mathbf{e}_\varphi \\
d\mathbf{e}_\theta &= -d\theta\, \mathbf{e}_r + d\varphi \cos\theta \mathbf{e}_\varphi \\
d\mathbf{e}_\varphi &= -d\varphi \sin\theta \mathbf{e}_r - d\varphi \cos\theta \mathbf{e}_\theta
\end{aligned} \tag{2.37}$$

In concluding this section we would like to mention that in a similar way one can determine the changes in the unit vectors of any orthogonal, curvilinear coordinate system. See Section 2.45.

2.23 Frames of Reference

In the preceding considerations the vector quantities were described with respect to a chosen origin. Such an origin is a point fixed in some *frame of reference*. By a frame of reference we mean a frame in space and time that will enable us, by suitable measurements, to describe physical phenomena such as the position of mass points and the passage of time. Although we may postulate the existence of a reference frame that is absolutely fixed in space and time, we are obliged, for practical reasons, to deal with reference frames that are in relative motion with each other. This being the case we should recognize that operations performed in one frame of reference on functions involving vectors and scalars will yield results that are, in general, different from those obtained by similar operations in another frame of reference. This means that an explicit mention of the reference frame in which the operations are being performed is essential unless it is understood that once and for all a particular reference frame is all that is employed.*

We shall now illustrate some of these ideas by working out a particular problem. Suppose that K and K_0 are two frames of reference such that the frame K rotates with an angular velocity $\boldsymbol{\omega} = \boldsymbol{\omega}(t)$ with respect to the frame K_0, t being time. Consider now a vector function of time, say $\mathbf{A} = \mathbf{A}(t)$. Our problem is to find the relation between the derivative of $\mathbf{A}$ with respect to t as computed in the frame K and the similar derivative computed in the frame K_0. First of all we must introduce an explicit notation to distinguish the derivative operations in the two frames. Accordingly, we shall denote by

$$\frac{{}^{K_0}d}{dt} \text{ the derivative operation in frame } K_0$$

and by

$$\frac{{}^{K}d}{dt} \text{ the derivative operation in frame } K$$

Let $\mathbf{e}_1$, $\mathbf{e}_2$, $\mathbf{e}_3$ represent a system of unit vectors *fixed* in the system K, and let a_1, a_2, a_3 be the respective scalar components of $\mathbf{A}$. We observe that the unit vectors $\mathbf{e}_1$, $\mathbf{e}_2$, $\mathbf{e}_3$ *are not* functions of the variable t in the frame K, while they *are* functions of t in the frame K_0. The scalar components are simply scalar functions of a scalar variable in either frame of reference; in this case the distinction between the reference frames becomes irrelevant.

---

* The differentiation of vectors with particular emphasis on the importance of reference frames is extensively discussed by Kane (1961).

Expressing $\mathbf{A}$ in component form we write

$$\frac{^{K_0}d\mathbf{A}}{dt} = \frac{^{K_0}d}{dt}(a_1\mathbf{e}_1 + a_2\mathbf{e}_2 + a_3\mathbf{e}_3)$$

$$= \left(\mathbf{e}_1 \frac{^{K_0}da_1}{dt} + \mathbf{e}_2 \frac{^{K_0}da_2}{dt} + \mathbf{e}_3 \frac{^{K_0}da_3}{dt}\right)$$

$$+ \left(a_1 \frac{^{K_0}d\mathbf{e}_1}{dt} + a_2 \frac{^{K_0}d\mathbf{e}_2}{dt} + a_3 \frac{^{K_0}d\mathbf{e}_3}{dt}\right) \qquad (2.38)$$

Consider the term $\dfrac{^{K_0}da_1}{dt}$. Since the derivative of a scalar function of a scalar variable does not depend on the reference frame, we note that

$$\frac{^{K_0}da_1}{dt} = \frac{^{K}da_1}{dt} = \frac{da_1}{dt}$$

We can, therefore, write

$$\mathbf{e}_1 \frac{^{K_0}da_1}{dt} = \mathbf{e}_1 \frac{^{K}da_1}{dt} = \frac{^{K}d(a_1\mathbf{e}_1)}{dt} \qquad (2.39a)$$

since $\mathbf{e}_1$ is independent of t in the frame K. Similarly, we have

$$\mathbf{e}_2 \frac{^{K_0}da_2}{dt} = \frac{^{K}d(a_2\mathbf{e}_2)}{dt} \qquad (2.39b)$$

and

$$\mathbf{e}_3 \frac{^{K_0}da_3}{dt} = \frac{^{K}d(a_3\mathbf{e}_3)}{dt} \qquad (2.39c)$$

Combining the relations (2.39), we arrive at the result

$$\mathbf{e}_1 \frac{^{K_0}da_1}{dt} + \mathbf{e}_2 \frac{^{K_0}da_2}{dt} + \mathbf{e}_3 \frac{^{K_0}da_3}{dt} = \frac{^{K}d}{dt}(a_1\mathbf{e}_1 + a_2\mathbf{e}_2 + a_3\mathbf{e}_3)$$

$$= \frac{^{K}d\mathbf{A}}{dt} \qquad (2.40)$$

Consider next the term

$$\frac{^{K_0}d\mathbf{e}_1}{dt}$$

Since $\mathbf{e}_1$ is a fixed vector in the frame K that is rotating with angular velocity $\boldsymbol{\omega}(t)$ with respect to frame K_0, it can be verified that

$$\frac{^{K_0}d\mathbf{e}_1}{dt} = \boldsymbol{\omega} \times \mathbf{e}_1$$

Then we can write

$$a_1 \frac{{}^{K_0}de_1}{dt} = a_1\boldsymbol{\omega} \times \mathbf{e}_1 = \boldsymbol{\omega} \times a_1\mathbf{e}_1 \tag{2.41a}$$

Similarly, we obtain

$$a_2 \frac{{}^{K_0}de_2}{dt} = \boldsymbol{\omega} \times a_2\mathbf{e}_2 \tag{2.41b}$$

and

$$a_3 \frac{{}^{K_0}de_3}{dt} = \boldsymbol{\omega} \times a_3\mathbf{e}_3 \tag{2.41c}$$

Combining the relations (2.41), we arrive at the result

$$a_1 \frac{{}^{K_0}de_1}{dt} + a_2 \frac{{}^{K_0}de_2}{dt} + a_3 \frac{{}^{K_0}de_3}{dt} = \boldsymbol{\omega} \times (a_1\mathbf{e}_1 + a_2\mathbf{e}_2 + a_3\mathbf{e}_3)$$

$$= \boldsymbol{\omega} \times \mathbf{A} \tag{2.42}$$

By using the relations (2.40) and (2.42), Eq. (2.38) may be rewritten as

$$\frac{{}^{K_0}d\mathbf{A}}{dt} = \frac{{}^{K}d\mathbf{A}}{dt} + \boldsymbol{\omega} \times \mathbf{A} \tag{2.43}$$

This gives the required relation between the derivatives of $\mathbf{A}(t)$ in the reference frames K and K_0.

2.24 Differentiation of a Scalar Function of a Vector: Concept of a Gradient

Let us consider specifically a scalar function of position $\phi = \phi(\mathbf{r})$ and ask for its spatial variation. The conclusions we arrive at for this function are equally applicable to other scalar functions of a vector variable. Since the independent variable in the function $\phi(\mathbf{r})$ is a vector, we have the choice of an infinite number of directions in which to take the increment $\Delta\mathbf{r}$ or, equivalently, the differential increment $d\mathbf{r}$. The differential increment in ϕ corresponding to $d\mathbf{r}$ would, in general, be different in different directions. This means that in describing the variation of ϕ we must specify the direction in which the variation is taken. We thus talk about the spatial derivative of ϕ in a particular direction and refer to it as a *directional derivative*. According to these ideas it would appear that to describe completely the spatial variation at any point of a function $\phi = \phi(\mathbf{r})$, we may have to specify the derivatives of ϕ in all possible directions at that point. Fortunately, however, this is not necessary. It turns out, as we shall see, that all that is necessary is to give the derivatives of ϕ in three independent directions. These three derivatives are then sufficient to determine the variation of ϕ in any other direction. Thus to specify completely the variation at $\mathbf{r}$ of a function $\phi(\mathbf{r})$, we need a set of three numbers. As will be seen, this set of numbers represents a vector. Such a vector is called a *gradient vector* or simply a *gradient*.

To discuss the nature and significance of a gradient, let us first choose Cartesian coordinates x, y, z and write

$$\mathbf{r} = (x, y, z) = x\mathbf{i} + y\mathbf{j} + z\mathbf{k}$$

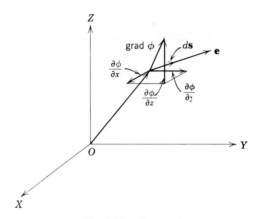

Fig. 2.30. Gradient.

and

$$\phi = \phi(\mathbf{r}) = \phi(x, y, z)$$

Let us denote by

$$d\mathbf{s} = ds\mathbf{e}$$

a small increment in $\mathbf{r}$ in some direction $\mathbf{e}$. Hereafter we adopt the notation $d\mathbf{s}$ for $d\mathbf{r}$ so as to avoid confusion between $|d\mathbf{r}|$ and $dr = d\,|\mathbf{r}|$, which are not the same. With this notation

$$|d\mathbf{r}| \equiv |d\mathbf{s}| = ds$$

We observe that

$$d\mathbf{s} = (dx, dy, dz) = dx\mathbf{i} + dy\mathbf{j} + dz\mathbf{k}$$

(See Fig. 2.30.)

The change in ϕ over the directed distance $d\mathbf{s}$ is given by

$$d\phi = \frac{\partial \phi}{\partial x}dx + \frac{\partial \phi}{\partial y}dy + \frac{\partial \phi}{\partial z}dz \tag{2.44}$$

Recalling that the scalar product of two vectors is the sum of the products of their corresponding components, we rewrite Eq. (2.44) as

$$d\phi = \left(\frac{\partial \phi}{\partial x}\mathbf{i} + \frac{\partial \phi}{\partial y}\mathbf{j} + \frac{\partial \phi}{\partial z}\mathbf{k}\right) \cdot (dx\mathbf{i} + dy\mathbf{j} + dz\mathbf{k})$$

$$= \left(\frac{\partial \phi}{\partial x}\mathbf{i} + \frac{\partial \phi}{\partial y}\mathbf{j} + \frac{\partial \phi}{\partial z}\mathbf{k}\right) \cdot d\mathbf{s} \qquad (2.45)$$

Dividing Eq. (2.45) by ds we obtain

$$\frac{d\phi}{ds} = \left(\frac{\partial \phi}{\partial x}\mathbf{i} + \frac{\partial \phi}{\partial y}\mathbf{j} + \frac{\partial \phi}{\partial z}\mathbf{k}\right) \cdot \frac{d\mathbf{s}}{ds}$$

$$= \left(\frac{\partial \phi}{\partial x}\mathbf{i} + \frac{\partial \phi}{\partial y}\mathbf{j} + \frac{\partial \phi}{\partial z}\mathbf{k}\right) \cdot \mathbf{e} \qquad (2.46)$$

We thus see that the directional derivative of $\phi(\mathbf{r})$ in any chosen direction is equal to the component in that direction of a particular vector. The components of that vector are the partial derivatives of ϕ with respect to distances along the three coordinate axes (see Fig. 2.30).

We arrive at the same conclusions if instead of Cartesians we choose any orthogonal, curvilinear coordinates. To see this, let us denote a point P in space by a set of orthogonal, curvilinear coordinates q_1, q_2, q_3 (see Section 2.17). Then the scalar function of position ϕ is expressed as

$$\phi = \phi(\mathbf{r}) = \phi(q_1, q_2, q_3)$$

Note that q_1, q_2, q_3 are coordinates and are not necessarily the components of $\mathbf{r}$. As before, let $d\mathbf{s} = ds\mathbf{e}$ denote a differential increment in $\mathbf{r}$ in some direction $\mathbf{e}$ from the point P. If dq_1, dq_2, dq_3 are the corresponding increments in the coordinates, the differential increment $d\phi$ over the divided distance $d\mathbf{s}$ is given by

$$d\phi = \frac{\partial \phi}{\partial q_1}dq_1 + \frac{\partial \phi}{\partial q_2}dq_2 + \frac{\partial \phi}{\partial q_3}dq_3 \qquad (2.47)$$

The partial derivatives appearing in Eq. (2.47) are not derivatives with respect to distances. Also, dq_1, dq_2, dq_3 are not components of $d\mathbf{s}$.

To express Eq. (2.47) in a form similar to Eq. (2.45), we proceed as follows. Let s_1, s_2, s_3 denote distances measured from P along the q_1, q_2, q_3 curves respectively. Let $\delta s_1, \delta s_2, \delta s_3$ denote the distances along q_1, q_2, q_3 curves corresponding to the increments dq_1, dq_2, dq_3. Let us say that the dq's and the δs's are related as follows

$$\delta s_1 = h_1 \, dq_1$$
$$\delta s_2 = h_2 \, dq_2$$
$$\delta s_3 = h_3 \, dq_3$$

where h_1, h_2, h_3 may be referred to as scale factors. These factors may vary from point to point; in other words they are, in general, functions of position. They can be determined in terms of q_1, q_2, q_3. This is shown in Section 2.45 where further details are given.

We may now write Eq. (2.47) as

$$
\begin{aligned}
d\phi &= \frac{1}{h_1}\frac{\partial \phi}{\partial q_1} h_1\, dq_1 + \frac{1}{h_2}\frac{\partial \phi}{\partial q_2} h_2\, dq_2 + \frac{1}{h_3}\frac{\partial \phi}{\partial q_3} h_3\, dq_3 \\
&= \frac{1}{h_1}\frac{\partial \phi}{\partial q_1}\, \delta s_1 + \frac{1}{h_2}\frac{\partial \phi}{\partial q_2}\, \delta s_2 + \frac{1}{h_3}\frac{\partial \phi}{\partial q_3}\, \delta s_3 \\
&= \left(\frac{1}{h_1}\frac{\partial \phi}{\partial q_1}\mathbf{e}_1 + \frac{1}{h_2}\frac{\partial \phi}{\partial q_2}\mathbf{e}_2 + \frac{1}{h_3}\frac{\partial \phi}{\partial q_3}\mathbf{e}_3 \right) \cdot (\delta s_1 \mathbf{e}_1 + \delta s_2 \mathbf{e}_2 + \delta s_3 \mathbf{e}_3)
\end{aligned}
$$
$$(2.48)$$

where $\mathbf{e}_1, \mathbf{e}_2, \mathbf{e}_3$ are the unit vectors at P in the directions of the coordinates q_1, q_2, q_3 (See Section 2.17 and Fig. 2.21).

Now since $\delta s_1, \delta s_2, \delta s_3$ are differential lengths along the q_1, q_2, q_3 curves, we have

$$\mathbf{ds} = \delta s_1 \mathbf{e}_1 + \delta s_2 \mathbf{e}_2 + \delta s_3 \mathbf{e}_3 \qquad (2.49)$$

By using this relation, Eq. (2.48) may be rewritten as

$$d\phi = \left(\frac{1}{h_1}\frac{\partial \phi}{\partial q_1}\mathbf{e}_1 + \frac{1}{h_2}\frac{\partial \phi}{\partial q_2}\mathbf{e}_2 + \frac{1}{h_3}\frac{\partial \phi}{\partial q_3}\mathbf{e}_3 \right) \cdot \mathbf{ds} \qquad (2.50)$$

which has the same form as Eq. (2.46). Dividing Eq. (2.50) by ds we obtain the spatial derivative of ϕ in the direction $\mathbf{e}$ as

$$
\begin{aligned}
\frac{d\phi}{ds} &= \frac{1}{h_1}\frac{\partial \phi}{\partial q_1}\mathbf{e}_1 + \frac{1}{h_2}\frac{\partial \phi}{\partial q_2}\mathbf{e}_2 + \frac{1}{h_3}\frac{\partial \phi}{\partial q_3}\mathbf{e}_3 \\
&= \left(\frac{\partial \phi}{\partial s_1}\mathbf{e}_1 + \frac{\partial \phi}{\partial s_2}\mathbf{e}_2 + \frac{\partial \phi}{\partial s_3}\mathbf{e}_3 \right) \cdot \mathbf{e} \qquad (2.51)
\end{aligned}
$$

We thus see that to every scalar function of position there corresponds at each point a particular vector which determines at that point the spatial variation of the scalar function. We call such a vector the *gradient* of the function and denote it by the word *grad*. With this notation Eq. (2.50) can be put in the form

$$d\phi = \text{grad } \phi \cdot \mathbf{ds} \qquad (2.52)$$

and (2.51) in the form

$$\frac{d\phi}{ds} = \text{grad } \phi \cdot \mathbf{e} \qquad (2.53)$$

The gradient is then *defined* by stating that *in any orthogonal, curvilinear* *coordinate system the components of the gradient of a scalar function of position are simply the partial derivatives of the function with respect to distances in the directions of the respective coordinate axes.* Symbolically, we write*

$$\text{grad } \phi = \left(\frac{\partial \phi}{\partial s_1} \mathbf{e}_1 + \frac{\partial \phi}{\partial s_2} \mathbf{e}_2 + \frac{\partial \phi}{\partial s_3} \mathbf{e}_3 \right)$$

$$= \left(\frac{1}{h_1} \frac{\partial \phi}{\partial q_1} \mathbf{e}_1 + \frac{1}{h_2} \frac{\partial \phi}{\partial q_2} \mathbf{e}_2 + \frac{1}{h_3} \frac{\partial \phi}{\partial q_3} \mathbf{e}_3 \right)$$

(2.54)

The components of grad ϕ in Cartesians are given by

$$\frac{\partial \phi}{\partial x}, \frac{\partial \phi}{\partial y}, \frac{\partial \phi}{\partial z}$$

in cylindrical coordinates r, θ, z by

$$\frac{\partial \phi}{\partial r}, \frac{1}{r} \frac{\partial \phi}{\partial \theta}, \frac{\partial \phi}{\partial z}$$

and in spherical coordinates r, θ, φ by

$$\frac{\partial \phi}{\partial r}, \frac{1}{r} \frac{\partial \phi}{\partial \theta}, \frac{1}{r \sin \theta} \frac{\partial \phi}{\partial \varphi}$$

Further significance of a gradient can be gathered from Eq. (2.53). If, at any point, the components of grad ϕ are constructed in different directions, the component that is numerically the greatest will be in the direction of the gradient itself and will be of the same magnitude as that of the gradient. This means (from Eq. 2.53) that the greatest value of the derivative $d\phi/ds$ at a given point occurs in the direction of grad ϕ and equals the magnitude of the gradient. Conversely, we may state that *at any point the gradient of a scalar function of position ϕ, is equal in magnitude and direction to the greatest derivative of ϕ with respect to distance at that point.*

In general, the gradient of a scalar function of position varies from point to point of the region of space in which the function is defined. Thus to every scalar function of position $\phi(\mathbf{r})$ there corresponds a certain vector function of position, grad $\phi(\mathbf{r})$, which describes the spatial variation of ϕ at all points of the space concerned.

Since a scalar function of position describes a scalar field, the concept of a gradient is directly applicable to a scalar field. Thus the spatial variation of a scalar field is given by a certain vector field, which is nothing

---

* Although for convenience we use the representation given in the top line of (2.54), we always mean by that the representation given in the lower line of (2.54).

but the field of the gradient of the scalar function in question. With this field interpretation, an interesting result follows from Eq. (2.53). Consider any level surface of the scalar field $\phi(\mathbf{r})$. By definition ϕ is a constant on such a surface. That is, at any point on a level surface

$$\frac{d\phi}{ds} = 0$$

for every direction lying in the surface. Then from Eq. (2.53) it follows that

$$\text{grad } \phi \cdot \mathbf{e} = 0$$

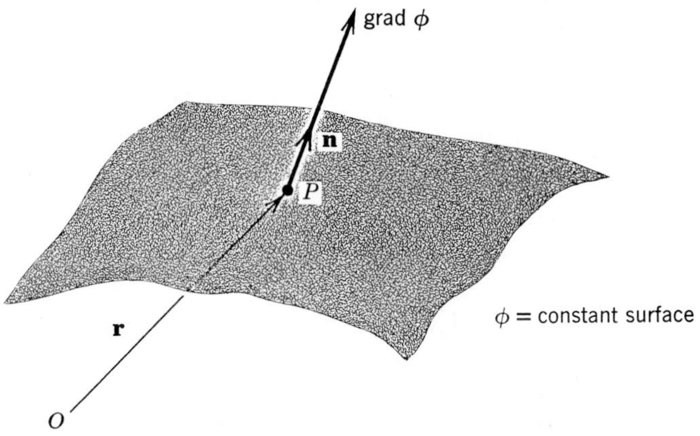

Fig. 2.31   Level surface.

when $\mathbf{e}$ *lies in a level surface.* Since a vector has no component in a direction normal to itself, we conclude that grad ϕ at any point is normal to the level surface passing through that point (Fig. 2.31). If we map the scalar field ϕ by means of its level surfaces and draw at the same time the field lines of grad ϕ, we find that the field lines intersect the level surfaces orthogonally.

We have defined here the gradient by means of a *differential operation*. A definition of the gradient by means of an *integral operation* will be given in Section 2.31.

2.25   Differentiation of a Vector Function of a Vector

Consider a vector function of position $\mathbf{A} = \mathbf{A}(\mathbf{r})$. Here again the independent variable is a vector. Therefore, when we speak about the spatial variation of $\mathbf{A}$, we must say in which direction that variation is constructed. To see what is required to describe completely the spatial rate of change at any point of the vector function $\mathbf{A(r)}$, let us choose Cartesians and write

$$\mathbf{A(r)} = A_x\mathbf{i} + A_y\mathbf{j} + A_z\mathbf{k}$$

As before, let $d\mathbf{s} = ds\mathbf{e}$ be a differential increment in $\mathbf{r}$ in some direction $\mathbf{e}$. Let $d\mathbf{A}$ be the differential increment in $\mathbf{A}$ over the directed distance ds, and let dA_x, dA_y, and dA_z be the corresponding increments in the scalar components of $\mathbf{A}$.

We therefore write

$$d\mathbf{A} = dA_x\mathbf{i} + dA_y\mathbf{j} + dA_z\mathbf{k} \tag{2.55}$$

Since A_x, A_y, A_z are scalar functions of position, according to Eq. (2.52), the increments dA_x, etc., are given by

$$dA_i = d\mathbf{s} \cdot \text{grad } A_i \tag{2.56}$$

where the subscript i may be x, y, or z. By using (2.56), Eq. (2.55) may be rewritten as

$$d\mathbf{A} = (d\mathbf{s} \cdot \text{grad } A_x)\mathbf{i} + (d\mathbf{s} \cdot \text{grad } A_y)\mathbf{j} + (d\mathbf{s} \cdot \text{grad } A_z)\mathbf{k} \tag{2.57}$$

Dividing this equation by ds we obtain, in Cartesians, the derivative of $\mathbf{A}$ with respect to distance in any direction $\mathbf{e}$ as

$$\frac{d\mathbf{A}}{ds} = (\mathbf{e} \cdot \text{grad } A_x)\mathbf{i} + (\mathbf{e} \cdot \text{grad } A_y)\mathbf{j} + (\mathbf{e} \cdot \text{grad } A_z)\mathbf{k} \tag{2.58}$$

Equations (2.57) and (2.58) show that to determine the spatial variation of $\mathbf{A(r)}$ in any direction from a given point, we need to know at that point a set of three vectors associated with $\mathbf{A}$, namely the gradients of A_x, A_y, A_z. Equivalently we need to know a set of nine numbers that constitute these three vectors. In Cartesians, the set of nine numbers is given by the array

$$\begin{pmatrix} \dfrac{\partial A_x}{\partial x} & \dfrac{\partial A_x}{\partial y} & \dfrac{\partial A_x}{\partial z} \\[2mm] \dfrac{\partial A_y}{\partial x} & \dfrac{\partial A_y}{\partial y} & \dfrac{\partial A_y}{\partial z} \\[2mm] \dfrac{\partial A_z}{\partial x} & \dfrac{\partial A_z}{\partial y} & \dfrac{\partial A_z}{\partial z} \end{pmatrix}$$

The elements of the array are the various partial derivatives of the components of $\mathbf{A}$ with respect to distances along the X, Y, Z axes. The elements obey the same rules as the elements of what is known as a *second-order tensor*. We may, therefore, describe the array of nine numbers as a second-order tensor and state that the spatial variation of $\mathbf{A(r)}$ is specified completely by a second-order tensor.

If instead of Cartesians we choose a system of orthogonal, curvilinear coordinates, we would arrive at the same conclusion. In this case, however, the elements of the tensor describing the spatial variation of $A(r)$ are not simply the partial derivatives of the components of A with respect to distances along the coordinate axes. They now include additional terms that arise due to the fact that the directions of the reference unit vectors change with change of position.

To see this, we choose some orthogonal, curvilinear coordinates q_1, q_2, q_3 and express $A(r)$ as

$$A(r) = A_1 e_1 + A_2 e_2 + A_3 e_3$$

where e_1, e_2, e_3 are the reference unit vectors associated with the point r and A_1, A_2, A_3 are the components of A with respect to the system e_1, e_2, e_3. The components and the unit vectors are functions of position. The differential increment in A over a directed distance $ds = ds e$ is then given by

$$dA = \{(dA_1)e_1 + (dA_2)e_2 + (dA_3)e_3\}$$
$$+ \{A_1(de_1) + A_2(de_2) + A_3(de_3)\} \qquad (2.59)$$

where dA_1, etc., are the changes in the components over the distance ds, and de_1, etc., are the corresponding changes in the unit vectors.

Now, as before, we can write

$$(dA_1)e_1 + (dA_2)e_2 + (dA_3)e_3$$
$$= (ds \cdot \text{grad } A_1)e_1 + (ds \cdot \text{grad } A_2)e_2 + (ds \cdot \text{grad } A_3)e_3 \qquad (2.60)$$

After evaluating de_1, etc., (see Section 2.45) it is possible to write

$$A_1(de_1) + A_2(de_2) + A_3(de_3)$$
$$= (ds \cdot \chi_1)e_1 + (ds \cdot \chi_2)e_2 + (ds \cdot \chi_3)e_3 \qquad (2.61)$$

where χ_1, χ_2, χ_3 are vectors involving A_1, A_2, A_3 and components of the vectors de_1, de_2, and de_3. Verify this for cylindrical and spherical coordinates.

Combining the relations (2.60) and (2.61) and introducing the notation

$$W_1 \equiv \text{grad } A_1 + \chi_1$$
$$W_2 \equiv \text{grad } A_2 + \chi_2 \qquad (2.62)$$
$$W_3 \equiv \text{grad } A_3 + \chi_3$$

equation (2.59) may be rewritten as

$$dA = (ds \cdot W_1)e_1 + (ds \cdot W_2)e_2 + (ds \cdot W_3)e_3 \qquad (2.63)$$

or, dividing by ds, as

$$\frac{dA}{ds} = (e \cdot W_1)e_1 + (e \cdot W_2)e_2 + (e \cdot W_3)e_3 \qquad (2.64)$$

where $\mathbf{e}$ is the direction of $d\mathbf{s}$. Thus we see that in any orthogonal, curvi-linear, coordinate system the spatial variation at any point of $\mathbf{A(r)}$ is given by three vectors or by a second order tensor. As seen from relation (2.62), the elements of the tensor are not simply the partial derivatives of the components of $\mathbf{A}$ with respect to distances along the coordinate axes.

The tensor that specifies completely the spatial variation of $\mathbf{A(r)}$ is known as the *tensor gradient* of $\mathbf{A}$, usually denoted by the symbol grad $\mathbf{A}$. We thus have

$$\text{grad } \mathbf{A} = \begin{pmatrix} W_{11} & W_{12} & W_{13} \\ W_{21} & W_{22} & W_{23} \\ W_{31} & W_{32} & W_{33} \end{pmatrix} \tag{2.65}$$

where the element W_{ij} (i and j may take any of the values 1, 2, or 3) denotes the jth component (i.e., in the direction of $\mathbf{e}_j$) of the vector $\mathbf{W}_i$, again i being 1, 2, or 3. The vectors $\mathbf{W}_1$, etc., are defined by the relation (2.62). In general, the tensor gradient varies from point to point. Hence we say that the spatial variation of a vector field is given by a tensor field.

In the vector description of physical problems, we are not directly concerned with the complete tensor gradient of $\mathbf{A(r)}$. Only certain combinations of the elements of the tensor are significant. Three such combinations are particularly important. One of them is a scalar quantity obtained by summing the diagonal terms W_{11}, W_{22}, and W_{33}. This sum is known as the *divergence* of $\mathbf{A}$ and is denoted by div $\mathbf{A}$. We thus have

$$\text{div } \mathbf{A} = W_{11} + W_{22} + W_{33}$$

The other two combinations are second-order tensors. One of them is given by the array

$$\frac{1}{2}\begin{pmatrix} 2W_{11} & W_{12} + W_{21} & W_{13} + W_{31} \\ W_{21} + W_{12} & 2W_{22} & W_{23} + W_{32} \\ W_{31} + W_{13} & W_{32} + W_{23} & 2W_{33} \end{pmatrix}$$

This tensor is usually known as the *strain* of $\mathbf{A}$. As seen, it is symmetrical about the diagonal formed by the elements W_{11}, W_{22}, and W_{33}.

The other second-order tensor is given by the array

$$\frac{1}{2}\begin{pmatrix} 0 & W_{12} - W_{21} & W_{13} - W_{31} \\ W_{21} - W_{12} & 0 & W_{23} - W_{32} \\ W_{31} - W_{13} & W_{32} - W_{23} & 0 \end{pmatrix}$$

This tensor is usually known as the *rotation* of **A**. As seen, it is anti-symmetrical about the diagonal formed by the elements that are zero.

The strain of **A**, being symmetric, actually contains six independent elements. The rotation of **A**, being antisymmetric, actually contains three independent elements. In other words, the rotation of **A** is actually specified by a set of three numbers. This set of three numbers, as may be verified, obeys the same rules as a set specifying a vector. This means the rotation of **A**, although it is a second-order antisymmetric tensor, can be represented by a vector. Denoting such a vector by **B**, its components in the directions of the reference unit vectors e_1, e_2, e_3 are given by

$$\mathbf{e_1} \cdot \mathbf{B} = \tfrac{1}{2}(W_{32} - W_{23})$$

$$\mathbf{e_2} \cdot \mathbf{B} = \tfrac{1}{2}(W_{13} - W_{31})$$

$$\mathbf{e_3} \cdot \mathbf{B} = \tfrac{1}{2}(W_{21} - W_{12})$$

For reasons that will become evident later, a vector equal to 2**B** is known as the *curl* of **A** denoted by curl **A**.

The physical significance of the names divergence, rotation, and curl will become apparent later on when we shall define divergence and curl of a vector by means of certain integral operations. For the significance of the name strain, reference may be made to Section 9.1.

The reader may verify that in Cartesians the following results are obtained:

$$\text{div } \mathbf{A} = \frac{\partial A_x}{\partial x} + \frac{\partial A_y}{\partial y} + \frac{\partial A_z}{\partial z}$$

$$\text{strain of } \mathbf{A} = \frac{1}{2}\begin{pmatrix} 2\dfrac{\partial A_x}{\partial x} & \dfrac{\partial A_x}{\partial y} + \dfrac{\partial A_y}{\partial x} & \dfrac{\partial A_x}{\partial z} + \dfrac{\partial A_z}{\partial x} \\[2ex] \dfrac{\partial A_y}{\partial x} + \dfrac{\partial A_x}{\partial y} & 2\dfrac{\partial A_y}{\partial y} & \dfrac{\partial A_y}{\partial z} + \dfrac{\partial A_z}{\partial y} \\[2ex] \dfrac{\partial A_z}{\partial x} + \dfrac{\partial A_x}{\partial z} & \dfrac{\partial A_z}{\partial y} + \dfrac{\partial A_y}{\partial z} & 2\dfrac{\partial A_z}{\partial z} \end{pmatrix}$$

and

$$\text{curl } \mathbf{A} = \begin{vmatrix} \mathbf{i} & \mathbf{j} & \mathbf{k} \\[1ex] \dfrac{\partial}{\partial x} & \dfrac{\partial}{\partial y} & \dfrac{\partial}{\partial z} \\[1ex] A_x & A_y & A_z \end{vmatrix}$$

Rotation of **A** is simply $\tfrac{1}{2}$ curl **A**.

2.26 Del, the Vector Differential Operator

Consider the expression (2.54) for the gradient of $\phi(\mathbf{r})$ in any orthogonal, curvilinear, coordinate system:

$$\text{grad } \phi = \mathbf{e}_1 \frac{\partial \phi}{\partial s_1} + \mathbf{e}_2 \frac{\partial \phi}{\partial s_2} + \mathbf{e}_3 \frac{\partial \phi}{\partial s_3}$$

This expression may be rewritten as

$$\text{grad } \phi = \left(\mathbf{e}_1 \frac{\partial}{\partial s_1} + \mathbf{e}_2 \frac{\partial}{\partial s_2} + \mathbf{e}_3 \frac{\partial}{\partial s_3} \right) \phi$$

This means to obtain the gradient of ϕ we operate on ϕ by the *operator*

$$\mathbf{e}_1 \frac{\partial}{\partial s_1} + \mathbf{e}_2 \frac{\partial}{\partial s_2} + \mathbf{e}_3 \frac{\partial}{\partial s_3}$$

This is a *vector differential operator* and is usually denoted by the symbol $\boldsymbol{\nabla}$, called *del*. We thus define*

$$\begin{aligned}
\boldsymbol{\nabla} &\equiv \mathbf{e}_1 \frac{\partial}{\partial s_1} + \mathbf{e}_2 \frac{\partial}{\partial s_2} + \mathbf{e}_3 \frac{\partial}{\partial s_3} \\
&\equiv \mathbf{e}_1 \frac{1}{h_1} \frac{\partial}{\partial q_1} + \mathbf{e}_2 \frac{1}{h_2} \frac{\partial}{\partial q_2} + \mathbf{e}_3 \frac{1}{h_3} \frac{\partial}{\partial q_3} \mathbf{e}_3
\end{aligned} \qquad (2.66)$$

The expression for del in Cartesians is

$$\boldsymbol{\nabla} = \mathbf{i} \frac{\partial}{\partial x} + \mathbf{j} \frac{\partial}{\partial y} + \mathbf{j} \frac{\partial}{\partial z}$$

in cylindrical coordinates r, θ, z is

$$\boldsymbol{\nabla} = \mathbf{e}_r \frac{\partial}{\partial r} + \mathbf{e}_\theta \frac{1}{r} \frac{\partial}{\partial \theta} + \mathbf{e}_z \frac{\partial}{\partial z}$$

and in spherical coordinates r, θ, φ is

$$\boldsymbol{\nabla} = \mathbf{e}_r \frac{\partial}{\partial r} + \mathbf{e}_\theta \frac{1}{r} \frac{\partial}{\partial \theta} + \mathbf{e}_\varphi \frac{1}{r \sin \theta} \frac{\partial}{\partial \varphi}$$

The Operators $\boldsymbol{\nabla} \cdot$ and $\boldsymbol{\nabla} \times$. Since del is a vector operator, the operators $\boldsymbol{\nabla} \cdot$ and $\boldsymbol{\nabla} \times$ may be introduced and applied on any vector field. If $\mathbf{A}(\mathbf{r})$ is any vector field, the scalar product $\boldsymbol{\nabla} \cdot \mathbf{A}$ and the vector product $\boldsymbol{\nabla} \times \mathbf{A}$ are formed as follows:

$$\boldsymbol{\nabla} \cdot \mathbf{A} = \left(\mathbf{e}_1 \frac{\partial}{\partial s_1} + \mathbf{e}_2 \frac{\partial}{\partial s_2} + \mathbf{e}_3 \frac{\partial}{\partial s_3} \right) \cdot (\mathbf{e}_1 A_1 + \mathbf{e}_2 A_2 + \mathbf{e}_3 A_3) \quad (2.67)$$

---

* Recall that although for convenience we use the representation in the top line of (2.66), we always mean by that the representation in the lower line of (2.66).

and

$$\nabla \times \mathbf{A} = \left(\mathbf{e}_1 \frac{\partial}{\partial s_2} + \mathbf{e}_2 \frac{\partial}{\partial s_2} + \mathbf{e}_3 \frac{\partial}{\partial s_3} \right) \times (\mathbf{e}_1 A_1 + \mathbf{e}_2 A_2 + \mathbf{e}_3 A_3) \quad (2.68)$$

In carrying out these operations in coordinate systems other than the Cartesians, we should take proper account of the fact that the system of reference units $\mathbf{e}_1$, $\mathbf{e}_2$, $\mathbf{e}_3$ changes with change of position.

It may be verified that in Cartesians we obtain the following results:

$$\nabla \cdot \mathbf{A} = \frac{\partial A_x}{\partial x} + \frac{\partial A_y}{\partial y} + \frac{\partial A_z}{\partial z}$$

and

$$\nabla \times \mathbf{A} = \begin{vmatrix} \mathbf{i} & \mathbf{j} & \mathbf{k} \\ \dfrac{\partial}{\partial x} & \dfrac{\partial}{\partial y} & \dfrac{\partial}{\partial z} \\ A_x & A_y & A_z \end{vmatrix}$$

These results are identical, respectively, with the divergence and curl of the vector $\mathbf{A}$ (see Section 2.25). Thus it is usual to set

$$\nabla \cdot \mathbf{A} \equiv \text{div } \mathbf{A} \qquad (2.69)$$

and

$$\nabla \times \mathbf{A} \equiv \text{curl } \mathbf{A} \qquad (2.70)$$

The Operator $\mathbf{B} \cdot \nabla$. If $\mathbf{B}$ is any vector, an operator $\mathbf{B} \cdot \nabla$ can be defined as

$$\mathbf{B} \cdot \nabla = (\mathbf{e}_1 B_1 + \mathbf{e}_2 B_2 + \mathbf{e}_3 B_3) \cdot \left(\mathbf{e}_1 \frac{\partial}{\partial s_1} + \mathbf{e}_2 \frac{\partial}{\partial s_2} + \mathbf{e}_3 \frac{\partial}{\partial s_3} \right)$$

$$= B_1 \frac{\partial}{\partial s_1} + B_2 \frac{\partial}{\partial s_2} + B_3 \frac{\partial}{\partial s_3} \qquad (2.71)$$

It is seen that $\mathbf{B} \cdot \nabla$ is a scalar differential operator. In Cartesians we have

$$\mathbf{B} \cdot \nabla = B_x \frac{\partial}{\partial x} + B_y \frac{\partial}{\partial y} + B_z \frac{\partial}{\partial z}$$

Applying the operator to a scalar function $\phi(\mathbf{r})$ we have

$$(\mathbf{B} \cdot \nabla)\phi = \left(B_1 \frac{\partial}{\partial s_1} + B_2 \frac{\partial}{\partial s_2} + B_3 \frac{\partial}{\partial s_3} \right) \phi$$

$$= \mathbf{B} \cdot \nabla \phi \qquad (2.72)$$

Applying it to a vector function $A(\mathbf{r})$ we have

$$(\mathbf{B} \cdot \nabla)\mathbf{A} = \left(B_1 \frac{\partial}{\partial s_1} + B_2 \frac{\partial}{\partial s_2} + B_3 \frac{\partial}{\partial s_3} \right)(A_1 \mathbf{e}_1 + A_2 \mathbf{e}_2 + A_3 \mathbf{e}_3) \quad (2.73)$$

In carrying out the operation, account must be taken of the fact that the unit vectors $\mathbf{e}_1$, $\mathbf{e}_2$, $\mathbf{e}_3$ change with change of position.

Of particular significance are the operators $d\mathbf{s} \cdot \nabla$ and $\mathbf{e} \cdot \nabla$, where $d\mathbf{s} = ds\,\mathbf{e}$ is, at any given point, a small increment in some direction $\mathbf{e}$ of the position vector $\mathbf{r}$. We recall [see Eq. (2.52)] that the differential change in a function $\phi(\mathbf{r})$ over the directed distance $d\mathbf{s}$ from a given point is given by

$$d\phi = d\mathbf{s} \cdot \operatorname{grad} \phi$$

This may be rewritten as

$$d\phi = (d\mathbf{s} \cdot \operatorname{grad})\phi$$
$$= (d\mathbf{s} \cdot \nabla)\phi \qquad (2.74)$$

This means that the operator $d\mathbf{s} \cdot \nabla$, when applied to a scalar function $\phi(\mathbf{r})$, yields the differential change in ϕ over the distance $d\mathbf{s}$. Similarly, we have

$$\frac{d\phi}{ds} = (\mathbf{e} \cdot \nabla)\phi \qquad (2.75)$$

showing that the operator $\mathbf{e} \cdot \nabla$ when applied to $\phi(\mathbf{r})$ yields the derivative of ϕ with respect to distance in the direction $\mathbf{e}$.

Consider now the expression (2.57) we have obtained, in Cartesians, for the change $d\mathbf{A}$ in the function $\mathbf{A}(\mathbf{r})$ over the directed distance $d\mathbf{s} = ds\,\mathbf{e}$ from a given point. We have

$$d\mathbf{A} = (d\mathbf{s} \cdot \operatorname{grad} A_x)\mathbf{i} + (d\mathbf{s} \cdot \operatorname{grad} A_y)\mathbf{j} + (d\mathbf{s} \cdot \operatorname{grad} A_z)\mathbf{k}$$

This may be rewritten (noting that $\mathbf{i}$, $\mathbf{j}$, $\mathbf{k}$ are fixed directions) as

$$d\mathbf{A} = (d\mathbf{s} \cdot \operatorname{grad})(A_x \mathbf{i}) + (d\mathbf{s} \cdot \operatorname{grad})(A_y \mathbf{j}) + (d\mathbf{s} \cdot \operatorname{grad})(A_z \mathbf{k})$$
$$= (d\mathbf{s} \cdot \operatorname{grad})(A_x \mathbf{i} + A_y \mathbf{j} + A_z \mathbf{k})$$
$$= (d\mathbf{s} \cdot \operatorname{grad})\mathbf{A}$$
$$= (d\mathbf{s} \cdot \nabla)\mathbf{A} \qquad (2.76)$$

This means *the operator $d\mathbf{s} \cdot \nabla$ when applied to a vector function $\mathbf{A}(\mathbf{r})$ yields, just as when applied to a scalar function $\phi(\mathbf{r})$, the differential change in $\mathbf{A}$ over the directed distance $d\mathbf{s}$.* Similarly, it can be seen that

$$\frac{d\mathbf{A}}{ds} = (\mathbf{e} \cdot \nabla)\mathbf{A} \qquad (2.77)$$

showing that the operator $\mathbf{e} \cdot \nabla$ applied to $\mathbf{A}(\mathbf{r})$ yields the derivative of $\mathbf{A}$

with respect to distance in the direction **e**. In applying Eqs. (2.76) and (2.77) to coordinate systems other than Cartesians we should take note of the fact that the reference unit vectors change with change of position.

The utility of the operator ∇ lies in the fact that in working out problems *we can treat ∇ formally as a vector and apply to it the rules of vector algebra and calculus.* In doing this, however, we should bear in mind that ∇ is an operator and not an actual vector. Therefore, in the formal application of vector rules to ∇, it is necessary *to preserve the order in which del appears with respect to the other factors* involved. For instance, even though $\mathbf{A} \cdot \mathbf{B} = \mathbf{B} \cdot \mathbf{A}$, the operation $\nabla \cdot \mathbf{A}$ is not equal to the operator $\mathbf{A} \cdot \nabla$.

2.27 Integration of a Vector Function of a Scalar

If a vector **A** is a function of a scalar variable t, we can form the so-called indefinite integral

$$\int \mathbf{A}(t) \, dt \tag{2.78}$$

in the same manner as is done in scalar integration (i.e., integration of a scalar function of a scalar variable). The result of the integration (2.78) is another vector function of the scalar t and is determined to within an additive constant which, in general, is a vector. We thus write

$$\int \mathbf{A}(t) \, dt = \mathbf{B}(t) + \mathbf{C}$$

It follows that

$$\frac{d\mathbf{B}}{dt} = \mathbf{A}(t)$$

If the variable t changes continuously from a particular value t_1 to another particular value t_2, the integral

$$\int_{t_1}^{t_2} \mathbf{A}(t) \, dt$$

is the definite integral of **A** between the limits t_1 and t_2.

2.28 Line Integrals: Circulation

Consider a scalar function of position $\phi = \phi(\mathbf{r})$ and the field described by it. In such a field let $\mathscr{C}$ represent a space curve drawn from a point a to another point b. We assign a direction to the curve as that of travel along the curve from a to b. Let **r** denote the position from some origin of a point P on the curve and $d\mathbf{s}$ an element of length along the curve from the point P (see Fig. 2.32). If e_s is a unit vector tangential to the curve at the point P, we have $d\mathbf{s} = ds\, e_s$. The integral

$$\oint_a^b \phi(\mathbf{r})\, d\mathbf{s}$$

or, equivalently,

$$\oint_a^b \phi(\mathbf{r})e_s\, ds$$

taken along the curve $\mathscr{C}$ is called the *line integral* of ϕ along the path $\mathscr{C}$. The value of this integral is a vector.

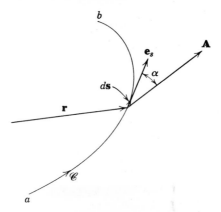

Fig. 2.32 Line integral.

Consider now a vector function of position $\mathbf{A}(\mathbf{r})$. If $\mathscr{C}$ is a space curve as before, we can form the line integral

$$\oint_a^b \mathbf{A} \cdot d\mathbf{s} \quad \text{or} \quad \oint_a^b \mathbf{A} \cdot e_s\, ds \tag{2.79}$$

along the curve $\mathscr{C}$ between the given endpoints (see Fig. 2.32). The integral is simply the integral along $\mathscr{C}$ of the component of $\mathbf{A}$ tangential to the curve. If, as shown in the figure, α is the angle between e_s and $\mathbf{A}$, the integral (2.79) may be written as

$$\oint_a^b A \cos \alpha\, ds$$

where A is the magnitude of $\mathbf{A}$. The result of this integral is a scalar. In general, this line integral, like any other line integral, depends on the function $\mathbf{A}(\mathbf{r})$, the path along which the integration is carried out and on the endpoints of the path. Under certain conditions, however, the value of the integral (2.79) depends only on the endpoints and becomes independent of the path that joins them. We shall look into those conditions later.

We note that another line integral of $\mathbf{A}(\mathbf{r})$ may be formed as follows:

$$\oint_a^b \mathbf{A}(\mathbf{r}) \times d\mathbf{s} \quad \text{or} \quad \oint_a^b \mathbf{A}(\mathbf{r}) \times \mathbf{e}_s \, ds$$

The result of this integral is a vector.

Circulation. Line integrals of the type described above may be formed around closed curves. Of particular interest is the integral

$$\oint \mathbf{A} \cdot d\mathbf{s} \quad \text{or} \quad \oint \mathbf{A} \cdot \mathbf{e}_s d_s \quad \text{or} \quad \oint A \cos \alpha \, ds$$

around a closed space curve $\mathscr{C}$. Such an integral is known as the *circulation* of the vector $\mathbf{A}$ around the curve $\mathscr{C}$. In general, the value of the circulation is nonzero and depends on the function $\mathbf{A}(\mathbf{r})$ and the closed curve $\mathscr{C}$. In certain circumstances, however, the circulation vanishes and becomes independent of the curve. We shall look into these details later.

2.29 Surface Integrals

Consider an open surface S drawn in the field described by a scalar function of position $\phi(\mathbf{r})$. Let the surface be divided into a number of infinitesimal elements. Each of the surface elements may be considered as a plane area and denoted as a vector

$$d\mathbf{S} = \mathbf{n} \, dS$$

$\mathbf{n}$ is a unit vector normal to the surface element (see Fig. 2.33). The unit vector $\mathbf{n}$ is drawn arbitrarily from one side or the other of the surface S. If, however, a direction of travel is first assigned along the boundary curve $\mathscr{C}$ of the surface, the direction of $\mathbf{n}$ is chosen according to the right-hand rule with respect to the direction of travel along $\mathscr{C}$. Using these notations we form the integral

$$\iint_S \phi(\mathbf{r}) \, d\mathbf{S} \quad \text{or} \quad \iint_S \phi(\mathbf{r})\mathbf{n} \, ds$$

over the entire surface S. Such an integral is called the *surface integral* of ϕ over the surface S. The result of the integral is a vector.

Consider next a vector function of position $\mathbf{A}(\mathbf{r})$ and let S be an open surface drawn in its field. Then the integral

$$\iint_S \mathbf{A} \cdot d\mathbf{S} \quad \text{or} \quad \iint_S \mathbf{A} \cdot \mathbf{n} \, dS \qquad (2.80)$$

taken over the surface S is called the surface integral of $\mathbf{A}$ over the surface S. Since $\mathbf{A} \cdot \mathbf{n}$ is the component of $\mathbf{A}$ in the direction of the normal to the surface element (see Fig. 2.33), the integral (2.80) is simply the surface integral of this component. The value of the integral is thus a scalar.

The quantity $\mathbf{A} \cdot d\mathbf{S}$ or $A \cdot \mathbf{n}\,dS$ is usually called the *outflow* of *vector* $\mathbf{A}$ *through the surface element* $d\mathbf{S}$. By outflow of $\mathbf{A}$ we mean the flow of $\mathbf{A}$

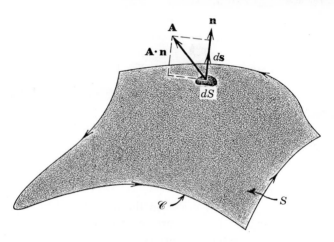

Fig. 2.33 Surface integral.

into the region that contains the normal to $d\mathbf{S}$. This is the case when the component $\mathbf{A} \cdot \mathbf{n}$ is positive. With this interpretation, the surface integral (2.80) is called the *outflow of vector* $\mathbf{A}$ *through the surface* S. Since $\mathbf{A} \cdot \mathbf{n}$ may be positive at some points and negative at other points of the surface S, *by outflow of* $\mathbf{A}$ *through* S *we mean actually the net outflow of* $\mathbf{A}$.

For the vector field $\mathbf{A}(\mathbf{r})$, we can form another surface integral expressed by

$$\iint_S \mathbf{A}(\mathbf{r}) \times d\mathbf{S} \quad \text{or} \quad \iint_S \mathbf{A}(\mathbf{r}) \times \mathbf{n}\,dS$$

The result of such an integral is a vector.

Surface integrals of the type described above may be formed also with closed surfaces. As shown in Fig. 2.34, let S be a closed surface and, as before, let $d\mathbf{S} = \mathbf{n}\,dS$ denote an element of S. *For a closed surface, we shall always draw the normal so as to point outward from the region enclosed by the surface and refer to it as the outward normal.* Using this convention we form the following surface integrals.

For a scalar field $\phi = \phi(\mathbf{r})$, we have the integral

$$\oiint_S \phi \, d\mathbf{S} \quad \text{or} \quad \oiint_S \phi \mathbf{n} \, dS \tag{2.81}$$

The result of this integral is a vector.

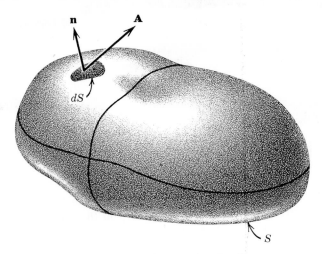

Fig. 2.34 Integral over a closed surface.

For a vector field $\mathbf{A} = \mathbf{A}(\mathbf{r})$, we have two integrals. One of them is

$$\oiint_S \mathbf{A} \cdot d\mathbf{S} \quad \text{or} \quad \oiint_S \mathbf{A} \cdot \mathbf{n} \, dS \tag{2.82}$$

the result of which is a scalar. The other integral is

$$\oiint_S \mathbf{A} \times d\mathbf{S} \quad \text{or} \quad \oiint_S \mathbf{A} \times \mathbf{n} \, dS \tag{2.83}$$

the result of which is a vector.

The three integrals (2.81), (2.82), and (2.83) appear frequently in the analysis of physical problems.

The integral (2.82) is called the outflow of $\mathbf{A}$ through the surface S. It actually gives the *net outflow of $\mathbf{A}$ through S from the region enclosed by S.*

2.30   Volume Integrals

Consider a region of space $\mathscr{R}$ in the field of a scalar function of position $\phi(\mathbf{r})$. Let the region be divided into a number of infinitesimal volume elements of magnitude $d\tau$. Then the integral

$$\iiint_{\mathscr{R}} \phi(\mathbf{r})\, d\tau$$

taken throughout the volume $\mathscr{R}$ is known as the *volume integral* of ϕ over the region $\mathscr{R}$. The result of the integral is a scalar.

Similarly, if $\mathscr{R}$ is a region of space in the field of a vector function of position $\mathbf{A}(\mathbf{r})$, the integral

$$\iiint_{\mathscr{R}} \mathbf{A}(\mathbf{r})\, d\tau$$

is known as the volume integral of $\mathbf{A}$ over the region $\mathscr{R}$. The result of such an integral is a vector.

In subsequent chapters we will find many examples of the line, surface, and volume integrals introduced in the preceding three sections. There are certain transformation relations that enable us to convert these integrals into one another. These relations follow directly from the integral definitions of gradient, divergence, and curl that will be given in the following sections. The operations involved in those definitions are the ones that usually arise in the setting up of physical problems.

2.31   Integral Definition of the Gradient

Consider a point P in a scalar field described by $\phi = \phi(\mathbf{r})$. Surround the point by a small closed surface ΔS. Let $\Delta \tau$ be the volume enclosed by ΔS (see Fig. 2.35). The shape of the volume element is arbitrary. If $d\mathbf{S} = \mathbf{n}\, dS$ is an elemental area on the surface ΔS ($\mathbf{n}$, according to our convention, is the outward normal), the quantity $\phi \mathbf{n}\, dS$ is a vector at the element $d\mathbf{S}$ of magnitude $\phi\, dS$ pointing in the direction $\mathbf{n}$. We form the integral

$$\oiint_{\Delta S} \phi \mathbf{n}\, dS$$

divide it by the volume $\Delta \tau$ and take the limit of the resulting ratio for vanishing $\Delta \tau$. This limit, when it exists, represents a certain vector associated with the point P. The vector, as obtained, is derived from the scalar function $\phi(\mathbf{r})$. Denoting this vector tentatively by $\mathbf{U}$ we write

$$\mathbf{U} = \lim_{\Delta \tau \to 0} \frac{1}{\Delta \tau} \oiint_{\Delta S} \phi \mathbf{n}\, dS \qquad (2.84)$$

To see the significance of the vector $\mathbf{U}$, we seek at the point P the component of $\mathbf{U}$ in some direction $\mathbf{e}$. We have

$$\mathbf{U} \cdot \mathbf{e} = \lim_{\Delta\tau \to 0} \frac{1}{\Delta\tau} \oiint_{\Delta S} \phi(\mathbf{n} \cdot \mathbf{e}) \, dS \tag{2.85}$$

since $\mathbf{e}$ is a given direction. The right side of Eq. (2.85) is evaluated as follows. Since the shape of the volume element $\Delta\tau$ is arbitrary, we set it

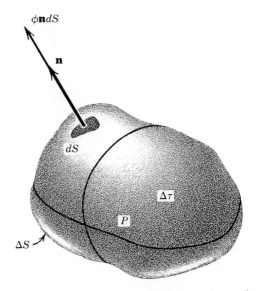

$\phi\mathbf{n}dS$

$\mathbf{n}$

dS

$\Delta\tau$

P

ΔS

Fig. 2.35 Illustrating the integral definition of a gradient.

up at the point P as a cylinder with axis along $\mathbf{e}$, base $\Delta\sigma$, and length Δs (see Fig. 2.36). Then for the cylinder we have

$$\oiint_{\Delta S} \phi(\mathbf{n} \cdot \mathbf{e}) \, dS = \iint_{\text{wall}} \phi(\mathbf{n} \cdot \mathbf{e}) \, dS + \iint_{\text{face 1}} \phi(\mathbf{n} \cdot \mathbf{e}) \, dS + \iint_{\text{face 2}} \phi(\mathbf{n} \cdot \mathbf{e}) \, dS \tag{2.86}$$

Since, at every point of the cylinder wall $\mathbf{n}$ is normal to $\mathbf{e}$, $\mathbf{n} \cdot \mathbf{e}$ vanishes on the wall, and consequently

$$\iint_{\text{wall}} \phi(\mathbf{n} \cdot \mathbf{e}) \, dS = 0 \tag{2.87}$$

On the face 1, $\mathbf{n} = -\mathbf{e}$. We assume that ϕ is uniform over the face and is equal to the value of ϕ at the point P, that is, at $\mathbf{r}$. We then have

$$\iint_{\text{face 1}} \phi(\mathbf{n} \cdot \mathbf{e})\, dS = -\phi(\mathbf{r})\, \Delta\sigma \tag{2.88}$$

On face 2, $\mathbf{n} = \mathbf{e}$ and assume that ϕ is uniform over the face with the value

$$\phi(\mathbf{r} + \Delta s\mathbf{e}) = \phi(\mathbf{r}) + \frac{d\phi}{ds}\, \Delta s \tag{2.89}$$

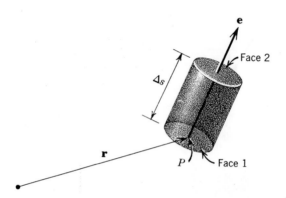

Fig. 2.36   Volume element to compute the gradient.

where $d\phi/ds$ is the derivative at P of ϕ with respect to distance in the direction $\mathbf{e}$. In relation (2.89) terms involving higher derivatives are neglected. This and the assumption of uniformity of ϕ over the faces are permissible in view of the ensuing limit operation. Using (2.89) we obtain

$$\iint_{\text{face 2}} \phi(\mathbf{n} \cdot \mathbf{e})\, dS = \phi(\mathbf{r})\, \Delta\sigma + \frac{d\phi}{ds}\, \Delta s\, \Delta\sigma \tag{2.90}$$

By using relations (2.87), (2.88), and (2.90), Eq. (2.86) may be reduced to

$$\oiint_{\Delta S} \phi(\mathbf{n} \cdot \mathbf{e})\, dS = \frac{d\phi}{ds}\, \Delta s\, \Delta\sigma$$

$$= \frac{d\phi}{ds}\, \Delta\tau \tag{2.91}$$

since $\Delta s\, \Delta\sigma$ is the volume of the cylinder.

Using Eqs. (2.91) and (2.85) we have

$$\mathbf{U} \cdot \mathbf{e} = \lim_{\Delta\tau \to 0} \frac{1}{\Delta\tau} \frac{d\phi}{ds} \Delta\tau$$

$$= \frac{d\phi}{ds} \tag{2.92}$$

This shows that the vector $\mathbf{U}$ at the point P is such that its component in any direction $\mathbf{e}$ gives the derivative at P of ϕ with respect to distance in that direction. Such a vector has been named previously the gradient of ϕ (see Section 2.24). Therefore we identify grad ϕ with $\mathbf{U}$ and give the following integral definition:

$$\text{grad } \phi \equiv \lim_{\Delta\tau \to 0} \frac{1}{\Delta\tau} \oiint_{\Delta S} \phi\mathbf{n} \, dS \tag{2.93}$$

In closing this section we draw attention to the fact that $\oiint_{\Delta S} \phi\mathbf{n} \, dS$ is small like $\Delta\tau$ (see Eq. 2.91).

2.32 Divergence of a Vector Field

Consider next a vector field $\mathbf{A} = \mathbf{A}(\mathbf{r})$. Let P be a point in that field and let ΔS be a small closed surface surrounding the point and enclosing a volume $\Delta\tau$ (see Fig. 2.37). As explained before (see Section 2.29), the quantity $\mathbf{A} \cdot \mathbf{n} \, dS$ is the outflow of $\mathbf{A}$ through the elemental area $\mathbf{n} \, dS$ of

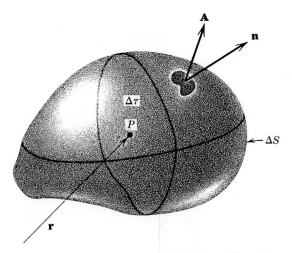

Fig. 2.37 Illustrating the integral definition of divergence.

ΔS and the integral $\displaystyle\oiint_{\Delta S} \mathbf{A} \cdot \mathbf{n}\, dS$ is the net outflow of vector $\mathbf{A}$ through the

surface ΔS. Forming $(1/\Delta\tau)\displaystyle\oiint_{\Delta S} \mathbf{A} \cdot \mathbf{n}\, dS$ we observe that the limit of this

ratio for $\Delta\tau \to 0$ is the *net outflow at P of* $\mathbf{A}$ *per unit volume of the region enclosing the point*. In other words, the limit denotes simply the *divergence* of the vector $\mathbf{A}$ from the point considered. Accordingly, such a limit is called the *divergence of the vector field* $\mathbf{A}$ and denoted by div $\mathbf{A}$. We thus define

$$\operatorname{div} \mathbf{A} \equiv \lim_{\Delta\tau \to 0} \frac{1}{\Delta\tau} \oiint_{\Delta S} \mathbf{A} \cdot \mathbf{n}\, dS \tag{2.94}$$

To obtain the expression for the divergence in any orthogonal, curvilinear coordinate system, we carry out, choosing suitably the volume element $\Delta\tau$, the operation shown on the right side of Eq. (2.94). In this way we obtain the following results:
(1) in Cartesians x, y, z

$$\operatorname{div} \mathbf{A} = \frac{\partial A_x}{\partial x} + \frac{\partial A_y}{\partial y} + \frac{\partial A_z}{\partial z} \tag{2.95}$$

(2) in cylindrical coordinates r, θ, z

$$\operatorname{div} \mathbf{A} = \frac{1}{r}\frac{\partial}{\partial r}(r A_r) + \frac{1}{r}\frac{\partial A_\theta}{\partial \theta} + \frac{\partial A_z}{\partial z} \tag{2.96}$$

(3) in spherical coordinates r, θ, φ

$$\operatorname{div} \mathbf{A} = \frac{1}{r^2}\frac{\partial}{\partial r}(r^2 A_r) + \frac{1}{r \sin\theta}\frac{\partial}{\partial \theta}(A_\theta \sin\theta) + \frac{1}{r \sin\theta}\frac{\partial A_\varphi}{\partial \varphi} \tag{2.97}$$

2.33 Curl of a Vector Field

The divergence operation, as seen above, derives a *scalar from* a vector. We can define, in an analogous manner, an operation that will derive a *vector from a vector*. To do this, we consider, as before (see Fig. 2.37), a point P in a vector field $\mathbf{A}(\mathbf{r})$ and form the limit

$$\lim_{\Delta\tau \to 0} \frac{1}{\Delta\tau} \oiint_{\Delta S} \mathbf{A} \times \mathbf{n}\, dS \tag{2.98}$$

The result of this limit is a vector, which we shall denote tentatively by the vector $\mathbf{B}$.

To see the *physical significance of the vector* $\mathbf{B}$ we choose the vector $\mathbf{A}$ in the limit (2.98) to be the velocity field $\mathbf{V}(\mathbf{r})$ of a rigid body rotating with

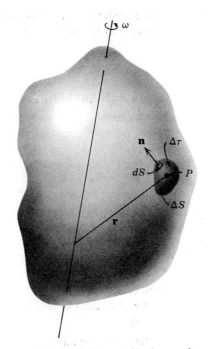

Fig. 2.38 Curl of velocity as twice the angular velocity.

an angular velocity $\boldsymbol{\omega}$. According to Eq. (2.5), we have

$$\mathbf{A}(\mathbf{r}) = \mathbf{V}(\mathbf{r}) = \boldsymbol{\omega} \times \mathbf{r}$$

Then the vector **B** associated with **V** is given by

$$\mathbf{B} \equiv \lim_{\Delta \tau \to 0} \frac{1}{\Delta \tau} \oiint_{\Delta S} \mathbf{V} \times \mathbf{n} \, dS$$

$$= \lim_{\Delta \tau \to 0} \frac{1}{\Delta \tau} \oiint_{\Delta S} (\boldsymbol{\omega} \times \mathbf{r}) \times \mathbf{n} \, dS \qquad (2.99)$$

(see Fig. 2.38.) Since

$$(\boldsymbol{\omega} \times \mathbf{r}) \times \mathbf{n} = (\mathbf{n} \cdot \boldsymbol{\omega})\mathbf{r} - (\mathbf{n} \cdot \mathbf{r})\boldsymbol{\omega}$$

Equation (2.99) may be written as

$$B = \lim_{\Delta \tau \to 0} \frac{1}{\Delta \tau} \oiint_{\Delta S} [(\mathbf{n} \cdot \boldsymbol{\omega})\mathbf{r} - (\mathbf{n} \cdot \mathbf{r})\boldsymbol{\omega}] \, dS. \qquad (2.100)$$

To evaluate the integral in Eq. (2.100) we proceed as follows. At the point

P we choose $\Delta\tau$ as a circular cylinder with its axis parallel to $\boldsymbol{\omega}$, its base $\Delta\sigma$ equal to πa^2 and its length equal to Δs (see Fig. 2.39). We may then write Eq. 2.100) as

$$
\mathbf{B} = \lim_{\Delta\tau \to 0} \frac{1}{\Delta\tau} \left\{ \iint_{\substack{\text{wall of} \\ \text{cylinder}}} (\mathbf{n} \cdot \boldsymbol{\omega})\mathbf{r}\, dS + \iint_{\text{face 1}} (\mathbf{n} \cdot \boldsymbol{\omega})\mathbf{r}\, dS + \iint_{\text{face 2}} (\mathbf{n} \cdot \boldsymbol{\omega})\mathbf{r}\, dS \right.
$$

$$
\left. - \boldsymbol{\omega} \left[\iint_{\substack{\text{wall of} \\ \text{cylinder}}} \mathbf{n} \cdot \mathbf{r}\, dS + \iint_{\text{face 1}} \mathbf{n} \cdot \mathbf{r}\, dS + \iint_{\text{face 2}} \mathbf{n} \cdot \mathbf{r}\, dS \right] \right\} \quad (2.101)
$$

To evaluate the various integrals in Eq. (2.101) we assume (which is permissible in view of the ensuing limit) that

$\mathbf{r}$ on face 1 is uniform and equal to
$\mathbf{r}$ at P which we denote by $\mathbf{r}(P)$;

$\mathbf{r}$ on face 2 is uniform and equal to
$\mathbf{r}(P) + \Delta s\mathbf{e}_\omega$, the value of $\mathbf{r}$ at the center of that face. Here $\mathbf{e}_\omega$ is a unit vector in the direction of $\boldsymbol{\omega}$;

$\mathbf{r}$ on the wall of the cylinder is uniform and equal to $\mathbf{r}(P) + \mathbf{a}$.

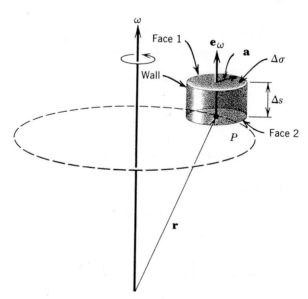

Fig. 2.39   Volume element to compute the curl of velocity.

Now, on the wall of the cylinder $\mathbf{n}$ is normal to $\boldsymbol{\omega}$. Therefore $\mathbf{n} \cdot \boldsymbol{\omega}$ vanishes on the wall of the cylinder and we have

$$\iint_{\text{wall}} (\mathbf{n} \cdot \boldsymbol{\omega}) \mathbf{r} \, dS = 0 \tag{2.102a}$$

On the face 1, $\mathbf{n} \cdot \boldsymbol{\omega}$ is equal to $-\omega$, and on the face 2, it is equal to ω. Therefore we obtain

$$\iint_{\text{face 1}} (\mathbf{n} \cdot \boldsymbol{\omega}) \mathbf{r} \, dS + \iint_{\text{face 2}} (\mathbf{n} \cdot \boldsymbol{\omega}) \mathbf{r} \, dS$$

$$= -\omega \mathbf{r}(P) \Delta\sigma + \omega[\mathbf{r}(P) + \Delta s \mathbf{e}_\omega] \Delta\sigma$$

$$= \boldsymbol{\omega} \, \Delta s \, \Delta\sigma$$

$$= \boldsymbol{\omega} \, \Delta\tau \tag{2.102b}$$

since $\Delta\tau = \Delta s \Delta\sigma$ is the volume of the cylinder. Next we observe that

$$\iint_{\text{wall}} \mathbf{n} \cdot \mathbf{r} \, dS = \iint_{\text{wall}} \frac{\mathbf{a}}{a} \cdot [\mathbf{r}(P) + \mathbf{a}] \, dS$$

$$= \frac{\mathbf{r}(P)}{a} \cdot \iint_{\text{wall}} \mathbf{a} \, dS + \iint_{\text{wall}} \frac{\mathbf{a} \cdot \mathbf{a}}{a} \, dS \tag{2.102c}$$

Now, $\displaystyle\iint_{\text{wall}} \mathbf{a} \, dS$ is zero. Further, we have

$$\iint_{\text{wall}} \frac{\mathbf{a} \cdot \mathbf{a}}{a} \, dS = \iint_{\text{wall}} \frac{a^2}{a} \, dS$$

$$= a \iint_{\text{wall}} dS = a 2\pi a \, \Delta s$$

$$= 2(\pi a^2) \, \Delta s$$

$$= 2\Delta\sigma \, \Delta s$$

$$= 2\Delta\tau$$

Therefore Eq. (2.102c) becomes

$$\iint_{\text{wall}} \mathbf{n} \cdot \mathbf{r} \, dS = 2\Delta\tau \tag{2.102d}$$

For the integral $\iint \mathbf{n} \cdot \mathbf{r} \, dS$ over the face 1 and 2 we have

$$\underset{\text{face 1}}{\iint} \mathbf{n} \cdot \mathbf{r} \, dS + \underset{\text{face 2}}{\iint} \mathbf{n} \cdot \mathbf{r} \, dS = -\mathbf{e}_\omega \cdot \mathbf{r}(P)\Delta s + \mathbf{e}_\omega \cdot [\mathbf{r}(P) + \mathbf{e}_\omega \, \Delta s] \, \Delta \sigma$$

$$= \Delta s \, \Delta \sigma = \Delta \tau \qquad (2.102e)$$

Using Eqs. (2.102), Eq. (101) may be reduced to

$$\mathbf{B} = \lim_{\Delta \tau \to 0} \frac{1}{\Delta \tau} \left\{ \boldsymbol{\omega} \, \Delta \tau - \boldsymbol{\omega} [2\Delta \tau + \Delta \tau] \right\}$$

$$= -2\boldsymbol{\omega}$$

or

$$-\mathbf{B} = 2\boldsymbol{\omega} \qquad (2.103)$$

In terms of Eq. (2.99), since $\mathbf{V} \times \mathbf{n} = -\mathbf{n} \times \mathbf{V}$, we have

$$\lim_{\Delta \tau \to 0} \frac{1}{\Delta \tau} \oiint_{\Delta S} \mathbf{n} \times \mathbf{V} \, dS \equiv -\mathbf{B} = 2\,\boldsymbol{\omega} \qquad (2.104)$$

This shows that for the velocity field $\mathbf{V}(\mathbf{r})$ of a rigid body rotating with the angular velocity $\boldsymbol{\omega}$, the limit in Eq. (2.104) *derives a vector*, namely $-\mathbf{B}$, which is simply equal to twice the angular velocity or the rotation associated with $\mathbf{V}(\mathbf{r})$. For this reason, we speak of the derived vector, $-\mathbf{B}$, as the *curl* or the *rotation* of $\mathbf{V}$. Extending this notion, the limit in Eq. (2.104) for any vector field $\mathbf{A}(\mathbf{r})$ is called the *curl of* $\mathbf{A}$ (or *rotation of* $\mathbf{A}$) and denoted by *curl* $\mathbf{A}$ (or *rot* $\mathbf{A}$). Thus we define

$$\text{curl } \mathbf{A} \equiv \lim_{\Delta \tau \to 0} \frac{1}{\Delta \tau} \oiint_{\Delta S} \mathbf{n} \times \mathbf{A} \, dS$$

or

$$\qquad (2.105)$$

$$\text{curl } \mathbf{A} \equiv \lim_{\Delta \tau \to 0} \frac{1}{\Delta \tau} \oiint_{\Delta S} - \mathbf{A} \times \mathbf{n} \, dS$$

When we are dealing with the velocity field of a rigid body, or of a deformable body such as a fluid or a solid, the curl of the velocity field is known as the *vortex vector* or the *vorticity*. In the general motion of a deformable body, the curl of the velocity at any point is equal to twice the angular velocity at that point (see Sections 9.1 and 9.2).

To derive the expressions for curl $\mathbf{A}$ in any orthogonal, curvilinear coordinate system, we choose the volume element $\Delta \tau$ suitably and carry out the limit specified in Eq. (2.105). To express curl $\mathbf{A}$ in a given coordinate system means giving the components of curl $\mathbf{A}$ in the directions of the three reference unit vectors. We shall now obtain the component of curl **A** in any direction **e**. This task at the same time establishes a relation between the curl of a vector field and the circulation of the vector field. Such a relation, as we shall learn, is of great importance in the study of many aspects of fluid motion.

2.34   Component of Curl as Circulation

Consider a point P in a vector field $\mathbf{A(r)}$, and let **e** denote a unit vector at P in some direction. Associated with **A** there is a vector curl **A** at the point P (see Fig. 2.40). We seek the component of **A** in the direction **e**.

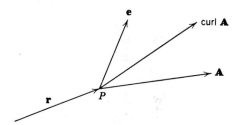

Fig. 2.40   A vector and its curl.

Using the definition (2.105), we have

$$\mathbf{e} \cdot \operatorname{curl} \mathbf{A} = \lim_{\Delta\tau \to 0} \frac{1}{\Delta\tau} \oiint_{\Delta S} \mathbf{e} \cdot \mathbf{n} \times \mathbf{A} \, dS \qquad (2.106)$$

since **e** is a given vector. Noting that

$$\mathbf{e} \cdot \mathbf{n} \times \mathbf{A} = \mathbf{A} \cdot \mathbf{e} \times \mathbf{n}$$

Eq. (2.106) may be written as

$$\mathbf{e} \cdot \operatorname{curl} \mathbf{A} = \lim_{\Delta\tau \to 0} \frac{1}{\Delta\tau} \oiint_{\Delta S} \mathbf{A} \cdot \mathbf{e} \times \mathbf{n} \, dS \qquad (2.107)$$

To evaluate the integral in Eq. (2.107) we choose the volume element $\Delta\tau$ as a cylinder with its axis parallel to **e**, its base equal to $\Delta\sigma$, and its height equal to Δh (see Fig. 2.41). Then we have

$$\oiint_{\Delta S} \mathbf{A} \cdot \mathbf{e} \times \mathbf{n} \, dS = \iint_{\text{wall}} \mathbf{A} \cdot \mathbf{e} \times \mathbf{n} \, dS$$

$$+ \iint_{\text{face 1}} \mathbf{A} \cdot \mathbf{e} \times \mathbf{n} \, dS + \iint_{\text{face 2}} \mathbf{A} \cdot \mathbf{e} \times \mathbf{n} \, dS \qquad (2.108)$$

Since the normals on the faces 1 and 2 are parallel to $\mathbf{e}$, the product $\mathbf{e} \times \mathbf{n}$ vanishes on these faces; consequently the integrals in Eq. (2.108) over the faces also vanish. To evaluate the remaining integral over the wall, we introduce h the distance measured along $\mathbf{e}$, and s the distance measured along C_e, the curve of intersection of the cylinder with the plane

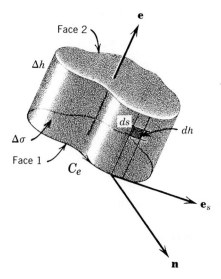

Fig. 2.41   Volume element for showing the relation between curl and circulation.

normal to $\mathbf{e}$. The direction along the curve C_e is assigned according to the rule of right-hand rotation about $\mathbf{e}$ (see Fig. 2.41). We then have

$$\oiint_{\Delta S} \mathbf{A} \cdot \mathbf{e} \times \mathbf{n} \, dS = \iint_{\text{wall}} \mathbf{A} \cdot \mathbf{e} \times \mathbf{n} \, dS$$

$$= \oint_{C_e} \left[\int_0^{\Delta h} \mathbf{A} \cdot \mathbf{e} \times \mathbf{n} \, dh \right] ds \qquad (2.109)$$

We observe that $\mathbf{e} \times \mathbf{n}$ does not change with h and is equal to a unit vector $\mathbf{e}_s$ in the direction of curve C_e at the point under consideration. We further assume (which is permissible in view of the ensuing limit) that at a given s the vector $\mathbf{A}$ is uniform over the height Δh and is equal to the value at $h = 0$. Equation (2.109) then becomes

$$\oiint_{\Delta S} \mathbf{A} \cdot \mathbf{e} \times \mathbf{n} \, dS = \Delta h \oint_{C_e} \mathbf{A} \cdot \mathbf{e}_s \, ds \qquad (2.110)$$

Using Eq. (2.110) with Eq. (2.107) we obtain

$$\mathbf{e} \cdot \text{curl } \mathbf{A} = \lim_{\Delta \tau \to 0} \frac{\Delta h}{\Delta \tau} \oint_{C_e} \mathbf{A} \cdot \mathbf{e}_s \, ds$$

Since $\Delta \tau = \Delta h \, \Delta \sigma$, this takes the form

$$\mathbf{e} \cdot \text{curl } \mathbf{A} = \lim_{\Delta \sigma \to 0} \frac{1}{\Delta \sigma} \oint_{C_e} \mathbf{A} \cdot \mathbf{e}_s \, ds \qquad (2.111)$$

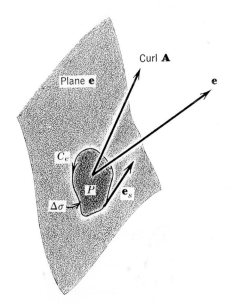

Fig. 2.42 Component of curl as the limit of circulation.

Recalling that $\oint_{C_e} \mathbf{A} \cdot \mathbf{e}_s \, ds$ is the circulation of $\mathbf{A}$ around curve C_e, we see that Eq. (2.111) gives the component of curl $\mathbf{A}$ in terms of the circulation of $\mathbf{A}$ around a certain curve. *Equation (2.111) is an alternative definition for curl* $\mathbf{A}$. It states that *curl* $\mathbf{A}$ *is a vector whose component in any direction* $\mathbf{e}$ *at the point P is equal to the limit of the circulation per unit area of* $\mathbf{A}$ *around a small curve enclosing P and lying in the plane* $\mathbf{e}$ (see Fig. 2.42).

The direction of integration of **A** *around the curve is according to the rule of right-hand rotation about the given direction* **e**.

Using Eq. (2.111) we can obtain in any orthogonal, curvilinear coordinate system the expression for curl **A**. In Cartesians x, y, z we have

$$\text{curl } \mathbf{A} = \begin{vmatrix} \mathbf{i} & \mathbf{j} & \mathbf{k} \\ \dfrac{\partial}{\partial x} & \dfrac{\partial}{\partial y} & \dfrac{\partial}{\partial z} \\ A_x & A_y & A_z \end{vmatrix} \tag{2.112}$$

In cylindrical coordinates r, θ, z we have

$$\text{curl } \mathbf{A} = \frac{1}{r} \begin{vmatrix} \mathbf{e}_r & r\mathbf{e}_\theta & \mathbf{e}_z \\ \dfrac{\partial}{\partial r} & \dfrac{\partial}{\partial \theta} & \dfrac{\partial}{\partial z} \\ A_r & rA_\theta & A_z \end{vmatrix} \tag{2.113}$$

In spherical coordinates r, θ, φ we have

$$\text{curl } \mathbf{A} = \frac{1}{r^2 \sin \theta} \begin{vmatrix} \mathbf{e}_r & r\mathbf{e}_\theta & r \sin \theta \mathbf{e}_\varphi \\ \dfrac{\partial}{\partial r} & \dfrac{\partial}{\partial \theta} & \dfrac{\partial}{\partial \varphi} \\ A_r & rA_\theta & r \sin \theta A_\varphi \end{vmatrix} \tag{2.114}$$

2.35 Some Related Remarks

1. The expressions for grad ϕ, div **A** and curl **A** in Cartesian, cylindrical, and spherical coordinates show that in terms of the differential operator ∇ we have

$$\text{grad } \phi \equiv \nabla \phi$$

$$\text{div } \mathbf{A} \equiv \nabla \cdot \mathbf{A}$$

and

$$\text{curl } \mathbf{A} \equiv \nabla \times \mathbf{A}$$

2. Using the operator ∇, the integral definitions (2.93), (2.94), and (2.105) for the gradient, divergence, and curl, respectively, may all be grouped in the form

$$\nabla \begin{pmatrix} \phi \\ \cdot \mathbf{A} \\ \times \mathbf{A} \end{pmatrix} = \lim_{\Delta \tau \to 0} \frac{1}{\Delta \tau} \oiint_{\Delta S} \begin{pmatrix} \phi \\ \mathbf{A} \cdot \\ -\mathbf{A} \times \end{pmatrix} \mathbf{n} \, dS \tag{2.115}$$

3. If the volume element in Eqs. (2.93), (2.94), and (2.105) defining the gradient, divergence, and curl is taken equal to $d\tau$, a differential volume

element, these equations can be written in the approximate forms:

$$\text{grad } \phi = \frac{1}{d\tau} \oiint_{\Delta S} \phi \mathbf{n} \, dS \qquad (2.116)$$

$$\text{div } \mathbf{A} = \frac{1}{d\tau} \oiint_{\Delta S} \mathbf{A} \cdot \mathbf{n} \, dS \qquad (2.117)$$

$$\text{curl } \mathbf{A} = \frac{1}{d\tau} \oiint_{\Delta S} -\mathbf{A} \times \mathbf{n} \, dS \qquad (2.118)$$

where now ΔS denotes the surface of the differential element $d\tau$. Equivalently (2.116), (2.117), and (2.118) may also be expressed in the forms:

$$d\tau \, \text{grad } \phi = \oiint_{\Delta S} \phi \mathbf{n} \, dS \qquad (2.116a)$$

$$d\tau \, \text{div } \mathbf{A} = \oiint_{\Delta S} \mathbf{A} \cdot \mathbf{n} \, dS \qquad (2.117a)$$

$$d\tau \, \text{curl } \mathbf{A} = \oiint_{\Delta S} -\mathbf{A} \times \mathbf{n} \, dS \qquad (2.118a)$$

The approximation implied in (2.116), (2.117), and (2.118) may be made as close as we please, since $d\tau$ may be taken as small as we please.

4. Consider a surface element $\mathbf{n} \, \Delta S$ in a vector field $\mathbf{A(r)}$. Let C_n denote the boundary curve of the surface element, and let $d\mathbf{s}$ denote a differential element of the curve. The direction of $d\mathbf{s}$ is that of right-hand rotation about $\mathbf{n}$ (see Fig. 2.43). According to the integral definition (2.111) for curl $\mathbf{A}$, we have

$$\mathbf{n} \cdot \text{curl } \mathbf{A} = \lim_{\Delta S \to 0} \frac{1}{\Delta S} \oint_{C_n} \mathbf{A} \cdot d\mathbf{s} \qquad (2.119)$$

If the surface element $\mathbf{n} \, dS$ is taken equal to $\mathbf{n} \, \Delta S$, a differential surface element, Eq. (2.119) can be written in the approximate form

$$\mathbf{n} \cdot \text{curl } \mathbf{A} = \frac{1}{dS} \oint_{C_n} \mathbf{A} \cdot d\mathbf{s} \qquad (2.120)$$

or in the form

$$\mathbf{n} \, dS \cdot \text{curl } \mathbf{A} = \oint_{C_n} \mathbf{A} \cdot d\mathbf{s} \qquad (2.120a)$$

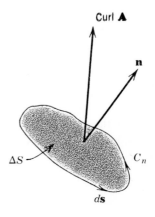

Curl **A**

n

ΔS

C_n

$d\mathbf{s}$

Fig. 2.43 Outflow of curl through a surface element is equal to the circulation around its boundary.

The approximation implied in Eqs. (2.120) may be made as close as we please, since dS may be taken as small as possible.

2.36   Relations between Surface and Volume Integrals

The integral definitions of grad ϕ, div $\mathbf{A}$, and curl $\mathbf{A}$ give rise to some important relations between surface and volume integrals that occur frequently in the analysis of scalar and vector fields. We shall now derive these relations.

Let us first consider a scalar field $\phi(\mathbf{r})$. In this field let $\mathscr{R}$ denote the volume of a finite region of space enclosed by a closed surface S. Subdivide $\mathscr{R}$ into a number of small volume elements. Let the volume of element k be denoted by $\delta\tau_k$ and its surface by ΔS_k. Then, for the element $\delta\tau_k$, according to Eq. (2.116a), we have

$$\oiint_{\Delta S_k} \phi\,\mathbf{n}_k\,dS_k = (\text{grad }\phi)_k\,\delta\tau_k \qquad (2.121)$$

where $\mathbf{n}_k\,dS_k$ is an element of area on the surface ΔS_k, and $(\text{grad }\phi)_k$ is grad ϕ taken at an arbitrary point within the element $\delta\tau_k$. The approximation implied in Eq. (2.121) becomes closer as $\delta\tau_k$ becomes smaller.

We sum the Eq. (2.121) for all elements in the region and proceed to the limit as the number of elements becomes indefinitely large. We thus have

$$\lim_{\substack{k\to\infty \\ \delta\tau_k\to 0}} \sum_{k=1}^{k} \oiint_{\Delta S_k} \phi\,\mathbf{n}_k\,dS_k = \lim_{\substack{k\to\infty \\ \delta\tau_k\to 0}} \sum_{k=1}^{k}(\text{grad }\phi)_k\,\delta\tau_k \qquad (2.122)$$

In carrying out the summation in the left side of Eq. (2.122), we observe that on the common surface between any two adjoining elements, the normal $\mathbf{n}$ for one element is opposite to that for the other element. This means the sum of the integrals $\iint\phi\mathbf{n}_k\,dS_k$ over the common surface for the two elements vanishes. In this way, considering all the elements, we will be left for the sum with contributions from only the surface elements that lie on the surface S. This result is independent of the way the region $\mathscr{R}$ is divided into elements $\delta\tau_k$. We therefore obtain

$$\lim_{k\to\infty} \sum_{k=1}^{k} \oiint_{\Delta S_k} \phi\,\mathbf{n}_k\,dS_k = \sum_{k=1}^{k} \oiint_{\Delta S_k} \phi\,\mathbf{n}_k\,dS_k$$

$$= \oiint_{S} \phi\,\mathbf{n}\,dS \qquad (2.123)$$

The sum in the right side of Eq. (2.122) is by definition the integral of grad ϕ in the region $\mathscr{R}$. We thus have

$$\lim_{k \to \infty} \sum_{k=1}^{k} (\text{grad } \phi)_k \, \delta\tau_k \equiv \iiint_{\mathscr{R}} \text{grad } \phi \, d\tau \qquad (2.124)$$

With the relations (2.123) and (2.124) Eq. (2.122) becomes

$$\oiint_{S} \phi \mathbf{n} \, dS = \iiint_{R} \text{grad } \phi \, d\tau \qquad (2.125)$$

which gives a relation between surface and volume integrals in a scalar field. Equation (2.125) is sometimes called the *gradient theorem*.

Consider next a vector field $\mathbf{A}(\mathbf{r})$ and, as before, let S be a closed surface enclosing a finite region of space, denoted by $\mathscr{R}$. Using relations (2.117a) and (2.118a), and proceeding on the same lines as above, we arrive at the following relations:

$$\oiint_{S} \mathbf{A} \cdot \mathbf{n} \, dS = \iiint_{R} \text{div } \mathbf{A} \, d\tau \qquad (2.126)$$

and

$$\oiint_{S} -\mathbf{A} \times \mathbf{n} \, dS = \iiint_{R} \text{curl } \mathbf{A} \, d\tau \qquad (2.127)$$

The relation (2.126) is usually known as *the divergence theorem* or *the theorem of Gauss*. It states that *the outflow of a vector field* $\mathbf{A}$ *through a closed surface S is equal to the volume integral of the divergence of the vector field over the region enclosed by S.*

Using the operator ∇, the integral relations (2.125), (2.126), and (2.127) may be grouped in the form

$$\oiint_{S} \begin{pmatrix} \phi \\ \mathbf{A} \cdot \\ -\mathbf{A} \times \end{pmatrix} \mathbf{n} \, dS = \iiint_{R} \nabla \begin{pmatrix} \phi \\ \cdot \mathbf{A} \\ \times \mathbf{A} \end{pmatrix} d\tau \qquad (2.128)$$

2.37 Theorem of Stokes

Equation (2.120a) gives rise to a relation between a line integral and a surface integral in a vector field. Let $\mathscr{C}$ be a closed curve in a vector field $\mathbf{A}(\mathbf{r})$ and let S be an arbitrary (in general, curved) surface bounded by the curve (see Fig. 2.44). We assign arbitrarily one or the other side of the surface as the positive side and divide that side of the surface into a large number of elements δS_k by a network of small curves C_k. At each element the normal $\mathbf{n}_k$ is set up from the positive side of S. A direction of travel along any C_k is chosen according to the rule of right-hand rotation about the corresponding normal $\mathbf{n}_k$. This procedure also fixes the direction of travel along the curve $\mathscr{C}$. For the surface element $\mathbf{n}_k\,\delta S_k$, according to Eq. (2.120a), we have

$$\oint_{C_k} \mathbf{A} \cdot d\mathbf{s}_k = (\text{curl } \mathbf{A})_k \cdot \mathbf{n}_k\,\delta S_k \tag{2.129}$$

where $d\mathbf{s}_k$ is a differential length along C_k, and $(\text{curl } \mathbf{A})_k$ is curl $\mathbf{A}$ taken at an arbitrary point within the element δS_k. The approximation in Eq. (2.129) becomes closer as δS_k becomes smaller.

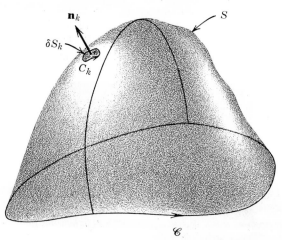

Fig. 2.44 Illustrating the theorem of Stokes.

We sum (2.129) for all the elements on the surface S and proceed to the limit as their number becomes indefinitely large. We thus have

$$\lim_{\substack{k \to \infty \\ \delta S_k \to \infty}} \sum_{k=1}^{k} \oint_{C_k} \mathbf{A} \cdot d\mathbf{s}_k = \lim_{\substack{k \to \infty \\ \delta S_k \to \infty}} \sum_{k=1}^{k} (\text{curl } \mathbf{A})_k \cdot \mathbf{n}_k\,\delta S_k \tag{2.130}$$

It is easily verified that

$$\lim_{\substack{k \to \infty \\ \delta S_k \to 0}} \sum_{k=1}^{k} \oint_{C_k} \mathbf{A} \cdot d\mathbf{s}_k = \sum_{k=1}^{k} \oint_{C_k} \mathbf{A} \cdot d\mathbf{s}_k = \oint_{\mathscr{C}} \mathbf{A} \cdot d\mathbf{s}$$

where $d\mathbf{s}$ is an elemental length along $\mathscr{C}$. The right of Eq. (2.130) is, by definition, the integral of curl $\mathbf{A} \cdot \mathbf{n}$ over the surface S. Therefore Eq. (2.130) becomes

$$\oint_{\mathscr{C}} \mathbf{A} \cdot d\mathbf{s} = \iint_{S} \text{curl } \mathbf{A} \cdot \mathbf{n}\, dS \tag{2.131}$$

This is known as *Stokes' theorem*. It states that *the circulation of a vector* **A** *around a curve* $\mathscr{C}$ *is equal to the outflow of curl* **A** *through an arbitrary surface S bounded by the curve* $\mathscr{C}$. Note that if we consider different surfaces drawn with the same boundary curve, the outflow of curl **A** is the same through all the surfaces.

In terms of fluid flow, if **A** represents the velocity field **V**, curl **A** becomes the *vorticity* and Eq. (2.131) states that the *circulation* around a curve $\mathscr{C}$ is equal to the *outflow of vorticity* through an arbitrary surface S bounded by $\mathscr{C}$.

A simple result that follows immediately from Eq. (2.131) is that if we consider a closed surface, the integral taken over the boundary curve vanishes giving

$$\oiint_S \text{curl } \mathbf{A} \cdot \mathbf{n} \, dS = 0 \qquad (2.132)$$

2.38 Further Operations

It may be recalled that gradient, divergence, and curl are operations that involve *first-order partial derivatives* with respect to space. Repeated application of gradient, divergence, and curl lead to expressions that involve spatial *derivatives* of *an order higher than the first*. We shall now look into those cases that involve only the second-order derivatives.

Consider first a scalar field $\phi(\mathbf{r})$. Associated with ϕ there is only one first-order differential expression, namely grad ϕ. Since grad ϕ is a vector field, we can form from it the following two second-order differential expressions:

$$\text{div (grad } \phi) \quad \text{or} \quad \boldsymbol{\nabla} \cdot \boldsymbol{\nabla}\phi$$

and

$$\text{curl (grad } \phi) \quad \text{or} \quad \boldsymbol{\nabla} \times \boldsymbol{\nabla}\phi$$

Consider next a vector field $\mathbf{A}(\mathbf{r})$. The first-order differential expressions associated with **A** are only two, the div **A** and the curl **A**. From these we can form the following second-order differential expressions:

$$\text{grad (div } \mathbf{A}) \quad \text{or} \quad \boldsymbol{\nabla}(\boldsymbol{\nabla} \cdot \mathbf{A})$$
$$\text{div (curl } \mathbf{A}) \quad \text{or} \quad \boldsymbol{\nabla} \cdot \boldsymbol{\nabla} \times \mathbf{A}$$

and

$$\text{curl (curl } \mathbf{A}) \quad \text{or} \quad \boldsymbol{\nabla} \times (\boldsymbol{\nabla} \times \mathbf{A})$$

2.39 Laplace Operator

In terms of the del operator, for div grad ϕ we can write

$$\text{div grad } \phi \equiv \boldsymbol{\nabla} \cdot \boldsymbol{\nabla}\phi = (\boldsymbol{\nabla} \cdot \boldsymbol{\nabla})\phi$$

The operator

$$\mathbf{\nabla} \cdot \mathbf{\nabla} \equiv \text{div grad}$$

is a *scalar second-order differential operator*. It is known as the *Laplace operator* or *Laplacian* and denoted by the symbol ∇^2. We thus have

$$\nabla^2 \equiv \mathbf{\nabla} \cdot \mathbf{\nabla} \equiv \text{div grad} \qquad (2.133)$$

The expression for the Laplacian in general orthogonal, curvilinear coordinates is given in Section 2.45. In Cartesians x, y, z it has the form

$$\nabla^2 = \frac{\partial^2}{\partial x^2} + \frac{\partial^2}{\partial y^2} + \frac{\partial^2}{\partial z^2} \qquad (2.134)$$

In cylindrical coordinates r, θ, z it has the form

$$\nabla^2 = \frac{1}{r}\frac{\partial}{\partial r}\left(r\frac{\partial}{\partial r}\right) + \frac{1}{r^2}\frac{\partial^2}{\partial \theta^2} + \frac{\partial^2}{\partial z^2} \qquad (2.135)$$

In spherical coordinates r, θ, φ it has the form

$$\nabla^2 = \frac{1}{r^2 \sin\theta}\left[\frac{\partial}{\partial r}\left(r^2 \sin\theta \frac{\partial}{\partial r}\right) + \frac{\partial}{\partial \theta}\left(\sin\theta \frac{\partial}{\partial \theta}\right) + \frac{\partial}{\partial}\left(\frac{1}{\sin\theta}\frac{\partial}{\partial}\right)\right] \qquad (2.136)$$

The Laplacian being a scalar operator may be applied to a vector field, say $\mathbf{A}(\mathbf{r})$, and we can speak of $\nabla^2\mathbf{A}$. In obtaining the expression for $\nabla^2\mathbf{A}$ in an orthogonal, curvilinear, coordinate system, account must be taken of the fact that not only the components of $\mathbf{A}$ but also the reference unit vectors are functions of position. In Cartesians, since the unit vectors are constant, the components of $\nabla^2\mathbf{A}$ are simply the Laplacians of the corresponding components of $\mathbf{A}$ (i.e., $\nabla^2 A_x$, etc.). This, however, is not true in the case of a general orthogonal, curvilinear system. To avoid any possible misinterpretation, in general, of the components of $\nabla^2\mathbf{A}$ as the Laplacians of the corresponding components, we express $\nabla^2\mathbf{A}$ in a form involving grad, div, and curl only. Such a form is the vector identity

$$\nabla^2\mathbf{A} = \text{grad div } \mathbf{A} - \text{curl (curl } \mathbf{A})$$
$$= \mathbf{\nabla}(\mathbf{\nabla} \cdot \mathbf{A}) - \mathbf{\nabla} \times (\mathbf{\nabla} \times \mathbf{A}) \qquad (2.137)$$

This identity may be readily verified by expansion in Cartesians or by expanding $\mathbf{\nabla} \times (\mathbf{\nabla} \times \mathbf{A})$ according to the formula for a vector triple product.

2.40 Green's Theorem

Having introduced the Laplacian, we shall now derive another integral relation involving it and known as *Green's theorem*. Green's theorem occupies an important position in mathematics and in various branches of mathematical physics. Its applications are numerous and may be found in the theory of gravitational fields, fluid mechanics, electrodynamics and optics, and in the theory of differential and integral equations.

To derive Green's theorem we start with the divergence theorem, Eq. (2.126):

$$\iiint_R \operatorname{div} \mathbf{A}\, d\tau = \oiint_S \mathbf{A} \cdot \mathbf{n}\, dS$$

Let $\psi = \psi(\mathbf{r})$ and $\phi = \phi(\mathbf{r})$ be two scalar functions of position. Form the vector field

$$\psi \operatorname{grad} \phi \equiv \psi \nabla \phi$$

and substitute it for the vector field $\mathbf{A}(\mathbf{r})$ in the divergence theorem (2.126). We thus have

$$\iiint_R \operatorname{div}(\psi \operatorname{grad} \phi)\, d\tau = \oiint_S \psi \operatorname{grad} \phi \cdot \mathbf{n}\, dS \qquad (2.138)$$

The integrand in the left side of Eq. (2.138) may be expanded as

$$\operatorname{div}(\psi \operatorname{grad} \phi) = \psi \operatorname{div} \operatorname{grad} \phi + \operatorname{grad} \psi \cdot \operatorname{grad} \phi$$

With this, Eq. (2.138) takes the form

$$\iiint_R (\psi \operatorname{div} \operatorname{grad} \phi + \operatorname{grad} \psi \cdot \operatorname{grad} \phi)\, d\tau = \oiint_S \psi \operatorname{grad} \phi \cdot \mathbf{n}\, dS$$

or, using the Laplacian, in the form

$$\iiint_R (\psi \nabla^2 \phi + \operatorname{grad} \psi \cdot \operatorname{grad} \phi)\, d\tau = \oiint_S \psi \operatorname{grad} \phi \cdot \mathbf{n}\, dS$$

$$(2.139)$$

Introducing $\partial\phi/\partial n$ to denote the derivative of ϕ with respect to distance in the direction of the *outward* normal $\mathbf{n}$, we have

$$\operatorname{grad} \phi \cdot \mathbf{n} = \frac{\partial \phi}{\partial n}$$

Equation (2.139) therefore may also be written in the form

$$\iiint_R (\psi \nabla^2 \phi + \operatorname{grad} \psi \cdot \operatorname{grad} \phi)\, d\tau = \oiint_S \psi \frac{\partial \phi}{\partial n}\, dS$$

Equation (2.139) is known as *Green's theorem in the first form.*

Now, consider the vector function

$$\psi \operatorname{grad} \phi - \phi \operatorname{grad} \psi$$

and substitute it for the vector $\mathbf{A}$ in the divergence theorem (2.126). We thus have

$$\iiint_R \text{div}(\psi \text{ grad } \phi - \phi \text{ grad } \psi)\, d\tau = \oiint_S (\psi \text{ grad } \phi - \phi \text{ grad } \psi) \cdot \mathbf{n}\, dS$$

(2.140)

The integrand in the left side of Eq. (2.140) may be expanded and shown to be equal to

$$\psi \nabla^2 \phi - \phi \nabla^2 \psi$$

Therefore Eq. (2.140) takes the form

$$\iiint_R (\psi \nabla^2 \phi - \phi \nabla^2 \psi)\, d\tau = \oiint_S (\psi \text{ grad } \phi - \phi \text{ grad } \psi) \cdot \mathbf{n}\, dS$$

or, equivalently, the form

(2.141)

$$\iiint_R (\psi \nabla^2 \phi - \phi \nabla^2 \psi)\, d\tau = \oiint_S \left(\psi \frac{\partial \phi}{\partial n} - \phi \frac{\partial \psi}{\partial n} \right) dS$$

This equation is known as *Green's theorem in the second form.*[*]

2.41 Irrotational Field

Let $\mathbf{n}\, dS$ be a differential surface element containing a point P in a scalar field $\phi(\mathbf{r})$. Consider the vector field grad ϕ and apply to it, at P, Eq. (2.120a). We thus have

$$\mathbf{n}\, dS \cdot \text{curl (grad } \phi) = \oint_{C_n} \text{grad } \phi \cdot d\mathbf{s}$$

(2.142)

where C_n is the boundary curve of the surface element $\mathbf{n}\, dS$. Since grad $\phi \cdot d\mathbf{s}$ is equal to $d\phi$, the total differential along C_n, we have

$$\oint_{C_n} \text{grad } \phi \cdot d\mathbf{s} = \oint_{C_n} d\phi = 0$$

since C_n is a closed curve. Therefore Eq. (2.142) becomes

$$\mathbf{n}\, dS \cdot \text{curl (grad } \phi) = 0$$

Since this equation is true for any arbitrary surface element $\mathbf{n}\, ds$ through P, it follows that

$$\text{curl (grad } \phi) = 0$$

(2.143)

This states that *the curl or rotation of any gradient vector is zero.* The

---

[*] The nomenclature regarding the first or second form is not uniform. The nomenclature used here seems to be preferred in America.

identity (2.143) may be demonstrated by working out the differential operation $\nabla \times \nabla \phi$ in any orthogonal, curvilinear coordinate. In particular, the identity is easily verified in Cartesians.

Now, suppose that a vector field is such that in certain regions of space its curl vanishes. Then in those regions, on the basis of Eq. (2.143), the vector field may be represented as the gradient of a scalar field. *Thus when*

$$\text{curl } \mathbf{A} = 0$$

we can write

$$\mathbf{A} = \text{grad } \phi \qquad (2.144)$$

where $\phi = \phi(\mathbf{r})$ is some scalar field. If $\mathbf{A}$ is known, ϕ can be determined from Eq. (2.144), which is a first-order partial differential (vector) equation. When $\mathbf{A}$ is not known, an equation for ϕ is to be developed from the equations that·govern $\mathbf{A}$. In physical problems, replacing an unknown irrotational vector field by an unknown scalar field reduces the number of scalar unknowns from three to one and consequently introduces some simplification in the analyses of the problems.

If the curl of a vector field vanishes in certain regions of space, we say it is an *irrotational field* in those regions. The scalar field, the gradient of which represents an irrotational vector field, is usually known as a *scalar potential* or simply a *potential*. This name results from the fact that the scalar field representing an irrotational force field is simply (except perhaps for the sign) the *potential energy* of the force field.

2.42   Solenoidal Field

Let $d\tau$ be a differential volume element containing a point P in a vector field $\mathbf{V} = \mathbf{V}(\mathbf{r})$. Consider the vector field curl $\mathbf{V}$ and apply to it, at P, Eq. (2.117a). We thus have

$$d\tau \text{ div (curl } \mathbf{V}) = \oiint_{\Delta S} \text{curl } \mathbf{V} \cdot \mathbf{n} \, dS \qquad (2.145)$$

where ΔS is the surface of the volume element $d\tau$. According to Eq. (2.132), the right side of Eq. (2.145) is zero. Therefore Eq. (2.145) becomes

$$d\tau \text{ div (curl } \mathbf{V}) = 0$$

Since this relation is true for any arbitrary volume element $d\tau$ enclosing P, it follows that

$$\text{div (curl } \mathbf{V}) = 0 \qquad (2.146)$$

This states that the *divergence of any curl vector is zero*. The identity (2.146) may be demonstrated by working out the differential operation $\nabla \cdot (\nabla \times \mathbf{V})$ in any orthogonal curvilinear coordinates. In particular, it is easily verified in Cartesians.

Now, suppose that $\mathbf{A}(\mathbf{r})$ is a vector field such that in certain regions of space its divergence vanishes. Then in those regions, on the basis of Eq. (2.146), $\mathbf{A}(\mathbf{r})$ may be represented as the curl of some other vector field, say $\mathbf{B}(\mathbf{r})$. *Thus when*

$$\operatorname{div} \mathbf{A} = 0$$

we can write

$$\mathbf{A} = \operatorname{curl} \mathbf{B} \tag{2.147}$$

In the analysis of physical problems, replacing a divergenceless vector $\mathbf{A}$ by curl $\mathbf{B}$ leads to certain advantages.

A vector field whose divergence is zero is known as a *solenoidal field.* In analogy with a scalar potential a vector such as $\mathbf{B}$, defined by Eq. (2.147), is known as a *vector potential.*

2.43 Laplace's Equation

Suppose $\mathbf{A}(\mathbf{r})$ is a vector field whose divergence and curl are both zero. We then have

$$\operatorname{div} \mathbf{A} = 0 \tag{2.148}$$

$$\operatorname{curl} \mathbf{A} = 0 \tag{2.149}$$

On the basis of Eq. (2.148) we can represent $\mathbf{A}$ as the gradient of a scalar function, say $\phi(\mathbf{r})$. Alternatively, on the basis of Eq. (2.149), we may represent $\mathbf{A}$ as the curl of a vector function $\mathbf{B}(\mathbf{r})$.

First let us write

$$\mathbf{A} = \operatorname{grad} \phi \tag{2.150}$$

satisfying identically Eq. (2.149). Substituting Eq. (2.150) into Eq. (2.148) we have

$$\nabla^2 \phi \equiv \operatorname{div} (\operatorname{grad} \phi) = 0 \tag{2.151}$$

as the equation to determine ϕ. Such an equation is called *Laplace's equation.*

Instead of setting $\mathbf{A}$ equal to grad ϕ, suppose we write

$$\mathbf{A} = \operatorname{curl} \mathbf{B} \tag{2.152}$$

satisfying identically Eq. (2.148). Substituting Eq. (2.152) into Eq. (2.149) we have

$$\operatorname{curl} (\operatorname{curl} \mathbf{B}) = 0 \tag{2.153}$$

According to the vector identity (2.137)

$$\operatorname{curl} (\operatorname{curl} \mathbf{B}) = \operatorname{grad} \operatorname{div} \mathbf{B} - \nabla^2 \mathbf{B}$$

Since div $\mathbf{B}$ is unspecified, we stipulate that div $\mathbf{B}$ is zero. Then Eq. (2.153) takes the form

$$\nabla^2 \mathbf{B} = 0 \tag{2.154}$$

which, again, is Laplace's equation.

2.44 Poisson's Equation

Let us consider the situation where the rotation of a vector field $A(r)$ is zero but its divergence is not zero. In fact, let div A be described by a scalar field $q(r)$. We thus have

$$\text{curl } A = 0 \tag{2.155}$$

$$\text{div } A = q(r) \tag{2.156}$$

We set

$$A = \text{grad } \phi \tag{2.157}$$

thus identically satisfying Eq. (2.155). Using Eqs. (2.156) and (2.157) we obtain

$$\nabla^2\phi \equiv \text{div grad } \phi = q(r) \tag{2.158}$$

as the equation to determine $\phi(r)$. Equation (2.158) is an inhomogeneous Laplace equation. Such an equation is called *Poisson's equation.*

Consider next a vector field $V(r)$ whose divergence is zero but whose rotation is not zero. Let curl V be described by a vector field $\Omega(r)$. Thus $V(r)$ is characterized by the equations

$$\text{div } V = 0 \tag{2.159}$$

$$\text{curl } V = \Omega(r) \tag{2.160}$$

We set

$$V = \text{curl } B \tag{2.161}$$

thus satisfying identically Eq. (2.159). Using Eqs. (2.160) and (2.161) we obtain

$$\text{curl (curl } B) = \Omega \tag{2.162}$$

On using the vector identity (2.137) and stipulating that div B is equal to zero, Eq. (2.162) becomes

$$\nabla^2 B = -\Omega(r) \tag{2.163}$$

This is Poisson's equation governing the vector field $B(r)$.

Recalling that the Laplacian is a second-order differential expression, the equations of Laplace and Poisson are second-order partial differential equations. They govern many physical phenomena, for example, steady heat conduction, electrostatics, magneto-statics, gravitational fields, flow of an ideal fluid. In many technically interesting problems involving these phenomena, the source term [i.e., $q(r)$ or $\Omega(r)$ in Eq. (2.158) or (2.163) in Poisson's equation] vanishes outside certain limited regions of space. In such a case Poisson's equation holds only inside those regions, whereas Laplace's equation holds outside those regions.

The theory of the motion of an ideal fluid, as we shall see, is identical with the pursuit of constructing solutions to the equations of Laplace and Poisson.

2.45 Expressions in General Orthogonal, Curvilinear Coordinates

We now complete our study of vector analysis with the expressions in orthogonal, curvilinear coordinates for the directed distance between two neighboring points, the differential operators—grad, div, curl, and Laplacian—and the changes in the reference unit vectors.

Let q_1, q_2, q_3 be the general curvilinear coordinates of a point P in space. They are defined by the transformation

$$q_1 = q_1(x, y, z)$$
$$q_2 = q_2(x, y, z)$$
$$q_3 = q_3(x, y, z)$$

or, inversely, by

$$x = x(q_1, q_2, q_3)$$
$$y = y(q_1, q_2, q_3) \tag{2.164}$$
$$z = z(q_1, q_2, q_3)$$

where x, y, z are, as usual, the Cartesian coordinates of P. Let e_1, e_2, e_3 denote, as always, the system of reference unit vectors, at P, corresponding to the coordinates q_1, q_2, q_3 (see Section 2.16). We first determine the unit vectors from the transformation relations (2.164). It should be noted that until the unit vectors are determined it is not possible to say whether or not they (or equivalently the chosen curvilinear coordinates) are orthogonal.

Unit Vectors. Consider the q_1-curve through the point P(see Section 2.16). Let **r** denote the position vector to P from a fixed origin and s_1 the distance along the q_1-curve (being positive when measured in the direction of increasing q_1). Then, according to Eq. (2.17) and the definition of e_1, we have

$$\frac{\partial \mathbf{r}}{\partial q_1} = \lim_{\Delta q_1 \to 0} \left(\frac{\delta \mathbf{r}}{\delta q_1} \right)_{q_2, q_3 \text{ kept const.}}$$

$$= \lim_{\Delta q_1 \to 0} \left[\frac{\delta \mathbf{r}}{\delta s_1} \cdot \frac{\delta s_1}{\delta q_1} \right]_{q_2, q_3 \text{ kept const.}}$$

$$= \left(\frac{\delta s_1}{\delta q_1} \right)_{\text{at the point } q_1, q_2, q_3} \lim_{\Delta q_1 \to 0} \left(\frac{\delta \mathbf{r}}{\delta s_1} \right)$$

$$= \left(\frac{\delta s_1}{\delta q_1} \right)_{\text{at } q_1, q_2, q_3} (\mathbf{e}_1)_{\text{at } q_1, q_2, q_3} \tag{2.165}$$

Similarly, considering the q_2 and q_3 curves, we obtain

$$\frac{\partial \mathbf{r}}{\partial q_2} = \left(\frac{\delta s_2}{\delta q_2}\right)_{\text{at } q_1,q_2,q_3} (\mathbf{e}_2)_{\text{at } q_1,q_2,q_3} \tag{2.166}$$

and,

$$\frac{\partial \mathbf{r}}{\partial q_3} = \left(\frac{\delta s_3}{\delta q_3}\right)_{\text{at } q_1,q_2,q_3} (\mathbf{e}_3)_{\text{at } q_1,q_2,q_3} \tag{2.167}$$

where s_2 and s_3 are distances measured, respectively, along the q_2 and q_3 curves. Equations (2.165), (2.166), and (2.167) enable us to determine $\mathbf{e}_1$, and so forth and $\delta s_1/\delta q_1$, etc., once we know what $\partial \mathbf{r}/\partial q_1$, etc., are. The derivatives $\partial \mathbf{r}/\partial q_1$, and so forth, are obtained as follows in terms of the transformation relations (2.164) and the $\mathbf{i}$, $\mathbf{j}$, $\mathbf{k}$ unit vectors. Introduce the notation

$$h_1 \equiv \frac{\delta s_1}{\delta q_1}$$

$$h_2 \equiv \frac{\delta s_2}{\delta q_2} \tag{2.168}$$

and

$$h_3 \equiv \frac{\delta s_3}{\delta q_3}$$

Equations (2.165), (2.166) and (2.167) may be grouped in the convenient form

$$h_m \mathbf{e}_m = \frac{\partial \mathbf{r}}{\partial q_m} \tag{2.169}$$

where the subscript m may be 1, 2, or 3.

Expressing the position vector as

$$\mathbf{r} = x(q_1, q_2, q_3)\mathbf{i} + y(q_1, q_2, q_3)\mathbf{j} + z(q_1, q_2, q_3)\mathbf{k} \tag{2.170}$$

we obtain

$$h_m \mathbf{e}_m = \frac{\partial \mathbf{r}}{\partial q_m} = \frac{\partial x}{\partial q_m}\mathbf{i} + \frac{\partial y}{\partial q_m}\mathbf{j} + \frac{\partial z}{\partial q_m}\mathbf{k} \tag{2.171}$$

where, again, the subscript m may be 1, 2, or 3. The relations (2.171) determine completely the h_m's and the $\mathbf{e}_m$'s. In fact, from (2.171) we have

$$h_m^2 = \left(\frac{\partial x}{\partial q_m}\right)^2 + \left(\frac{\partial y}{\partial q_m}\right)^2 + \left(\frac{\partial z}{\partial q_m}\right)^2 \tag{2.172}$$

and

$$\mathbf{e}_m = \frac{1}{h_m}\left(\frac{\partial x}{\partial q_m}\mathbf{i} + \frac{\partial y}{\partial q_m}\mathbf{j} + \frac{\partial z}{\partial q_m}\mathbf{k}\right) \tag{2.173}$$

Equations (2.172) and (2.173) hold whether or not the curvilinear co-ordinates, defined by the relations (2.164), are orthogonal. Having determined the e_m's, if we find that the products

$$\mathbf{e}_1 \cdot \mathbf{e}_2, \mathbf{e}_2 \cdot \mathbf{e}_3, \mathbf{e}_3 \cdot \mathbf{e}_1$$

vanish at all points, then the unit vectors or, equivalently, the curvilinear coordinates are orthogonal.

Infinitesimal Distance between Two Neighboring Points. Let Q be a neighboring point of P, and let

$$q_1 + dq_1, q_2 + dq_2, q_3 + dq_3$$

be the curvilinear coordinates of Q while

$$x + dx, y + dy, z + dz$$

are its Cartesian coordinates. We denote by $\mathbf{ds} = ds\,\mathbf{e}_s$ the directed distance $\overrightarrow{PQ}$. If ds_1, ds_2, ds_3 are differential arc lengths along the q_1, q_2, q_3 curves, respectively, we have

$$\mathbf{ds} = \mathbf{e}_1 \, ds_1 + \mathbf{e}_2 \, ds_2 + \mathbf{e}_3 \, ds_3 \qquad (2.174)$$

This relation does not assume that the coordinates are orthogonal. The arc lengths ds_1, etc., are determined as follows.

As always $\mathbf{ds}$ is identical with $\mathbf{dr}$, where $\mathbf{r}$ is the position vector of point P. Now, since

$$\mathbf{r} = \mathbf{r}(q_1, q_2, q_3)$$

we obtain

$$\mathbf{ds} \equiv \mathbf{dr} = \frac{\partial \mathbf{r}}{\partial q_1} \, \partial q_1 + \frac{\partial \mathbf{r}}{\partial q_2} \, dq_2 + \frac{\partial \mathbf{r}}{\partial q_3} \, dq_3$$

$$= \mathbf{e}_1 h_1 \, dq_1 + \mathbf{e}_2 h_2 \, dq_2 + \mathbf{e}_3 h_3 \, dq_3 \qquad (2.175)$$

where the h's and $\mathbf{e}$'s are given by Eqs. (2.172) and (2.173). From Eqs. (2.174) and (2.175) it follows that

$$\begin{aligned} ds_1 &= h_1 \, dq_1 \\ ds_2 &= h_2 \, dq_2 \\ ds_3 &= h_3 \, dq_3 \end{aligned} \qquad (2.176)$$

giving the differential distances along the coordinate curves in terms of the coordinates. From Eq. (2.176) we see that h_1, h_2, h_3 are of the nature of *scale factors* which relate the differential distances to the differentials of the coordinates. *These scale factors vary from point to point and thus are, in general, functions of position.*

From Eq. (2.175), for the square of the infinitesimal distance ds, we obtain

$$(ds)^2 \equiv \mathbf{ds} \cdot \mathbf{ds}$$

$$= \sum_{m=1}^{3} \sum_{n=1}^{3} (h_m \mathbf{e}_m \cdot h_m \mathbf{e}_n) \, dq_m \, dq_n \tag{2.177}$$

This equation is true for any curvilinear coordinate system, orthogonal or not.

For an *orthogonal* system we have the result*

$$\mathbf{e}_m \cdot \mathbf{e}_n = 0 \quad \text{when} \quad m \neq n$$

$$= 1 \quad \text{when} \quad m = n$$

Therefore, *for an orthogonal, curvilinear, coordinate system*, Eq. (2.177) becomes

$$(ds)^2 = (h_1 \, dq_1)^2 + (h_2 \, dq_2)^2 + (h_3 \, dq_3)^2$$

$$= (ds_1)^2 + (ds_2)^2 + (ds_3)^2 \tag{2.178}$$

Differential Volume and Surface Elements. We concern ourselves hereafter with an *orthogonal, curvilinear, coordinate system* only. Consider two points P and Q such that the coordinates of P are q_1, q_2, q_3 and those of Q are $q_1 + dq_1$, $q_2 + dq_2$ and $q_3 + dq_3$. The coordinate surfaces passing through P and Q describe a volume element that we shall denote by $d\tau$. The sides of the element are curvilinear, and, in general, the element is not a parallelepiped (see Fig. 2.45). For illustration, the volume elements in cylindrical and spherical coordinates are shown in Fig. 2.45. To the first order the volume of the element in curvilinear, orthogonal coordinates is given by

$$d\tau = ds_1 \, ds_2 \, ds_3$$

$$= h_1 h_2 h_3 \, dq_1 \, dq_2 \, dq_3 \tag{2.179}$$

where the scale factors h_1, h_2, h_3 refer to the point q_1, q_2, q_3.

The area of the surface element forming the q_1 face (i.e., the face normal to q_1 coordinate) of the volume element $d\tau$ is given by

$$dS_1 = h_2 h_3 \, dq_2 \, dq_3 \tag{2.180}$$

Similarly, the areas of the q_2 and q_3 faces of the element $d\tau$ are given, respectively, by

$$dS_2 = h_3 h_1 \, dq_3 \, dq_1 \tag{2.181}$$

and

$$dS_3 = h_1 h_2 \, dq_1 \, dq_2 \tag{2.182}$$

---

* For a *nonorthogonal* system $\mathbf{e}_m \cdot \mathbf{e}_n \neq 0$ when $m \neq n$.

144   Ideal-Fluid Aerodynamics

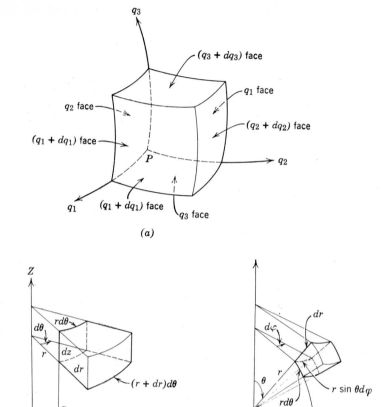

Fig. 2.45 Volume element in curvilinear coordinates: (a) general coordinates; (b) cylindrical coordinates; (c) spherical coordinates.

Gradient. The expression for the gradient of a scalar function $\phi(q_1, q_2, q_3)$ is conveniently obtained as described in Section (2.24). We have

$$\operatorname{grad}\phi = \frac{1}{h_1}\frac{\partial\phi}{\partial q_1}\mathbf{e}_1 + \frac{1}{h_2}\frac{\partial\phi}{\partial q_2}\mathbf{e}_2 + \frac{1}{h_3}\frac{\partial\phi}{\partial q_3}\mathbf{e}_3$$

The del operator is given by

$$\nabla \equiv e_1 \frac{1}{h_1} \frac{\partial}{\partial q_1} + e_2 \frac{1}{h_2} \frac{\partial}{\partial q_2} + e_3 \frac{1}{h_3} \frac{\partial}{\partial q_3}$$

The expression for grad ϕ may be derived also on the basis of the integral definition (2.93). This is left as an exercise.

Divergence. The expression for the divergence of a vector field $A(q_1, q_2, q_3)$ is conveniently obtained by using the integral definition (2.94). As may be verified we obtain

$$\text{div } \mathbf{A} = \frac{1}{h_1 h_2 h_3} \left[\frac{\partial}{\partial q_1} (h_2 h_3 A_1) + \frac{\partial}{\partial q_2} (h_3 h_1 A_2) + \frac{\partial}{\partial q_3} (h_1 h_2 A_3) \right] \quad (2.183)$$

Curl. The expression for the curl of a vector field $A(q_1, q_2, q_3)$ is conveniently obtained by constructing its components in the directions of the reference unit vectors e_1, e_2, e_3 on the basis of the definition (2.111). As may be verified, we obtain a result that can be exhibited as

$$\text{Curl } \mathbf{A} = \frac{1}{h_1 h_2 h_3} \begin{vmatrix} h_1 e_1 & h_2 e_2 & h_3 e_3 \\ \dfrac{\partial}{\partial q_1} & \dfrac{\partial}{\partial q_2} & \dfrac{\partial}{\partial q_3} \\ h_1 A_1 & h_2 A_2 & h_3 A_3 \end{vmatrix} \quad (2.184)$$

Laplacian. The expression for the Laplacian of a scalar function of position $\phi(q_1, q_2, q_3)$ is readily obtained by substituting grad ϕ for **A** and the corresponding components of grad ϕ for A_1, A_2, and A_3 in the relation (2.183). We thus obtain

$$\nabla^2 \phi = \text{div grad } \phi = \frac{1}{h_1 h_2 h_3} \left[\frac{\partial}{\partial q_1} \left(\frac{h_2 h_3}{h_1} \frac{\partial \phi}{\partial q_1} \right) \right.$$
$$\left. + \frac{\partial}{\partial q_2} \left(\frac{h_3 h_1}{h_2} \frac{\partial \phi}{\partial q_2} \right) + \frac{\partial}{\partial q_3} \left(\frac{h_1 h_2}{h_3} \frac{\partial \phi}{\partial q_3} \right) \right] \quad (2.185)$$

It follows that the Laplacian operator is given by

$$\nabla^2 \equiv \text{div grad}$$
$$= \frac{1}{h_1 h_2 h_3} \left[\frac{\partial}{\partial q_1} \left(\frac{h_2 h_3}{h_1} \frac{\partial}{\partial q_1} \right) + \frac{\partial}{\partial q_2} \left(\frac{h_3 h_1}{h_2} \frac{\partial}{\partial q_2} \right) + \frac{\partial}{\partial q_3} \left(\frac{h_1 h_2}{h_3} \frac{\partial}{\partial q_3} \right) \right]$$

Changes in the Reference Unit Vectors. The relation (2.173) enables us to determine the change in any of the unit vectors e_1, e_2, e_3 resulting from a change in the coordinates q_1, q_2, q_3. Such a determination, however, leaves the result in terms of the $\mathbf{i}$, $\mathbf{j}$, $\mathbf{k}$ systems of unit vectors, but we are usually interested in having the result in terms of the coordinates q_1, q_2, q_3 and the corresponding $\mathbf{e}_1$, $\mathbf{e}_2$, $\mathbf{e}_3$ unit vectors. Such a result is readily obtained by the following method.

Denote by $\mathbf{e}_1'$, $\mathbf{e}_2'$, $\mathbf{e}_3'$ the unit vectors at the point $q_1 + \delta q_1$, $q_2 + \delta q_2$, and $q_3 + \delta q_3$, where δq_1, δq_2, δq_3 are infinitesimal changes (for clarity we use temporarily the symbol δ instead of d). Denote by $\boldsymbol{\delta}\mathbf{s}$ the change in position corresponding to the coordinate changes δq_1, δq_2, δq_3. The unit vectors $\mathbf{e}_1'$, $\mathbf{e}_2'$, $\mathbf{e}_3'$ may be regarded as resulting from a translation of the vectors $\mathbf{e}_1$, $\mathbf{e}_2$, $\mathbf{e}_3$ over the directed distance $\boldsymbol{\delta}\mathbf{s}$ and a rotation of them through an infinitesimal angle $\delta\psi$ about a certain axis passing through their common origin. Note that we are concerned here with orthogonal curvilinear coordinates only. Let $\boldsymbol{\delta}\boldsymbol{\psi}$ represent vectorially the infinitesimal angular rotation. Once $\boldsymbol{\delta}\boldsymbol{\psi}$ is known, the changes in the unit vectors are given (since translation causes no change)

$$\delta\mathbf{e}_1 = \mathbf{e}_1' - \mathbf{e}_1 = \boldsymbol{\delta}\boldsymbol{\psi} \times \mathbf{e}_1$$
$$\delta\mathbf{e}_2 = \mathbf{e}_2' - \mathbf{e}_2 = \boldsymbol{\delta}\boldsymbol{\psi} \times \mathbf{e}_2 \qquad (2.187)$$
$$\delta\mathbf{e}_3 = \mathbf{e}_3' - \mathbf{e}_3 = \boldsymbol{\delta}\boldsymbol{\psi} \times \mathbf{e}_3$$

Now, we know that the angular displacement $\boldsymbol{\delta}\boldsymbol{\psi}$ and the associated displacement $\boldsymbol{\delta}\mathbf{s}$ are related. In fact, we have

$$\boldsymbol{\delta}\boldsymbol{\psi} = \tfrac{1}{2} \operatorname{curl} \boldsymbol{\delta}\mathbf{s} \qquad (2.188)$$

To express $\boldsymbol{\delta}\boldsymbol{\psi}$ in terms of the coordinates q_1, q_2, q_3 and the unit vectors $\mathbf{e}_1$, $\mathbf{e}_2$, $\mathbf{e}_3$ we write

$$\boldsymbol{\delta}\mathbf{s} = \delta s_1 \mathbf{e}_1 + \delta s_2 \mathbf{e}_2 + \delta s_3 \mathbf{e}_3$$
$$= h_1\,\delta q_1 \mathbf{e}_1 + h_2\,\delta q_2 \mathbf{e}_2 + h_3\,\delta q_3 \mathbf{e}_3$$

Using this and (2.184), we express relation (2.188) as

$$\boldsymbol{\delta}\boldsymbol{\psi} = \frac{1}{2h_1 h_2 h_3} \begin{vmatrix} h_1 \mathbf{e}_1 & h_2 \mathbf{e}_2 & h_3 \mathbf{e}_3 \\ \dfrac{\partial}{\partial q_1} & \dfrac{\partial}{\partial q_2} & \dfrac{\partial}{\partial q_3} \\ h_1^{\,2}\,\delta q_1 & h_2^{\,2}\,\delta q_2 & h_3^{\,2}\,\delta q_3 \end{vmatrix} \qquad (2.189)$$

2.46 Some Useful Relations

The following formulas are of frequent use in applications. They can be verified either by expansion in Cartesian coordinates or by treating ∇ as a vector (while retaining its actual meaning as an operator) in the appropriate vector formulas for products. In the following the scalars and vectors are all functions of position.

$$\nabla(\psi\phi) = \psi\nabla\phi + \phi\nabla\psi$$

$$\nabla \cdot (\phi\mathbf{A}) = \phi\nabla \cdot \mathbf{A} + \nabla\phi \cdot \mathbf{A}$$

$$\nabla \times (\phi\mathbf{A}) = \phi\nabla \times \mathbf{A} + \nabla\phi \times \mathbf{A}$$

$$\nabla(\mathbf{A} \cdot \mathbf{B}) = (\mathbf{A} \cdot \nabla)\mathbf{B} + (\mathbf{B} \cdot \nabla)\mathbf{A} + \mathbf{A} \times (\nabla \times \mathbf{B}) + \mathbf{B} \times (\nabla \times \mathbf{A})$$

$$\nabla \cdot (\mathbf{A} \times \mathbf{B}) = \mathbf{B} \cdot \nabla \times \mathbf{A} - \mathbf{A} \cdot \nabla \times \mathbf{B}$$

$$\nabla \times (\mathbf{A} \times \mathbf{B}) = \mathbf{A}(\nabla \cdot \mathbf{B}) + (\mathbf{B} \cdot \nabla)\mathbf{A} - \mathbf{B}(\nabla \cdot \mathbf{A}) - (\mathbf{A} \cdot \nabla)\mathbf{B}$$

$$\nabla \times (\nabla \times \mathbf{A}) = \nabla(\nabla \cdot \mathbf{A}) - \nabla^2\mathbf{A}$$

Chapter 3

Stress in a Fluid

Theory of the motion of a fluid, like the theory of the motion of a system of masses, is based on Newton's laws of mechanics. Similarly, the study of the state of rest of a fluid is based on the laws of static equilibrium as applied to a mechanical system of masses. In adapting these laws to a fluid we usually choose as our system a certain finite or elemental region of the fluid. Thus for a fluid in motion we require, according to Newton's second law, that the rate of change of momentum of the fluid contained within any chosen region be equal to the resultant of all the forces acting on it. If the fluid is at rest, the momentum is zero and we require that the resultant of all the forces acting on any portion of the fluid be zero. If the fluid is in a state of uniform motion (i.e., all fluid elements have the same velocity for all times), the rate of change of its momentum is zero. Therefore the laws of static equilibrium also apply to a state of uniform motion. To the law of change of momentum we add the companion law of moments that the sum of all the moments of the forces acting on any part of a fluid at rest is zero. The moments are all taken with respect to a single reference point.

As a preliminary step in the formulation of these laws, we shall consider in this chapter the nature of the forces that act on any region of a fluid and the method of their specification. This involves the concept of stress in a fluid. Following these considerations, we shall develop the law of static equilibrium for a fluid. To formulate completely the law of motion for a fluid we must first decide about a method for describing fluid motion. This we do in Chapter 4. Finally, in Chapter 5, we shall formulate the law of motion along with other equations necessary for the analysis of fluid motion.

3.1 Surface Forces and Body Forces

In the space occupied by a fluid, which is in motion or at rest, let us imagine a surface enclosing some part of the fluid. The portions of the fluid close to the surface on its two sides exert forces on each other. These

148

forces are in the nature of actions and reactions. Such forces are commonly referred to as *internal forces*. Since they act across a surface that is imagined to separate the fluid, they are also called *surface forces*. The complete specification of these forces is based on the concept of stress, which we shall take up presently (see Section 3.2).

In addition to the surface forces the fluid may, in general, be subjected to forces that act throughout the body of the fluid as such. The simplest example of such a force is the force due to gravity, that is, the so-called weight of the fluid. Forces of this type are called *body forces*. They are proportional to the volume or mass of the fluid considered. Thus the body forces can be specified as so much per unit volume or per unit mass of the fluid, that is, as an intensity. In general, the body force may vary from point to point of the fluid, and at any point it may have different values at different instances of time. Thus the body force is a vector function of position and time.

3.2 Concept of Stress and the Specification of Stress at a Point

The concept of stress and the method of its specification, as described below, are applicable to any continuous deformable medium whether it be a fluid or a solid.*

Let us consider a plane surface S drawn through some point P in a fluid. The fluid may be either in motion or at rest. We denote by **n** the normal to the surface and consider the forces exerted across S by the portion of the fluid which lies on the side of **n**. These internal forces are not, in general, distributed uniformly across S. We represent these forces as equivalent to a force **F** acting at the point P and a moment **M** about some axis through P (Fig. 3.1a). If we gradually shrink the area of the surface to the point P, both **F** and **M** will tend to zero. However, for a vanishingly small area the equivalent force **F** may be assumed to be proportional to the area. Then the ratio of force to area may be assumed to tend to a definite limit as the area shrinks to a point. It can be shown that the ratio of the moment to the area vanishes as the area goes to zero. This means the action of the internal forces in the immediate vicinity of point P across the surface **n** (i.e., normal to **n**) can be specified by the limit of the ratio of force to area, that is, by a *force per unit area* at that point. This is called *the stress at the point P across a plane* **n**. It is a vector, and its direction is, in general, different from that of **n**. Denoting such a *stress vector* by $\boldsymbol{\sigma}_n$ we define symbolically

$$\boldsymbol{\sigma}_n = \lim_{S_n \to 0} \frac{\mathbf{F}}{S_n} \qquad (3.1)$$

* In this connection see Love (1944).

where S_n is a plane area normal to **n**. The definition of a stress vector involves two directions—that of the normal to the surface and that of the stress itself. Now if dS is an elemental area normal to **n**, the force exerted across dS by the fluid that is on the side of **n** is (Fig. 3.1b)

$$\boldsymbol{\sigma}_n \, ds$$

The stress at a point P across an elemental surface **n** can be specified either by a stress vector $\boldsymbol{\sigma}_n$ or, equivalently, by its three components

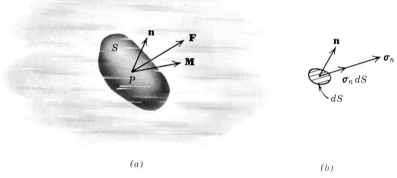

(a)   (b)

Fig. 3.1   Representation of internal forces in a fluid.

referred to a system of three unit vectors. Denoting the unit vectors by $\mathbf{e}_1$, $\mathbf{e}_2$, $\mathbf{e}_3$ and the respective components of $\boldsymbol{\sigma}_n$ by σ_{n1}, σ_{n2}, σ_{n3}, we write

$$\boldsymbol{\sigma}_n = \sigma_{n1}\mathbf{e}_1 + \sigma_{n2}\mathbf{e}_2 + \sigma_{n3}\mathbf{e}_3 = (\sigma_{n1}, \sigma_{n2}, \sigma_{n3}) \qquad (3.2)$$

The meaning of the double subscript is apparent. The first subscript denotes the plane across which the stress component acts, whereas the second subscript denotes the direction of the component.

The decomposition of a stress vector is usually done with respect to an orthogonal system of unit vectors. Generally the unit vectors selected are those associated with the (orthogonal) coordinate system employed (Fig. 3.2a). Another way of choosing the unit vectors is to select one of them in the direction of the normal **n** and the other two in two mutually perpendicular directions lying in the surface **n**. Denoting the latter directions by $\mathbf{e}_t$ and $\mathbf{e}_s$ we write

$$\boldsymbol{\sigma}_n = \sigma_{nn}\mathbf{n} + \sigma_{nt}\mathbf{e}_t + \sigma_{ns}\mathbf{e}_s \qquad (3.3)$$

The component σ_{nn} is known as the *normal stress* at P on the **n** surface (Fig. 3.2b). Its positive direction is that of **n**. It is then known as a

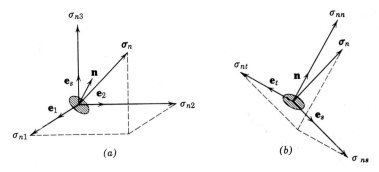

Fig. 3.2 Decomposition of a stress vector.

tension or *a tensile stress*. The components σ_{nt} and σ_{ns} are known as the *tangential stresses* at P on the **n** surface. They are *shearing stresses*.

The definition (3.1) of a stress vector enables us to specify the internal forces at a point P across only a certain plane **n**. For the complete specification of the internal forces at P we should know the stress vector at P across all the planes, infinite in number, that can be drawn through P. It is easily seen, by considering the equilibrium of an infinitesimal tetrahedron such as shown in Fig. 3.3, that *if the stress vector across three independent planes passing through a point is given, the stress vector across any other plane passing through that point is determined.* It thus follows that the state of the internal forces, also known as the *state of stress*, at any point is completely specified by giving a set of three stress vectors. Equivalently we can give the components of these stress vectors. Thus if $\boldsymbol{\sigma}_l$, $\boldsymbol{\sigma}_m$, $\boldsymbol{\sigma}_n$ denote the stress vectors at P across the planes $\mathbf{e}_l$, $\mathbf{e}_m$, $\mathbf{e}_n$, and if

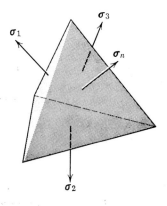

Fig. 3.3 Equilibrium of a tetrahedron.

e_1, e_2, e_3 are a set of unit vectors at P, the stress components can be given in the form

$$\begin{pmatrix} \sigma_{l1} & \sigma_{l2} & \sigma_{l3} \\ \sigma_{m1} & \sigma_{m2} & \sigma_{m3} \\ \sigma_{n1} & \sigma_{n2} & \sigma_{n3} \end{pmatrix} \tag{3.4}$$

For purposes of calculation it is convenient to take the unit vectors e_1, e_2, e_3 as those defined by the (orthogonal) coordinate system used and the

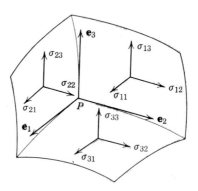

Fig. 3.4 State of stress at a point.

planes e_l, e_m, e_n as the corresponding coordinate planes e_1, e_2, e_3. With this notation the stress components (3.4) become

$$\begin{pmatrix} \sigma_{11} & \sigma_{12} & \sigma_{13} \\ \sigma_{21} & \sigma_{22} & \sigma_{23} \\ \sigma_{31} & \sigma_{32} & \sigma_{33} \end{pmatrix} \tag{3.5}$$

This array of nine numbers is known as a *stress tensor*. It specifies completely the state of stress or simply the stress at a point. The diagonal terms σ_{11}, etc., represent the normal stresses, while the nondiagonal terms σ_{12}, etc., represent tangential or shear stresses (Fig. 3.4).

The equilibrium of the moments on an infinitesimal cube shows that

$$\sigma_{12} = \sigma_{21}$$
$$\sigma_{23} = \sigma_{32} \tag{3.6}$$
$$\sigma_{31} = \sigma_{13}$$

That is, the stress tensor is symmetrical about its diagonal. This means to specify completely the state of stress at a point we actually need six stress components.

For the complete specification of the state of stress in the region of interest of a fluid we must give at each point the stress tensor. The tensor will generally be different at different points, and at any point it may vary from instant to instant. This means that the state of stress within a fluid is specified by a tensor function of position and time, that is, by a tensor field.

3.3 Stress in a Fluid at Rest: Hydrostatic Pressure

The stress tensor for a fluid takes on a particularly simple form when the fluid is at rest. *It is a fact of experience that tangential stresses do not exist in a fluid at rest.* This means the *stress vector at any point of the fluid at rest is wholly normal to any surface element passing through that point* (see Eq. 3.3). Symbolically we write that

$$\boldsymbol{\sigma}_n = \sigma_{nn}\mathbf{n} \tag{3.7}$$

In such a case the stress tensor takes the form

$$\begin{pmatrix} \sigma_{11} & 0 & 0 \\ 0 & \sigma_{22} & 0 \\ 0 & 0 & \sigma_{33} \end{pmatrix} \tag{3.8}$$

For such a state of stress we can draw a fundamental conclusion. Considering the equilibrium of an infinitesimal tetrahedron we can show that *when the stress vector at a point is wholly normal in all directions, its magnitude is the same for all elemental planes passing through the point.*[*] In such a case all that is required to specify the stress at a point is simply a single number. Denoting this number by σ we write

$$\boldsymbol{\sigma}_n = \sigma\mathbf{n} \tag{3.9}$$

for *all directions* of $\mathbf{n}$.

If σ is positive, $\boldsymbol{\sigma}_n$ represents a tensile stress. But from experience we find that *no tensile stresses occur in the interior of a fluid.*[†] This means it would be more appropriate to represent $\boldsymbol{\sigma}_n$ as a compression. This we do by replacing the number σ by a negative number, say $-p$. Then *the stress in a fluid at rest* is represented by

$$\boldsymbol{\sigma}_n = -p\mathbf{n} \tag{3.10}$$

---

[*] This conclusion is known as *Pascal's law*. In this case the stress vector describes a sphere.

[†] Tensile stresses are exhibited at the so-called "free surfaces" and in "thin fluid films."

for *all directions of* **n**. The corresponding stress tensor is given by

$$\begin{pmatrix} -p & 0 & 0 \\ 0 & -p & 0 \\ 0 & 0 & -p \end{pmatrix} \tag{3.11}$$

The scalar p is called the *pressure* or, more precisely, the *hydrostatic** *pressure* at the point considered. Equation (3.10) or (3.11) can be taken as the definition of a fluid. Note that p is a positive number.

The state of stress within the whole region of a fluid at rest is specified by giving the pressure as a function of position and time, that is, by a scalar field, denoted by $p = p(\mathbf{r}, t)$.

The hydrostatic pressure is generally related to the density and temperature of the fluid. For instance, we know that in gases that obey Charles' and Boyle's laws the pressure, the density ρ, and the temperature T are connected by the relation $p = \rho RT$, where R is a constant for the gas considered.

3.4 Stress in a Fluid in Motion

When a fluid is in motion the phenomenon of viscosity or of internal friction manifests itself, and at any point in the fluid both tangential and normal stresses occur. In this case we deal with the complete stress tensor and express it as made up of two parts—one that represents *viscous* or *frictional stresses* only, and the other that represents compressive stresses equal in all directions, that is, a pressure and thus not related to friction. This pressure, also denoted by p, is similar but not identical to the hydrostatic pressure. The viscous stresses, which occur both as tangential and normal components, are usually denoted by τ_{11}, τ_{12}, etc. Thus the state of stress at any point in a moving fluid is given in the form

$$\begin{pmatrix} -p & 0 & 0 \\ 0 & -p & 0 \\ 0 & 0 & -p \end{pmatrix} + \begin{pmatrix} \tau_{11} & \tau_{12} & \tau_{13} \\ \tau_{21} & \tau_{22} & \tau_{23} \\ \tau_{31} & \tau_{32} & \tau_{33} \end{pmatrix} \tag{3.12}$$

The viscous stresses are assumed to be proportional to the *rates of strain*† occurring at the point considered. The proportionality constants, known as viscosity coefficients, depend on the nature of the fluid. For any given fluid the viscous stresses are small when the rates of strain are small, and

---

* "Hydrostatic" signifies a fluid at rest.

† The rates of strain at a point are in turn given by certain combinations of the partial derivatives of the components of the velocity at that point (see Section 9.1).

they disappear when the rates of strain disappear, thus leaving the stress tensor as that of a uniform pressure in all directions. Similarly, if the velocity of the fluid is zero everywhere, the viscous stresses become zero, and we realise again uniform pressure at a point. Because of these interpretations, the splitting of the stress tensor in a moving fluid into the uniform pressure p and the viscous stresses is convenient.

According to these ideas, the state of stress within the whole region of a moving fluid is expressed by the combination of a scalar field and a tensor field.

In developing the theories of fluid motion we assume generally that thermodynamic considerations, which are strictly applicable to equilibrium situations, are valid for moving fluids and use them. In such a case we must talk about a thermodynamic pressure that is related to the density and temperature at a point in the fluid. It is then assumed that the thermodynamic pressure is the same as the pressure p occurring in the stress tensor. The justification for these assumptions is that the theories based on them seem to give results that are in good agreement with experiments.

3.5   State of Stress in a Nonviscous Fluid in Motion

A large part of the theory of fluid motion, as pointed out in Chapter 1, is developed on the assumption that the *fluid is frictionless or nonviscous.* In such a case the coefficients of viscosity, and consequently the viscous stresses, are set to zero and one takes the *state of stress* as that of a *uniform pressure.* The ultimate justification for such an assumption lies again in the comparison of its consequences with experiments.

The study of fluid motion treated in this book, as previously stated, is based on the assumption of a nonviscous fluid.

3.6   Pressure Distribution in a Fluid at Rest

For the static equilibrium of a fluid, the sum of all the forces acting on any part of the fluid should be zero. To apply this law we consider at any point $\mathbf{r}$ an infinitesimal region of the fluid and write that

the resultant of the body forces acting on the region + the
resultant of the surface forces acting on it $= 0$ (3.13)

Let $\delta\tau$ denote the volume of the region considered and δS the surface enclosing it. If $\mathbf{f} = \mathbf{f}(\mathbf{r})$ represents the distribution of the body forces per unit mass of the fluid, the body force acting on the element $\delta\tau$ is equal to

$$\rho\mathbf{f}\,\delta\tau \qquad\qquad (3.14)$$

where $\rho = \rho(\mathbf{r})$ is the density of the fluid. The density may vary from point to point of the fluid.

To determine the resultant of the surface forces let us first consider an infinitesimal area $\mathbf{n}\,dS$ on the surface δS (Fig. 3.5). Since the state of stress in a fluid at rest is given by a pressure $p = p(\mathbf{r})$, the surface force acting on the fluid within δS across the surface element $\mathbf{n}\,dS$ is

$$-p\mathbf{n}\,dS$$

The resultant of the surface forces is then obtained simply by adding vectorially the pressure forces acting on all the elemental areas of the surface δS. It is thus equal to

$$-\oiint_{\delta S} p\mathbf{n}\,dS \tag{3.15}$$

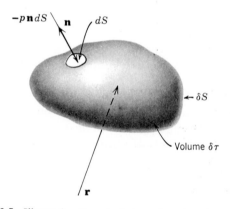

Fig. 3.5 Illustrating the calculation of resultant pressure.

For an infinitesimal volume element $\delta\tau$ we have, according to Eq. (2.116a),

$$-\oiint_{\delta S} p\mathbf{n}\,dS = -\delta\tau\,\text{grad}\,p$$

This shows that the resultant of the pressure forces acting on the surface of an infinitesimal fluid element $\delta\tau$ is given by

$$-\delta\tau\,\text{grad}\,p \tag{3.16}$$

With the expressions (3.14) and (3.16), the condition (3.13) becomes

$$\rho\mathbf{f}\,\delta\tau - \delta\tau\,\text{grad}\,p = 0$$

or, expressing it per unit volume,

$$\rho\mathbf{f} - \text{grad}\,p = 0$$

Equivalently we may write this equation as

$$\text{grad } p = \rho \mathbf{f} \tag{3.17}$$

This equation is thus the analytical form of the condition for static equilibrium of a fluid. Therefore it is the basis for analysing the statics of incompressible and compressible fluids. The content of Eq. (3.17) is significant. It states that *static equilibrium of a fluid is possible only if the body force ($\rho\mathbf{f}$) per unit volume can be expressed as the gradient of a scalar function*. This implies that $\rho\mathbf{f}$ should be irrotational. If the density of the fluid is uniform throughout space, we have

$$\text{grad}\left(\frac{p}{\rho}\right) = \mathbf{f}$$

Equation (3.17) assures us that the resultant of the forces acting on any part of a fluid at rest is zero. For the fluid to be at rest the condition that the resultant of the moments acting on a fluid element should also be zero. That this condition is automatically satisfied may be verified by the reader.

3.7 Concluding Remarks

To pass on from the statics to the dynamics of a fluid, we may consider a certain portion of the fluid and apply to it the second law of Newton. Thus we write that

the rate of change of momentum of the region of fluid considered = the resultant of the body forces acting on the region + the resultant of the pressure forces acting at its surface + the resultant of the viscous forces acting at its surface. (3.18)

For the complete analytical formulation of this law we must first decide about a method for describing the motion of the fluid, whether we wish to talk about the fate of each individual element of the fluid or about the whole fluid as such. This we shall do in the next chapter.

Equation (3.18) involves the density, the pressure, and the velocity as unknowns if we assume that the body forces are given and that the viscous stresses are related to the derivatives of the velocity. It is thus apparent that additional equations have to be formulated. This we do in Chapter 5.

Chapter 4

Description of Fluid Motion

We shall now begin to set up analytically the problem of fluid motion. The fluid, as remarked before, is regarded as matter distributed continuously. At various points of the distribution, we may choose infinitesimally small regions of fluid and regard them as possessing individuality. We may thus refer to them as *fluid elements* or *fluid particles*. To describe fluid motion, two methods are possible. One possibility is to describe the fate of each individual fluid particle, that is, to adopt a *particle point of view*. The other possibility is to forget about the individuality of the various fluid elements and concern ourselves only with the state of motion in the space filled by the fluid, that is, to adopt a so-called *field point of view*. The particle description is known as the *Lagrangian method* (after Lagrange, 1736–1813), while the field description is known as the *Eulerian method* (after Euler, 1707–1783).* In this chapter we shall discuss these two methods and also consider additional concepts related to the description of fluid motion.

4.1 Lagrangian Method

In this method we ask about what happens to the individual fluid elements in the course of time, what are their paths, their velocities, densities, or any other properties associated with them. Thus in this case the fluid motion is described by describing the fate of each individual fluid particle. Such a procedure follows the practice usually adopted in the mechanics of mass particles where one seeks the trajectories of the individual particles.

Since in this method we are concerned with each particle, we must, so to speak, label the various elements (i.e., name them) in order to distinguish between them. We do this by assigning, at some instant of time, three scalar parameters, say a, b, c, to each element. A convenient choice of these parameters, for instance, is to give the position coordinates of an

---

* Historically, equations of fluid motion based on Lagrangian method occur in Euler's papers.

158

element with respect to any chosen coordinate system. Since the fluid elements are continuously distributed, the values that the parameters a, b, c will assume for the various elements are continuous. Now, the motion of the fluid can be described by giving as a function of time, t, any quantity Q that is associated with each element a, b, c. Considering the whole fluid, Q becomes a function of time and the *particle* parameters a, b, c. Thus we write

$$Q = Q(a, b, c, t) \tag{4.1}$$

For instance, the position $\mathbf{r}$ of the various elements at any time t is expressed by the functional form

$$\mathbf{r} = \mathbf{r}(a, b, c, t) \tag{4.2}$$

If x, y, z are the Cartesian components of $\mathbf{r}$, the scalar form of (4.2) is shown by

$$x = x(a, b, c, t)$$
$$y = y(a, b, c, t)$$
$$z = z(a, b, c, t)$$

Thus in Lagrangian description the particle parameters a, b, c, and time t are the independent variables. The unknown variables are the position coordinates (or, equivalently, the velocity or acceleration components) of an element and other quantities such as the density, giving its state. To determine the unknowns we set up a necessary number of equations between them by applying to each fluid particle natural laws, such as Newton's second law of motion, and conservation of mass. We shall not enter into the derivation of these equations, which are usually known as the *Lagrangian equations of fluid motion*. For these equations reference may be made to Lamb (1932), or Sommerfeld (1950).

Although the Lagrangian description appears to be a natural way to set up problems of fluid motion, generally it is not as convenient and meaningful as the Eulerian description, which we shall follow exclusively in our studies here. The Lagrangian method gives more information than one needs, for often one is not interested in the fate of each fluid particle. There are, however, specific instances, such as certain one-dimensional (involving one space coordinate) problems, where the Lagrangian point of view is fruitful.

4.2 Eulerian Method

In this method we focus our attention on the various points of the space filled by the flowing fluid and what is happening at each of these points as time goes on. What is happening is, of course, to be given in terms of quantities such as velocity, density, and pressure, which are of interest in the motion of the fluid. Thus, in the Eulerian point of view, we give the values that a fluid quantity Q assumes at various points of the space at different instances of time. In other words, Q is described as a function of position and time. If $\mathbf{r}$ denotes the position of a space point with respect to a chosen coordinate system and t denotes the time, we write

$$Q = Q(\mathbf{r}, t) \qquad (4.3)$$

Recalling the notion of a *field* we observe that in Eulerian description fluid motion is specified by various scalar and vector fields. Thus we talk about velocity field, acceleration field, density field, and so on. In such a description, the identity of the fluid particles that occupy the various points in space at various times is irrelevant. At any instant each point of the region may be associated with a fluid element for which the velocity, density, etc., may be taken as those occurring at the point at the instant considered.

In the Eulerian description, the independent variables are the position coordinates (i.e., components of $\mathbf{r}$) of a point in space and time t. The dependent variables are the fields of velocity, density, etc. To solve for these fields we set up a necessary system of equations by again using laws such as Newton's second law of motion and conservation of mass. The derivation of these *Eulerian* equations will occupy us in the next chapter.

4.3 Connection between the Lagrangian and Eulerian Descriptions

To relate these two descriptions we need to know the identity of the fluid element that occupies a certain position in space at a given time. In other words, we need the relations between points in space and the particle parameters.

We consider first the transition from the Lagrangian to the Eulerian description. Let some quantity Q be given in terms of the Lagrangian variables a, b, c, t. We then have

$$Q = Q(a, b, c, t) \qquad (4.4)$$

To pass to the Eulerian description we must express a, b, c in terms of the coordinates, say x, y, z, of a point in space. In the Lagrangian method, they are given by relations of the form

$$x = f_1(a, b, c, t)$$
$$y = f_2(a, b, c, t) \qquad (4.5)$$
$$z = f_3(a, b, c, t) .$$

These may be solved to obtain a, b, c in terms of x, y, z, and t. We shall then have

$$a = g_1(x, y, z, t)$$
$$b = g_2(x, y, z, t) \qquad (4.6)$$
$$c = g_3(x, y, z, t)$$

With (4.6), Eq. (4.4) may be expressed in terms of the Eulerian variables x, y, z, t. We write

$$Q = Q[a = g_1(x, y, z, t), b = g_2(x, y, z, t), c = g_3(x, y, z, t), t] \quad (4.7)$$

The transition from the Eulerian to the Lagrangian description is not so simple. In terms of Eulerian variables we have

$$Q = Q(x, y, z, t) \quad (4.8)$$

To pass to the Lagrangian description we must express x, y, z in terms of the particle parameters a, b, c. This we do as follows. Let the velocity components at the point x, y, z at time t be given (in terms of Eulerian description) by

$$u = F_1(x, y, z, t)$$
$$v = F_2(x, y, z, t) \quad (4.9)$$
$$w = F_3(x, y, z, t)$$

In terms of the Lagrangian description we shall have

$$u = \frac{\partial x}{\partial t}, \qquad v = \frac{\partial y}{\partial t}, \qquad w = \frac{\partial z}{\partial t}, \quad (4.10)$$

where x, y, z are functions of the variables a, b, c, t. Equations (4.9) and (4.10) both describe the velocity components of a fluid element. Combining them we obtain

$$\frac{\partial x}{\partial t} = F_1(x, y, z, t)$$

$$\frac{\partial y}{\partial t} = F_2(x, y, z, t) \quad (4.11)$$

$$\frac{\partial z}{\partial t} = F_3(x, y, z, t)$$

These are (first order) differential equations for the position coordinates of an element as described by the Lagrangian method. Integration of these equations leads to solutions of the form

$$x = f_1(x_0, y_0, z_0, t)$$
$$y = f_2(x_0, y_0, z_0, t) \quad (4.12)$$
$$z = f_3(x_0, y_0, z_0, t)$$

where x_0, y_0, z_0, the constants of integration, are chosen as the initial values of x, y, z at an initial instant $t = t_0$. We may set the particle parameters a, b, c equal to x_0, y_0, z_0 respectively and rewrite Eq. (4.12) as

$$x = f_1(a, b, c, t)$$
$$y = f_2(a, b, c, t) \quad (4.13)$$
$$z = f_3(a, b, c, t)$$

Equation (4.13) enables us to express (4.8) in terms of the Lagrangian variables. Thus we obtain

$$Q = Q[x = f_1(a, b, c, t), y = f_2(a, b, c, t), z = f_3(a, b, c, t), t] \quad (4.14)$$

4.4 Steady and Unsteady Motions

If at various points of the flow field all quantities (such as velocity, density, pressure) associated with the fluid flow remain unchanged with time, the motion is said to be *steady*; otherwise it is called *unsteady*. Thus in steady motion time drops out of the independent variables, and the various field quantities simply become functions of the space coordinates.

4.5 Path Line

The curve described in space by a moving fluid element is known as its *trajectory* or *path line*. Such a line is obtained by giving the position of an element as a function of time.

4.6 Streamlines

We can describe another set of curves in the space filled by a flowing fluid.* At a certain instant of time, mark out the direction of velocity at each point of space. Then draw a family of curves such that each curve is tangent at each point to the velocity direction at that point. Such curves, as mentioned before, are called *streamlines*. They are nothing but the *field lines* of the vector field of velocity (see 2.20).

To express analytically the equations for the streamlines we proceed as follows. Considering a certain instant of time, at any point **r**, let **ds** be an element of the streamline passing through the point and let **V** denote the velocity vector ·at that point at that instant (Fig. 4.1). Then, in view of the definition of a streamline, we state that the direction of **ds** is the same as that of **V**, that is,

$$\textbf{ds is parallel to V} \quad (4.15)$$

Since the cross product of two parallel vectors is zero, we express (4.15) by writing

$$\textbf{ds} \times \textbf{V} = 0 \quad (4.16)$$

which then is the differential equation for a streamline. Equation (4.16) can be readily specialized to any coordinate description. Choosing Cartesians, for instance, let the components of **ds** be denoted by dx, dy, dz, and those of **V** by u, v, w. Then (4.16) becomes

$$(\textbf{i} \, dx + \textbf{j} \, dy + \textbf{k} \, dz) \times (\textbf{i}u + \textbf{j}v + \textbf{k}w) = 0 \quad (4.17a)$$

* For detailed considerations regarding the relation between path lines, streamlines, and the so-called streak lines consult Prandtl and Tietjens (1934).

Carrying out the cross product, we obtain

$$
\begin{vmatrix}
\mathbf{i} & \mathbf{j} & \mathbf{k} \\
dx & dy & dz \\
u & v & w
\end{vmatrix}
= \mathbf{i}(w\,dy - v\,dz) + \mathbf{j}(u\,dz - w\,dx) + \mathbf{k}(v\,dx - u\,dy)
$$
$$
= 0
$$

(4.17b)

This vector equation is equivalent to three scalar equations

$$w\,dy - v\,dz = 0 \qquad (4.17c)$$

$$u\,dz - w\,dx = 0 \qquad (4.17d)$$

$$v\,dx - u\,dy = 0 \qquad (4.17e)$$

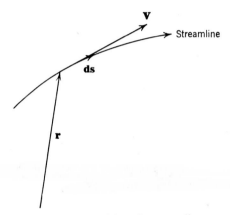

Fig. 4.1   Definition of a streamline.

which can further be put in the symmetric form

$$\frac{dx}{u} = \frac{dy}{v} = \frac{dz}{w} \qquad (4.18)$$

Note that u, v, w are functions of x, y, z, and t. Equation (4.17) or (4.18) is the Cartesian form of the differential equations for a streamline. We consider the integration of such equations in Section 4.9.

Thus at each instant of time we can construct a picture of the stream-lines. If the motion is unsteady, the streamline picture will change from instant to instant. If the motion is steady, the picture remains the same for all times. *In this case path lines and streamlines are identical.* A picture of the streamlines helps us to see, as it were, the flow field and therefore plays an important part in the analysis and understanding of fluid flow problems.

Experimentally, various methods are available by which we can make fluid flow visible and obtain photographs of the streamlines. We cannot enter into a discussion of these techniques here.* Examples of experimentally obtained streamline pictures have already been given. In our studies here we shall compare such pictures with those obtained analytically. This would help us to evaluate our theoretical ideas in the light of experimental facts.

4.7 Stream Surfaces and Stream Tubes

If at any instant of time we draw an arbitrary line in the fluid region and draw the streamlines passing through the line, a surface is formed. Such a surface is called a *streamline surface* or simply a *stream surface* (Fig. 4.2).

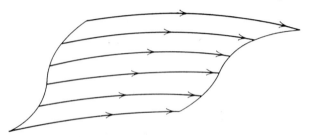

Fig. 4.2   Stream surface.

If we consider a closed curve and draw all the streamlines passing through it, a tube called a *stream tube* is formed (Fig. 4.3). In unsteady motion, the shape of a stream surface drawn through an arbitrarily chosen curve and the shape of a stream tube drawn through a chosen closed curve will change with time. In steady motion stream surfaces and stream tubes once drawn remain unchanged.

No fluid can cross a stream surface or the walls of a stream tube, for the walls and the stream surface are always parallel to the fluid velocity

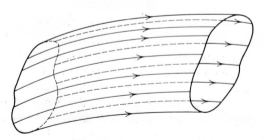

Fig. 4.3   Stream tube.

* Consult, for instance Pankhurst and Holder (1952) or Prandtl and Tietjens (1934).

everywhere. *For steady flow*, a stream tube behaves like one with rigid boundaries inside which fluid flows. It follows that in steady flow, an impermeable solid-fluid boundary, across which there cannot be any flow of fluid, would be a stream surface.

4.8   Reference Frame and Streamline Pattern

It is to be noted that a streamline pattern would appear differently from different reference frames. This is illustrated in Plates 12 and 13 for flow around an airfoil that is moving through an otherwise undisturbed fluid. The streamline pattern shown in (12) is obtained in a reference frame fixed with respect to the airfoil, while the pattern shown in (13) is obtained in a reference frame fixed with respect to the undisturbed fluid far away from the airfoil.

4.9   Stream Functions

We now consider the integration of the Eq. (4.16) for a streamline

$$\mathbf{ds} \times \mathbf{V} = 0$$

Although $\mathbf{V} = \mathbf{V}(\mathbf{r}, t)$, the integration of (4.16) involves only the space variables. As such, in what follows, we shall not exhibit explicitly the dependence on time. Bear in mind, however, that the results obtained hold at every instant of time.

First consider the Cartesian form of (4.16) as given by (4.17).

$$
\begin{aligned}
v(x, y, z)\, dx - u(x, y, z)\, dy &= 0 \\
w(x, y, z)\, dy - v(x, y, z)\, dz &= 0 \qquad (4.19) \\
w(x, y, z)\, dx - u(x, y, z)\, dz &= 0
\end{aligned}
$$

These are a set of differential equations for the variables x, y, z. As may be readily verified, (4.19) is actually a set of two independent equations. We may, therefore, represent (4.19) by two equations of the form

$$
\begin{aligned}
a_1(x, y, z)\, dx + b_1(x, y, z)\, dy + c_1(x, y, z)\, dz &= 0 \\
a_2(x, y, z)\, dx + b_2(x, y, z)\, dy + c_2(x, y, z)\, dz &= 0
\end{aligned} \qquad (4.20)
$$

Equations of this type are known as *Pfaffian differential equations*. For the theory of such equations, reference may be made to any suitable book, for instance that by Margenau and Murphy (1956) or by Sneddon (1957).

The solution of each equation in (4.20) may be represented by an equation of the form

$$f(x, y, z) = c \qquad (4.21)$$

where c is a constant. As is well known, Eq. (4.21) describes a one-parameter family of surfaces. We thus conclude that *the solution* of

(4.20) or equivalently *of the differential equation for the streamline is given by two independent functions*

$$\psi_1(x, y, z) = c_1$$
$$\psi_2(x, y, z) = c_2 \tag{4.22}$$

where c_1 and c_2 are constants. Such a conclusion could have been reached immediately from the observation that a line in space, such as a streamline, may be described as the curve of intersection of two surfaces.

The functions ψ_1 and ψ_2 should naturally be related to the velocity

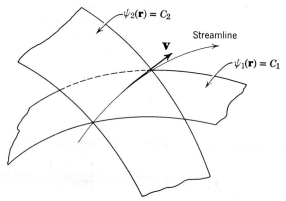

Fig. 4.4 Stream functions and streamline.

components u, v, w. We now establish the relation. For generality and convenience we now switch to vector representation. We write

$$\mathbf{V} = \mathbf{V(r)}$$
$$\psi_1 = \psi_1(\mathbf{r}) = c_1 \tag{4.23}$$
$$\psi_2 = \psi_2(\mathbf{r}) = c_2$$

Now, along the streamline, the velocity vector $\mathbf{V}$ lies in both the surfaces $\psi_1 = c_1$ and $\psi_2 = c_2$ (Fig. 4.4). This being the case, along the streamline $\mathbf{V}$ is normal to both grad ψ_1 and grad ψ_2, the gradients being parallel to the normals to their respective surfaces. Symbolically we have

$$\mathbf{V} \cdot \text{grad } \psi_1 = 0$$
$$\mathbf{V} \cdot \text{grad } \psi_2 = 0$$

This shows that $\mathbf{V}$ is normal to the plane formed by the vectors grad ψ_1 and grad ψ_2. In other words $\mathbf{V}$ is parallel to the cross product grad ψ_1 × grad ψ_2. Hence we write

$$\mu(\mathbf{r})\mathbf{V} = \text{grad } \psi_1 \times \text{grad } \psi_2 \tag{4.24}$$

where $\mu(\mathbf{r})$ is a scalar function of position.

The function $\mu(\mathbf{r})$ is arbitrary except that it should satisfy a certain condition that follows directly from Eq. (4.24). As may be verified it is a vector identity that if $f_1(\mathbf{r})$ and $f_2(\mathbf{r})$ are two scalar functions of position, then

$$\text{div} (\text{grad} f_1 \times \text{grad} f_2) = 0$$

It, therefore, follows that $\mu(\mathbf{r})$ should be such that it satisfies the condition

$$\text{div} (\mu \mathbf{V}) = 0 \qquad (4.25)$$

To sum up, we state that the solution of the equation

$$\mathbf{ds} \times \mathbf{V} = 0$$

for a streamline is given by two functions

$$\psi_1(\mathbf{r}) = c_1$$
$$\psi_2(\mathbf{r}) = c_2$$

where c_1 and c_2 are constants. These functions and the velocity are related by

$$\mu \mathbf{V} = \text{grad} \ \psi_1 \times \text{grad} \ \psi_2$$

where μ is a scalar function of position that satisfies the condition

$$\text{div} \ \mu \mathbf{V} = 0$$

The functions ψ_1 and ψ_2 are known as *stream functions.** If the function $\mu(\mathbf{r})$ is known, then the vector field $\mathbf{V}(\mathbf{r})$ may be replaced by two scalar fields $\psi_1(\mathbf{r})$ and $\psi_2(\mathbf{r})$ gaining possible mathematical advantage. In general, a function μ satisfying the condition (4.25) without at the same time imposing physically untenable conditions on the velocity field, is not readily obtained. Under certain circumstances, however, independent considerations relating to the fluid motion lead to relations of the form

$$\text{div} f(\mathbf{r})\mathbf{V} = 0$$

where f is a certain scalar function of position. In such cases we identify $\mu(\mathbf{r})$ with $f(\mathbf{r})$ and replace $\mathbf{V}(\mathbf{r})$ by $\psi_1(\mathbf{r})$ and $\psi_2(\mathbf{r})$ by use of Eq. (4.24).

Examples of such situations are the motion of an incompressible fluid and the steady motion of a compressible fluid. As we shall see later, for an incompressible fluid we shall have

$$\text{div} \ \mathbf{V} = 0$$

then we can set

$$\mu(\mathbf{r}) = 1$$

* Note that we have suppressed showing explicitly the dependence on time.

For steady compressible flow, we shall have

$$\text{div } \rho(\mathbf{r})\mathbf{V} = 0$$

where $\rho(\mathbf{r})$ is the density field. Then we can set

$$\mu(\mathbf{r}) = \rho(\mathbf{r})$$

Solving fluid flow problems by use of two stream functions when such use is possible has not received much attention, and at this stage no general comments can be made about it. In certain problems where only two velocity components appear, the other being zero, there would be only one single unknown stream function instead of two. We now consider such situations.

4.10 Stream Function for Two-Dimensional Flow

When the motion of a fluid is such that the flow pattern and the various flow quantities are independent of distance along a certain fixed direction, the motion is said to be *two-dimensional* or *planar*. Thus, if we designate such a direction as the Z axis, we shall have

$$\frac{\partial}{\partial z}(\quad) \equiv \frac{\partial}{\partial z} \text{ of any quantity} = 0$$

The motion in all planes normal to Z will appear the same.

Introducing Cartesians x, y, z, we write

$$\mathbf{ds} = (dx, dy, dz)$$
$$\mathbf{V} = (u, v, 0)$$
$$u = u(x, y)$$
$$v = v(x, y)$$

We consider the equation for a streamline in the form

$$\frac{dx}{u(x, y)} = \frac{dy}{v(x, y)} = \frac{dz}{0}$$

It immediately follows that

$$dz = 0$$

or

$$z = \text{constant}$$

Thus, one of the stream functions is simply z. The other is the only unknown function. It is a function of x and y only. Denote it by $\psi(x, y)$.

To relate $\psi(x, y)$ to the velocity components u, v we use the Eq. (4.24). Set*

$$\psi_1 = \psi(x, y)$$
$$\psi_2 = z$$

---

* We may as well set $\psi_1 = z$ and $\psi_2 = \psi$. The choice made leads to the usual form of the results.

We then have

$$\mu(x, y)\mathbf{V} = \begin{vmatrix} \mathbf{i} & \mathbf{j} & \mathbf{k} \\ \dfrac{\partial \psi}{\partial x} & \dfrac{\partial \psi}{\partial y} & 0 \\ 0 & 0 & 1 \end{vmatrix}$$

Equivalently, we obtain

$$\mu u = \frac{\partial \psi}{\partial y}, \qquad \mu v = -\frac{\partial \psi}{\partial x} \tag{4.26}$$

The function μ should be such that the following relation is satisfied:

$$\frac{\partial \mu u}{\partial x} + \frac{\partial \mu v}{\partial y} = \operatorname{div} \mu \mathbf{V} = 0 \tag{4.27}$$

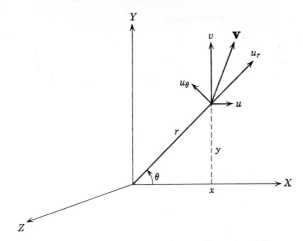

Fig. 4.5 Velocity components in two-dimensional flow.

We now express the results in terms of cylindrical polar coordinates r, θ, z (see Fig. 4.5). We write

$$\mathbf{ds} = (dr, r\,d\theta, dz)$$
$$\mathbf{V} = (u_r, u_\theta, 0)$$
$$u_r = u_r(r, \theta)$$
$$u_\theta = u_\theta(r, \theta)$$

We then have

$$\frac{dr}{u_r(r, \theta)} = \frac{r\,d\theta}{u_\theta(r, \theta)} = \frac{dz}{0}$$

As before, one of the stream functions is z. For the other, say $\psi = \psi(r, \theta)$ we have, using Eq. (4.24),

$$\mu(r, \theta)\mathbf{V} = \begin{vmatrix} \mathbf{e}_r & \mathbf{e}_\theta & \mathbf{e}_z \\ \dfrac{\partial \psi}{\partial r} & \dfrac{1}{r}\dfrac{\partial \psi}{\partial \theta} & 0 \\ 0 & 0 & 1 \end{vmatrix}$$

or, equivalently,

$$\mu u_r = \frac{1}{r}\frac{\partial \psi}{\partial \theta}, \qquad \mu u_\theta = -\frac{\partial \psi}{\partial r} \tag{4.28}$$

The function $\mu(r, \theta)$ should satisfy the relation

$$\frac{1}{r}\left[\frac{\partial}{\partial r}(r\mu u_r) + \frac{\partial}{\partial \theta}(\mu u_\theta)\right] = \operatorname{div}\mu\mathbf{V} = 0 \tag{4.29}$$

4.11   Stream Function for Axisymmetric Motion

When the motion of a fluid is such that the flow pattern and the flow quantities at corresponding points are the same in all planes passing

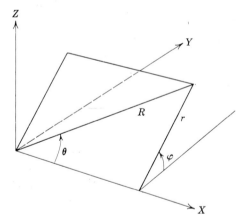

Fig. 4.6   Coordinates for axisymmetric motion.

through a certain fixed axis, the motion is said to be axisymmetric. Introduce spherical coordinates R, θ, φ or cylindrical coordinates r, φ, x as shown in Fig. 4.6. Let the motion be axisymmetric about the X axis. We then have

$$\frac{\partial}{\partial \varphi}(\ ) = \frac{\partial}{\partial \varphi} \text{ of any quantity} = 0$$

In terms of the spherical coordinates R, θ, φ, we write

$$\mathbf{ds} = (dR, \, R\, d\theta, \, R \sin \theta \, d\varphi)$$
$$\mathbf{V} = (u_R, u_\theta, 0)$$
$$u_R = u_R(R, \theta)$$
$$u_\theta = u_\theta(R, \theta)$$

Consider the equation for a streamline in the form

$$\frac{dR}{u_R(R, \theta)} = \frac{R\, d\theta}{u_\theta(R, \theta)} = \frac{R \sin \theta \, d\varphi}{0}$$

It immediately follows that
$$d\varphi = 0$$
or
$$\varphi = \text{constant}$$

Thus, one of the stream functions is simply φ. The other, say $\psi = \psi(R, \theta)$, is the only unknown function. To relate $\psi(R, \theta)$ to the velocity components u_R, u_θ, we set

$$\psi_1 = \psi(R, \theta)$$
$$\psi_2 = \varphi$$

and use Eq. (4.24). We thus have

$$\mu(R, \theta)\mathbf{V} = \begin{vmatrix} \mathbf{e}_R & \mathbf{e}_\theta & \mathbf{e}_\varphi \\[2mm] \dfrac{\partial \psi}{\partial R} & \dfrac{1}{R}\dfrac{\partial \psi}{\partial \theta} & 0 \\[3mm] 0 & 0 & \dfrac{1}{R \sin \theta} \end{vmatrix}$$

Equivalently, we have

$$\mu u_R = \frac{1}{R^2 \sin \theta}\frac{\partial \psi}{\partial \theta}$$

$$\mu u_\theta = -\frac{1}{R \sin \theta}\cdot\frac{\partial \psi}{\partial R} \qquad\qquad (4.30)$$

The function $\mu(R, \theta)$ should satisfy the relation

$$\frac{1}{R^2 \sin \theta}\left[\frac{\partial}{\partial R}(R^2 \sin \theta \mu u_R) + \frac{\partial}{\partial \theta}(R \sin \theta \mu u_\theta)\right] = \text{div } \mu\mathbf{V}$$
$$= 0 \qquad\qquad (4.31)$$

We now express the results in terms of cylindrical coordinates r, φ, x (see Fig. 4.6). We have

$$\mathbf{ds} = (dx, dr, r\,d\varphi)$$

$$\mathbf{V} = (u_x, u_r, 0)$$

$$u_x = u_x(x, r)$$

$$u_r = u_r(x, r)$$

$$\frac{dx}{u_x(x, r)} = \frac{dr}{u_r(x, r)} = \frac{r\,d\varphi}{0}$$

$$\psi_1 = \psi(x, r)$$

$$\psi_2 = \varphi$$

$$\mu(x, r)\mathbf{V} = \begin{vmatrix} \mathbf{e}_x & \mathbf{e}_r & \mathbf{e}_\varphi \\[2mm] \dfrac{\partial \psi}{\partial x} & \dfrac{\partial \psi}{\partial r} & 0 \\[4mm] 0 & 0 & \dfrac{1}{r} \end{vmatrix}$$

$$\mu u_x = \frac{1}{r}\frac{\partial \psi}{\partial r}$$

$$\mu u_r = -\frac{1}{r}\frac{\partial \psi}{\partial x} \tag{4.32}$$

$$\frac{1}{r}\left[\frac{\partial}{\partial x}(r\mu u_x) + \frac{\partial}{\partial r}(r\mu u_r)\right] = \operatorname{div}\mu\mathbf{V} = 0 \tag{4.33}$$

The stream function for axisymmetric flow is known as the *Stokes stream function*.

4.12 Stagnation Points

Points in the flow field where the velocity becomes zero are known as *stagnation points*. In terms of the components of the velocity, this means that at a stagnation point all the velocity components are zero.

Consider the differential equations for a streamline in the Cartesian form

$$\frac{dy}{dx} = \frac{v(x, y, z)}{u(x, y, z)}$$

$$\frac{dz}{dx} = \frac{w(x, y, z)}{u(x, y, z)} \tag{4.34}$$

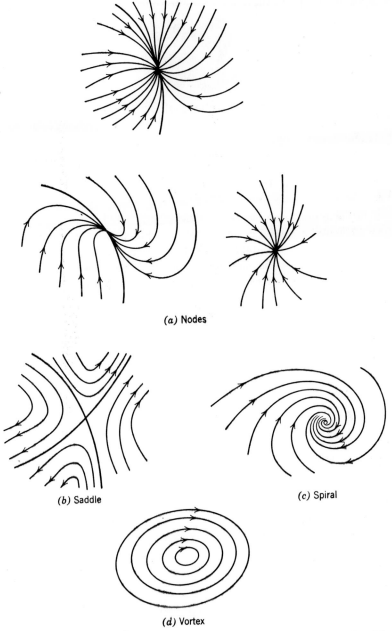

(a) Nodes

(b) Saddle   (c) Spiral

(d) Vortex

Fig. 4.7   Examples of streamline patterns at a stagnation point in two-dimensional flow.

At a stagnation point these equations become

$$\frac{dy}{dx} = \frac{0}{0}$$

$$\frac{dz}{dx} = \frac{0}{0}$$

(4.35)

Points at which differential equations of the type (4.34) take the form (4.35) are called *singular points* of the equations. Thus *stagnation points are such singular points.*

We shall not enter here into the theory of integration of Eqs. (4.34) at the singular points. For such theory, reference may be made to books on differential equations, for instance see Goursat (1959).

Without explicit analysis we state that the streamline pattern at a stagnation point is not the simple picture of a single streamline passing through a given point. Examples of possible streamline patterns at a stagnation point in two-dimensional flow are given in Fig. 4.7. In this context, reference may be made to Karman and Biot (1940) where analysis is given for determining the shape of the integral curves for the equation

$$\frac{dy}{dx} = \frac{f(x, y)}{g(x, y)}$$

at its singular points.

Chapter 5

Eulerian Equations
for the Motion of an Ideal Fluid

The Eulerian method of description defines, for any particular value of time, the state of motion at all points of the space occupied by the fluid, while for a given position the method gives the history of what goes on at that place. Thus in this point of view the fluid flow is characterized by the fields of velocity, pressure, density, and so on, and a fluid element or particle occupying a certain point at a certain time assumes for its properties the values that are appropriate to that point at that instant. The object of our investigations in fluid flow problems, then, is to determine these fields. We attempt this by first formulating possible relationships that should be satisfied between the field quantities. These relationships are established on the basis of certain natural laws such as Newton's second law of motion.

We refer to the relationships between the field quantities as the *equations of fluid motion*. The equations may be set up either in *differential form* or in *integral form*. Furthermore, they may be developed either from the point of view of a certain "fluid region" that contains the same fluid elements (or element) for all times or from the point of view of a "fixed volume in space" through which different fluid elements flow through.

We shall concern ourselves first with the derivation of the equations in the differential form from the point of view of an infinitesimal fluid region. We shall then naturally be involved with the calculation of the time rate of change of any quantity following a fluid element. We shall discuss this in the next section and then formulate the basic equations.

5.1 Local, Convective, and Material Derivatives

Consider the fluid element situated at the point $\mathbf{r}$ at time t. Let $Q(\mathbf{r}, t)$ denote some fluid property Q (density, velocity, etc.) associated with the point $\mathbf{r}$ at that instant. The fluid element situated there will, therefore, assume for its corresponding property Q the value $Q(\mathbf{r}, t)$. In a short time interval Δt, the element moves through a directed distance $\mathbf{\Delta s} = \mathbf{V}\,\Delta t$, where $\mathbf{V}$ is its velocity at $\mathbf{r}$ and t (Fig. 5.1). In the new situation the element will assume for Q the value appropriate to the position $(\mathbf{r} + \mathbf{V}\,\Delta t)$ at the time $(t + \Delta t)$. This we denote as $Q(\mathbf{r} + \mathbf{V}\,\Delta t, t + \Delta t)$. Then the change in Q for that moving element in the time interval Δt is given by

$$\Delta Q = Q(\mathbf{r} + V\,\Delta t, t + \Delta t) - Q(\mathbf{r}, t)$$

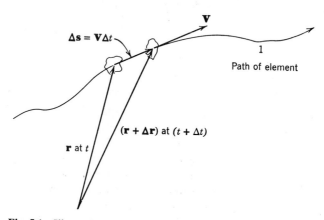

Fig. 5.1 Illustrating the local, convective and material derivatives.

and the rate of change of Q following the element is simply the limit of $\Delta Q/\Delta t$ as Δt vanishes. This rate of change is usually denoted by DQ/Dt. Therefore we have

$$\frac{DQ}{Dt} \equiv \lim_{\Delta t \to 0} \frac{\Delta Q}{\Delta t} = \lim_{\Delta t \to 0} \frac{Q(\mathbf{r} + V\,\Delta t, t + \Delta t) - Q(\mathbf{r}, t)}{\Delta t} \tag{5.1}$$

This rate of change can be expressed as made up of two parts, one a change due to local variation with time of the fluid property at a given position and the other due to a change of position at a given time. Formally we write

$$Q(\mathbf{r} + V\,\Delta t, t + \Delta t) = Q(\mathbf{r}, t) + \left(\frac{\partial Q}{\partial t}\right)_{\mathbf{r},t} \Delta t + \left(\frac{\partial^2 Q}{\partial t^2}\right)_{\mathbf{r},t} (\Delta t)^2 + \cdots$$

$$+ \left(\frac{\partial Q}{\partial s}\right)_{\mathbf{r},t} V\,\Delta t + \left(\frac{\partial^2 Q}{\partial s^2}\right)_{\mathbf{r},t} (V\,\Delta t)^2 + \cdots \tag{5.2}$$

where s *denotes distance in the direction of the velocity* $\mathbf{V}$ at the point $\mathbf{r}$ at time t. Using the Taylor expansion (5.2) in relation (5.1) we obtain

$$\frac{DQ}{Dt} = \frac{\partial Q}{\partial t} + \frac{\partial Q}{\partial s} V \tag{5.3}$$

where the derivatives are evaluated at **r** and t. The derivative $\partial Q/\partial t$ denotes local variation with time at a given position, while the derivative $\partial Q/\partial s$ denotes variation with change of position at a given time. The term $(\partial Q/\partial s)V$ itself denotes the time rate of change of Q due to change of position. A physical interpretation of (5.3) is as follows.

Consider a flow field which at any given instant is uniform throughout the space but varies from instant to instant. If a fluid element moves in such a field from the point **r**, the change in any property Q for that element in a small time interval δt is $(\partial Q/\partial t)\,\delta t$ (correct to the first order in δt), where $\partial Q/\partial t$ denotes the rate at which Q is changing locally at the point **r**. This change is called the *local change.*

Suppose the flow field is steady but not uniform, that is, the fluid property Q though varying from point to point does not change with time at any point. Now, if in such a flow field an element moves from a point **r** to a new point **r** + **V** δt, a change in Q must take place for that element so as to adjust the element to the new location. This change, which results from the fact that the element is getting into a new environment of the flow field, is called the *convective* change. This change is equal to $(\partial Q/\partial s)V\,\delta t$ (to the first order in $V\,\delta t$), the change in Q over the directed distance Δs equal to **V** δt.

When a flow field is neither steady nor uniform, the change in any property Q for a particular fluid element will be made up of both the local and convective changes and is equal to $[(\partial Q/\partial t)\,\delta t + (\partial Q/\partial s)V\,\delta t]$. Hence it follows that the rate of change of Q following a fluid element is given by Eq. (5.3).

The local rate of change $\partial Q/\partial t$ is known as the *local derivative*, and the convective rate of change $(\partial Q/\partial s)V$ is known as the *convective derivative.* The total rate of change DQ/Dt is usually known as the *substantial derivative.* It is sometimes referred to as the *particle* or *material derivative.* This is a more descriptive name, for the derivative is constructed following a certain fluid element.

Equation (5.3) may be used to compute the material derivative of any quantity whether it be a scalar field, a vector field, or a more general tensor field. For scalar and vector fields, the convective derivative in (5.3) may be put, as follows, into a more explicit form. We recall that the operator **e** · grad (or **e** · ∇) applied to a scalar or vector field yields the derivative of that field with respect to distance in the direction **e** (see 2.75 and 2.77). Now, in (5.3) the term $\partial Q/\partial s$ represents the derivative of Q with respect to distance in the direction of the velocity. Denoting this direction by $\mathbf{e}_V$ we may therefore write

$$\frac{\partial Q}{\partial s} = \mathbf{e}_V \cdot \text{grad } Q$$

and

$$\frac{\partial Q}{\partial s} V = \mathbf{V} \cdot \text{grad } Q$$

whether Q is a scalar or a vector. Then for the material derivative we have

$$\frac{DQ}{Dt} = \frac{\partial Q}{\partial t} + \mathbf{V} \cdot \text{grad } Q \tag{5.4}$$

If the quantity Q is a vector field, denoted, say, by $\mathbf{A}$, the convective derivative of $\mathbf{A}$ may be further expanded according to the following formula:

$$\mathbf{V} \cdot \text{grad } \mathbf{A} = \tfrac{1}{2}[\text{grad } (\mathbf{V} \cdot \mathbf{A}) - \mathbf{V} \times \text{curl } \mathbf{A}$$
$$- \mathbf{A} \times \text{curl } \mathbf{V} - \text{curl } (\mathbf{V} \times \mathbf{A})$$
$$+ \mathbf{V}(\text{div } \mathbf{A}) - \mathbf{A}(\text{div } \mathbf{V})] \tag{5.5}$$

We will now set up the basic equations that govern the motion of an inviscid incompressible fluid. At the start, however, we will not regard the fluid as incompressible. The condition of incompressibility will be introduced at the appropriate time so as to bring out clearly the role of incompressibility in the formulation of fluid problems. Initially, we regard as our unknowns the velocity field $\mathbf{V}(\mathbf{r}, t)$, the pressure field $p(\mathbf{r}, t)$, and the density field $\rho(\mathbf{r}, t)$. We seek to establish relationships between these fields by applying to a certain fluid element the basic laws of nature: *Newton's second law of motion, law of conservation of mass, law of conservation of energy*. It might turn out that these laws are actually inadequate in setting up sufficient relations between the unknowns. In such a case one must invoke additional relations suggested by experience.

5.2 Euler's Equation

The fundamental equation governing the motion of any mechanical system expresses Newton's second law of motion, which states that *at any instant, the rate of change of momentum of a system is equal to the force acting on it at that instant*. By force is understood the resultant of all the forces that are acting. To obtain the equation governing the motion of a fluid we apply Newton's second law to a moving fluid element.

Let us consider an infinitesimally small element of fluid situated at the position $\mathbf{r}$ at time t (Fig. 5.2). If $\mathbf{V}$ and ρ are the velocity and density, respectively, at $\mathbf{r}$ and t, and if $\delta\tau$ denotes the volume of the element, the mass and momentum of the element are $\rho \, \delta\tau$ and $\rho \, \delta\tau\mathbf{V}$, respectively. We denote by $\mathbf{F}$ the force acting on the element at that instant and equate it, according to Newton's law, to the rate of change of momentum of the element, which is simply the material derivative of the momentum. Thus we have

$$\frac{D}{Dt}(\rho\,\delta\tau\mathbf{V}) = \mathbf{F} \tag{5.6}$$

as the equation of motion of the fluid.

As the element moves, its shape, volume, and density will, in general, change. *Its mass*, however, *remains constant*. Thus (5.6) takes the form

$$\rho\,\delta\tau\,\frac{D\mathbf{V}}{Dt} = \mathbf{F} \tag{5.7}$$

where $D\mathbf{V}/Dt$ is the acceleration of the fluid element situated at $\mathbf{r}$ at time

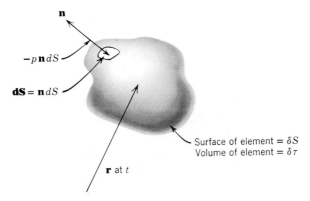

$\mathbf{n}$

$-p\,\mathbf{n}\,dS$

$\mathbf{dS} = \mathbf{n}\,dS$

Surface of element $= \delta S$
Volume of element $= \delta\tau$

$\mathbf{r}$ at t

Fig. 5.2   Fluid element.

t. Equation (5.7) expresses the usual acceleration form of Newton's law: mass times acceleration is equal to force.

The acceleration $D\mathbf{V}/Dt$ may be determined by use of Eqs. (5.1) or (5.4). According to (5.4) we have

$$\frac{D\mathbf{V}}{Dt} = \frac{\partial\mathbf{V}}{\partial t} + \mathbf{V}\cdot\operatorname{grad}\mathbf{V} \tag{5.8}$$

Using (5.5) we obtain

$$\mathbf{V}\cdot\operatorname{grad}\mathbf{V} = \operatorname{grad}\frac{V^2}{2} - \mathbf{V}\times\operatorname{curl}\mathbf{V}$$

It then follows that the acceleration may also be expressed in the form

$$\frac{D\mathbf{V}}{Dt} = \frac{\partial\mathbf{V}}{\partial t} + \operatorname{grad}\frac{V^2}{2} - \mathbf{V}\times\operatorname{curl}\mathbf{V} \tag{5.9}$$

This form, which may be easily specialized to any orthogonal, curvilinear coordinate system, is very useful when we investigate the integration of the equations of fluid motion (see Section 8.4). In Eq. (5.9) the vorticity, namely curl $\mathbf{V}$, which is twice the angular velocity of the fluid, appears explicitly.

In determining the total force acting on the fluid element we distinguish between the so-called body forces and surface forces (Sections 3.2 and 3.7). We denote by $\mathbf{f}$ the body force per unit mass of the fluid. *The body force acting on the fluid element* is, therefore, $\rho\, \delta\tau \mathbf{f}$.

To obtain the resultant of the surface forces acting on the element we note that, since *we are concerned only with a nonviscous fluid*, the surface forces are simply pressure forces that act normal to the surface of the element. Let us denote by p the pressure, by δS the total surface area of the fluid element, and by $\mathbf{n}\, dS$ an elemental area on δS, $\mathbf{n}$ being an outward normal (Fig. 5.2). The *resultant of the pressure forces acting on the element is then equal* to

$$-\oiint_{\delta S} p\mathbf{n}\, dS$$

According to the integral definition of the gradient of a scalar function (Eq. 2.116a) we have

$$-\oiint_{\delta S} p\mathbf{n}\, dS = -\delta\tau\, \mathrm{grad}\, p$$

This shows that *in an inviscid fluid,* $-\,\mathrm{grad}\,p$ *represents the resultant surface force acting on unit volume of the fluid.*

The total force on the fluid element is thus given by

$$\mathbf{F} = \rho\, \delta\tau \mathbf{f} - \delta\tau\, \mathrm{grad}\, p \qquad (5.10)$$

The equation of motion of the fluid (Eq. 5.6) now takes the form

$$\rho\, \delta\tau\, \frac{D\mathbf{V}}{Dt} = \rho\, \delta\tau \mathbf{f} - \delta\tau\, \mathrm{grad}\, p$$

or

$$\rho\, \frac{D\mathbf{V}}{Dt} = \rho\mathbf{f} - \mathrm{grad}\, p \qquad (5.11a)$$

when expressed per unit volume of the fluid. Using Eqs. (5.8) or (5.9) for the acceleration we may write (5.11a) in the following alternate form:

$$\rho\left(\frac{\partial\mathbf{V}}{\partial t} + \mathbf{V}\cdot\mathrm{grad}\,\mathbf{V}\right) = \rho\mathbf{f} - \mathrm{grad}\, p \qquad (5.11b)$$

or

$$\rho\left(\frac{\partial\mathbf{V}}{\partial t} + \mathrm{grad}\,\frac{V^2}{2} - \mathbf{V}\times\mathrm{curl}\,\mathbf{V}\right) = \rho\mathbf{f} - \mathrm{grad}\, p \qquad (5.11c)$$

This equation of motion is one of the fundamental equations of fluid dynamics. It was first obtained by Euler in 1755 and is called *Euler's Equation*. In deriving this equation no account has been taken of the viscous nature of a fluid. Therefore it holds good only for the motions of fluids in which viscosity is assumed to be zero, that is, for the motions of an *inviscid* fluid.

Euler's equation (5.11) represents a system of three scalar equations for the five unknowns—the three scalar components of the velocity, the pressure, and the density. Hence it is necessary to obtain additional equations. We do this by the application of the laws of conservation of mass and energy. We derive first the equation that expresses the conservation of mass.

5.3   Equation of Conservation of Mass

We assert that the *mass of any fluid element remains constant* as it moves about even though, in general, its shape, volume, and density may change. Consider a fluid element of volume $\delta\tau$ that is situated at the point $\mathbf{r}$ at time t. The mass of the element is $\rho\,\delta\tau$. Since the mass of the element is a constant, it follows that $\rho\,\delta\tau$ *is a constant if one follows the same element.* In other words, the material derivative of $\rho\,\delta\tau$ is zero.

$$\frac{D}{Dt}(\rho\,\delta\tau) = 0 \tag{5.12}$$

This is the equation of conservation of mass in its simplest form.

Equation (5.12) may be rewritten as

$$\delta\tau\,\frac{D\rho}{Dt} + \rho\,\frac{D}{Dt}(\delta\tau) = 0 \tag{5.13}$$

The material derivative of the density is given by

$$\frac{D\rho}{Dt} = \frac{\partial\rho}{\partial t} + \mathbf{V}\cdot\operatorname{grad}\rho \tag{5.14}$$

The material derivative $D(\delta\tau)/Dt$ of the volume of the element may be expressed in terms of the velocity field. To do this we observe that

$$\frac{D}{Dt}(\delta\tau) = \lim_{\delta t\to 0}\frac{\delta\tau(t + \delta t) - \delta\tau(t)}{\delta t}$$

where $\delta\tau(t)$ is the volume of the element at time t and $\delta\tau(t + \delta\tau)$ is its volume after a small time interval δt. The change in volume of the element during the time interval δt may be determined as follows. Let us consider the element as situated in the position $\mathbf{r}$ at time t and assume, for clarity,

that at that instant the velocities at the various points on the surface of the
element are all directed *outward*, that is, toward the region that contains
the outward normal (see Fig. 5.3). Then, as shown in the figure, the
surface of the element that is δS at time t grows into the surface δS_1 during
the interval δt. Both the surfaces δS and δS_1 contain the same fluid
element but at different times. The change in volume of the element is
simply equal to the volume swept by the surface of the element during

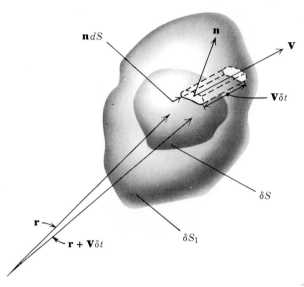

Fig. 5.3 Illustrating the change in volume of a fluid element.

the time δt. Now, if the velocities at various points of the surface of the
element are not all directed outward, the change in volume is equal to the
net volume swept outward by the surface of the element. If $\mathbf{n}\,dS$ is an
elemental area on the surface δS, $\mathbf{n}$ being the outward normal (Fig. 5.3), the
net volume swept outward by δS in time δt is given (to the first order in
δt) by

$$\oiint_{\delta S} \mathbf{V}\,\delta t \cdot \mathbf{n}\,dS$$

where $\mathbf{V}$ denotes velocity. Thus it follows that

$$\frac{D}{Dt}(\delta\tau) = \oiint_{\delta S} \mathbf{V} \cdot \mathbf{n}\,dS$$

Now, for an infinitesimally small volume element according to the integral definition of the divergence of a vector (Eq. 2.117) we have

$$\oiint_{\delta S} \mathbf{V} \cdot \mathbf{n} \, dS = \delta\tau \, \text{div} \, \mathbf{V}$$

Using this relation we obtain

$$\frac{D}{Dt}(\delta\tau) = \delta\tau \, \text{div} \, \mathbf{V} \tag{5.15}$$

This may be rewritten as

$$\text{div} \, \mathbf{V} = \frac{1}{\delta\tau} \frac{D}{Dt}(\delta\tau) \tag{5.16}$$

This equation shows that *the divergence of the velocity field gives the rate of change of volume of a fluid element per unit volume.* It thus represents the rate of volumetric strain. It is also known as the *dilatation* or *extension*.
   From Eqs. (5.12) and (5.15) we obtain

$$\delta\tau \frac{D\rho}{Dt} + \rho \, \delta\tau \, \text{div} \, \mathbf{V} = 0$$

or

$$\frac{D\rho}{Dt} + \rho \, \text{div} \, \mathbf{V} = 0 \tag{5.17}$$

when expressed per unit volume of the fluid. This is the required *equation of conservation of mass* or, simply, the equation of mass. It is a relation between the velocity and density fields only. A relation that does not involve any dynamical quantities, such as forces and pressures, is known as a *kinematical relation*. Thus the equation of conservation of mass is purely a kinematical relation. Consequently, it holds good for the motion of all fluids.
   Substituting from Eq. (5.14) for the material derivative of the density we may rewrite Eq. (5.17) as

$$\frac{\partial\rho}{\partial t} + \mathbf{V} \cdot \text{grad} \, \rho + \rho \, \text{div} \, \mathbf{V} = 0$$

or as

$$\frac{\partial\rho}{\partial t} + \text{div} \, \rho\mathbf{V} = 0 \tag{5.18}$$

This equation has an interesting interpretation which we shall see later (Section 6.2). The equation of conservation of mass is also known as the *equation of continuity*. By continuity we mean physical continuity, implying that the fluid always remains a continuum, that is, as continuously distributed matter.

5.4 Equation of Energy

The law of conservation of energy expresses the balance of energy exchanges that take place between a system and its surroundings. In applying the law to a fluid in motion we should regard the fluid as a thermodynamic system, assume the existence of the usual thermodynamic variables and take into account all the several processes that may contribute to the exchange of energy between the different fluid elements themselves and between the fluid and its surroundings. Some examples of such processes are: work on the element by the body and surface forces, heat conduction, chemical reaction, radiation and electromagnetic action. For our purposes *we shall assume that the fluid is a nonheat-conducting medium and that no processes of energy exchange other than the work by body and surface forces take place.* Under these assumptions the law of conservation of energy for a fluid element may be expressed as

rate of increase of energy E of a fluid element
= rate W_1 at which work is done on the element by the body forces
+ rate W_2 at which work is done on the element, at its surface, by the surface forces

This is expressed symbolically as

$$\frac{DE}{Dt} = W_1 + W_2 \tag{5.19}$$

The energy of the fluid is the sum of two parts, one the *kinetic energy*, a mechanical quantity, and the other the *internal energy*, a thermodynamic quantity. We specify the internal energy of the fluid by the scalar field $e = e(\mathbf{r}, t)$, which denotes the internal energy per unit mass at any point $\mathbf{r}$ at time t. The kinetic energy of the fluid per unit mass is given by the scalar field $V^2/2$. Thus the energy of the fluid per unit mass is equal to

$$e + \frac{V^2}{2}$$

Consequently, the energy of the fluid element is given by

$$E = \rho \, \delta\tau \left(e + \frac{V^2}{2} \right) \tag{5.20}$$

The rate at which work is done by the body forces is given by

$$W_1 = \rho \, \delta\tau \mathbf{f} \cdot \mathbf{V} \tag{5.21}$$

To obtain the rate at which work is done by the surface forces we assume that the fluid is inviscid and, consequently, that the surface forces are only pressure forces. Accordingly, we have

$$W_2 = -\oiint_{\delta S} p\mathbf{n}\, dS \cdot \mathbf{V} = -\oiint_{\delta S} p\mathbf{V} \cdot \mathbf{n}\, dS \qquad (5.22)$$

where δS, as before, is the surface of the element. We have assumed, as before, that the fluid is inviscid and, consequently, that the surface forces are only pressure forces. For an infinitesimally small volume element Eq. (5.22) becomes

$$W_2 = -\delta\tau \operatorname{div} p\mathbf{V}$$
$$= -\delta\tau\,(\operatorname{grad} p \cdot \mathbf{V} + p \operatorname{div} \mathbf{V}) \qquad (5.23)$$

This equation shows that the rate at which work is done by the surface forces is equal to $-\operatorname{div} p\mathbf{V}$ per unit volume of the fluid and that this rate of work is made up of two parts—one part given by $-\operatorname{grad} p \cdot \mathbf{V}$ represents the rate, per unit volume, at which work is done by the resultant of the surface forces, and the other part given by $-p \operatorname{div} \mathbf{V}$ represents the rate, per unit volume, at which work is done on the element due to an increase in its volume. Recall that $\operatorname{div} \mathbf{V}$ represents the dilatation.

Using Eqs. (5.19), (5.20), (5.21), and (5.23), we obtain

$$\frac{D}{Dt}\left[\rho\,\delta\tau\left(e + \frac{V^2}{2}\right)\right] = \rho\,\delta\tau\mathbf{f} \cdot \mathbf{V} - \delta\tau \operatorname{grad} p \cdot \mathbf{V} - \delta\tau p \operatorname{div} \mathbf{V}$$

or, since the material derivative of $\rho\,\delta\tau$ is zero,

$$\rho\,\delta\tau\frac{D}{Dt}\left(e + \frac{V^2}{2}\right) = \rho\,\delta\tau\mathbf{f} \cdot \mathbf{V} - \delta\tau \operatorname{grad} p \cdot \mathbf{V} - \delta\tau p \operatorname{div} \mathbf{V}$$

Expressing it per unit volume, we may write this equation as

$$\rho\frac{D}{Dt}\left(e + \frac{V^2}{2}\right) = \rho\mathbf{f} \cdot \mathbf{V} - \operatorname{grad} p \cdot \mathbf{V} - p \operatorname{div} \mathbf{V} \qquad (5.24)$$

This is the required *equation of conservation of energy* or, simply, the *equation of energy*. It is also referred to as the *equation of total energy*.

Equation (5.24) may further be reduced to a simpler form as follows. We form the scalar product of the equation of motion (5.11a) with $\mathbf{V}$ and obtain

$$\rho\frac{D}{Dt}\left(\frac{V^2}{2}\right) = \mathbf{f} \cdot \mathbf{V} - \operatorname{grad} p \cdot \mathbf{V} \qquad (5.25)$$

[handwritten: should be $\rho\vec{f} \cdot \vec{V}$]

This equation, known as the *equation of mechanical energy*, states that the rate of change of kinetic energy of a fluid element is equal to the rate at which work is done by the body forces and the resultant of the surface forces. Subtracting (5.25) from (5.24) we obtain

$$\rho\frac{De}{Dt} = -p \operatorname{div} \mathbf{V} \qquad (5.26)$$

which states that the rate of increase of internal energy of a fluid element is equal to the rate at which work is done on the element during an increase in its volume. This simple form of the energy equation is an expression of the first law of thermodynamics applied to a system undergoing an adiabatic change of volume, that is, without any heat addition. Equation (5.26) is thus known as the *thermodynamic form of the energy equation.* It holds good only for a nonviscous, nonheat-conducting fluid.

The energy equation introduces the internal energy as an additional unknown in the formulation of the equations of fluid motion. The list of unknowns is thus increased from five to six, namely $\mathbf{V}$, p, ρ, e, while the number of equations is brought up to five. It therefore follows that an additional relation between some of these unknowns is needed. Such a relation is provided by the so-called equation of state of the fluid.

5.5 Equation of State

The thermodynamic state of a system is described by a certain number of thermodynamic variables, such as pressure, density, temperature, internal energy, enthalpy, and entropy. Such variables depend only on the state of the system and are called variables of state. For the complete description of the thermodynamic state only a few, instead of all, of the state variables are necessary. The number of variables required depends on the nature of the system: it may be a *simple system* composed of a single fluid or a homogeneous mixture of *inert* fluids, or it may be a nonsimple system composed of a mixture of reacting fluids. From measurements we realize that for a simple system the thermodynamic state is completely described by two state variables, whereas for a nonsimple system a greater number of variables is required. Once the number of variables required for a complete thermodynamic description is known, we can express any of the state variables as a function of a few others chosen as independent variables. The number of these independent variables is simply the number required for complete thermodynamic description. Such functional relations between state variables are called *equations of state.* A material medium may be characterized by its equations of state. The actual functional relation implied in an equation of state cannot be determined from thermodynamics. It is usually obtained from measurements.

For our purposes we shall assume that the *fluid* in motion is a *simple thermodynamic system.* In such a case the fluid may be characterized by equations of state such as

$$e = e(p, \rho)$$

$$p = p(e, \rho)$$

$$p = p(\rho, T)$$

where T is the temperature. Since we are presently concerned with the variables e, p, and ρ, we shall choose

$$e = e(p, \rho) \qquad (5.27)$$

as the equation of state.

5.6 Equations for an Inviscid Compressible Fluid

Summarizing the preceding considerations we may state that the motion of an inviscid, nonheat-conducting, compressible fluid, which is assumed to be a simple thermodynamic system, involves three scalar fields ρ, p, e and one vector field $\mathbf{V}$ as the unknown functions and is governed by the following system of equations:

(1) *Equation of motion*

$$\rho \frac{D\mathbf{V}}{Dt} = \rho \mathbf{f} - \text{grad } p \qquad (5.11a)$$

(2) *Equation of conservation of mass*

$$\frac{D\rho}{Dt} + \rho \text{ div } \mathbf{V} = 0 \qquad (5.17)$$

(3) *Equation of energy*

$$\rho \frac{De}{Dt} = -p \text{ div } \mathbf{V} \qquad (5.26)$$

(4) *Equation of state*

$$e = e(p, \rho)$$

or (5.27)

$$p = p(e, \rho)$$

These equations are known as the equations of inviscid compressible flow. The integration of these equations forms the subject matter of the *mechanics of an inviscid compressible fluid*.

Our business is with the mechanics of an inviscid fluid that is incompressible. We shall now introduce the condition of incompressibility and deduce the equations that govern the motion of an inviscid incompressible fluid.

5.7 Condition of Incompressibility

A fluid is said to be incompressible if the volume of every element of the fluid is a constant for all times. Thus, if $\delta\tau$ denotes, as before, the volume of a fluid element, the *condition of incompressibility* may be expressed in the differential form as

$$\frac{D}{Dt}(\delta\tau) = \delta\tau \text{ div } \mathbf{V} = 0$$

or simply as

$$\text{div } \mathbf{V} = 0 \tag{5.28}$$

This equation, which is purely a kinematical relation, states that the divergence of the velocity field is zero.

5.8 Consequences of Incompressibility

The assumption of incompressibility gives rise to much simplification of the equations of fluid mechanics. We observe that if the fluid is assumed incompressible, the condition of incompressibility may be added to the equations of motion and mass conservation to form a system of equations sufficient to solve for the unknown functions ρ, p, and $\mathbf{V}$. This means we need not invoke any thermodynamical considerations and additional equations, such as the equation of energy and the equation of state. It thus appears that the assumption of incompressibility reduces the problem of the motion of an inviscid fluid from one in mechanics and thermodynamics to one in mechanics only.

As a consequence of Eq. (5.28), the equation of conservation of mass (5.17) becomes

$$\frac{D\rho}{Dt} = \frac{\partial \rho}{\partial t} + \mathbf{V} \cdot \text{grad } \rho = 0 \tag{5.29}$$

This equation expresses the fact that, *in an incompressible fluid, the density of any fluid element is a constant for all times.* The equation does not imply that the density is a constant throughout the flow field for all times. From Eq. (5.29) it follows that if, at some initial instant, the fluid is *inhomogeneous* (i.e., one whose fluid elements are of different densities), it will remain inhomogeneous for all other times and that if the fluid is homogeneous (i.e., one whose elements are all of the same density) at some initial instant, it will remain homogeneous for all other times. Thus for an incompressible inhomogeneous fluid there will be both local and spatial variations of density although the material rate of change of density is zero. For an incompressible homogeneous fluid, however, the local and spatial variation vanish independently, and the density is constant in time as well as in space. Thus for *an incompressible homogeneous fluid* Eq. (5.29) takes the simple form

$$\rho = \text{const.} \tag{5.30}$$

We consider next Euler's equation (5.11a). For an inhomogeneous fluid, the assumption of incompressibility leaves Euler's equation unchanged. For a homogeneous fluid, however, the assumption of incompressibility allows one to set ρ equal to a constant and to write Euler's equation in the simpler form

$$\frac{D\mathbf{V}}{Dt} = \mathbf{f} - \text{grad} \left(\frac{p}{\rho}\right) \tag{5.31}$$

For an incompressible fluid, the energy equation (5.26) becomes

$$\frac{De}{Dt} = \frac{\partial e}{\partial t} + \mathbf{V} \cdot \text{grad } e = 0 \qquad (5.32)$$

which states that *in an incompressible fluid the internal energy of any fluid element is a constant for all times.* From (5.32) it follows that if at some initial time e is nonuniform in space (whence we say the fluid is again inhomogeneous), then for all other times there will exist both local and spatial variations of e although the material derivative of e is zero. However, if at some instant e is uniform in space (whence we say that the fluid is homogeneous), then for all other times both the local and spatial variations of e vanish independently and (5.32) reduces to

$$e = \text{const.} \qquad (5.33)$$

Because of the condition of incompressibility (5.28) and the consequent energy equation (5.32), the total energy equation (5.24) becomes

$$\rho \frac{D}{Dt} \frac{V^2}{2} = \rho \mathbf{f} \cdot \mathbf{V} - \text{grad } p \cdot \mathbf{V}$$

which is simply the equation of mechanical energy.

We see that for an inviscid, incompressible, inhomogeneous fluid we may solve for ρ, p, and $\mathbf{V}$ from the condition of incompressibility (5.28), the equation of mass conservation (5.29), and the equation of motion (5.11a) independently of the equation of energy (5.32). Once $\mathbf{V}$ is known, we may solve the energy equation (5.32) to obtain e. Equations (5.28), (5.29), and (5.32) are purely mechanical in nature, and their solution does not involve any thermodynamical considerations. This situation becomes strikingly clear in the case of a homogeneous fluid. For a homogeneous fluid the density is constant and drops out of the unknown functions, thus reducing the initial list from ρ, p, $\mathbf{V}$ to p and $\mathbf{V}$ only. We may solve for these from the condition of incompressibility (5.28) and the equation of motion (5.31). For a homogeneous fluid the internal energy is also a constant for all times. Consequently, the internal energy and along with it all thermodynamic considerations disappear, as it were, from the problem of the motion of an inviscid, incompressible, homogeneous fluid.

Once the fluid is assumed incompressible, the pressure becomes purely a mechanical variable, and no thermodynamic significance may be given to it. This is clearly seen in the case of the homogeneous fluid. For an inviscid, incompressible, homogeneous fluid both the density and the internal energy are constant although the pressure varies both in time and in space. There is thus no functional relation between the variable p and the density and the internal energy. For an illuminating explanation of the concept of pressure in an incompressible fluid, the reader may consult Sommerfeld (1950).

5.9 Equations for an Ideal Fluid

Our concern here is with the study of the motion of an inviscid, incompressible, homogeneous fluid, the so-called *ideal fluid*. Naturally no real fluid has the properties of an ideal fluid. However, as pointed out in Chapter 1, under certain circumstances many aspects of the motion of a real fluid may be analyzed on the basis of an ideal fluid. As we have seen, the motion of an ideal fluid is characterized by pressure and velocity fields only and is governed by the following equations:
Condition of incompressibility

$$\text{div } \mathbf{V} = 0 \tag{5.28}$$

and *Euler's equation*

$$\frac{D\mathbf{V}}{Dt} = \mathbf{f} - \text{grad}\left(\frac{p}{\rho}\right) \tag{5.31}$$

where

$$\frac{D\mathbf{V}}{Dt} = \frac{\partial \mathbf{V}}{\partial t} + \mathbf{V} \cdot \text{grad } \mathbf{V} = \frac{\partial \mathbf{V}}{\partial t} + \text{grad}\,\frac{V^2}{2} - \mathbf{V} \times \text{curl } \mathbf{V}$$

These equations are the basis of all investigations in the mechanics of an ideal fluid. To specify a given problem we must add to these differential equations the so-called initial and boundary conditions. We shall now see how to formulate such conditions.

5.10 Initial Conditions

Initial conditions are conditions that describe completely the state of the fluid at some instant of time which is referred to as the initial time. *For an ideal fluid* a complete set of initial conditions is obtained if the velocity $\mathbf{V}$ and the pressure p are specified at the initial time.

5.11 Boundary Conditions for an Ideal Fluid

Physical conditions that should be satisfied on given boundaries of the fluid are known as boundary conditions. There are several types of boundaries and as such there are various possibilities for the boundary conditions. We consider two types of boundaries, (1) *a solid-fluid boundary*, where the fluid is bounded by a solid surface and (2) a *fluid-fluid boundary*, where the fluid is bounded by another fluid or the same fluid in a different state of motion; this boundary is usually referred to as a *free surface* or a *free boundary*.

The nature and number of the boundary conditions depend also on the assumptions made with regard to the nature of the fluid, more specifically on the nature of the differential equations that are assumed to govern the motion of the fluid. In this sense, for instance, differences exist between the boundary conditions for a viscous fluid and a nonviscous fluid. *We shall concern ourselves with only an ideal fluid.*

Condition at a Solid-fluid Boundary. We assume that the fluid is bounded by an impermeable solid and require that no fluid should therefore penetrate the solid surface. Since, in general, each element of the solid surface may be in motion relative to the fluid, this requirement may be expressed as the following condition:

At each point of the solid-fluid surface, at every instant, the component normal to the surface of the relative velocity between the fluid and the solid must vanish.

If at any point on the surface V_R represents the relative velocity and $\mathbf{n}$ the normal to the surface, this condition may be written symbolically as

$$\mathbf{V}_R \cdot \mathbf{n} = 0 \quad \text{on a solid-fluid surface} \tag{5.34}$$

If $\mathbf{V}$ and $\mathbf{V}_S$ denote, respectively, the velocities of the fluid and the surface, condition (5.34) takes the form

$$(\mathbf{V} - \mathbf{V}_S) \cdot \mathbf{n} = 0$$

or

$$\mathbf{V} \cdot \mathbf{n} = \mathbf{V}_S \cdot \mathbf{n} \quad \text{on a solid-fluid-surface} \tag{5.35}$$

Equation (5.35) states that at each point of a solid-fluid boundary, at every instant, the normal component of the velocity of the fluid is equal to the normal component of the velocity of the boundary. If the boundary is formed by a *fixed* (stationary with respect to a frame fixed in space), *rigid* solid surface, the velocity of the surface at every point is zero and we obtain $\quad \mathbf{V} \cdot \mathbf{n} = 0 \quad$ on a fixed, rigid solid-fluid surface $\tag{5.36}$

Now, consider a solid-fluid surface, each element of which is in motion relative to the fluid.* We represent the surface by the equation

$$F(\mathbf{r}, t) = 0$$

where $F(\mathbf{r}, t)$ is a scalar function of position and time. The normal to the surface at any point on the surface is then given by

$$\mathbf{n} = \pm \frac{\text{grad } F}{|\text{grad } F|}$$

---

* For example, such a surface may be that of a deformable solid body which is either stationary or moving through a flowing fluid. Similarly, the surface may be that of a rigid solid body moving through a fluid.

With this relation for $\mathbf{n}$, the boundary condition (5.35) becomes

$$\mathbf{V} \cdot \operatorname{grad} F = \mathbf{V}_S \cdot \operatorname{grad} F \text{ on } F(\mathbf{r}, t) = 0 \qquad (5.37)$$

This condition (5.37) may be transformed further into a more convenient form. To do this we first show that $\mathbf{V}_S \cdot \operatorname{grad} F$ is equal to $-\partial F/\partial t$, the local rate of change of the function $F(\mathbf{r}, t)$. Consider the scalar field $F = F(\mathbf{r}, t)$ and imagine the surface particles of the solid to move in such a field. At each instant of time, the surface of the solid is given by the surface formed by those points that have a value of F equal to zero at that instant.* Thus if we move with the surface particles of the solid, we observe no change in the function $F(\mathbf{r}, t)$. In other words, the change or, equivalently, *the total time rate of change in $F(\mathbf{r}, t)$ following a surface particle of the solid is zero.*

Since $\mathbf{V}_S$ is the velocity of any surface particle of the solid, we therefore obtain

$$\frac{\partial F}{\partial t} + \mathbf{V}_S \cdot \operatorname{grad} F = 0$$

or

$$\mathbf{V}_S \cdot \operatorname{grad} F = -\frac{\partial F}{\partial t} \qquad (5.38)$$

Using this relation, we express the condition (5.37) as

$$\frac{DF}{Dt} = \frac{\partial F}{\partial t} + \mathbf{V} \cdot \operatorname{grad} F = 0 \quad \text{on} \quad F(\mathbf{r}, t) = 0 \qquad (5.39)$$

If the solid-fluid boundary is formed by the surface of a stationary rigid solid, the equation of the boundary becomes

$$F(\mathbf{r}) = 0$$

and the boundary condition (5.39) reduces to

$$\mathbf{V} \cdot \operatorname{grad} F = 0 \quad \text{on} \quad F(\mathbf{r}) = 0 \qquad (5.40)$$

The boundary condition at a solid-fluid boundary is purely a kinematical condition. The condition that we have formulated refers only to the normal component of the relative velocity between the fluid and the solid. Nothing has been said about the tangential component of the relative velocity. For an inviscid fluid, the one we have assumed, there is no *a priori* physical requirement that may be used to stipulate a condition

---

* According to the function $F(\mathbf{r}, t)$, a value of F is attached to each point of space at any given instant. However, only certain points at that instant will have a value of F equal to zero. A surface drawn through such points gives at that instant the position of the surface of the solid.

on the tangential component of the relative velocity between the fluid and a solid surface bounding the fluid. This means *in an inviscid fluid, at a solid-fluid boundary, the tangential component of the relative velocity may assume any value that is consistent with the solution of the flow field obtained on the basis of other specified conditions. The fluid thus may slip past the solid boundary.* This is in contrast to the situation in a real viscous fluid. For a viscous fluid, on the basis of experience, we require (*for the most part*) that at a solid-fluid boundary, the relative velocity should be zero at all times, that is, both the normal and tangential components of the relative velocity should vanish.

Condition at a Free Surface. Kinematical conditions similar to those applicable at a solid-fluid boundary should be satisfied also at any surface that forms the boundary between two different *immiscible* fluids or between two different states of motion of the same fluid. We describe such a surface (which may be deformable and moving), as before, by the equation

$$F(\mathbf{r}, t) = 0$$

Let $\mathbf{V}_S$ denote the velocity of any element of the surface, and let $\mathbf{V}_1$ and $\mathbf{V}_2$ denote the velocities of the fluid on either side of the surface. Requiring that there be no fluid flow across the surface, we obtain the condition

$$\mathbf{V}_1 \cdot \mathbf{n} = \mathbf{V}_S \cdot \mathbf{n} = \mathbf{V}_2 \cdot \mathbf{n} \text{ at each point on the surface } F(\mathbf{r}, t) = 0 \quad (5.41)$$

where $\mathbf{n}$ is the normal to the surface. Expressing $\mathbf{n}$ in terms of grad F, we have

$$\mathbf{V}_1 \cdot \text{grad } F = \mathbf{V}_S \cdot \text{grad } F = \mathbf{V}_2 \cdot \text{grad } F \text{ on } F(\mathbf{r}, t) = 0 \quad (5.42)$$

By using relation (5.38), this condition may be expressed by the equations

and

$$\left.\begin{array}{c} \dfrac{\partial F}{\partial t} + \mathbf{V}_1 \cdot \text{grad } F = 0 \\[2em] \dfrac{\partial F}{\partial t} + \mathbf{V}_2 \cdot \text{grad } F = 0 \end{array}\right\} \quad \text{on} \quad F(\mathbf{r}, t) = 0 \quad (5.43)$$

If the surface is fixed in space, $\mathbf{V}_S$ is zero, and this surface condition reduces to

$$\mathbf{V}_1 \cdot \mathbf{n} = \mathbf{V}_2 \cdot \mathbf{n} = 0 \text{ on } F(\mathbf{r}) = 0 \quad (5.44)$$

In addition to this kinematic condition, a dynamic condition has to be satisfied at a free surface. This condition is formulated as follows. Consider any infinitesimal element $\mathbf{n}\, dS$ of the surface and let p_1 and p_2 denote the pressures on either side of the element. We assume that the fluid is inviscid, as before, and that *there are no cohesive forces* (such as those arising from surface tension). Then, since the surface element is of zero thickness, the condition for dynamic equilibrium of the element becomes

$$(p_1 - p_2)\mathbf{n}\, dS = 0$$

It therefore follows that *at a free surface*

$$p_1 = p_2 \tag{5.45}$$

at every point of the surface at any time.

The kinematic condition at a free surface refers, on either side of the surface, only to the normal component of the relative velocity between the fluid and the surface. As in the case of the solid-fluid boundary no *a priori* condition can be stipulated for the tangential component of the relative velocity on either side of a free surface. It thus follows that for a inviscid fluid the tangential component of the fluid velocity at any point on one side of a free surface is not equal to the tangential component, at that point, of the fluid velocity on the other side of the surface, and that neither of these components is equal to the tangential component of the velocity of the surface at that point. This means *the tangential component of the fluid velocity is discontinuous across a free surface.* In this sense, a free surface is known as a *surface of discontinuity*, specifically, a *surface of tangential discontinuity.*

The possibility of such a surface of discontinuity arises solely from the assumption of an inviscid fluid. In a real viscous fluid, although there may be narrow regions of space where the fluid velocity may change very rapidly, a strict discontinuity is impossible.

5.12 Conditions at Infinity

In this book we shall consider mostly fluid flows related to the motion of a body through a fluid which is otherwise undisturbed. In such flows, the boundaries of the flow which are at large distances from the body are regarded as being at infinity. Boundary conditions at infinity are usually given by prescribing the velocity and pressure at infinity. We may require, for instance, that the velocity due to the disturbance of the body be zero, or at least finite at infinity. A more precise specification of the condition depends on the nature of the equations governing the fluid motion under consideration.

5.13 Stream Functions for Incompressible Flow

We had seen that a flow field may be described by two stream functions $\psi_1(\mathbf{r}, t)$ and $\psi_2(\mathbf{r}, t)$ that are connected to the velocity field $\mathbf{V}(\mathbf{r}, t)$ by the relation (see Section 4.9)

$$\mu(\mathbf{r}, t)\mathbf{V} = \operatorname{grad} \psi_1 \times \operatorname{grad} \psi_2$$

The function μ should be such that

$$\operatorname{div} \mu \mathbf{V} = 0$$

Now, according to the condition of incompressibility we have

$$\operatorname{div} \mathbf{V} = 0$$

It therefore follows that in incompressible flow we may set

$$\mu(\mathbf{r}, t) = 1$$

and express

$$\mathbf{V} = \operatorname{grad} \psi_1 \times \operatorname{grad} \psi_2 \tag{5.46}$$

Thus, in incompressible flow, the vector field $\mathbf{V}(r, t)$ may be replaced by two scalar fields, $\psi_1(\mathbf{r}, t)$ and $\psi_2(\mathbf{r}, t)$.

It is readily seen that for *two-dimensional incompressible flow*, Eqs. (4.26) and (4.28) become respectively

$$u(x, y, t) = \frac{\partial}{\partial y} \psi(x, y, t)$$

$$v(x, y, t) = -\frac{\partial}{\partial x} \psi(x, y, t) \tag{5.47}$$

and

$$u(r, \theta, t) = \frac{1}{r} \frac{\partial}{\partial \theta} \psi(r, \theta, t)$$

$$v(r, \theta, t) = -\frac{\partial}{\partial r} \psi(r, \theta, t) \tag{5.48}$$

The corresponding relations for *axisymmetric incompressible flow* are obtained from Eqs. (4.30) and (4.32) by setting $\mu = 1$. In terms of spherical coordinates R, θ, φ we have

$$u_R(R, \theta, t) = \frac{1}{R^2 \sin \theta} \frac{\partial}{\partial \theta} \psi(R, \theta, t)$$

$$u_\Theta(R, \theta, t) = -\frac{1}{R \sin \theta} \frac{\partial}{\partial R} \psi(R, \theta, t) \tag{5.49}$$

In terms of cylindrical coordinates x, r, φ we have

$$u_x(x, r, t) = \frac{1}{r} \frac{\partial}{\partial r} \psi(x, r, t)$$

$$u_r(x, r, t) = -\frac{1}{r} \frac{\partial}{\partial x} \psi(x, r, t) \tag{5.50}$$

5.14 Vector Potential for Incompressible Flow and Its Relation to the Stream Functions

Since the divergence of any curl vector is identically zero, we may express the velocity field in an incompressible flow for which

$$\text{div } \mathbf{V} = 0$$

in terms of another vector field $\mathbf{A}(\mathbf{r}, t)$ such that

$$\mathbf{V} = \text{curl } \mathbf{A} \tag{5.51}$$

We speak of $\mathbf{A}$ as a "vector potential" (see Section 2.42).

Such a vector potential is naturally related to the stream functions. From Eqs. (5.46) and (5.51) we have

$$\text{Curl } \mathbf{A} = \mathbf{V} = \text{grad } \psi_1 \times \text{grad } \psi_2 \tag{5.52}$$

Consider two-dimensional motion. Using Cartesians we have

$$\mathbf{V} = [u(x, y, t), v(x, y, t), 0]$$
$$\psi_1 = \psi(x, y, t)$$
$$\psi_2 = z$$

Let the components of $\mathbf{A}$ be denoted by A_x, A_y, A_z. Equation (5.52) then yields the relations

$$\frac{\partial A_z}{\partial y} - \frac{\partial A_y}{\partial z} = u = \frac{\partial \psi}{\partial y}$$

$$\frac{\partial A_x}{\partial z} - \frac{\partial A_z}{\partial x} = v = -\frac{\partial \psi}{\partial x} \tag{5.53}$$

$$\frac{\partial A_y}{\partial x} - \frac{\partial A_x}{\partial y} = 0$$

It is readily verified that since $\mathbf{A} = A(x, y, t)$ only, the system (5.53) is satisfied by having

$$\mathbf{A}(x, y, t) = (0, 0, \psi(x, y, t)) \tag{5.54}$$
$$= \mathbf{k}\psi$$

In cylindrical coordinates r, θ, and z, Eq. (5.54) takes the form

$$\mathbf{A}(r, \theta, t) = \mathbf{k}\psi(r, \theta, t)$$

In a similar way the relation between the vector potential and the stream function in axisymmetric incompressible flow may be established. In terms of spherical coordinates R, θ, φ, we obtain

$$\mathbf{A}(R, \theta, t) = \mathbf{e}_\varphi \frac{1}{R \sin \theta} \psi(R, \theta, t) \tag{5.55}$$

In terms of cylindrical coordinates x, r, φ we have

$$A(x, r, t) = \mathbf{e}_\varphi \frac{1}{r}\, \psi(x, r, t) \tag{5.56}$$

5.15 Elimination of the Body Force from the Equation of Motion for a Certain Incompressible Flow Problem

Consider the problem of flow due to a body moving through an infinitely extending incompressible homogeneous fluid, a problem of most interest to us. Let us assume that initially the fluid with the body immersed in it is at rest under the action of body forces $\mathbf{f}$ per unit mass. The fluid is later set into motion by letting the body move through it. The body forces continue to act on the fluid. Denote by p_h the pressure when the fluid is at rest and by p the pressure when the fluid is in motion. For static equilibrium of the fluid we require that*

* Note that Eq. (5.57) shows that $\mathbf{f}$ should be irrotational.

$$\rho \mathbf{f} = \operatorname{grad} p_h \tag{5.57}$$

For dynamic equilibrium of the fluid we require that

$$\rho \frac{D\mathbf{V}}{Dt} = \mathbf{f} - \operatorname{grad} p + \mathbf{f}_\mu \tag{5.58}$$

where, for generality, we have included a term $\mathbf{f}_\mu$ to represent the viscous force on the fluid element.

Using Eq. (5.57) and noting that ρ does not vary when the fluid moves, we may eliminate the body force from Eq. (5.58) and obtain

$$\rho \frac{D\mathbf{V}}{Dt} = -\operatorname{grad}(p - p_h) + \mathbf{f}_\mu$$

$$= -\operatorname{grad} p' + \mathbf{f}_\mu \tag{5.59}$$

where p' is the difference of the pressure in the dynamical situation from the hydrostatic pressure p_h. It thus follows that *for the problem under consideration the body force may be dropped from the equation of motion if the pressure in the resulting equation is regarded as the difference of the actual pressure from the hydrostatic.* The effect of the body force may be accounted for independently by solving the corresponding static problem. It should be borne in mind that the present conclusions are strictly valid for the particular problem considered here.

Chapter 6

Alternate Forms of the Equations

In the preceding chapter the differential equations that govern the motion of an inviscid fluid were developed from the point of view of a moving fluid element. We shall now derive an alternate but equivalent form of these equations from a different point of view, namely that of a *definite region of space* enclosed by a surface *fixed* in the flowing fluid. Following this, we shall obtain, from the two points of view, *integral forms* of the equations. We begin by formulating an equation for the change of any fluid quantity contained in a fixed region of space.

6.1 Equation of Change

Consider a certain region R *fixed* in the space occupied by a flowing fluid and bounded by a surface S. Let G denote the amount of any fluid quantity, such as mass, momentum, energy, angular momentum, contained in the region R at any time t. We set up an equation for the change in G by equating $\partial G / \partial t$, the rate of increase of G in R, to the sum of all the changes per unit time in G due to different causes. One of the causes by which a change in G may occur is that of production of G in R. When this happens we say that there are sources of G in R. *We shall assume that there are no sources of G.* Another cause by which an increase in G will occur is that of *net transport* of G into the region R by the fluid flowing into and out of R through the bounding surface. We shall refer to this increase in G as the (*net*) *rate of inflow* of G into R. Changes in G may be brought about by causes other than transport by the flowing fluid. Depending on the actual quantities under consideration, such changes are accounted for by certain laws of mechanics and thermodynamics. Calling such changes as "other changes" we express the *equation of change* for G as

$$\frac{\partial G}{\partial t} = \text{rate of inflow of } G \text{ into } R$$

$$+ \text{ other changes of } G \text{ per unit time} \qquad (6.1)$$

198

Alternate Forms of the Equations

An equation such as (6.1) is usually referred to as the *equation of conservation of the quantity G*. We shall now specialize this equation to obtain the differential form of the so-called *conservation equations* for the motion of an inviscid fluid.

6.2 Conservation of Mass

We choose a small element of volume $\delta\tau$ fixed in space and enclosed by the surface δS. To set up the equation of change for the mass contained in $\delta\tau$ we assume that the changes in that mass are brought about solely by inflow of mass through δS. Then Eq. (6.1) takes the form

$$\frac{\partial \rho \delta\tau}{\partial t} = - \oiint_{\delta S} \rho \mathbf{V} \cdot \mathbf{n}\, dS \qquad (6.2)$$

where $\rho\mathbf{V}$ is the so-called *mass flux vector* and $\mathbf{n}\, dS$ is an elemental area on the surface δS, $\mathbf{n}$ being the outward normal. Since $\delta\tau$ is a fixed volume element, it may be taken out of the differentiation in the left-hand side of (6.2). Also, for an infinitesimal volume element, the surface integral on the right-hand side may be replaced by $\delta\tau \operatorname{div} \rho\mathbf{V}$. With these changes (6.2) may be rewritten as

$$\delta\tau \frac{\partial \rho}{\partial t} = -\delta\tau \operatorname{div} \rho\mathbf{V}$$

or, writing it per unit volume, as

$$\frac{\partial \rho}{\partial t} = -\operatorname{div} \rho\mathbf{V} \qquad (6.3)$$

As we have already seen (Section 5.3), this form of the equation of conservation of mass is equivalent to the equation

$$\frac{D\rho}{Dt} + \rho \operatorname{div} \mathbf{V} = 0$$

which was obtained from the point of view of a moving fluid element.

6.3 Conservation of Momentum

To write the equation of change for the momentum of the fluid, besides the rate of change due to inflow of momentum, one must take into account the rate of change of momentum of the fluid contained at any instant in the fixed volume $\delta\tau$ due to the body and surfaces acting on that fluid at the instant considered. We denote, as before, by $\mathbf{f}$ the body force per unit mass of the fluid and assume that the fluid is inviscid. Then the total force acting on the fluid in $\delta\tau$ is simply

$$\overline{F}_{total} = \rho\mathbf{f}\, \delta\tau - \oiint_{\delta S} p\mathbf{n}\, dS$$

As before, p is the pressure. The net rate of inflow of momentum into the volume $\delta\tau$ is given by

$$-\oiint_{\delta S} \mathbf{V}(\rho\mathbf{V}\cdot\mathbf{n})\,dS \tag{6.4}$$

Therefore, for the momentum of the fluid, the equation of change (6.1) takes the form

$$\frac{\partial\rho\mathbf{V}\,\delta\tau}{\partial t} = -\oiint_{\delta S}\mathbf{V}(\rho\mathbf{V}\cdot\mathbf{n})\,dS + \rho\mathbf{f}\,\delta\tau - \oiint_{\delta S}p\mathbf{n}\,dS \tag{6.5}$$

The term in (6.5) representing the rate of inflow of momentum may be expressed in different forms. To show this we introduce a Cartesian system with $\mathbf{e}_1$, $\mathbf{e}_2$, $\mathbf{e}_3$ as the reference unit vectors and form the ith component of the term (6.4). We thus obtain

$$\mathbf{e}_i \cdot \oiint_{\delta S}\mathbf{V}(\rho\mathbf{V}\cdot\mathbf{n})\,dS = \oiint_{\delta S}u_i(\rho\mathbf{V}\cdot\mathbf{n})\,dS$$

$$= \oiint_{\delta S}(\rho u_i\mathbf{V})\cdot\mathbf{n}\,dS$$

where u_i is the ith component of $\mathbf{V}$. For an infinitesimal volume element we have

$$\oiint_{\delta S}(\rho u_i\mathbf{V})\cdot\mathbf{n}\,dS = \delta\tau\,\text{div}\,(\rho u_i\mathbf{V})$$

It then follows that

$$\oiint_{\delta S}\mathbf{V}(\rho\mathbf{V}\cdot\mathbf{n})\,dS = \delta\tau[\mathbf{e}_1\,\text{div}\,(\rho u_1\mathbf{V}) + \mathbf{e}_2\,\text{div}\,(\rho u_2\mathbf{V}) + \mathbf{e}_3\,\text{div}\,(\rho u_3\mathbf{V})]$$

Introducing the *notation that* $\text{div}(\rho\mathbf{V}\mathbf{V})$ *represents a vector** such that

$$\mathbf{e}_i \cdot \text{div}\,(\rho\mathbf{V}\mathbf{V}) \equiv \text{div}\,(\rho u_i\mathbf{V})$$

we write

$$\oiint_{\delta S}\mathbf{V}(\rho\mathbf{V}\cdot\mathbf{n})\,dS = \delta\tau\,\text{div}\,(\rho\mathbf{V}\mathbf{V}) \tag{6.6}$$

* A product such as $\mathbf{AB}$ between two vectors is known as a *dyadic product*. A dyadic product $\mathbf{AB}$ is a tensor, the elements of which are given by

$$A_iB_j$$

where A_i and B_j are ith and jth components of $\mathbf{A}$ and $\mathbf{B}$, respectively. The subscripts refer to the directions of the axes of the chosen coordinate system and may take any of the values 1, 2, 3. The divergence of a dyadic product is a vector.

The vector div $(\rho \mathbf{VV})$ may further be expressed in a form that contains only familiar vector operations. Thus we observe that

$$\text{div} \,(\rho u_i \mathbf{V}) = \rho \mathbf{V} \cdot \text{grad} \, u_i + u_i \, \text{div} \, \rho \mathbf{V}$$

From this we obtain

$$\text{div} \,(\rho \mathbf{VV}) = \rho \mathbf{V} \cdot \text{grad} \, \mathbf{V} + \mathbf{V} \, \text{div} \, \rho \mathbf{V} \tag{6.7}$$

Using relation (6.6) and remembering that for an infinitesimal volume element

$$\oiint_{\delta S} p\mathbf{n} \, dS = \delta\tau \, \text{grad} \, p$$

Eq. (6.5) may be expressed, per unit volume of the fluid, as

$$\frac{\partial \rho \mathbf{V}}{\partial t} = -\text{div} \,(\rho \mathbf{VV}) + \rho \mathbf{f} - \text{grad} \, p \tag{6.8}$$

This equation is usually referred to as the equation of *conservation of momentum*. The dyadic $\rho \mathbf{VV}$ *is called the momentum flux tensor.* Using the relation (6.7) and the equation of change for mass (6.3), Eq. (6·8) may be reduced to the form

$$\rho \left(\frac{\partial \mathbf{V}}{\partial t} + \mathbf{V} \cdot \text{grad} \, \mathbf{V} \right) = \rho \mathbf{f} - \text{grad} \, p$$

which is the equation of motion for a moving fluid element.

6.4 Conservation of Energy

To write the equation of change for the energy of the fluid, besides the rate of inflow of energy, we must account for addition of energy to the fluid in the fixed volume $\delta\tau$ due to several other processes. We assume, as before, that the fluid is nonviscous and nonheat-conducting and that no processes other than the work by body and surface forces take place. The rate at which work is done by the body and surfaces forces on the fluid contained in $\delta\tau$ is

$$\delta\tau\rho\mathbf{f} \cdot \mathbf{V} - \oiint_{\delta S} p\mathbf{V} \cdot \mathbf{n} \, dS$$

The rate of inflow of energy into the volume is

$$-\oiint_{\delta S} \rho \left(e + \frac{V^2}{2} \right) \mathbf{V} \cdot \mathbf{n} \, dS$$

where e, as before, is the internal energy per unit mass of the fluid. Therefore, for the energy of the fluid, the equation of change (6.1) takes the form

$$\frac{\partial}{\partial t} \rho\left(e + \frac{V^2}{2}\right) \delta\tau = -\iint_{\delta S} \rho\left(e + \frac{V^2}{2}\right) \mathbf{V} \cdot \mathbf{n} \, dS$$

$$+ \delta\tau\rho\mathbf{f} \cdot \mathbf{V} - \iint_{\delta S} p\mathbf{V} \cdot \mathbf{n} \, dS \qquad (6.9)$$

Recalling that for an infinitesimal volume $\delta\tau$

$$\iint_{\delta S} \mathbf{A} \cdot \mathbf{n} \, dS = \delta\tau \, \mathrm{div} \, \mathbf{A}$$

Eq. (6.9) may be expressed, per unit volume of the fluid, as

$$\frac{\partial}{\partial t} \rho\left(e + \frac{V^2}{2}\right) = -\mathrm{div} \, \rho\left(e + \frac{V^2}{2}\right)\mathbf{V} + \rho\mathbf{f} \cdot \mathbf{V} - \mathrm{div} \, p\mathbf{V} \qquad (6.10)$$

This equation is usually referred to as the *equation of conservation of energy*. The vector $\rho[e + (V^2/2)]\mathbf{V}$ is known as the *energy flux vector*. Using the equation of change for mass (6.1), Eq. (6.10) may be reduced to the form

$$\rho \frac{D}{Dt}\left(e + \frac{V^2}{2}\right) = \rho\mathbf{f} \cdot \mathbf{V} - \mathrm{div} \, p\mathbf{V}$$

which is the equation of energy derived from the point of view of a moving fluid element.

6.5 Integral Form of the Equations from the Point of View of a Fixed Region of Space

In the preceding sections we obtained the differential form of the equations of change for mass, momentum, and energy by considering a small fixed volume of space in the flowing fluid. Instead of a small region we now consider a finite region R of space enclosed by a surface S fixed in the flowing fluid. The equations of change then take the following form:

(1) *Mass*

$$\frac{\partial}{\partial t} \iiint_{R} \rho \, d\tau = -\iint_{S} \rho\mathbf{V} \cdot \mathbf{n} \, dS \qquad (6.11)$$

(2) *Momentum*

$$\frac{\partial}{\partial t} \iiint_{R} \rho\mathbf{V} \, d\tau = -\iint_{S} \mathbf{V}(\rho\mathbf{V} \cdot \mathbf{n}) \, dS + \iiint_{R} \rho\mathbf{f} \, d\tau - \iint_{S} p\mathbf{n} \, dS \qquad (6.12)$$

(3) *Energy*

$$\frac{\partial}{\partial t} \iiint_{R} \rho\left(e + \frac{V^2}{2}\right) d\tau = -\oiint_{S} \rho\left(e + \frac{V^2}{2}\right) \mathbf{V} \cdot \mathbf{n} \, dS$$

$$+ \iiint_{R} \rho\mathbf{f} \cdot \mathbf{V} \, d\tau - \oiint_{S} p\mathbf{V} \cdot \mathbf{n} \, dS \qquad (6.13)$$

Equations (6.11) to (6.13) are referred to as the *integral form of the conservation equations for an inviscid fluid.* In the left-hand side of each of these equations, the partial derivative $\partial/\partial t$ may be taken inside the integral sign, for R is a fixed volume.

The integral relations (6.11) to (6.13) may be reduced to the differential form by assuming that the functions involved have sufficiently many derivatives in the region R and letting the volume of the region R tend to zero or, as is usually done, by applying the general integral relations

$$\oiint_{S} \phi\mathbf{n} \, dS = \iiint_{R} \operatorname{grad} \phi \, d\tau$$

$$\oiint_{S} \mathbf{A} \cdot \mathbf{n} \, dS = \iiint_{R} \operatorname{div} \mathbf{A} \, d\tau$$

6.6 Integral Form of the Equations from the Point of View of a Finite Fluid Region

In the last chapter we derived the equations of mass, motion, and energy from the point of view of a moving fluid element. If, instead of a fluid element, we follow a finite *fluid region*, we obtain these equations in an integral form. *By a fluid region we mean a region that is composed of the same fluid elements for all times.* Suppose that at some instant of time we mark out a certain region R of the fluid as that enclosed by a surface S. If the surface moves with the fluid such that it always passes through the same fluid elements, it will always enclose the same fluid elements. Thus the surface S is a fluid surface and the region R it encloses is a fluid region. In general, the shape and size of a fluid region change as the region moves with the fluid.

Now, applying the law of constancy of mass, the second law of Newton, and the law of energy conservation to a fluid region R enclosed by a fluid surface S, we obtain the following equations:

(1) *Mass*

$$\frac{D}{Dt} \iiint_{R(t)} \rho \, d\tau = 0 \qquad (6.14)$$

(2) *Motion*

$$\frac{D}{Dt}\iiint_{R(t)} \rho \mathbf{V}\, d\tau = \iiint_{R(t)} \rho \mathbf{f}\, d\tau - \oiint_{S(t)} p\mathbf{n}\, dS \qquad (6.15)$$

(3) *Energy*

$$\frac{D}{Dt}\iiint_{R(t)} \rho\left(e + \frac{V^2}{2}\right) d\tau = \iiint_{R(t)} \rho \mathbf{f}\cdot\mathbf{V}\, d\tau - \oiint_{S(t)} p\mathbf{V}\cdot\mathbf{n}\, dS \qquad (6.16)$$

The notation $R(t)$ and $S(t)$ are used to emphasize that we are dealing with a fluid region and a fluid surface.

In each of these equations the left-hand side represents the rate of change of the total amount of a quantity associated with the fluid that is contained in the fluid surface S at some instant; the rate of change is computed following that portion of fluid, that is, following the fluid surface S or, equivalently, the fluid region R. Such a rate of change may generally be represented as

$$\frac{D}{Dt}\iiint_{R(t)} \rho Q\, d\tau \qquad (6.17)$$

where $Q = Q(\mathbf{r}, t)$ is any fluid quantity (scalar or vector) specified per unit mass of the fluid. Now, although the volume $R(t)$ changes with time, the mass contained in that volume is a constant for all times. Therefore we temporarily change the volume integral in (6.17) into an integral over the mass by writing $\rho\, d\tau = dm$ and obtain

$$\frac{D}{Dt}\iiint_{R(t)} \rho Q\, d\tau = \frac{D}{Dt}\int_{\substack{\text{mass} \\ \text{in } R(t)}} Q\, dm = \int_{\substack{\text{mass} \\ \text{in } R(t)}} \frac{D}{Dt}(Q\, dm)$$

Since the mass dm of a fluid element is a constant for all times, this relation becomes

$$\frac{D}{Dt}\iiint_{R(t)} \rho Q\, d\tau = \int_{\substack{\text{mass} \\ \text{in } R(t)}} \frac{DQ}{Dt}\, dm = \iiint_{R(t)} \rho \frac{DQ}{Dt}\, d\tau \qquad (6.18)$$

By using this relation, the left-hand side of each of the equations (6.14) to (6.16) may be represented as a volume integral.

The equations obtained in this section are equivalent to the equations obtained from the point of view of a fixed region of space (see Section 6.5). One way to show this is by means of the general relation

$$\rho \frac{DQ}{Dt} = \frac{\partial}{\partial t}(\rho Q) + \operatorname{div} \rho Q\mathbf{V} \qquad (6.19)$$

where, as before, $Q = Q(\mathbf{r}, t)$ denotes per unit mass of the fluid any quantity associated with the fluid. Another way, which is more direct and simple, is based on the following interpretation of the material derivative of

$$\iiint_{R(t)} \rho Q \, d\tau.$$

6.7 Rate of Change of a Quantity Following a Fluid Region

Let S denote the bounding surface of a certain fluid region at the time t (Fig. 6.1). In a short time interval δt thereafter the different fluid elements composing the fluid region move to new locations. Let S' denote the bounding surface of the fluid region at time $t + \delta t$ (Fig. 6.1). Denote by

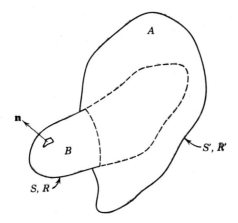

Fig. 6.1 Illustrating the computation of the rate of change of a quantity following a fluid region.

R the volume enclosed by the surface S and by R' the volume enclosed by S'. If $Q = Q(\mathbf{r}, t)$ denotes per unit mass any quantity (scalar or vector) associated with the fluid, by definition we have

$$\frac{D}{Dt} \iiint_{R(t)} \rho Q \, d\tau = \lim_{\delta t \to 0} \frac{1}{\delta t} \left[\iiint_{R'} \rho Q \, d\tau - \iiint_{R} \rho Q \, d\tau \right]$$

where, as before, $R(t)$ denotes the volume of the moving fluid region. We introduce temporarily the notation

$$G(R, t) \equiv \iiint_{R} \rho Q \, d\tau$$

$$G(R', t + \delta t) \equiv \iiint_{R'} \rho Q \, d\tau$$

with the obvious meaning that G represents the total amount of the quantity concerned in the volume under consideration. With this meaning for G, and referring to Fig. 6.1, we write

$$G(R', t + \delta t) = G(R, t + \delta t) + G(A, t + \delta t) - G(B, t + \delta t)$$

Now, to the first order in δt, we have

$$G(R, t + \delta t) = G(R, t) + \left[\frac{\partial}{\partial t} G(R, t) \right]_{R \text{ fixed}} \delta t$$

and

$G(A, t + \delta t) - G(B, t + \delta t) =$ the *net* outflow, during the time δt, of G through the surface S fixed in the flow field

$$= \delta t \oiint_{S \text{ fixed}} Q(\rho \mathbf{V} \cdot \mathbf{n}) \, dS$$

It therefore follows that

$$\frac{D}{Dt} \iiint_{R(t)} \rho Q \, d\tau = \frac{\partial}{\partial t} \iiint_{\substack{R \\ \text{fixed}}} \rho Q \, d\tau + \oiint_{\substack{S \\ \text{fixed}}} Q(\rho \mathbf{V} \cdot \mathbf{n}) \, dS$$

$$= \iiint_{R} \frac{\partial \rho Q}{\partial t} \, d\tau + \oiint_{S} Q(\rho \mathbf{V} \cdot \mathbf{n}) \, dS \qquad (6.20)$$

This equation expresses the significant result that the material rate of change of the total amount of any quantity associated with the fluid which at any instant is contained in a surface S fixed in space is equal to the sum of two parts, (1) a part that is the local rate of change of the total amount of the quantity in the fixed volume R enclosed by the surface S and (2) a part that is the net outflow per unit time of the quantity concerned through the surface S fixed in the flowing fluid. The differential form of (6.20) is simply

$$\rho \frac{DQ}{Dt} = \frac{\partial \rho Q}{\partial t} + \text{div } \rho Q \mathbf{V}$$

which is Eq. (6.19).

Using (6.19) and (6.20) we may readily show that the equations derived from the point of view of a moving fluid element or fluid region are equivalent to the equations of change (the so-called conservation equations) derived from the point of view of a fixed, infinitesimal or finite, volume of space. In particular, we observe that the equation of change for the momentum is indeed an expression of Newton's second law of motion.

6.8 Equations of Change for an Ideal Fluid

For an ideal fluid, the density and internal energy are constant for all times and the equations of change take the following forms:

(1) *Differential form of the equations* (see Eqs. 6.3, 6.8, 6.10, respectively)

Mass: $\operatorname{div} \mathbf{V} = 0$

Momentum:

$$\rho \frac{\partial \mathbf{V}}{\partial t} = -\rho \operatorname{div}(\mathbf{V}\mathbf{V}) + \rho \mathbf{f} - \operatorname{grad} p$$

$$= -\rho \mathbf{V} \cdot \operatorname{grad} \mathbf{V} + \rho \mathbf{f} - \operatorname{grad} p$$

Energy:

$$\rho \frac{\partial}{\partial t}\frac{V^2}{2} = -\rho \operatorname{div}\left(\frac{V^2}{2}\mathbf{V}\right) + \rho \mathbf{f} \cdot \mathbf{V} - \operatorname{div} p\mathbf{V}$$

$$= \rho \mathbf{V} \cdot \operatorname{grad}\frac{V^2}{2} + \rho \mathbf{f} \cdot \mathbf{V} - \mathbf{V} \cdot \operatorname{grad} p$$

(2) *Integral form of the equations* (see Eqs. 6.11, 6.12, 6.13, respectively)

Mass:

$$\oiint_S \mathbf{V} \cdot \mathbf{n}\, dS = 0$$

Momentum:

$$\rho \iiint_R \frac{\partial \mathbf{V}}{\partial t}\, d\tau = -\rho \oiint_S \mathbf{V}(\mathbf{V} \cdot \mathbf{n})\, dS + \rho \iiint_R \mathbf{f}\, d\tau - \oiint_S p\mathbf{n}\, dS$$

Energy:

$$\rho \iiint_R \frac{\partial}{\partial t}\left(\frac{V^2}{2}\right) d\tau = -\rho \oiint_S \frac{V^2}{2}\mathbf{V} \cdot \mathbf{n}\, dS$$

$$+ \rho \iiint_R \mathbf{f} \cdot \mathbf{V}\, d\tau - \oiint_S p\mathbf{V} \cdot \mathbf{n}\, dS$$

6.9 Rate of Change of a Quantity Following a Moving Region of Space

In Section 6.7 we considered the total time rate of change of a quantity following a fluid region. We now inquire into the total time rate of change of a quantity following a region that moves with an arbitrary velocity different from that of the fluid. In other words, let us consider the situation in which the moving region is not a fluid region.

Let S denote the surface of a region R which moves through a scalar or vector field $A = A(\mathbf{r}, t)$. Let the velocity of an element of the surface S be denoted by $\boldsymbol{\xi}$. In general $\boldsymbol{\xi}$ varies over S. The region R and the enclosing surface S are functions of time, and they do not in general retain their shape or size:

$$R = R(t) \quad \text{and} \quad S = S(t)$$

We ask for the total time rate of change of the quantity $\iiint\limits_{R(t)} A(\mathbf{r}, t)\, d\tau$. This rate is given by

$$\frac{d}{dt} \iiint\limits_{R(t)} A\, d\tau = \lim_{\delta t \to 0} \frac{1}{\delta t} \left(\iiint\limits_{R(t+\delta t)} A\, d\tau - \iiint\limits_{R(t)} A\, d\tau \right) \tag{6.21}$$

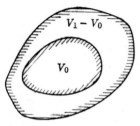

Denote by V_0 the volume of space occupied by R at time t and by V_1 that occupied by R at time $t + \delta t$. We have

$$\iiint\limits_{R(t)} A\, d\tau = \iiint\limits_{V_0} A(\mathbf{r}, t)\, d\tau$$

$$\iiint\limits_{R(t+\delta t)} A\, d\tau = \iiint\limits_{V_1} A(\mathbf{r}, t + \delta t)\, d\tau$$

Fig. 6.2 Illustrating the rate of change of a quantity following a moving region.

For simplicity, suppose that at time t the velocity $\boldsymbol{\xi}$ is directed outward all around the surface of R; then V_1 encloses V_0 as shown in Fig. 6.2. We write

$$\iiint\limits_{V_1} A(\mathbf{r}, t + \delta t)\, d\tau = \iiint\limits_{V_0} A(\mathbf{r}, t + \delta t)\, d\tau + \iiint\limits_{V_1-V_0} A(\mathbf{r}, t + \delta t)\, d\tau$$

Furthermore, we have

$$\iiint\limits_{V_0} A(\mathbf{r}, t + \delta t)\, d\tau = \iiint\limits_{V_0} \left[A(\mathbf{r}, t) + \frac{\partial A}{\partial t}\, \delta t \right] d\tau$$

to first order in δt and

$$\iiint\limits_{V_1-V_0} A(\mathbf{r}, t + \delta t)\, d\tau = \text{the amount of } A \text{ swept by the surface of } R \text{ during the time } \delta t$$

$$= \delta t \text{ times the rate at which the surface of } R \text{ sweeps the field at the instant } t$$

$$= \delta t \left(\oiint\limits_S A\boldsymbol{\xi} \cdot \mathbf{n}\, dS \right)$$

to first order in δt. With these results it follows that

$$\frac{d}{dt} \iiint\limits_{R(t)} A \, d\tau = \iiint\limits_{V_0} \frac{\partial A}{\partial t} \, d\tau + \oiint\limits_{S} A \boldsymbol{\xi} \cdot \mathbf{n} \, dS$$

$$= \iiint\limits_{R} \frac{\partial A}{\partial t} \, d\tau + \oiint\limits_{S} A \boldsymbol{\xi} \cdot \mathbf{n} \, dS \qquad (6.22)$$

If the region is a fluid region, the surface moves with the fluid; $\boldsymbol{\xi}$ equals the fluid velocity $\mathbf{V}$. Equation 6.22 then reduces, as it should, to (6.20) for the material derivative of $\iiint\limits_{R(t)} A \, d\tau$.

Chapter 7

Equations of Discontinuous Motion

In Chapter 5 we saw that at a free surface the tangential component of the fluid velocity is discontinuous across the surface. In this sense the motion is discontinuous and the free surface is a discontinuity surface. In general, *a discontinuity surface (or simply a discontinuity) is any surface across which the motion is discontinuous*, that is, *across which jumps may occur in any of the fluid properties*. We now investigate in a general way the type of discontinuous motion that is possible in the flow of an ideal fluid. For this purpose we assume that a discontinuity is present in the flow and inquire into the relations that must be satisfied by the fluid properties across the discontinuity so that the motion obeys the basic laws of fluid flow. The discontinuity surface may be stationary (i.e., fixed relative to a fixed observer) or it may be in motion, the velocity of any element of the surface being in general different from that of the fluid on either side of it. We shall consider both these cases. In the following we treat first the motion of an ideal fluid, and in that context we shall assume that the density and the internal energy are constant throughout the flow field for all times. The case of an inhomogeneous incompressible inviscid fluid is then considered.

7.1 A Stationary Discontinuity in a Steady Flow

Let the surface D denote a *stationary discontinuity* in a *steady flow field* (see Fig. 7.1). We designate one side of the surface as the positive side and the other as the negative side and introduce the following notation:

D_1 denotes the positive side of the surface
D_2 denotes the negative side of the surface
$\mathbf{n}_d \, dS$ denotes any surface element on D_1
$-\mathbf{n}_d \, dS$ denotes the corresponding surface element on D_2
$\mathbf{V}_1$ and p_1 denote the velocity and pressure at any point on D_1
$\mathbf{V}_2$ and p_2 denote the velocity and pressure at the corresponding point on D_2

Our problem is to derive, according to the basic laws of fluid motion, the relations between $\mathbf{V}_1, p_1$ and $\mathbf{V}_2, p_2$.

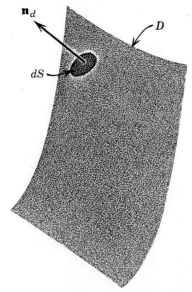

Fig. 7.1 Stationary discontinuity in steady flow.

For this purpose we choose, as shown in Fig. 7.2, a region R that is *fixed* in space and contains the discontinuity such that the discontinuity separates R into two parts. According to the integral form of the conservation equations for an ideal fluid (see Section 6.5) we have

Conservation of mass

$$\oiint_S \mathbf{V} \cdot \mathbf{n} \, dS = 0 \tag{7.1}$$

Conservation of momentum

$$\rho \oiint_S \mathbf{V}(\mathbf{V} \cdot \mathbf{n}) \, dS = -\oiint_S p\mathbf{n} \, dS + \rho \iiint_R \mathbf{f} \, d\tau \tag{7.2}$$

Conservation of energy

$$\rho \oiint_S \frac{V^2}{2} (\mathbf{V} \cdot \mathbf{n}) \, dS = -\oiint_S p\mathbf{V} \cdot \mathbf{n} \, dS + \rho \iiint_R \mathbf{f} \cdot \mathbf{V} \, d\tau \tag{7.3}$$

where S is the fixed surface enclosing R.

The boundary of the discontinuity marks out the surface S into two distinct parts, which we denote by Σ_1 and Σ_2. Introducing Σ_1 and Σ_2 we express each of the surface integrals in the Eqs. (7.1) to (7.3) as the sum of two surface integrals:

$$\oiint_S (\quad)\, dS = \iint_{\Sigma_1} (\quad)\, dS + \iint_{\Sigma_2} (\quad)\, dS$$

The conservation equations are valid no matter how small the region R is. We therefore perform a limiting process by letting the region R shrink to the discontinuity surface D. In such a limiting process Σ_1 tends to D_1,

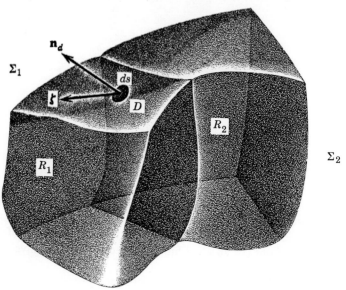

Fig. 7.2   Fluid region containing a discontinuity.

Σ_2 tends to D_2, and the volume integrals in (7.2) and (7.3) tend to zero. Thus we obtain the following relations:

Conservation of mass

$$\iint_D (\mathbf{V}_1 - \mathbf{V}_2) \cdot \mathbf{n}_d\, dS = 0 \tag{7.4}$$

Conservation of momentum

$$\rho \iint_D [\mathbf{V}_1(\mathbf{V}_1 \cdot \mathbf{n}_d) - \mathbf{V}_2(\mathbf{V}_2 \cdot \mathbf{n}_d)]\, dS = -\iint_D (p_1 - p_2)\mathbf{n}_d\, dS \tag{7.5}$$

Conservation of energy

$$\frac{\rho}{2} \iint_D [V_1{}^2(\mathbf{V}_1 \cdot \mathbf{n}_d) - V_2{}^2(\mathbf{V}_2 \cdot \mathbf{n}_d)]\, dS = -\iint_D (p_1\mathbf{V}_1 - p_2\mathbf{V}_2) \cdot \mathbf{n}_d\, dS \tag{7.6}$$

Now, because the relations (7.4) to (7.6) hold for any arbitrary portion of the surface D, they reduce to the following conditions:

Conservation of mass

$$(\mathbf{V}_1 - \mathbf{V}_2) \cdot n_d = 0$$

or
$$\mathbf{V}_1 \cdot \mathbf{n}_d = \mathbf{V}_2 \cdot \mathbf{n}_d = \lambda, \text{ say.}$$
(7.7)

Conservation of momentum

$$\rho\lambda(\mathbf{V}_1 - \mathbf{V}_2) = (p_2 - p_1)\mathbf{n}_d$$
(7.8)

Conservation of energy

$$\frac{\rho}{2}(\lambda V_1{}^2 - \lambda V_2{}^2) = (\lambda p_2 - \lambda p_1)$$
(7.9)

These conditions must be satisfied at each point of the discontinuity. Equation (7.7) states that

$$(\mathbf{V}_1 - \mathbf{V}_2) \text{ is perpendicular to } \mathbf{n}_d$$

whereas Eq. (7.8) states that if $\lambda \neq 0$, $(\mathbf{V}_1 - \mathbf{V}_2)$ is parallel to $\mathbf{n}_d$. Therefore it follows that the solution of (7.7) and (7.8) is

$$\lambda = \mathbf{V}_1 \cdot \mathbf{n}_d = \mathbf{V}_2 \cdot \mathbf{n}_d = 0$$
(7.10)

$$p_1 = p_2$$
(7.11)

Equation (7.9) is then automatically satisfied. From (7.10) we infer that *there is no flow across the surface and that the tangential component of the fluid velocity is discontinuous across it*; in other words, the surface D is a tangential discontinuity. We thus conclude that *only a tangential discontinuity is possible in the steady flow of an ideal fluid.*

7.2 A Moving Discontinuity in the Unsteady Flow of an Ideal Fluid

We now take up the more general case where the discontinuity surface is moving through the flow field which is unsteady. Let ζ denote the velocity of any element of the surface. This velocity may, in general, be nonuniform over the surface. We use the same scheme as before to describe the two sides of the discontinuity and the fluid quantities on its two sides.

We choose, as shown in Fig. 7.2, a *fluid* region* R that contains the discontinuity D such that it separates R into two distinct parts, R_1 and R_2. Also, the boundary of D marks out the surface S enclosing R into two distinct parts, Σ_1 and Σ_2. We note that R, R_1, R_2, S, Σ_1, Σ_2 are all functions of time. Now, according to the integral form of the basic equations

* Note that now R is not fixed in space; being a fluid region it moves with the fluid.

governing the motion of an ideal fluid (see Section 6.6), we have

Condition of incompressibility

$$\frac{D}{Dt} \iiint\limits_{R(t)} d\tau = 0 \tag{7.12}$$

Mass

$$\frac{D}{Dt} \iiint\limits_{R(t)} \rho \, d\tau = \rho \frac{D}{Dt} \iiint\limits_{R(t)} d\tau = 0$$

This reduces to Eq. (7.12).

Motion

$$\rho \frac{D}{Dt} \iiint\limits_{R(t)} \mathbf{V} \, d\tau = \rho \iiint\limits_{R(t)} \mathbf{f} \, d\tau - \oiint\limits_{S(t)} p\mathbf{n} \, dS$$

$$= \rho \iiint\limits_{R(t)} \mathbf{f} \, d\tau + \iint\limits_{\Sigma_1(t)} p\mathbf{n} \, dS - \iint\limits_{\Sigma_2(t)} p\mathbf{n} \, dS \tag{7.13}$$

Energy

$$\frac{\rho}{2} \frac{D}{Dt} \iiint\limits_{R(t)} V^2 \, d\tau = \rho \iiint\limits_{R(t)} \mathbf{f} \cdot \mathbf{V} \, d\tau - \oiint\limits_{S(t)} p\mathbf{V} \cdot \mathbf{n} \, dS$$

$$= \rho \iiint\limits_{R(t)} \mathbf{f} \cdot \mathbf{V} \, d\tau - \iint\limits_{\Sigma_1(t)} p\mathbf{V} \cdot \mathbf{n} \, dS - \iint\limits_{\Sigma_2(t)} p\mathbf{V} \cdot \mathbf{n} \, dS \tag{7.14}$$

We now wish to express the left-hand side of each of these equations in terms of volume and surface integrals related to the regions R_1 and R_2. For this purpose we note that

$$\frac{D}{Dt} \iiint\limits_{R(t)} (\ ) \, d\tau \neq \frac{D}{Dt} \iiint\limits_{R_1(t)} (\ ) \, d\tau + \frac{D}{Dt} \iiint\limits_{R_2(t)} (\ ) \, d\tau$$

for although the region R is a fluid region, the regions R_1 and R_2 are not. This is so because the surface D_1 (or D_2), which forms a portion of the surface that encloses the region R_1 (or R_2), is not a fluid surface. In view of this we proceed as follows. We recall (see Section 6.9) that if $P(t)$ is a (moving) time-dependent region of space in a scalar or vector field $Q(\mathbf{r}, t)$, the total time rate of change of the integral $\iiint\limits_{P(t)} Q(\mathbf{r}, t) \, d\tau$ following the region $P(t)$ is given by

$$\frac{d}{dt} \iiint\limits_{P(t)} Q \, d\tau = \iiint\limits_{P(t)} \frac{\partial Q}{\partial t} \, d\tau + \oiint\limits_{\sigma(t)} Q\boldsymbol{\xi} \cdot \mathbf{n} \, dS \tag{7.15}$$

where $\sigma(t)$ is the time-dependent surface of the region $P(t)$ and $\boldsymbol{\xi}$ is the velocity of any element $\mathbf{n} \, dS$ of the surface $\sigma(t)$. Now, the left-hand side of each of the Eqs. (7.12) to (7.14) is of the form

$$\frac{D}{Dt} \iiint_{R(t)} Q \, d\tau$$

where Q may be 1 or $\rho \mathbf{V}$ or $\rho(V^2/2)$. This is simply the total time rate of change of $\iiint_{R(t)} Q \, d\tau$ following the time-dependent region $R(t)$. Hence we write*

$$\frac{D}{Dt} \iiint_{R(t)} Q \, d\tau = \frac{d}{dt} \iiint_{R(t)} Q \, d\tau = \frac{d}{dt} \iiint_{R_1(t)} Q \, d\tau + \frac{d}{dt} \iiint_{R_2(t)} Q \, d\tau \quad (7.16)$$

Using Eq. (7.15) and denoting by $S_1(t)$ the surface enclosing R_1 we obtain

$$\frac{d}{dt} \iiint_{R_1(t)} Q \, d\tau = \iiint_{R_1(t)} \frac{\partial Q}{\partial t} \, d\tau + \iint_{S_1(t)} Q \boldsymbol{\zeta} \cdot \mathbf{n} \, dS$$

$$= \iiint_{R_1(t)} \frac{\partial Q}{\partial t} \, d\tau + \iint_{\Sigma_1(t)} Q \mathbf{V} \cdot \mathbf{n} \, dS - \iint_{D_1(t)} Q_1 \boldsymbol{\zeta} \cdot \mathbf{n}_d \, dS \quad (7.17)$$

where Q_1 signifies a fluid quantity on the positive side D_1 of the discontinuity. Similarly, we obtain

$$\frac{d}{dt} \iiint_{R_2(t)} Q \, d\tau = \iiint_{R_2(t)} \frac{\partial Q}{\partial t} \, d\tau + \iint_{\Sigma_2(t)} Q \mathbf{V} \cdot \mathbf{n} \, dS + \iint_{D_2(t)} Q_2 \boldsymbol{\zeta} \cdot \mathbf{n}_d \, dS \quad (7.18)$$

where Q_2 signifies a fluid quantity on the negative side D_2 of the discontinuity. Combining (7.17) and (7.18), we obtain the result

$$\frac{D}{Dt} \iiint_{R(t)} Q \, d\tau = \iiint_{R(t)} \frac{\partial Q}{\partial t} \, d\tau + \iint_{\Sigma_1(t)} Q \mathbf{V} \cdot \mathbf{n} \, dS - \iint_{D_1(t)} Q_1 \boldsymbol{\zeta} \cdot \mathbf{n}_d \, dS$$

$$+ \iint_{\Sigma_2(t)} Q \mathbf{V} \cdot \mathbf{n} \, dS + \iint_{D_2(t)} Q_2 \boldsymbol{\zeta} \cdot \mathbf{n}_d \, dS \quad (7.19)$$

* Although

$$\frac{D}{Dt} \iiint_{R(t)} Q \, d\tau = \frac{d}{dt} \iiint_{R(t)} Q \, d\tau$$

$$\frac{d}{dt} \iiint_{R_1(t)} Q \, d\tau \neq \frac{D}{Dt} \iiint_{R_1(t)} Q \, d\tau$$

for $R_1(t)$ is not a fluid region. Similarly,

$$\frac{d}{dt} \iiint_{R_2(t)} Q \, d\tau \neq \frac{D}{Dt} \iiint_{R_2(t)} Q \, d\tau$$

Now, we rewrite the left-hand sides of (7.12) to (7.14) in the form expressed by (7.19) and perform a limiting process in which the region R is allowed to shrink to the discontinuity surface D. The surface Σ_1 tends to D_1, the surface Σ_2 tends to D_2, and the volume integrals tend to zero. In this way we obtain the following relations:

Condition of incompressibility

$$\iint_D [(\mathbf{V}_1 - \boldsymbol{\zeta}) \cdot \mathbf{n}_d - (\mathbf{V}_2 - \boldsymbol{\zeta}) \cdot \mathbf{n}_d] \, dS = 0 \qquad (7.20)$$

Motion

$$\rho \iint_D \{\mathbf{V}_1[(\mathbf{V}_1 - \boldsymbol{\zeta}) \cdot \mathbf{n}_d] - \mathbf{V}_2[(\mathbf{V}_2 - \boldsymbol{\zeta}) \cdot \mathbf{n}_d]\} \, ds$$
$$= \iint_D (p_2 - p_1)\mathbf{n}_d \, dS \quad (7.21)$$

Energy

$$\frac{\rho}{2} \iint_D \{V_1{}^2[(\mathbf{V}_1 - \boldsymbol{\zeta}) \cdot \mathbf{n}_d] - V_2{}^2[(\mathbf{V}_2 - \boldsymbol{\zeta}) \cdot \mathbf{n}_d]\} \, dS$$
$$= \iint_D (p_2\mathbf{V}_2 - p_1\mathbf{V}_1) \cdot \mathbf{n}_d \, dS \quad (7.22)$$

Since the relations (7.20) to (7.22) hold for any arbitrary portion of the surface D, they reduce to the following conditions:

Condition of incompressibility

$$(\mathbf{V}_1 - \boldsymbol{\zeta}) \cdot \mathbf{n}_d = (\mathbf{V}_2 - \boldsymbol{\zeta}) \cdot \mathbf{n}_d = \lambda, \text{ say}$$

or

$$(\mathbf{V}_1 - \mathbf{V}_2) \cdot \mathbf{n}_d = 0$$

(7.23)

Motion

$$\rho\lambda(\mathbf{V}_1 - \mathbf{V}_2) = (p_2 - p_1)\mathbf{n}_d \qquad (7.24)$$

Energy

$$\frac{\rho}{2}(V_1{}^2 - V_2{}^2) = (p_2\mathbf{V}_2 - p_1\mathbf{V}_1) \cdot \mathbf{n}_d \qquad (7.25)$$

These conditions must be satisfied at each point of the discontinuity. From (7.23) and (7.24) we infer that

$$\lambda = (\mathbf{V}_1 - \boldsymbol{\zeta}) \cdot \mathbf{n}_d = (\mathbf{V}_2 - \boldsymbol{\zeta}) \cdot \mathbf{n}_d = 0$$

or

$$\mathbf{V}_1 \cdot \mathbf{n}_d = \mathbf{V}_2 \cdot \mathbf{n}_d = \boldsymbol{\zeta} \cdot \mathbf{n}_d$$

(7.26)

and that

$$p_1 = p_2 \tag{7.27}$$

Equation (7.25) is then automatically satisfied.

Equations (7.26) and (7.27) show that there is no flow across the discontinuity surface, that the tangential component of the fluid velocity is discontinuous across this surface and is different on either side of the surface from the tangential component of the velocity of the surface. Thus the discontinuity is a tangential discontinuity. We therefore conclude that *in the motion of an ideal fluid only a tangential discontinuity* (whether stationary or moving) *is all that is possible.* For a stationary discontinuity, that is, for ζ equal to zero, (7.26) reduces to

$$\mathbf{V}_1 \cdot \mathbf{n}_d = \mathbf{V}_2 \cdot \mathbf{n}_d = 0$$

which is precisely Eq. (7.10) derived in the previous section. Equation (7.10) applies to a moving discontinuity if we replace the fluid velocity $\mathbf{V}$ by the relative velocity $(\mathbf{V} - \zeta)$ between the fluid and the moving discontinuity.

7.3 Discontinuity in the Flow of an Inhomogeneous Incompressible Fluid

We now investigate the possibility of discontinuous motion in the flow of an inhomogeneous, incompressible, inviscid fluid. We consider a moving discontinuity in the flow of such a fluid and adopt the same notation as in the foregoing section to describe the necessary details of the flow field (see also Fig. 7.2).

In the motion of an inhomogeneous, incompressible, inviscid fluid, the density and the internal energy vary, in general, with time and position although the volume of any fluid region is preserved for all times (recall Section 5.8). The equations governing the motion of such a fluid region $R(t)$ containing the discontinuity (Fig. 7.2) are

Condition of incompressibility

$$\frac{D}{Dt} \iiint_{R(t)} d\tau = 0$$

Mass

$$\frac{D}{Dt} \iiint_{R(t)} \rho \, d\tau = 0$$

Motion

$$\frac{D}{Dt} \iiint_{R(t)} \rho \mathbf{V} \, d\tau = \iiint_{R(t)} \rho \mathbf{f} \, d\tau - \oiint_{S(t)} p\mathbf{n} \, dS$$

Energy

$$\frac{D}{Dt} \iiint_{R(t)} \rho \left(e + \frac{V^2}{2} \right) d\tau = \iiint_{R(t)} \rho \mathbf{f} \cdot \mathbf{V} \, d\tau - \oiint_{S(t)} p\mathbf{V} \cdot \mathbf{n} \, dS$$

We rewrite the left-hand side of each of these equations in the form expressed by Eq. (7.19) and perform the limiting procedure, as before, in which the region R is allowed to shrink to the discontinuity surface D. We thus obtain the following relations:

Condition of incompressibility

$$\iint_{D} [(\mathbf{V}_1 - \boldsymbol{\zeta}) \cdot \mathbf{n}_d - (\mathbf{V}_2 - \boldsymbol{\zeta}) \cdot \mathbf{n}_d] \, dS = 0$$

Mass

$$\iint_{D} [\rho_1(\mathbf{V}_1 - \boldsymbol{\zeta}) \cdot \mathbf{n}_d - \rho_2(\mathbf{V}_2 - \boldsymbol{\zeta}) \cdot \mathbf{n}_d] \, dS = 0$$

Motion

$$\iint_{D} [\rho_1\mathbf{V}_1(\mathbf{V}_1 - \boldsymbol{\zeta}) \cdot \mathbf{n}_d - \rho_2\mathbf{V}_2(\mathbf{V}_2 - \boldsymbol{\zeta}) \cdot \mathbf{n}_d] \, dS = \iint_{D} (p_2 - p_1)\mathbf{n}_d \, dS$$

Energy

$$\iint_{D} \left\{ \rho_1 \left(e_1 + \frac{V_1^2}{2} \right)[(\mathbf{V}_1 - \boldsymbol{\zeta}) \cdot \mathbf{n}_d] - \rho_2 \left(e_2 + \frac{V_2^2}{2} \right)[(\mathbf{V}_2 - \boldsymbol{\zeta}) \cdot \mathbf{n}_d] \right\} dS$$

$$= \iint_{D} (p_2\mathbf{V}_2 - p_1\mathbf{V}_1) \cdot \mathbf{n}_d \, dS$$

Since these relations hold for any arbitrary portion of the surface D, they reduce to the following conditions:

Condition of incompressibility

$$(\mathbf{V}_1 - \boldsymbol{\zeta}) \cdot \mathbf{n}_d = (\mathbf{V}_2 - \boldsymbol{\zeta}) \cdot \mathbf{n}_d = 0$$

or

$$\mathbf{V}_1 \cdot \mathbf{n}_d = \mathbf{V}_2 \cdot \mathbf{n}_d = \boldsymbol{\zeta} \cdot \mathbf{n}_d$$

Motion

$$p_2 = p_1$$

The equations of mass and energy are automatically satisfied. We thus conclude that *in the motion of an inviscid incompressible fluid* (*whether it is homogeneous or inhomogeneous*), *a tangential discontinuity* (*whether stationary or moving*) *is all that is possible*. The conditions to be satisfied across the discontinuity are given in general by Eqs. (7.26) and (7.27).

7.4   Remarks

In view of the considerations given in this chapter we conclude that in the flow of an inviscid, incompressible fluid, the only possible type of a free surface (i.e., a fluid-fluid boundary) is one of tangential discontinuity. Sharp discontinuities as envisaged here are a consequence of the assumption that the fluid is inviscid (i.e., the viscosity if zero) and, therefore, should not be expected to appear in the flow of a viscous fluid. When

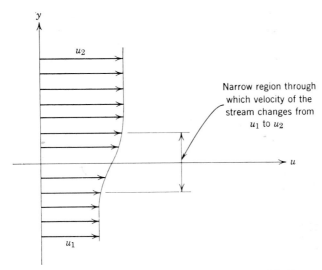

Fig. 7.3   Example of a region of vorticity in a real fluid.

finite viscosity, no matter how small, is taken into account, a sharp discontinuity is impossible. In the motion of a viscous fluid one, however, observes regions through which the velocity changes rapidly. An example of such a region is illustrated in Fig. 7.3. The region is a region of rotational motion or, equivalently, a region of voticity. In the formulation of theoretical models for real-fluid flows involving such regions of vorticity, one generally idealizes them as sharp discontinuities. In this sense, a discontinuity is a surface of concentrated vorticity and we refer to it as a *vortex surface* or *vortex sheet*. In ideal fluids *vortex surfaces and discontinuities are in in fact identical.* Vortex surfaces are employed in the analysis of many physical problems, in particular, of the *problem of lift.* We shall thus be concerned a great deal with flows involving vortex sheets. There are many properties associated with such flows and we shall learn about them in their proper context (see Chapters 17 to 19).

In real flows the narrow regions of vorticity originate generally at the solid surfaces present in the flow. The so-called boundary-layer region close to the solid surface and the extension of that region after it leaves the surface are usually narrow regions of vorticity. The region of vorticity that extends from the solid surface does not retain its shape permanently. It usually rolls up and forms so-called discrete vortices. The appearance of such vortices changes the flow field from that associated with the presence of only a narrow vortex region. This should be borne in mind when one formulates the theoretical models of real flows.

It can be shown that a discontinuity surface in an ideal fluid is unstable to any small disturbance. By this we mean that if a small disturbance, in the shape of the surface for instance, is introduced, the disturbance will continue to grow with time and thus completely alter the shape of the discontinuity.

We conclude this chapter with a few remarks about the type of discontinuity surface possible in the motion of a nonviscous, but compressible, fluid. The investigation in this case may be carried out in the same manner as done for the incompressible fluid. We then find that there are two possible types of discontinuities, (1) a tangential discontinuity as defined before and (2) a *normal discontinuity*, where the tangential component of the fluid velocity is continuous across the discontinuity while the normal component of the velocity, the pressure, and the density (as well as the other thermodynamic quantities) are discontinuous. Furthermore, we find that, so as to satisfy the second law of thermodynamics, the normal discontinuity can occur only in a certain range of flow speeds (namely in the range of the so-called supersonic speeds) and only as a compression discontinuity, that is, as the so-called *shock*. Across a shock discontinuity, the pressure and density increase in the direction of the flow velocity measured relative to the discontinuity while the normal component of the relative velocity decreases.

Chapter 8

Integration of Euler's Equation in Special Cases

In this chapter we shall consider briefly the question of integration of the equations that govern the motion of an ideal fluid. First we shall outline some of the mathematical features of the equations and then shall see how under certain circumstances it is possible to form an integral of Euler's equation. We shall find that there are two cases when this is possible. (1) When the body forces are irrotational and the motion is steady, integration is possible along a streamline. (2) When the body forces are irrotational and the motion itself is without rotation, integration is possible in any direction and for unsteady motion. These results naturally suggest that we inquire into the conditions under which fluid motions are irrotational. This we shall do in the next chapter.

8.1 Mathematical Character of the Equations

The differential equations governing the motion of an ideal fluid are

(1) *Condition of incompressibility*

$$\text{div } \mathbf{V} = 0 \tag{8.1}$$

which in Cartesians takes the form

$$\frac{\partial u}{\partial x} + \frac{\partial v}{\partial y} + \frac{\partial w}{\partial z} = 0 \tag{8.1a}$$

(2) *Euler's equation*

$$\frac{\partial \mathbf{V}}{\partial t} + \text{grad } \frac{V^2}{2} - \mathbf{V} \times \text{curl } \mathbf{V} = \mathbf{f} - \frac{1}{\rho} \text{grad } p \tag{8.2}$$

which in Cartesians is expressed by the three scalar equations:

$$\frac{\partial u}{\partial t} + u\frac{\partial u}{\partial x} + v\frac{\partial u}{\partial y} + w\frac{\partial u}{\partial z} = f_x - \frac{1}{\rho}\frac{\partial p}{\partial x}$$

$$\frac{\partial v}{\partial t} + u\frac{\partial v}{\partial x} + v\frac{\partial v}{\partial y} + w\frac{\partial v}{\partial z} = f_y - \frac{1}{\rho}\frac{\partial p}{\partial y} \qquad (8.2a)$$

$$\frac{\partial w}{\partial t} + u\frac{\partial w}{\partial x} + v\frac{\partial w}{\partial y} + w\frac{\partial w}{\partial z} = f_z - \frac{1}{\rho}\frac{\partial p}{\partial z}$$

where u, v, w are the components of $\mathbf{V}$ and f_x, f_y, f_z are the components of $\mathbf{f}$.

The velocity field $\mathbf{V}(\mathbf{r}, t)$ and the pressure field $p(\mathbf{r}, t)$ constitute four scalar unknowns. To determine these the system of four equations expressed by (8.1) and (8.2) are to be solved simultaneously together with specified *initial* and *boundary* conditions. We characterize these equations with the following remarks.

Since the equations involve more than one independent variable, they form a system of *partial differential equations*. The incompressibility condition involves three dependent variables (i.e., the velocity components) and three independent variables (the space coordinates), while the equation of motion involves four dependent variables (i.e., the three velocity components and the pressure) with four independent variables (the space coordinates and time). The integration of partial differential equations is a more difficult problem mathematically than that of ordinary differential equations.

The *order* of a differential equation is given by the order of the highest derivative that appears in the equation. Thus all four equations we have are of the first order. A differential equation is said to be *linear* if no non-linear terms, that is, products or powers of the unknowns and/or their derivatives, occur in the equation. If such products and powers appear, the equation is *nonlinear*. We notice that the incompressibility condition (8.1) is *linear*, whereas the Euler's equation (8.2) is *nonlinear*. The *non-linearity* of this equation is due to the so-called *convection terms* in the acceleration:

$$\operatorname{grad}\left(\frac{V^2}{2}\right) - \mathbf{V} \times \operatorname{curl} \mathbf{V}$$

or in Cartesians

$$u\frac{\partial u}{\partial x} + v\frac{\partial u}{\partial y} + w\frac{\partial u}{\partial z}, \qquad \text{and so forth.}$$

The integration of nonlinear differential equations is a very difficult mathematical problem. In fact, there is no general theory yet available for such equations. In case of a linear equation we find that the sum of two

independent solutions is also a solution, and that, therefore, we can build a complete solution by simply superposing various particular solutions. This *principle of superposition* cannot be used in case of a nonlinear equation.
Thus there is no general way by which we may attempt to solve the equations (8.1) and (8.2). In such a situation we may adopt alternative methods involving numerical integration or transformation of variables or approximations leading to linearization of the equations. These procedures, however, will not concern us in our present studies. Fortunately, the problem of *motion of an ideal fluid* takes on a different aspect than that of solving simultaneously the condition of incompressibility and the nonlinear Euler's equation. It turns out, as we shall show in the next chapter, that *for an ideal fluid under the action of rotation free forces all motions started from a state of rest are permanently irrotational.* This is very significant mathematically. When the angular velocity field is permanently zero, curl **V** is zero and the term **V** × curl **V** in the acceleration vanishes leaving the equation of motion in a form that can be integrated once and for all (see equation below). The velocity field when irrotational may be replaced by a scalar field, $\Phi = \Phi(\mathbf{r}, t)$ say, by setting $\mathbf{V} = \text{grad } \Phi$. This permits the integration of the equation of motion, reducing the problem to that of determining Φ. A single scalar equation is all that is then necessary to solve for Φ, and this turns out to be, using the incompressibility condition (8.1), nothing but Laplace's equation

$$\nabla^2 \Phi = 0$$

This means that in the irrotational motion of an ideal fluid we can solve for the velocity field (i.e., for Φ) independently of the equation of motion! Another significant factor is that Laplace's equation is *linear*. This means that in the analysis of ideal-fluid motion mathematical difficulties associated with nonlinear partial differential equations drop out of the picture. In fact, a great amount of mathematical knowledge already gathered becomes available.

8.2 Integration of Euler's Equation in Steady Rotational Motion

For steady motion Euler's equation becomes

$$\text{grad}\left(\frac{V^2}{2} + \frac{p}{\rho}\right) - \mathbf{V} \times \text{curl } \mathbf{V} = \mathbf{f} \qquad (8.3)$$

We observe that integration of this equation becomes possible if it is carried out along a streamline and if the body force **f** is assumed irrotational. The vector given by the term **V** × curl **V** in the acceleration is normal to **V**. Since the direction of a streamline at any point is that of the velocity vector at that point, it follows that (**V** × curl **V**) · **ds** is zero if **ds** is taken along a streamline. Thus, taking the scalar product of (8.3) with $\mathbf{ds}$ an element of a streamline, we obtain

$$\text{grad}\left(\frac{V^2}{2} + \frac{p}{\rho}\right) \cdot \mathbf{ds} = \mathbf{f} \cdot \mathbf{ds} \quad \text{along a streamline} \qquad (8.4)$$

We *assume that the body force is an irrotational force field* and set

$$\mathbf{f} = \text{grad } \bar{U} \qquad (8.5)$$

where $\bar{U}$ is a scalar function of position. In the problems we shall deal with, the body force consists of only *gravity force* and this is irrotational. Using (8.4) and (8.5) we obtain

$$\text{grad}\left(\frac{V^2}{2} + \frac{p}{\rho} - \bar{U}\right) \cdot \mathbf{ds} = 0 \quad \text{along a streamline} \qquad (8.6)$$

This shows that

$$\text{grad}\left(\frac{V^2}{2} + \frac{p}{\rho} - \bar{U}\right)$$

is a vector normal to the streamline. Equation (8.6) integrates to

$$\frac{V^2}{2} + \frac{p}{\rho} - \bar{U} = \text{const.} = H_s, \quad \text{say, along a streamline} \qquad (8.7)$$

This is known as the *Bernoulli's equation along a streamline*. The constant H_s is *not* a constant that has the same value for the entire space filled by the fluid. In general, it changes from one streamline to another. In this sense H_s *is a function of position, although along a streamline it is a constant.* We now obtain an equation that gives the change of H_s in space.

8.3 Spatial Variation of H_s

For steady motion, and assuming as before that the body force is irrotational, Euler's equation may be written as

$$\text{grad}\left(\frac{V^2}{2} + \frac{p}{\rho} - \bar{U}\right) = \mathbf{V} \times \text{curl } \mathbf{V}$$

or as

$$\text{grad } H_s = \mathbf{V} \times \text{curl } \mathbf{V} \qquad (8.8)$$

where

$$H_s = \frac{V^2}{2} + \frac{p}{\rho} - \bar{U}$$

From equation (8.8) it immediately follows that H_s is a constant along a streamline and is indeed the constant introduced in (8.7). Equation (8.8) is the equation that determines the spatial variation of H_s. It shows that the *variation* of H_s is related to curl $\mathbf{V}$, the vorticity (or, equivalently, the angular velocity) field of the fluid motion. From this equation it follows that grad H_s is zero if curl $\mathbf{V}$ is zero or if curl $\mathbf{V}$ is parallel to $\mathbf{V}$. In either case H_s becomes a true constant throughout the flow field.

To obtain the *variation of H_s from streamline to streamline*, let us consider a streamline and form the component of (8.8) in a direction normal to

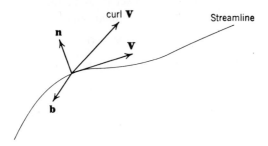

Fig. 8.1 Illustrating the spatial variation of H_s.

the streamline. Thus, if $\mathbf{n}$ denotes the normal to the streamline at any point on it (see Fig. 8.1), we obtain

$$\frac{\partial H_s}{\partial n} = \mathbf{n} \cdot \text{grad } H_s = \mathbf{n} \cdot \mathbf{V} \times \text{curl } \mathbf{V}$$

$$= \mathbf{n} \times \mathbf{V} \cdot \text{curl } \mathbf{V} \qquad (8.9)$$

where $\partial H_s/\partial n$ is the spatial rate of change of H_s in the direction $\mathbf{n}$. The unit vector $\mathbf{n}$ is normal to $\mathbf{V}$. Therefore the magnitude of $\mathbf{n} \times \mathbf{V}$ is simply V. Choosing a unit vector $\mathbf{b}$ such that the vectors $\mathbf{V}, \mathbf{n}$, and $\mathbf{b}$ in that order form a right-hand system (Fig. 8.1), we write

$$\mathbf{n} \times \mathbf{V} = -V\mathbf{b}$$

Equation (8.9) then takes the form

$$\frac{\partial H_s}{\partial n} = -V(\mathbf{b} \cdot \text{curl } \mathbf{V}) \qquad (8.10)$$

Since the angular velocity $\boldsymbol{\omega}$ is half of curl $\mathbf{V}$, this equation may also be expressed as

$$\frac{\partial H_s}{\partial n} = -2V(\mathbf{b} \cdot \boldsymbol{\omega}) \qquad (8.11)$$

This shows that there is no variation of H_s from streamline to streamline if the angular velocity is zero, that is, if the motion is irrotational. In other words, if the motion is irrotational, the Euler equation may be integrated once and for all. As we shall see in the next section, this is true even for unsteady motion.

8.4 Integration of Euler's Equation in Irrotational Motion

As we shall learn in the next chapter, for the motion of a fluid to be irrotational it is necessary that the body force $\mathbf{f}$ be irrotational. Thus we assume that $\mathbf{f}$ is irrotational and set it, as before, equal to grad $\bar{U}$. We *now assume that the motion is irrotational*, that is, that

$$\text{curl } \mathbf{V} = 0 \tag{8.12}$$

and, consequently, that the velocity may be expressed as

$$\mathbf{V} = \text{grad } \Phi \tag{8.13}$$

where $\Phi = \Phi(\mathbf{r}, t)$ is a scalar function of position and time.

With the assumption (8.12) and the relation (8.13), Euler's equation becomes

$$\text{grad} \left(\frac{\partial \Phi}{\partial t} + \frac{V^2}{2} + \frac{p}{\rho} - \bar{U} \right) = 0 \tag{8.14}$$

where

$$V^2 = (\text{grad } \Phi)^2 = \text{grad } \Phi \cdot \text{grad } \Phi \tag{8.15}$$

This equation readily integrates to

$$\frac{\partial \Phi}{\partial t} + \frac{V^2}{2} + \frac{p}{\rho} - \bar{U} = \text{const.} = F(t), \quad \text{a function of time} \tag{8.16}$$

Because the integration is only with respect to space, the constant is independent of space coordinates, whereas it may, in general, still depend on time. Hence it is expressed as $F(t)$. At any instant of time $F(t)$ has a uniform value for all points of the fluid.

If the motion is steady, variations with time do not exist and equation (8.16) reduces to

$$\frac{V^2}{2} + \frac{p}{\rho} - \bar{U} = \text{const.} = H \tag{8.17}$$

This is the famous *Bernoulli equation*. It was obtained by Daniel Bernoulli, before the discovery of Euler's equation, by considerations similar to the modern principle of energy conservation. The constant H in (8.17) is now truly a constant, that is, independent of both time and space. It has the same value for all points of the fluid for all times.

Equation (8.16) is a *generalization* of Bernoulli's equation for unsteady motions. Accordingly, it is referred to as the *unsteady Bernoulli's equation.*

Bernoulli's equation is the most important relation in elementary fluid dynamics and is used for the solution of numerous technical problems in aerodynamics, hydraulics, hydraulic machinery, and so on.

We wish to point out at this stage that the foregoing discussions in Sections 8.2 to 8.4 have been carried out without any recourse to a co-ordinate description, and the conclusions drawn are, therefore, independent of the choice of a coordinate system. This has been possible because of the use of vector notation and the associated concepts of scalar product, vector product, and the gradient. We recommend that the reader work out the preceding results by using Cartesians, for instance, and gain for himself an appreciation for the role of vector methods.

8.5 Remarks on an Irrotational Force Field

A force field **F** is said to be irrotational in a certain region of space when in that region curl **F** vanishes. Further, we can then represent the force field as the gradient of a scalar field. Denoting this by $\bar{U}$ we write, as before,

$$\mathbf{F} = \text{grad } \bar{U} \tag{8.18}$$

In Cartesians, for instance, the component form of this equation is expressed by

$$F_x = \frac{\partial \bar{U}}{\partial x}, \qquad F_y = \frac{\partial \bar{U}}{\partial y}, \qquad F_z = \frac{\partial \bar{U}}{\partial z} \tag{8.19}$$

where F_x, F_y, F_z are the components of **F**. Relation (8.18) forms the basis for determining $\bar{U}$ if **F** is known or, conversely, for determining **F** if $\bar{U}$ is known.

As the simplest example, let us consider the case of our "gravity" field. We choose a Cartesian coordinate system with the Z-axis pointing upward, that is, pointing in a direction opposite to that of the acceleration due to gravity. The force on a body of mass m in this force field is then given by

$$\mathbf{F} = (F_x, F_y, F_z) = (0, 0, -mg) \tag{8.20}$$

Curl **F** vanishes everywhere. Thus gravity field is an irrotational force field. To determine the scalar function $\bar{U}$, the gradient of which may represent **F**, we must solve the differential equation (8.18) or, equivalently, the component equations (8.19). From (8.19) and (8.20) we obtain

$$\frac{\partial \bar{U}}{\partial x} = 0, \quad \frac{\partial \bar{U}}{\partial y} = 0, \quad \frac{\partial \bar{U}}{\partial z} = -mg \tag{8.21}$$

which may readily be integrated to yield

$$\bar{U} = -mgz \tag{8.22}$$

$\bar{U}$ is determined only to within an additive constant which we have here set equal to zero so that $\bar{U}$ is zero at $z = 0$.

In case of an irrotational force field, such as the gravity field we have
been discussing, one can attach a certain physical significance to the
scalar function that represents the force field. To do this, consider the
motion of a particle acted on by a force field **F**. The work done by
the force field on the particle during its motion from a point P_0 to a
point P along a path $\mathscr{C}$ is expressed by

$$W = \oint_{P_0}^{P} \mathbf{F} \cdot \mathbf{ds} \qquad (8.23)$$

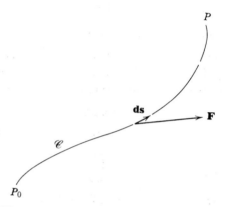

Fig. 8.2 Illustrating the work done by a force.

where **ds** is an element of length on the path $\mathscr{C}$ (Fig. 8.2). The symbol $\oint$
indicates that the integration is to be carried out along the given path $\mathscr{C}$.
Generally, the value of the integral

$$\oint_{P_0}^{P} \mathbf{F} \cdot \mathbf{ds}$$

would depend on the endpoints and the path connecting them. If the force
field is irrotational, **F** may be replaced by the gradient of a scalar function,
say $\bar{U}$. Then **F · ds** becomes an exact or a total differential of $\bar{U}$ and the
integral of (**F · ds**) between any two endpoints P_0 and P becomes independent of the path connecting them. Thus for the *work done by an
irrotational force field* we write

$$W = \int_{P_0}^{P} \mathbf{F} \cdot \mathbf{ds} = \int_{P_0}^{P} \operatorname{grad} U \cdot ds = \int_{P_0}^{P} dU$$
$$= \bar{U}(P) - \bar{U}(P_0) \qquad (8.24)$$

This states that the work done by an irrotational force field is simply a
function of position of the endpoints. This function is nothing but the

scalar function, $\bar{U}$, the gradient of which represents the force field. Thus, in an *irrotational force field, the scalar function $\bar{U}$ is physically equivalent to the work done by the force field.*

In problems of mechanics, the negative of the work done by an irro-. tational force field is defined as the "*potential energy*" or the "*potential*" of the system on which the force field acts. According to this definition a scalar function that represents an irrotational force field is essentially the negative of the potential energy.* In light of these considerations it is customary to refer to our scalar function, such as $\bar{U}$ in Eq. (8.18), as a "*force potential*" or simply as a "*potential.*" This nomenclature is further extended to all irrotational vector fields, and it is common to talk about "*velocity potential,*" "*acceleration potential,*" etc. On the same basis we refer to irrotational vector fields as "*potential fields.*"

For a mechanical system subjected to only irrotational forces, the total energy of the system (i.e., the sum of the kinetic and potential energies) is *conserved*, that is, remains a constant for all times. Hence an irrotational force field is also sometimes called a "*conservative field.*"

In concluding this section we add a few remarks about the independence of the integral $\oint_{P_0}^{P} d\bar{U}$ with respect to the path of integration. For this integral to be independent of the path and to be dependent only on the endpoints, we require that the value of $\bar{U}(P) - \bar{U}(P_0)$ be the same no matter how we approach the points P_0 and P, that is, by what path or direction. This means that the function $\bar{U}$ should assume at any point a definite single value irrespective of the path chosen to arrive at the point. *Thus, for the integral $\int_{P_0}^{P} d\bar{U}$ to be independent of the path connecting the endpoints P and P_0, $\bar{U}$ should be a single-valued function of position.* This is what we have considered here. In the analysis of physical problems, one does meet with *multiple valued* potentials for irrotational vector fields. In such a case an integral such as $\int_{P_0}^{P} d\bar{U}$ need not have the same value for *all* paths connecting P_0 and P. We shall learn in our studies that a multiple-valued potential plays an important part in the theory of lifting bodies.

8.6   Remarks on Bernoulli's Equation

In light of the preceding considerations we may *interpret Bernoulli's equation as simply a statement of the conservation of energy.* For steady,

---

* If we had initially written $\mathbf{F} = -\text{grad } \bar{U}$ instead of $\mathbf{F} = \text{grad } \bar{U}$ as done in equation (8.18) U would have become identical with the potential energy.

irrotational motion of an ideal fluid we have from Eq. (8.17)

$$\rho \frac{V^2}{2} + p - \bar{U} = \text{const.}$$

Here $\rho(V^2/2)$ is the kinetic energy of the fluid per unit volume and $-\bar{U}$ is, again per unit volume, the potential energy related to the body forces which are assumed irrotational. To interpret p let us recall that the field of the resultant pressure force, specified again per unit volume, is $-\text{grad } p$. This means the resultant pressure force is irrotational and $-p$ is its potential. It follows immediately that p is the potential energy (of course per unit volume again) related to the resultant pressure force. With the preceding interpretations, the term $(p - \bar{U})$ is simply the potential energy of the fluid. Thus Bernoulli's equation (8.17) reads

the kinetic energy + the potential energy = const.

It is not surprising that Bernoulli's equation is simply a statement of the law of conservation of energy. We have seen (refer to Section 5.8) that for an incompressible, inviscid fluid the equation of energy conservation (Eq. 5.24) reduces to the equation for mechanical energy (Eq. 5.25). In such a case, when the equation of mechanical energy is integrated (that is what we do to obtain Bernoulli's equation), one should obtain that the total energy of the system is conserved.

Before concluding this chapter it is interesting to put Bernoulli's equation (8.17) in the form commonly used in hydraulics. Considering the body force to be that due to gravity alone, we have from (8.22)

$$\bar{U} = -\rho g z$$

for $\bar{U}$ in (8.17). Thus we obtain

$$\rho \frac{V^2}{2} + p + \rho g z = \text{const.}$$

or writing per unit mass of the fluid

$$\frac{V^2}{2} + \frac{p}{\rho} + g z = \text{const.}$$

which is the form commonly met in hydraulics.

Chapter 9

Irrotational Motion

In the last chapter we saw that great mathematical simplicity could be achieved in the analysis of fluid motion if the motion can be treated as irrotational. We shall now examine the conditions under which the motion may be treated as irrotational. We begin with an analysis of the general motion of a fluid element with a view to recognizing particularly that the angular velocity of a fluid element is indeed half the curl of the velocity, which has been referred to as the vorticity. We then develop the important theorems of Helmholtz and Kelvin which describe the fate of vorticity and circulation in an inviscid fluid. These theorems show that if the motion of an ideal fluid is once irrotational, it is always irrotational. We then take up certain properties of irrotational motion.

9.1 Most General Motion of a Fluid Element

In the most general case, the motion of a fluid element consists of a *translation*, a *rotation*, and a *deformation*. We show this by considering the relative motion between two infinitely close points of a fluid element. At a certain instant of time *t*, let P and Q denote any two such points and let **r** and **r** + **dr** be their respective positions (Fig. 9.1). In a general motion of the element, the points P and Q both experience changes of position. Let ζ_0 denote the displacement of P in a small time interval dt and ζ the corresponding displacement of Q. If $V(r, t)$ is the velocity at the point P at the time t, the velocity at Q at the same moment is expressed to the first order by

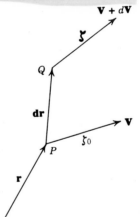

Fig. 9.1 Illustrating the general displacement of a fluid element.

$$V(\mathbf{r} + \mathbf{dr}, t) = V(\mathbf{r}, t) + dV \qquad (9.1)$$

where $d\mathbf{V}$ is the change in $\mathbf{V}$ over the directed distance $\mathbf{dr}$. Since $\mathbf{dr}$ is infinitesimal, $d\mathbf{V}$ is given by (see Section 2.26, Eq. 2.76)

$$d\mathbf{V} = (\mathbf{dr} \cdot \text{grad})\mathbf{V} \qquad (9.2)$$

where $\mathbf{V} = \mathbf{V}(\mathbf{r}, t)$. For the displacements of P and Q we have

$$\boldsymbol{\zeta}_0 = \mathbf{V}(\mathbf{r}, t)\, dt \qquad (9.3)$$

$$\boldsymbol{\zeta} = \mathbf{V}(\mathbf{r} + \mathbf{dr}, t)\, dt$$

$$= \boldsymbol{\zeta}_0 + d\mathbf{V}\, dt \qquad (9.4)$$

We notice immediately that in the displacement of the point Q, there is a part that is the same as that of P. This part, denoted by $\boldsymbol{\zeta}_0$, is the same for all points of the fluid element and, therefore, corresponds to a *translation* of the element as a whole.

The component $d\mathbf{V}$, which is the relative velocity between P and Q, can be shown to be made up of a rotation and a deformation. We carry out the proof in Cartesians. Accordingly, we denote the reference unit vectors by $\mathbf{e}_1$, $\mathbf{e}_2$, $\mathbf{e}_3$ and write*

$$\mathbf{r} = (x_1, x_2, x_3)$$

$$\mathbf{dr} = (dx_1, dx_2, dx_3)$$

$$\mathbf{V} = (u_1, u_2, u_3)$$

$$d\mathbf{V} = (du_1, du_2, du_3)$$

We then obtain

$$d\mathbf{V} = (\mathbf{dr} \cdot \text{grad})\mathbf{V} = \sum_{i=1}^{3} \left(\sum_{j=1}^{3} \frac{\partial u_i}{\partial x_j}\, dx_j \right) \mathbf{e}_i \qquad (9.5)$$

where the partial derivatives are evaluated at the point $\mathbf{r}$ at the time t. We now express each of the partial derivatives as the sum of a symmetric and an antisymmetric term and write

$$d\mathbf{V} = \sum_{i=1}^{3} \left\{ \sum_{j=1}^{3} \left[\frac{1}{2}\left(\frac{\partial u_i}{\partial x_j} + \frac{\partial u_j}{\partial x_i} \right) + \frac{1}{2}\left(\frac{\partial u_i}{\partial x_j} - \frac{\partial u_j}{\partial x_i} \right) \right] dx_j \right\} \mathbf{e}_i$$

$$= \sum_{i=1}^{3} \left(\sum_{j=1}^{3} \epsilon_{ij}\, dx_j \right) \mathbf{e}_i + \sum_{i=1}^{3} \left(\sum_{j=1}^{3} \varphi_{ij}\, dx_j \right) \mathbf{e}_i$$

$$= d\mathbf{V}_1 + d\mathbf{V}_2 \qquad (9.6)$$

where

$$\epsilon_{ij} = \frac{1}{2}\left(\frac{\partial u_i}{\partial x_j} + \frac{\partial u_j}{\partial x_i} \right) \qquad (9.7)$$

$$\varphi_{ij} = \frac{1}{2}\left(\frac{\partial u_i}{\partial x_j} - \frac{\partial u_j}{\partial x_i} \right) \qquad (9.8)$$

* We adopt this convention for simplicity, which will become apparent in the following.

and where $d\mathbf{V}_1$ and $d\mathbf{V}_2$ are velocity vectors given by the first and second terms respectively on the right-hand side of Eq. (9.6).

Since φ_{ij} is antisymmetric, as φ_{ij} is equal to $-\varphi_{ji}$, the vector $d\mathbf{V}_2$ becomes

$$d\mathbf{V}_2 \equiv \sum_{i=1}^{3}\left(\sum_{j=1}^{3}\varphi_{ij}\,dx_j\right)\mathbf{e}_i = \begin{vmatrix} \mathbf{e}_1 & \mathbf{e}_2 & \mathbf{e}_3 \\ -\varphi_{23} & -\varphi_{31} & -\varphi_{12} \\ dx_1 & dx_2 & dx_3 \end{vmatrix}$$

$$= \boldsymbol{\omega} \times d\mathbf{r} \tag{9.9}$$

where

$$\boldsymbol{\omega} = (-\varphi_{23},\ -\varphi_{31},\ -\varphi_{13}) \tag{9.10}$$

Equation (9.9) shows that $d\mathbf{V}_2$ is the velocity at the point Q due to a *rigid body rotation* of the fluid element as a whole about an instantaneous axis through the point P. The angular velocity is equal to $\boldsymbol{\omega}$.

The velocity $d\mathbf{V}_1$ is characterized by the set of nine numbers

$$\begin{pmatrix} \epsilon_{11} & \epsilon_{12} & \epsilon_{13} \\ \epsilon_{21} & \epsilon_{22} & \epsilon_{23} \\ \epsilon_{31} & \epsilon_{32} & \epsilon_{33} \end{pmatrix} \tag{9.11}$$

obtained by specializing ϵ_{ij} by giving i and j all the possible values from 1, 2, 3. This array forms a second-order symmetric tensor, symmetric because ϵ_{ij} is equal to ϵ_{ji}. The diagonal elements of this tensor represent *rates of normal strains*, while the nondiagonal terms represent *rates of shear* or *tangential strains*.* The tensor (9.11) is therefore known as the *rate of strain tensor*. Hence it follows that the velocity $d\mathbf{V}_1$ represents a *velocity due to deformation of the fluid element*.

We have thus shown that the general motion of a fluid element is made up of three parts—a translation, a rotation, and a deformation.

9.2 Rotation and Vorticity

From the preceding considerations it follows that at any point in the flow field of a fluid, associated with the velocity vector, there is a rate of strain tensor (denoted above by ϵ_{ij}) and a rate of rotation tensor (denoted above by φ_{ij}). Recalling the splitting of the tensor gradient of a vector field into a symmetric and an antisymmetric part (see Section 2.25), we observe that the symmetric part of the tensor gradient of the velocity measures the (instantaneous) rate of strain while the antisymmetric part measures the (instantaneous) rate of rotation at the point considered.

---

* Proof is left out as an exercise.

The rate of strain tensor being symmetric requires actually six components for its complete specification. The rate of rotation tensor being antisymmetric requires actually a set of three components for its complete specification. It can, therefore, be represented by a vector. This vector is nothing but the angular velocity vector **ω** introduced in Eq. (9.9). The

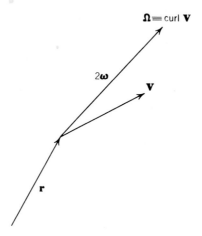

Fig. 9.2   Angular velocity and vorticity.

relation between the elements of the rotation tensor and those of **ω** is given by Eq. (9.10), which may be expressed as

$$\omega_i = -\varphi_{jk} \tag{9.12}$$

where ω_i is the component of **ω** in the direction $\mathbf{e}_i$ and where i, j, k take *successive values* from 1, 2, 3. By using Eq. (9.8), the vector **ω** may be expressed in Cartesians as

$$\boldsymbol{\omega} = \frac{1}{2} \begin{vmatrix} \mathbf{e}_1 & \mathbf{e}_2 & \mathbf{e}_3 \\ \dfrac{\partial}{\partial x_1} & \dfrac{\partial}{\partial x_2} & \dfrac{\partial}{\partial x_3} \\ u_1 & u_2 & u_3 \end{vmatrix}$$

The determinant in this relation is nothing but the curl of the velocity vector. Thus it follows that the vector representing the rate of rotation tensor is given by

$$\boldsymbol{\omega} = \tfrac{1}{2} \operatorname{curl} \mathbf{V} \tag{9.13}$$

which may be specialized easily to any orthogonal, curvilinear coordinate system. This equation states that the *instantaneous angular velocity (or the rate of rotation) at any point of the flowing fluid is equal to half the curl of the velocity at that point at the instant considered.*

The vector curl $\mathbf{V}$ is called the *vortex vector* or simply the *vorticity* of the fluid at the point considered. Denoting the vorticity by $\boldsymbol{\Omega}$, we write

$$\boldsymbol{\Omega} \equiv \operatorname{curl} \mathbf{V} \qquad (9.14)$$

It then follows that

$$\boldsymbol{\omega} = \tfrac{1}{2}\boldsymbol{\Omega} \qquad (9.15)$$

that is,

$$\text{angular velocity} = \tfrac{1}{2} \text{ vorticity}$$

We thus see that associated with the velocity field $\mathbf{V}(\mathbf{r}, t)$ of a flowing fluid there is a vorticity field $\boldsymbol{\Omega}(\mathbf{r}, t)$ and an angular velocity field $\boldsymbol{\omega}(\mathbf{r}, t)$. A schematic representation of the velocity, vorticity, and angular velocity at any point is shown in Fig. 9.2.

9.3 Circulation and Vorticity

We shall now apply to the velocity field of a flowing fluid the relations that exist between circulation and curl of any vector field. Following the definition for any vector field (see Section 2.28), we define the circulation of

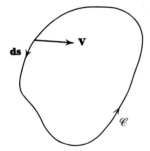

Fig. 9.3   Circulation.

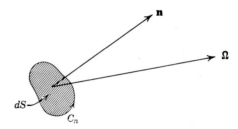

Fig. 9.4   Vorticity and circulation around an infinitesimal surface element

the velocity field as the line integral of $\mathbf{V}$ around any closed curve $\mathscr{C}$. We call this simply *circulation* and denote it by Γ or $\Gamma_{\mathscr{C}}$. We thus have

$$\Gamma_{\mathscr{C}} = \oint_{\mathscr{C}} \mathbf{V} \cdot \mathbf{ds} \qquad (9.16)$$

where $\mathbf{ds}$ is an element of the curve $\mathscr{C}$ (see Fig. 9.3).

Consider an infinitesimal surface element $\mathbf{n}\, dS$ situated at any point P. Let C_n denote the boundary curve of the element and $d\Gamma_{C_n}$ the circulation around it, the direction of integration along the curve being that given by the rule of right-hand rotation about $\mathbf{n}$ (see Fig. 9.4). Then, according to the relation (2.120a), we have

$$d\Gamma_{C_n} = \oint_{C_n} \mathbf{V} \cdot \mathbf{ds} = \text{curl } \mathbf{V} \cdot \mathbf{n} \, dS$$

$$= \mathbf{\Omega} \cdot \mathbf{n} \, dS \qquad (9.17)$$

where $\mathbf{\Omega}$ is the vortex vector at the point P. This equation states that the *circulation around any infinitesimal surface element* $\mathbf{n} \, dS$ *is equal to the flux of vorticity through that element.*

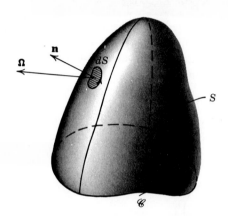

Fig. 9.5 Circulation around $\mathscr{C}$ is equal to the outflow of vorticity through S.

Consider now a finite closed curve $\mathscr{C}$ and let S represent any surface bounded by $\mathscr{C}$. Then, according to Stokes' theorem (Eq. 2.131), we obtain

$$\Gamma_{\mathscr{C}} = \oint_{\mathscr{C}} \mathbf{V} \cdot \mathbf{ds} = \iint_{S} \text{curl } \mathbf{V} \cdot \mathbf{n} \, dS$$

$$= \iint_{S} \mathbf{\Omega} \cdot \mathbf{n} \, dS \qquad (9.18)$$

This equation states that the *circulation around a closed curve $\mathscr{C}$ is equal to the outflow of vorticity through any surface S whose boundary is formed by $\mathscr{C}$* (Fig. 9.5). It also shows that if we consider different surfaces drawn with the same boundary curve, the outflow of vorticity is the same through all these surfaces.

9.4 Rate of Change of Vorticity

We now investigate the conditions under which the motion of an ideal fluid is irrotational. For this purpose we examine the fate of vorticity of a fluid element or, equivalently, the fate of circulation around a fluid curve. This examination may be carried out by working directly either with the vorticity or with the circulation. Here we work with the vorticity; in the next section we shall work with the circulation.

The equation governing the rate of change of vorticity of a fluid element may be obtained directly from the equation of motion. For this purpose, considering only an ideal fluid, we take the curl of Eq. (5.31), which is

$$\frac{\partial \mathbf{V}}{\partial t} + \text{grad} \frac{V^2}{2} - \mathbf{V} \times \mathbf{\Omega} = \mathbf{f} - \text{grad} \left(\frac{p}{\rho}\right)$$

and obtain

$$\frac{\partial \mathbf{\Omega}}{\partial t} - \text{curl} (\mathbf{V} \times \mathbf{\Omega}) = \text{curl } \mathbf{f} \tag{9.19}$$

Expanding the second term on the left-hand side, we have

$$\text{curl} (\mathbf{V} \times \mathbf{\Omega}) = (\text{div } \mathbf{\Omega})\mathbf{V} - \mathbf{V} \cdot \text{grad } \mathbf{\Omega} + \mathbf{\Omega} \cdot \text{grad } \mathbf{V} - (\text{div } \mathbf{V})\mathbf{\Omega}$$
$$= -\mathbf{V} \cdot \text{grad } \mathbf{\Omega} + \mathbf{\Omega} \cdot \text{grad } \mathbf{V} \tag{9.20}$$

since the *divergence of the vorticity* is always zero, and since the fluid is *assumed incompressible*, the divergence of the velocity is zero. Substituting (9.20) into (9.19), we obtain

$$\frac{\partial \mathbf{\Omega}}{\partial t} + \mathbf{V} \cdot \text{grad } \mathbf{\Omega} - \mathbf{\Omega} \cdot \text{grad } \mathbf{V} = \text{curl } \mathbf{f}$$

or

$$\frac{D\mathbf{\Omega}}{Dt} - \mathbf{\Omega} \cdot \text{grad } \mathbf{V} = \text{curl } \mathbf{f}$$

We assume that the *body force is irrotational* and set

$$\text{curl } \mathbf{f} = 0$$

Thus we obtain

$$\frac{D\mathbf{\Omega}}{Dt} = \mathbf{\Omega} \cdot \text{grad } \mathbf{V} \tag{9.21}$$

which gives the rate of change of vorticity of an element of an ideal fluid as it moves under the action of an irrotational body force. Recalling the meaning of a term such as $\mathbf{B} \cdot \text{grad } \mathbf{A}$ (see Section 2.26, Eqs. 2.76 and 2.77), we see that

$$\mathbf{\Omega} \cdot \text{grad } \mathbf{V} = \Omega \text{ (the rate of change of } \mathbf{V} \text{ with respect to distance}$$
$$\text{in the direction of } \mathbf{\Omega})$$

where Ω is the magnitude of $\mathbf{\Omega}$.

Equation (9.21) states that *the material rate of change of the vorticity is zero whenever the vorticity is zero.* On the basis of this result, we should not, however, proceed immediately to the conclusion that if the *vorticity of a fluid element is zero at some time it is zero at all times.* It is of course true that (9.21) leads to such a conclusion. To show this rigorously we first cite the following kinematical relation,* which is true whether the fluid is compressible or incompressible:

$$\frac{D}{Dt}(\mathbf{\Omega} \cdot \mathbf{n} \, dS) = \left[\frac{\partial \mathbf{\Omega}}{\partial t} - \text{curl}\,(\mathbf{V} \times \mathbf{\Omega})\right] \cdot \mathbf{n} \, dS$$

$$= \left[\frac{D\mathbf{\Omega}}{Dt} - \mathbf{\Omega} \cdot \text{grad}\,\mathbf{V} + (\text{div}\,\mathbf{V})\mathbf{\Omega}\right] \cdot \mathbf{n} \, dS$$

where $\mathbf{n} \, dS$ is a surface element moving with the fluid. If the fluid is incompressible, we have

$$\frac{D}{Dt}(\mathbf{\Omega} \cdot \mathbf{n} \, dS) = \left[\frac{D\mathbf{\Omega}}{Dt} - \mathbf{\Omega} \cdot \text{grad}\,\mathbf{V}\right] \cdot \mathbf{n} \, dS$$

With this relation (9.21) reduces to

$$\frac{D}{Dt}(\mathbf{\Omega} \cdot \mathbf{n} \, dS) = 0 \tag{9.22}$$

This equation expresses the important result: *In the motion of an ideal fluid subjected to irrotational body forces, the material rate of change of the outflow of vorticity through any surface element moving with the fluid is permanently zero, or, equivalently, the outflow of vorticity through any surface element moving with the fluid remains a constant for all times. In this sense vorticity is convected with the fluid.*
We may now draw the conclusion that if the vorticity of any surface element is zero at some time, it will remain so for all times as the element moves with the fluid, for according to (9.22) the outflow of vorticity through any surface element moving with the fluid should remain permanently zero if it is zero at some time.

Equation (9.22) may readily be expressed in terms of the circulation around the surface element. Denoting by $d\Gamma_{C_n}$ the circulation around the

---

* If $\mathbf{A}$ is any vector field, the rate of change of outflow of $\mathbf{A}$ through any surface element $\mathbf{n} \, dS$ moving with the fluid may be expressed as

$$\frac{D}{Dt}(\mathbf{A} \cdot \mathbf{n} \, dS) = \left[\frac{\partial \mathbf{A}}{\partial t} + \mathbf{V}\,\text{div}\,\mathbf{A} - \text{curl}\,(\mathbf{V} \times \mathbf{A})\right] \cdot \mathbf{n} \, dS$$

For a proof of this formula see Sommerfeld (1950).

boundary C_n of the element $\mathbf{n}\, dS$, we have

$$\boldsymbol{\Omega} \cdot \mathbf{n}\, dS = d\Gamma_{C_n}$$

Therefore Eq. (9.22) takes the form

$$\frac{D}{Dt}(d\Gamma_{C_n}) = 0 \qquad (9.23)$$

This expresses the result: *In the motion of an ideal fluid subjected to irrotational body forces, the material rate of change of the circulation around any surface element moving with the fluid is permanently zero, or, equivalently, the circulation around any surface element moving with the fluid remains a constant for all times.*

Equation (9.21), or, equivalently, (9.22), is referred to as *Helmholtz's theorem.** If forms the basis for the derivation of *Hemlholtz's theorem of vortex motion,* which we shall take up in the chapter on vortex motion. The above derivation of Helmholtz's theorem, which, following Helmholtz, starts from Euler's equation, seems to indicate that any conclusions drawn from the theorem are valid in the motion of only an ideal fluid, that is, of an incompressible inviscid fluid of constant density. We can, however, show that such conclusions hold also in the motion of a compressible inviscid fluid. This can be done readily if we start, following Kelvin (1869), from the concept of circulation. This we do in the next section.

We close this section with the observation that *in two-dimensional and but not in axisymmetric flows the term* $\boldsymbol{\Omega} \cdot$ grad $\mathbf{V}$ *is identically zero*[†] *and* (9.21) *therefore reduces to*

$$\frac{D\boldsymbol{\Omega}}{Dt} = 0 \qquad (9.24)$$

This equation expresses the result: *in two-dimensional flows of an ideal fluid under the action of irrotational body forces, the vorticity of a fluid element remains a constant for all times.*

9.5 Rate of Change of Circulation

We consider a closed *fluid curve* (i.e., a curve consisting of the same fluid particles at all times) and seek the rate of change of circulation around such a curve as it moves with the fluid. Let $\mathscr{C}$ represent any closed curve drawn at the instant t and let Γ denote the circulation around $\mathscr{C}$ at that instant. During a time interval Δt the different fluid particles that make up $\mathscr{C}$ move into new positions. Consequently, in the interval Δt the curve $\mathscr{C}$ will assume a different shape and occupy a new location. Let $\mathscr{C}_1$ be the

---

* See Helmholtz (1858).
† The reader should verify this as an exercise.

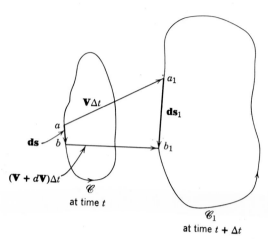

Fig. 9.6 Illustrating the change in circulation around a fluid curve.

curve at $t + \Delta t$ and let Γ_1 denote the circulation around $\mathscr{C}_1$ (see Fig. 9.6). The rate of change in circulation along $\mathscr{C}$ following its motion is then given by the material derivative

$$\frac{D\Gamma}{Dt} = \lim_{\Delta t \to 0} \frac{\Gamma_1 - \Gamma}{\Delta t} \qquad (9.25)$$

In terms of the line integrals of the velocity this equation takes the form

$$\frac{D\Gamma}{Dt} \equiv \frac{D}{Dt} \oint_{\mathscr{C}} \mathbf{V} \cdot \mathbf{ds} = \lim_{\Delta t \to 0} \frac{1}{\Delta t} \left[\oint_{\mathscr{C}_1} \mathbf{V}_1 \cdot \mathbf{ds}_1 - \oint_{\mathscr{C}} \mathbf{V} \cdot \mathbf{ds} \right] \qquad (9.26)$$

Since we are following the same set of fluid particles, the limit on the right side of this equation is the same as the integral $\oint_{\mathscr{C}} (D/Dt)(\mathbf{V} \cdot \mathbf{ds})$ along the curve $\mathscr{C}$ at the instant t. Therefore (9.26) becomes

$$\frac{D\Gamma}{Dt} \equiv \frac{D}{Dt} \oint_{\mathscr{C}} \mathbf{V} \cdot \mathbf{ds} = \oint_{\mathscr{C}} \frac{D}{Dt}(\mathbf{V} \cdot \mathbf{ds})$$

$$= \oint_{\mathscr{C}} \frac{D\mathbf{V}}{Dt} \cdot \mathbf{ds} + \oint_{\mathscr{C}} \mathbf{V} \cdot \frac{D}{Dt} \mathbf{ds} \qquad (9.27)$$

where $D\mathbf{V}/Dt$ is the acceleration at the instant t at the point considered on the curve $\mathscr{C}$. To interpret the expression $(D/Dt)\,\mathbf{ds}$ we proceed as follows. If $\mathbf{ds}$ represents the element $\mathbf{ab}$ of the curve $\mathscr{C}$ and $\mathbf{ds}_1$ represents the corresponding element $\mathbf{a}_1\mathbf{b}_1$ of $\mathscr{C}_1$ (see Fig. 9.6), we have

$$\frac{D}{Dt} \mathbf{ds} = \lim_{\Delta t \to 0} \frac{\mathbf{ds}_1 - \mathbf{ds}}{\Delta t} \qquad (9.28)$$

From the quadrilateral abb_1a_1 (see Fig. 9.6) we see that

$$\mathbf{aa}_1 + \mathbf{ds}_1 = \mathbf{ds} + \mathbf{bb}_1$$

or

$$\mathbf{ds}_1 - \mathbf{ds} = \mathbf{bb}_1 - \mathbf{aa}_1 \qquad (9.29)$$

Denoting the velocity at the point a by $\mathbf{V}$ and that at b by $\mathbf{V} + d\mathbf{V}$, where $d\mathbf{V}$ is the change in $\mathbf{V}$ from a over the directed distance $\mathbf{ds}$, the displacements $\mathbf{aa}_1$ and $\mathbf{bb}_1$ can be expressed as

$$\mathbf{aa}_1 = \mathbf{V}\,\Delta t$$

and

$$\mathbf{bb}_1 = (\mathbf{V} + d\mathbf{V})\,\Delta t \qquad (9.30)$$

With Eqs. (9.28), (9.29), and (9.30) we obtain

$$\frac{D}{Dt}\,\mathbf{ds} = d\mathbf{V} \qquad (9.31)$$

It then follows that

$$\oint_{\mathscr{C}} \mathbf{V} \cdot \frac{D}{Dt}\,\mathbf{ds} = \oint_{\mathscr{C}} \mathbf{V} \cdot d\mathbf{V}$$

$$= \oint_{\mathscr{C}} d\left(\frac{V^2}{2}\right) = 0 \qquad (9.32)$$

With this relation Eq. (9.27) becomes

$$\frac{D\Gamma}{Dt} = \oint_{\mathscr{C}} \frac{D\mathbf{V}}{Dt} \cdot \mathbf{ds} = \oint_{\mathscr{C}} \mathbf{a} \cdot \mathbf{ds} \qquad (9.33)$$

where $\mathbf{a}$ denotes the acceleration vector. *Equation (9.33) states that at any instant the rate of change of circulation around any fluid curve is equal to the line integral of the acceleration around that curve taken at the instant considered.* An immediate consequence of Eq. (9.33) is that if the acceleration is expressible as the gradient of some (single-valued) scalar function, that is, if the acceleration field is an irrotational vector field, then the integral $\oint_{\mathscr{C}} \mathbf{a} \cdot \mathbf{ds}$ is zero and consequently $D\Gamma/Dt$ is zero.

The considerations leading to Eq. (9.33) are still kinematical. To proceed further we have to bring in dynamical considerations. To evaluate the integral $\oint_{\mathscr{C}} \mathbf{a} \cdot \mathbf{ds}$ we restrict ourselves to an inviscid fluid and use Euler's equation (5.11). Thus we obtain

$$\frac{D\Gamma}{Dt} = \oint_{\mathscr{C}} \mathbf{a} \cdot \mathbf{ds} = \oint_{\mathscr{C}} \mathbf{f} \cdot \mathbf{ds} - \oint_{\mathscr{C}} \frac{\operatorname{grad} p}{\rho} \cdot \mathbf{ds} \qquad (9.34)$$

In this equation the density ρ is *not* assumed to be a constant. We now *assume* that the *body force is irrotational* and set

$$\mathbf{f} = \operatorname{grad} \overline{U}$$

where $\overline{U}$ is a single-valued scalar function of position and time. Then we have

$$\oint_{\mathscr{C}} \mathbf{f} \cdot \mathbf{ds} = \oint_{\mathscr{C}} d\overline{U} = 0$$

Equation (9.34) consequently reduces to

$$\frac{D\Gamma}{Dt} = -\oint_{\mathscr{C}} \frac{\operatorname{grad} p}{\rho} \cdot \mathbf{ds} \qquad (9.35)$$

If the fluid is assumed to be incompressible and homogeneous so that ρ is a constant, the integral on the right side of the above equation vanishes.

$$\oint_{\mathscr{C}} \frac{\operatorname{grad} p}{\rho} \cdot \mathbf{ds} = \frac{1}{\rho} \oint_{\mathscr{C}} dp = 0$$

If the fluid is assumed to be either compressible or incompressible and inhomogeneous so that ρ is a variable, the integral no longer vanishes automatically for any arbitrary curve $\mathscr{C}$. However, if there is a single-valued relation between the density and the pressure so that one can express*

$$\frac{\operatorname{grad} p}{\rho} = \operatorname{grad} P$$

where P is a single-valued scalar function of position and time, the integral again vanishes for any arbitrary curve $\mathscr{C}$. Then we have

$$\oint_{\mathscr{C}} \frac{\operatorname{grad} p}{\rho} \cdot \mathbf{ds} = \oint_{\mathscr{C}} \operatorname{grad} P \cdot \mathbf{ds}$$

$$= \oint_{\mathscr{C}} dP = 0$$

Thus, when we assume that ρ is a constant or that there is a single-valued relation between ρ and p, Eq. (9.35) becomes

$$\frac{D\Gamma}{Dt} = 0 \qquad (9.36)$$

This equation expresses the *theorem of conservation of circulation: In the*

---

* For instance, in the motion of an inviscid compressible fluid of constant entropy, $\operatorname{grad} p/\rho$ is simply $\operatorname{grad} h$, where h is the enthalpy per unit mass.

motion of an inviscid fluid, the rate of change of circulation around any fluid curve is permanently zero if the body forces are irrotational and if there is a single-valued pressure-density relation, or, equivalently under these conditions, the circulation around a fluid curve remains a constant for all times as the curve moves with the fluid. This theorem is known as *Kelvin's circulation theorem.*

Consider now any infinitesimal surface element $\mathbf{n}\,dS$ that moves with the fluid. Let $d\Gamma_{C_n}$ denote the circulation around the boundary C_n of such a surface element. Then Eq. (9.36) may be put in the form

$$\frac{D}{Dt}(d\Gamma_{C_n}) = 0$$

which is the same as Eq. (9.23). Furthermore, since

$$d\Gamma_{C_n} = \mathbf{\Omega} \cdot \mathbf{n}\,dS$$

we readily obtain the equation

$$\frac{D}{Dt}(\mathbf{\Omega} \cdot \mathbf{n}\,dS) = 0$$

which is exactly Helmoltz's theorem, namely, Eq. (9.22). Thus we see that the contents of the theorems of Helmholtz and Kelvin are identical. Kelvin's derivation shows clearly that the theorems are applicable not only to incompressible homogeneous fluids but also to compressible fluids in which there is a single-valued $\rho - p$ relation.

Considering at any instant a finite fluid curve $\mathscr{C}$, draw an arbitrary surface S such that $\mathscr{C}$ is the boundary of S. Then, since

$$\Gamma = \oint_{\mathscr{C}} \mathbf{V} \cdot \mathbf{ds} = \iint_{S} \mathbf{\Omega} \cdot \mathbf{n}\,dS$$

Eq. (9.36) may be put in the form

$$\frac{D}{Dt}\iint_{S} \mathbf{\Omega} \cdot \mathbf{n}\,dS = 0 \tag{9.37}$$

This equation states that *under the conditions considered in its derivation, the outflow of vorticity through any surface bounded by a closed fluid curve remains constant for all times as the curve moves with the fluid.*

In concluding this section we note that the conditions under which the theorems of Helmholtz and Kelvin are valid are precisely the conditions under which an *acceleration potential* can be defined.

9.6 Irrotational Motion

Several important results may be deduced from the theorems of Helmholtz and Kelvin. They are, for instance, the starting equations for developing the theorems of vortex motion that we shall consider in a later chapter. At present we use the theorems to deduce the conditions under which the motion of a fluid is irrotational.

Consider an inviscid fluid and let the body forces be irrotational. Also, let the fluid be such that there is a single-valued pressure-density relation. Such a fluid is known as a *barotropic fluid*. An incompressible homogeneous fluid is a special case of a barotropic fluid. In a state of rest or of uniform motion the circulation around every infinitesimal surface element of the fluid is zero; similarly, the vorticity (or equivalently the rotation) of every element of the fluid is also zero. If the fluid is then brought into a different state of motion, the circulation around every surface element and the vorticity of every. element, according to Kelvin's theorem, are still zero. This means that *under the action of potential body forces all motions of an inviscid barotropic fluid set up from a state of rest or of uniform motion are permanently rotation-free*, that is, *they are permanently irrotational.* Since irrotational motion implies the existence of a velocity potential, such motions are said to be *potential motions.*

Our main concern in this book is with the motion of an ideal fluid on which only irrotational body forces act. Generally the motions we shall consider originate from a state of rest or of uniform motion and are consequently irrotational.

Therefore our concern hereafter will be with the irrotational motion of an ideal fluid. We now proceed to set up the equations that govern such a motion.*

9.7 Velocity Potential

Throughout the region in which the fluid motion is irrotational we have

$$\boldsymbol{\Omega} \equiv \operatorname{curl} \mathbf{V} = 0 \tag{9.38}$$

which is the *condition of irrotationality.* Equation (9.38) ensures the existence of a scalar function of position and time, say

$$\Phi = \Phi(\mathbf{r}, t)$$

such that

$$\mathbf{V} = \operatorname{grad} \Phi \tag{9.39}$$

* For sake of continuity and convenience we gather in the following two sections ideas and results that have been introduced on previous occasions. See, for instance, Sections 8.1 and 8.4.

Since the gradient of a constant is zero, Φ is determined only up to an additive constant. The scalar function Φ is known as the *velocity potential*.

9.8 The Equations for Irrotational Motion of an Ideal Fluid

The motion of an ideal fluid is characterized by the velocity and pressure fields as the unknowns and is governed by (5.28) and (5.31). These are

(1) *Condition of incompressibility*

$$\text{div } \mathbf{V} = 0$$

(2) *Equation of motion*

$$\frac{D\mathbf{V}}{Dt} = \frac{\partial \mathbf{V}}{\partial t} + \text{grad } \frac{V^2}{2} - \mathbf{V} \times \text{curl } \mathbf{V} = \mathbf{f} - \text{grad}\left(\frac{p}{\rho}\right)$$

If the motion is irrotational, we add to these the condition of irrotationality (9.38) and replace, according to the definition (9.39), the velocity by the velocity potential. Furthermore, as we have seen, for the motion to be irrotational the body force must be irrotational. Therefore we set $\mathbf{f}$ equal to grad $\overline{U}$, where $\overline{U}(\mathbf{r}, t)$ is the scalar potential of the body force. In this way we obtain the following as *the equations governing the irrotational motion of an ideal fluid:*

(1) *Condition of incompressibility*

$$\text{div (grad } \Phi) = \text{div } \mathbf{V} = 0$$

or

$$\nabla^2 \Phi = 0 \qquad (9.40)$$

(2) *Equation of motion*

$$\frac{\partial}{\partial t}(\text{grad } \Phi) + \text{grad } \frac{V^2}{2} = \text{grad } \overline{U} - \text{grad}\left(\frac{p}{\rho}\right) \qquad (9.41)$$

or

$$\text{grad}\left(\frac{\partial \Phi}{\partial t} + \frac{V^2}{2} + \frac{p}{\rho} - \overline{U}\right) = 0$$

where

$$V^2 = (\text{grad } \Phi)^2$$

This equation immediately integrates to

$$\frac{\partial \Phi}{\partial t} + \frac{1}{2}(\text{grad } \Phi)^2 + \frac{p}{\rho} - \overline{U} = f(t) \qquad (9.42)$$

where $f(t)$ is a function of time that is uniform throughout space at any instant. This equation is the *unsteady Bernoulli's equation*, which was obtained before (see Eq. 8.16).

246

Ideal-Fluid Aerodynamics

We see that whereas the general (i.e., rotational) motion of an ideal fluid is characterized by four scalar unknowns, namely, the pressure and the three components of the velocity, the irrotational motion of the fluid involves only two scalar unknowns, namely, the pressure and the velocity potential. In the general motion of the fluid, to determine the unknowns we must solve simultaneously the condition of incompressibility and the equation of motion. In the irrotational motion, however, this is no longer necessary, for we may solve for the potential from the condition of incompressibility independently of the equation of motion. The equation of motion itself may be integrated once and for all. Once the potential is determined, the pressure is readily found from the integrated form of the equation of motion, namely, Bernoulli's equation. Also, we may then obtain the velocity field, if desired, from Eq. (9.39).

We thus conclude that *the entire problem of irrotational motion of an ideal fluid amounts to that of solving Laplace's equation, which now expresses the condition of incompressibility.* This is a great mathematical advantage, for there is considerable knowledge already available with regard to the solutions of Laplace's equation, which governs many physical phenomena.

It is instructive to recognize that Eq. (9.40) could have been written down by stipulating that the fluid in motion is incompressible and that the motion is irrotational; in other words, by specifying that the velocity field $\mathbf{V}$ is such that its divergence and curl are both zero (see Section 2.43). Then we can introduce a potential Φ on the basis of the condition that curl $\mathbf{V}$ is equal to zero and determine Φ from the condition

$$\nabla^2\Phi = \text{div } \mathbf{V} = 0$$

In such a case the velocity is determined without invoking any dynamic and thermodynamic considerations. Such a velocity field, we say, is only kinematically feasible. As we have seen, if in addition to the conditions on the velocity field we further assume that the fluid is homogeneous and inviscid, all kinematically feasible velocity fields are also dynamically feasible.

9.9 Irrotational Motion as an Impulsively Generated Motion: Velocity Potential as the Potential of an Impulse

The instantaneous state of irrotational motion of an ideal fluid or a change in such a state may be interpreted as brought about through the action of suddenly, or impulsively, applied forces. Such an interpretation is not only illuminating but is useful in applications.* To build this interpretation we first recall basic definitions related to impulsive motion.

---

* See, for instance, Chapter 10 on unsteady motions.

Suppose a mechanical system is subjected to very large forces for a very short time. Finite changes of velocity and momentum of the system will then occur suddenly. Such changes are said to be brought about *impulsively*, and the forces producing them are known as *impulsive forces*. Let the momentum of a system be changed impulsively at any instant t_0 from $\mathbf{P_0}$ to $\mathbf{P}$. Then the relation between the sudden change in momentum at t_0 and $\mathbf{F}$, the impulsive force causing the change, is given by

$$\delta \mathbf{P}(t_0) = \mathbf{P} - \mathbf{P_0} = \lim_{\delta t \to 0} \int_{t_0}^{t_0+\delta t} \mathbf{F}\, dt$$

The right member of this relation is known as the *impulse*.

Now consider an element of an ideal fluid in motion and let its momentum be changed impulsively at any instant t_0 by means of impulsive body forces and pressure forces. Impulsive pressure forces are generated by sudden changes in the boundary conditions as, for example, when a solid body immersed in the fluid is suddenly set in motion. The impulsive change in momentum of the fluid element is given by

$$\delta m[\mathbf{V} - \mathbf{V_0}] = \lim_{\delta t \to 0} \left[\int_{t_0}^{t_0+\delta t} \delta m \mathbf{f}\, dt - \int_{t_0}^{t_0+\delta t} \delta \tau \operatorname{grad} p\, dt \right] \quad (9.43)$$

where δm is the mass of the element, $\delta \tau$ its volume, $\mathbf{f}$ the body force per unit mass, and $\mathbf{V} - \mathbf{V_0}$ is the impulsive change in the velocity at t_0. This equation is simply an integral of the equation of motion (Eq. 5.11) with time over the interval δt, which is allowed to go to zero. Since the mass, the volume, and the density of the fluid element are constant, Eq. (9.43) takes the form

$$\mathbf{V} - \mathbf{V_0} = \lim_{\delta t \to 0} \left[\int_{t_0}^{t_0+\delta t} \mathbf{f}\, dt - \frac{1}{\rho} \int_{t_0}^{t_0+\delta t} \operatorname{grad} p\, dt \right] \quad (9.44)$$

We assume that the *body forces are irrotational* and set $\mathbf{f}$ equal to grad $\bar{U}$. Equation (9.44) may then be expressed as

$$\mathbf{V} - \mathbf{V_0} = \operatorname{grad} \chi - \frac{1}{\rho} \operatorname{grad} \varpi$$

$$= \operatorname{grad} \left(\chi - \frac{\varpi}{\rho} \right) \quad (9.45)$$

where

$$\chi \equiv \lim_{\delta t \to 0} \int_{t_0}^{t_0+\delta t} \bar{U}\, dt$$

and

$$\varpi \equiv \lim_{\delta t \to 0} \int_{t_0}^{t_0+\delta t} p\, dt$$

Grad χ is the impulse of the body forces and $-(\text{grad } \varpi/\rho)$ is the impulse of the pressure forces, both impulses being measured per unit mass of the fluid. We may call χ the potential of the impulse of the body forces, ϖ the potential of the impulse of the pressure forces, and $[\chi - (\varpi/\rho)]$ the potential of the impulse, that is, of the total impulse. The potential ϖ is known as the *impulsive pressure*.

If we set $\mathbf{V}_0$ as equal to zero, that is, if we consider that the state of motion just before the application of the impulse is a state of rest,* Eq. (9.45) takes the form

$$\mathbf{V} = \text{grad} \left(\chi - \frac{\varpi}{\rho} \right) \tag{9.46}$$

From this equation it immediately follows that curl $\mathbf{V}$ is zero, that is, the motion immediately after the impulse is irrotational and that the velocity potential† of this motion is given by

$$\Phi = \chi - \frac{\varpi}{\rho} \tag{9.47}$$

Furthermore, since both χ and ϖ are single-valued functions, the velocity potential is also a single-valued function. Equations (9.46) and (9.47) show that if impulsive irrotational forces (in the form of body and pressure forces) are applied on an ideal fluid that is initially in a state of rest, the fluid is instantaneously set into an irrotational motion and the resulting velocity potential is single-valued and equal to the potential of the impulse. Equivalently, any state of irrotational motion of an ideal fluid for which there is a single-valued potential may be brought to rest instantaneously by the application of suitable impulsive irrotational forces. We have thus shown that *any given instantaneous state of irrotational motion of an ideal fluid, for which there is a single-valued potential, may be interpreted as generated impulsively from a state of rest by the application of suitable impulsive irrotational forces and that the velocity potential is the potential of the impulse.* Any changes in the state of an irrotational motion for which there is a single-valued potential may also be interpreted as generated by impulsive pressure forces and forces that are irrotational.

Equations (9.46) and (9.47) show that the preceding conclusions hold equally when there are no body forces, and the impulse is only due to impulsive pressure forces.

On the basis of the foregoing conclusions we infer the following important properties of the irrotational motion of an ideal fluid for which there

---

* A state of rest is possible only if the body forces are irrotational.
† An arbitrary constant may always be added to the potential without altering Eq. (9.46) and any of the results that follow from it.

is a *single-valued* potential:* The state of motion at some instant of time is independent of the state of motion at any other instant of time. The state of motion at any given instant is determined solely by the state of the boundary conditions at that instant. In this sense the motion has no memory. Any changes in the flow field as brought about by changes in the boundary conditions are immediately felt throughout the flow field. One says that all changes are propagated instantaneously or with infinite speed in all directions. This means that at each instant the state of motion at any point of the flow field is intimately connected with the state of motion at another point at that instant. Since the flow has no memory, the notion of initial conditions becomes meaningless. In other words, boundary conditions are the only conditions that may be given to specialise the solution of the governing differential equation, which in this case is Laplace's equation. In this sense the mathematical problem representing the motion is a so-called *boundary value* problem.

9.10 Boundary Conditions

We now express the boundary conditions (see Section 5.11) for an ideal fluid in terms of the velocity potential.

Condition at a Solid-Fluid Boundary. The kinematical condition at such a boundary is expressed by Eq. (5.35) or, equivalently by Eq. (5.39). Equation (5.35) now becomes

$$\operatorname{grad} \Phi \cdot \mathbf{n} = \frac{\partial \Phi}{\partial n} = \mathbf{V}_s \cdot \mathbf{n} \text{ on the solid-fluid surface} \qquad (9.48)$$

Here $\partial \Phi / \partial n$ is the spatial derivative of Φ in the direction of the outward normal. Equation (5.39) takes the form

$$\frac{DF}{Dt} = \frac{\partial F}{\partial t} + \operatorname{grad} \Phi \cdot \operatorname{grad} F = 0 \quad \text{on} \quad F(\mathbf{r}, t) = 0 \quad (9.49)$$

Conditions at a Free Surface. Let Φ_1 denote the potential on one side of the surface and Φ_2 the potential on the other side of the surface. Then the kinematic condition expressed by Eq. (5.40) may be written as

$$\operatorname{grad} \Phi_1 \cdot \mathbf{n} = \mathbf{V}_s \cdot \mathbf{n} = \operatorname{grad} \Phi_2 \cdot \mathbf{n} \text{ on the surface} \qquad (9.50)$$

Equations (5.42), which express this condition in an equivalent manner,

---

* It is important to bear this in mind, for as it will be apparent later that important differences exist between flows with single-valued potentials and flows with multivalued potentials.

now become

and

$$
\left.
\begin{array}{l}
\dfrac{\partial F}{\partial t} + \operatorname{grad} \Phi_1 \cdot \operatorname{grad} F = 0 \\[2em]
\dfrac{\partial F}{\partial t} + \operatorname{grad} \Phi_2 \cdot \operatorname{grad} F = 0
\end{array}
\right\}
\qquad \text{on} \qquad F(\mathbf{r}, t) = 0 \qquad (9.51)
$$

The dynamic condition expressed by Eq. (5.44) requires that the pressure be continuous across the free surface. Using the unsteady Bernoulli's equation (9.42), we may write this dynamic condition as

$$
\rho_1 \left[\frac{\partial \Phi_1}{\partial t} + \frac{1}{2} (\operatorname{grad} \Phi_1)^2 - f_1(t) \right] = -p_1 = -p_2
$$

$$
= \rho_2 \left[\frac{\partial \Phi_2}{\partial t} + \frac{1}{2} (\operatorname{grad} \Phi_2)^2 - f_2(t) \right] \qquad \text{on} \qquad F(\mathbf{r}, t) = 0 \quad (9.52)
$$

We have yet to specify the conditions at infinity. To do this we need first to consider the problems we shall be interested in and some of the general properties of the solutions to such problems.

9.11 Problems of Concern

The problem of irrotational motion of an ideal fluid, as we have seen, reduces to that of solving Laplace's equation with specified auxiliary conditions. Solutions of Laplace's equation are known as *harmonic functions*. We shall, therefore, be concerned with harmonic functions and their properties. In learning about some of these properties we shall consider specifically problems that are of most interest to us.

There are two such problems which, although appearing distinct, are actually equivalent. One of the problems concerns the motion arising out of a solid body moving through an infinitely extending ideal fluid. The motion of the fluid is entirely due to that of the body (Fig. 9.7). The mathematical problem is to determine the velocity potential $\Phi_1(\mathbf{r}, t)$ as the solution of the equation

$$
\nabla^2 \Phi_1 = 0 \qquad\qquad (9.53)
$$

in the region R exterior to the body, such that the solution satisfies the boundary condition

$$
\frac{\partial \Phi_1}{\partial n} = \operatorname{grad} \Phi_1 \cdot \mathbf{n} = \mathbf{U}_b(\mathbf{r}, t) \cdot \mathbf{n} \qquad \text{on} \qquad S \qquad (9.54)
$$

where $S = S(t)$ denotes the surface of the body, $\mathbf{n}$ the outward normal, and $\mathbf{U}_b(\mathbf{r}, t)$ the velocity of the surface points of the body. The solution

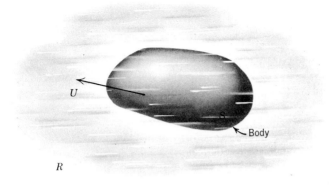

Fig. 9.7 A body moving with a constant velocity through a fluid.

should also satisfy an infinity condition that has yet to be completely specified.

The other problem of interest is that of uniform steady flow past a fixed rigid body* (Fig. 9.8). The mathematical problem is again to determine the velocity potential $\Phi_2(\mathbf{r})$ as the solution of the equation

$$\nabla^2\Phi_2 = 0 \qquad (9.55)$$

in the region R exterior to the body such that the solution satisfies the condition

$$\frac{\partial\Phi_2}{\partial n} = \text{grad } \Phi_2 \cdot \mathbf{n} = 0 \qquad \text{on} \quad S \qquad (9.56)$$

and an infinity condition yet to be specified. Now the surface S is not a function of time.

The two problems are of equivalent form. To see this, let us write

$$\text{grad } \Phi_2 = \mathbf{U} + \text{grad } \phi \qquad (9.57)$$

where, $\phi = \phi(\mathbf{r})$ is another scalar function of position defined by the

Fig. 9.8 Steady flow past a fixed body.

* Note that t̶ uniform flow could as well be time dependent and the following considerations ⋮ equally applicable to that situation.

Eq. (9.57). Equations (9.55) and (9.56) now take the form

$$\nabla^2\phi = 0 \qquad (9.58)$$

and

$$\frac{\partial \phi}{\partial n} = \text{grad } \phi \cdot \mathbf{n} = -\mathbf{U} \cdot \mathbf{n} \qquad \text{on} \quad S \qquad (9.59)$$

These equations are of the same form as (9.53) and (9.54).

We may, therefore, state that the problem is to determine Φ as the solution of the equation

$$\nabla^2\Phi = 0 \qquad (9.60)$$

in the region R exterior to the body such that the solution satisfies the boundary condition*

$$\frac{\partial \Phi}{\partial n} = \text{grad } \Phi \cdot \mathbf{n} = f(\mathbf{r}, t) \qquad \text{on} \quad S \qquad (9.61)$$

and certain conditions at infinity. For the first problem we have

$$\Phi = \Phi_1(\mathbf{r}, t)$$

and

$$f(\mathbf{r}, t) = \mathbf{U}_b(\mathbf{r}, t) \cdot \mathbf{n} \qquad (9.62)$$

For the second problem we have

$$\Phi = \phi(\mathbf{r})$$

and

$$f(\mathbf{r}, t) = -\mathbf{U} \cdot \mathbf{n} \qquad (9.63)$$

The mathematical problem represented by (9.60) and (9.61) is known as the *Neumann exterior problem.* We shall consider some general properties of the solution for such a problem and the specification of the conditions at infinity. To proceed we must first introduce a few simple topological notions.

9.12 Some Topological Notions

Connectivity. A region of space is said to be *connected if any two points of that space can be connected by a continuous line (or path) that does not leave the boundaries of the region.* For example, the regions interior and exterior to a closed surface, individually, are connected regions. The whole region of space consisting together of the interior and exterior regions is, however, not a connected region.

---

* The function f in the following should not be confused with $f(t)$ in the Bernoulli equation.

Reconcilable and Irreconcilable Paths.   A path is any line joining two points of a connected region. Consider the region exterior or interior to a finite closed surface (Fig. 9.9). Let P and Q be any two points and let C_1, C_2, C_3 be any three paths lying in that region and connecting P and Q. Any two of the paths can be made to coincide with each other by continuous variation of the paths concerned and without ever leaving the

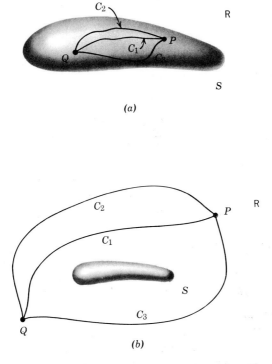

(a)

(b)

Fig. 9.9   Paths defined by a finite closed surface; (a) interior paths; (b) exterior paths.

region under consideration. The paths are then said to be *reconcilable*. We note that all paths between any two points in the region exterior or interior to a finite closed surface are reconcilable.

Now consider the regions exterior and interior to the surface of an infinite cylinder (Fig. 9.10). The cylinder extends to infinity in both the directions normal to the plane of the figure. It is immediately seen that in the interior region all paths are reconcilable. Let P and Q be any two points in the exterior region R and as shown let C_1, C_2, C_3 be any three paths lying in that region and connecting P and Q. We see that the paths C_1 and C_2 can be made to coincide by continuous variation without ever

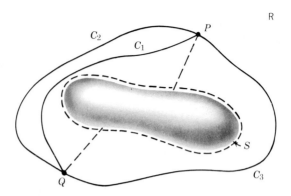

Fig. 9.10   Paths in the region exterior to the surface of an infinite cylinder.

leaving the region R, and they are therefore reconcilable. The paths C_1 and C_3, or equivalently C_2 and C_3, can never be made to coincide without leaving the region R. The best we can do without leaving the region is, as shown by dotted lines in Fig. 9.10, to approach the boundary of the cylinder with C_1 or C_2 brought to one side and C_3 to the other side. The paths C_1 and C_3, and C_2 and C_3 are said to be *irreconcilable*. We note that there are only two irreconcilable paths in the region exterior to the infinite cylinder.

Summing up, we state that *any two paths in a connected region are reconcilable if they can be made to coincide by continuous variation without ever passing out of that region. They are irreconcilable if they cannot be made to coincide without leaving the region.*

Reducible and Irreducible Circuits.   A circuit is any closed line in a connected region. Consider the region exterior or interior to a finite closed surface (Fig. 9.11). Let $\mathscr{C}_1$ and $\mathscr{C}_2$ be, as shown, any two circuits. By continuous variation (or deformation) each of the circuits can be contracted to a point without ever leaving the region under consideration.

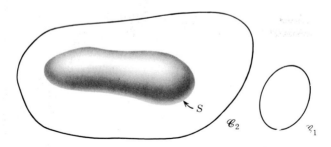

Fig. 9.11   Circuits in regions defined by a finite closed surface.

The circuits are then said to be *reducible*. We note that all circuits in the region exterior or interior to a finite closed surface are reducible.

Now consider the regions exterior and interior to the surface of an infinite cylinder (Fig. 9.12). It is immediately seen that all circuits in the interior region are reducible. Let $\mathscr{C}_1$ and $\mathscr{C}_2$ (or $\mathscr{C}_3$ and $\mathscr{C}_4$) be, as shown, any two circuits, one of them, namely $\mathscr{C}_1$, enclosing the cylinder, the other, namely $\mathscr{C}_2$, not enclosing the cylinder. The latter is reducible. The circuit $\mathscr{C}_1$ cannot be contracted to a point without leaving the region R.

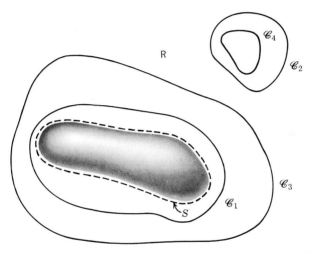

Fig. 9.12   Circuits in the region exterior to the surface of an infinite cylinder.

The best that can be done is to contract it, as shown by the dotted lines in the figure, to the boundary of the cylinder. The circuit $\mathscr{C}_1$ (or $\mathscr{C}_3$) is, therefore, said to be *irreducible*. We note that in the region exterior to the cylinder there is only one irreducible circuit, namely the circuit enclosing the cylinder.

Summing up, we state that *any circuit in a connected region is reducible if the circuit can be contracted to a point without ever leaving that region. A circuit that cannot be contracted to a point without leaving the region of interest is irreducible.*

We note that if P and Q are any two points in a connected region then any two reconcilable paths joining the points form a reducible circuit. Any two irreconcilable paths joining the points form an irreducible circuit.

Reconcilable and Irreconcilable Circuits.   Consider the region exterior or interior to a closed finite surface and any two circuits $\mathscr{C}_1$ and $\mathscr{C}_2$ in that region (Fig. 9.11). The circuits can be made to coincide by continuous variation without ever leaving the region under consideration. We say that the circuits are reconcilable. We note that all circuits are reconcilable in the region exterior or interior to a finite closed surface.

Now consider the regions exterior and interior to the surface of an infinite cylinder (Fig. 9.12). We note immediately that all circuits in the interior region are reconcilable. Any two circuits such as $\mathscr{C}_2$ and $\mathscr{C}_4$ in the exterior region R can be made to coincide by continuous deformation without ever leaving the region R. Similarly, circuits such as $\mathscr{C}_1$ and $\mathscr{C}_3$ can be made to coincide with each other. We say that $\mathscr{C}_1$ and $\mathscr{C}_3$, and $\mathscr{C}_2$ and $\mathscr{C}_4$ form reconcilable circuits. The circuits $\mathscr{C}_1$ and $\mathscr{C}_2$ (or equivalently $\mathscr{C}_3$ and $\mathscr{C}_4$) cannot be made to coincide without passing out of the region R. We say that $\mathscr{C}_1$ and $\mathscr{C}_2$ are irreconcilable circuits.

Summing up we state that *any two circuits in a connected region are reconcilable if they can be made to coincide by continuous variation without ever leaving that region. Two circuits are irreconcilable if they cannot be made to coincide without leaving the region under consideration.*

We note that if $\mathscr{C}_1$ and $\mathscr{C}_2$ are any two reconcilable circuits lying in a connected region, they can be connected by a continuous surface lying wholly in that region, the circuits forming the boundaries of the surface. For irreconcilable circuits no such surface is possible.

Simply Connected Region. *A connected region in which all paths connecting any two points of the region are reconcilable is said to be simply connected. In such a region all circuits are reducible and reconcilable.* The regions exterior and interior to a closed finite surface and the region interior to the surface of an infinite cylinder are simply connected regions. The region exterior to the surface of an infinite cylinder is not a simply-connected region.

Doubly Connected Region. Consider the region exterior to the surface of an infinite cylinder. Of all the paths that can be drawn connecting any two points of the region there are, as we had seen, only two paths (such as C_1 and C_3 of Fig. 9.10) that are irreconcilable with each other. The others are either reconcilable with one another or with one of the two irreconcilable paths. In view of this situation we speak of the region exterior to the surface of the infinite cylinder as doubly connected. We state that *a connected region in which there are only two irreconcilable paths and no more is a doubly connected region. In such a region, there is only one irreducible circuit.* In the case of the infinite cylinder it is the circuit enclosing the cylinder.

Multiply Connected Region. *A connected region in which there are irreconcilable paths or, equivalently, irreducible circuits, is said to be a multiply connected region.* If there are n irreconcilable paths or, equivalently, $n - 1$ irreducible circuits, the region is said to be n-ply connected.

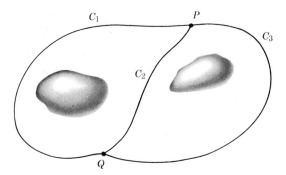

Fig. 9.13 A triply connected region.

The region exterior to the surfaces of two infinite cylinders is a triply connected region (Fig. 9.13).

Barriers. A multiply connected region may be rendered simply connected by introducing suitable *barriers* or boundaries that may not be crossed. Consider the doubly connected region exterior to an infinite cylinder. Insert a barrier, as shown in Fig. 9.14*a*, extending from the cylinder to infinity and agree not to cross it. Then in the region thus modified (exterior to the shaded part in the figure) all paths joining any two

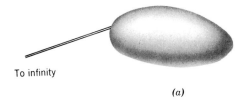

To infinity

(a)

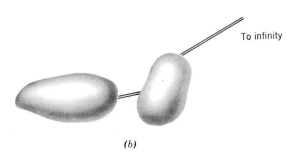

To infinity

(b)

Fig. 9.14 Barriers: *(a)* doubly connected region; *(b)* triply connected region.

points are reconcilable, and all circuits that can be drawn in that region are reducible. The modified region is, therefore, simply connected.

In a similar way by inserting a suitable number of barriers, a multiply connected region may be rendered simply connected. A triply connected region is made simply connected by inserting two barriers (see Fig. 9.14b).

9.13 Irrotational Motion in a Simply Connected Region

The important role played by the preceding notions in problems of irrotational motion will become apparent in this and succeeding sections.

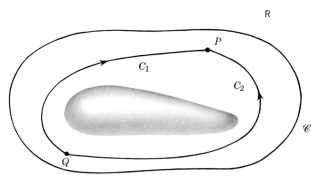

Fig. 9.15 Illustrating the properties of irrotational motion in a simply connected region.

Consider irrotational motion in a simply connected region R such as that exterior to a finite body moving through an infinitely extending fluid (Fig. 9.15). Let $\mathscr{C}$ be any circuit drawn in that region. The circulation around $\mathscr{C}$ is given by

$$\Gamma = \oint_{\mathscr{C}} \mathbf{V} \cdot \mathbf{ds} = \oint_{\mathscr{C}} d\Phi \qquad (9.64)$$

where, to recall, $\mathbf{V}$ is the velocity and Φ is the velocity potential. Let σ be any open surface lying in the region R (i.e., in the fluid) such that the circuit $\mathscr{C}$ forms the boundary of σ. Since the region is simply connected, it is always possible to draw such a surface. Now according to Stokes' theorem we have

$$\oint_{\mathscr{C}} \mathbf{V} \cdot \mathbf{ds} = \iint_{\sigma} \operatorname{curl} \mathbf{V} \cdot \mathbf{n} \, dS$$

$$= \iint_{\sigma} \mathbf{\Omega} \cdot \mathbf{n} \, dS$$

where $\mathbf{\Omega} = \operatorname{curl} \mathbf{V}$ is the vorticity. But the vorticity is zero everywhere in the region R. Hence it follows that

$$\Gamma = \oint_{\mathscr{C}} d\Phi = \oint_{\mathscr{C}} \mathbf{V} \cdot \mathbf{ds} = 0 \qquad (9.65)$$

We conclude that *the circulation around every circuit in a simply connected region is zero.* Thus, the irrotational motion of a fluid in a simply connected region is motion without any circulation. We refer to it as *acyclic motion.*

Now consider any two points P and Q in the region R. Let C_1 and C_2 be any two paths joining Q to P. Since the circulation around the circuit formed by C_1 and C_2 is zero, we have

$$\int_{Q \text{ via } C_1}^{P} \mathbf{V} \cdot \mathbf{ds} - \int_{Q \text{ via } C_2}^{P} \mathbf{V} \cdot \mathbf{ds} = 0$$

In terms of the potential this becomes

$$\int_{Q \text{ via } C_1}^{P} d\Phi - \int_{Q \text{ via } C_2.}^{P} d\Phi = 0 \qquad (9.66)$$

$$[\Phi(P) - \Phi(Q)]_{\text{via } C_1} = [\Phi(P) - \Phi(Q)]_{\text{via } C_2}$$

We conclude that the integral in (9.66) is independent of the path connecting Q to P and that consequently $\Phi(Q)$ and $\Phi(P)$ should be single valued. We state that the irrotational motion of a fluid in a simply connected region is characterized by a single valued velocity potential.

9.14 Irrotational Motion in a Doubly Connected Region

Consider the irrotational motion in a doubly connected region R such as that exterior to the surface of an infinite cylinder moving through an infinitely extending fluid, the generators of the cylinder being normal to the plane of motion.* Let $\mathscr{C}_1$ be any reducible circuit (Fig. 9.16). A surface σ_1 whose boundary is $\mathscr{C}_1$ can always be drawn such that σ_1 lies wholly in the region R (i.e., the fluid) where the vorticity $\mathbf{\Omega}$ is equal to zero. Therefore, for the circulation around the circuit $\mathscr{C}_1$ we obtain

$$\Gamma_{\mathscr{C}_1} = \oint_{\mathscr{C}_1} \mathbf{V} \cdot \mathbf{ds} = \iint_{\sigma_1} \mathbf{\Omega} \cdot \mathbf{n} \, dS = 0 \qquad (9.67)$$

We conclude *that in a doubly connected region the circulation around all reducible circuits is zero.*

Let $\mathscr{C}_2$ be an irreducible circuit. Now, it is impossible to draw a surface whose boundary is $\mathscr{C}_2$ such that the surface will lie entirely in the region R. This means we can no longer determine by application of Stokes' theorem whether $\Gamma_{\mathscr{C}_2}$ vanishes or not. Furthermore, if it does not vanish there is

* We are considering two-dimensional motion.

no way of telling what its value is. We thus conclude that *in a doubly connected region, the circulation around any irreducible circuit may not vanish and its value remains undetermined.*

Consider next another irreducible circuit such as $\mathscr{C}_3$ shown in Fig. (9.16). It is now possible to draw a surface, say σ, whose boundaries

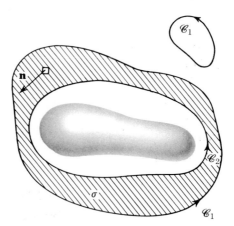

Fig. 9.16 Illustrating the properties of irrotational motion in a doubly connected region.

are the circuits $\mathscr{C}_2$ and $\mathscr{C}_3$ such that the surface lies wholly in the region R (Fig. 9.16). We then have

$$\oint_{\mathscr{C}_2} \mathbf{V} \cdot d\mathbf{s} - \oint_{\mathscr{C}_3} \mathbf{V} \cdot d\mathbf{s} = \iint_{\sigma} \mathbf{\Omega} \cdot \mathbf{n}\, dS = 0 \qquad (9.68)$$

where the directions of integration along $\mathscr{C}_2$, $\mathscr{C}_3$ are consistent with the indicated direction for the normal $\mathbf{n}$. It follows that

$$\Gamma_{\mathscr{C}_3} = \oint_{\mathscr{C}_3} \mathbf{V} \cdot d\mathbf{s} = \oint_{\mathscr{C}_2} \mathbf{V} \cdot d\mathbf{s} = \Gamma_{\mathscr{C}_2} \qquad (9.69)$$

We conclude *that the circulation around all irreducible circuits has the same value.*

Similar results hold also for a multiply connected region. We shall not enter into a derivation of those results. It is thus seen that the irrotational motion of a fluid in multiply connected regions may be a motion characterized by non-zero circulations around irreducible circuits. The present considerations leave the values of the circulations undetermined. Additional independent considerations need to be invoked if the circulations are to be determined. Irrotational motion with non-zero circulations around irreducible circuits is referred to as *cyclic motion.*

We now consider the properties of the velocity potential in a doubly connected region. Let P and Q be any two points in the region R (Fig. 9.17). Let C_1 and C_2 be any two reconcilable paths joining Q and P. Then since the circulation around the reducible circuit formed by C_1 and C_2 is zero, we obtain

$$\int_{Q\,\text{via}\,C_1}^{P} \mathbf{V}\cdot\mathbf{ds} - \int_{Q\,\text{via}\,C_2}^{P} \mathbf{V}\cdot\mathbf{ds} = 0$$

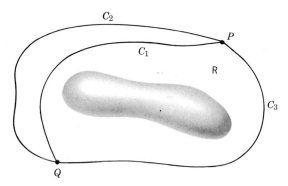

Fig. 9.17   Illustrating the multivaluedness of the potential in a doubly connected region.

In terms of the potential we have

$$\int_{Q\,\text{via}\,C_1}^{P} d\Phi - \int_{Q\,\text{via}\,C_2}^{P} d\Phi = 0 \tag{9.70}$$

$$[\Phi(P) - \Phi(Q)]_{\text{via}\,C_1} = [\Phi(P) - \Phi(Q)]_{\text{via}\,C_2}$$

We conclude that *the integral* $\displaystyle\int_{Q}^{P} d\Phi$ *has the same value for all reconcilable paths connecting Q and P.* It further follows that *along reducible circuits the velocity potential Φ is single valued.*

The situation is different for irreconcilable paths and irreducible circuits. Let C_3 be a path that is irreconcilable with C_1 and consequently also with C_2. Now the circulation around the irreducible circuit formed by C_1 and C_3 need not be zero; for example, it may be Γ. Then we obtain (taking proper account of the direction of integration for Γ)

$$\int_{Q\,\text{via}\,C_3}^{P} d\Phi - \int_{Q\,\text{via}\,C_1}^{P} d\Phi = \Gamma \tag{9.71}$$

$$[\Phi(P) - \Phi(Q)]_{\text{via}\,C_3} - [\Phi(P) - \Phi(Q)]_{\text{via}\,C_1} = \Gamma$$

We conclude that *the integral* $\int_Q^P d\Phi$ *may have different values along ir-reconcilable paths connecting P and Q*. It further follows that *along irreducible circuits the velocity potential Φ may be multivalued*. From Eqs. (9.70) and (9.71) we obtain

$$[\Phi(P) - \Phi(Q)]_{\text{via } C_3} - [\Phi(P) - \Phi(Q)]_{\text{via } C_2} = \Gamma$$

From (9.71) we have

$$[\Phi(P)]_{\text{via } C_3} = [\Phi(P)]_{\text{via } C_1} + \Gamma + [\Phi(Q)]_{\text{via } C_3} - [\Phi(Q)]_{\text{via } C_1} \quad (9.72)$$

Now consider another path, irreconcilable to C_1, such as C_4 in Fig. 9.18. The paths C_3 and C_4 form an irreducible circuit. Hence, we obtain

$$[\Phi(P) - \Phi(Q)]_{\text{via } C_4} - [\Phi(P) - \Phi(Q)]_{\text{via } C_3} = \Gamma \quad (9.73)$$

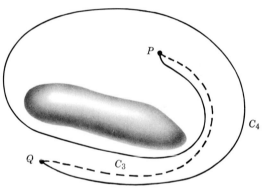

Fig. 9.18 Illustrating further the multivaluedness of the potential in a doubly connected region.

From (9.72) and (9.73) it follows that

$$[\Phi(P)]_{\text{via } C_4} = [\Phi(P)]_{\text{via } C_1} + 2\Gamma + [\Phi(Q)]_{\text{via } C_4} - [\Phi(Q)]_{\text{via } C_1} \quad (9.74)$$

In this, depending on the path from Q to P, the potential at P may assume many values that will differ by integer multiples of the circulation* Γ.

Similar results hold for irrotational motion in multiply connected regions. We shall not enter into a derivation of those results. Reference may be made to Lamb (1932).

Consider again the doubly connected region R. Let us now render the region into a simply connected region by inserting a barrier as shown in Fig. 9.19. Consider two adjacent points P and P_1 on either side of the barrier and the integral of $d\Phi$ along any path C connecting P to P_1. We

---

* Assume that $\Phi(Q)$ is zero for all paths.

then have

$$\lim_{P_1 \to P} \int_P^{P_1} d\Phi = \lim_{P_1 \to P} [\Phi(P_1) - \Phi(P)] = \Gamma \qquad (9.75)$$

where Γ is the circulation around an irreducible circuit. We thus conclude that *across the barrier the potential jumps by an amount* Γ.

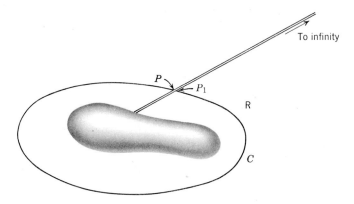

Fig. 9.19 Illustrating the jump in the potential across a barrier.

9.15   Summary

We summarize briefly the main results of the preceding two sections. For the irrotational motion of a fluid in a simply connected region, the circulation around every circuit is zero and the velocity potential is single valued.

For the irrotational motion of a fluid in a doubly connected region the following results hold: the circulation around any reducible circuit is zero; the circulation around an irreducible circuit may or may not be zero, it remains unknown; the circulation has the same value for all irreducible circuits; the velocity potential may be many valued over irreducible circuits; the various values of the potential differ by multiples of the circulation; if a barrier is inserted to make the region simply connected there is a jump in the potential across the barrier, the jump being equal to the value of the circulation.

9.16   Conditions at Infinity

We now turn to the specification of the conditions at infinity for the Neumann exterior problem. To recall, the problem concerns the irrotational motion of fluid in the region R exterior to the surface S of a body (see Section 9.11). We are to determine the potential Φ as the solution of the equation

$$\nabla^2 \Phi = 0$$

such that the solution satisfies the condition

$$\frac{\partial \Phi}{\partial n} = \text{grad } \Phi \cdot \mathbf{n} = f(\mathbf{r}, t) \quad \text{on } S$$

and some conditions at infinity.

To determine the nature of the conditions at infinity we proceed as follows. Let Σ be an arbitrary surface drawn such that it encloses the

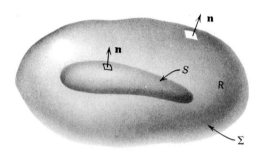

Fig. 9.20   Illustrating the determination of infinity conditions.

surface S of the body (see Fig. 9.20). Now, according to the divergence theorem we have

$$\oiint_\Sigma \text{grad } \Phi \cdot \mathbf{n} \, dS - \oiint_S \text{grad } \Phi \cdot \mathbf{n} \, dS$$

$$= \iiint_R \text{div grad } \Phi \, d\tau$$

$$= \iiint_R \nabla^2 \Phi \, d\tau = 0 \tag{9.76}$$

where R is the volume of the region included between the surfaces S and Σ and, as shown in Fig. 9.20, $\mathbf{n}$ denotes the outward normal on the surfaces S and Σ. It follows that

$$\oiint_\Sigma \text{grad } \Phi \cdot \mathbf{n} \, dS = \oiint_S \text{grad } \Phi \cdot \mathbf{n} \, dS = \oiint_S f(\mathbf{r}, t) \, dS \tag{9.77}$$

Now from Eqs. (9.62) and (9.63) we obtain

$$\oiint_S f(\mathbf{r}, t) \, dS = \oiint_S \mathbf{U}_b(\mathbf{r}, t) \cdot \mathbf{n} \, dS \tag{9.78}$$

for a body moving through a fluid and

$$\oiint_S f(\mathbf{r}, t)\, dS = -\mathbf{U} \cdot \oiint_S \mathbf{n}\, dS$$
$$= 0 \qquad\qquad (9.79)$$

for *steady flow past a fixed rigid body*.

Consider the case of a body moving through the fluid. Suppose that the motion of the body is purely a rigid body motion; that is, the motion consists of translation and rotation only but no deformation. Let $\mathbf{V}(t)$ denote the velocity of translation and $\boldsymbol{\omega}(t)$ the angular velocity of the body, the axis of rotation passing through some point of the body. With $\mathbf{r}$ measured from a point on the axis of rotation we have

$$\mathbf{U}_b(\mathbf{r}, t) = \mathbf{V}(t) + \boldsymbol{\omega}(t) \times \mathbf{r}$$

Equation (9.78) then becomes

$$\oiint_S f(\mathbf{r}, t)\, dS = \mathbf{V}(t) \cdot \oiint_S \mathbf{n}\, dS + \boldsymbol{\omega}(t) \cdot \oiint_S \mathbf{r} \times \mathbf{n}\, dS$$
$$= 0 \qquad\qquad (9.80)$$

If the body is also deforming, the integral

$$\oiint_S \mathbf{U}_b(\mathbf{r}, t) \cdot \mathbf{n}\, dS$$

need not vanish, there being in general a contribution due to the deformation part of the body motion. *The contribution is non-zero only if the deformation is such that the volume of the body is altering.*

With these considerations in mind, we shall distinguish two cases: one that of a rigid body, whether in motion through the fluid or fixed in a flow past it, and the other that of a deforming body. In the latter case by deformation we shall mean only volume changes. We conclude that

$$\oiint_S f(\mathbf{r}, t)\, dS = 0, \qquad \text{for a rigid body}$$
$$\neq 0, \qquad \text{for a deforming body} \qquad (9.81)$$

It follows that

$$\oiint_\Sigma \operatorname{grad} \Phi \cdot \mathbf{n}\, dS = 0, \qquad \text{for a rigid body}$$
$$\neq 0, \qquad \text{for a deforming body} \qquad (9.82)$$

The integral over Σ will therefore vanish if the body is rigid or will be finite and equal to a definite function of time if the body is undergoing volume changes. Now, the integral over Σ must be independent of the shape and location of the arbitrary surface Σ. This means grad Φ should behave in a certain way, with distance from the body such that Eq. (9.82) is fulfilled.

To examine this behavior, consider first a finite three-dimensional body. The region exterior to the body surface S is then simply connected. Let us choose Σ to be the surface of a sphere with its center at some point close to the body. Choose the center of the sphere as the origin of coordinates and introduce spherical coordinates r, θ, φ. Equation (9.82) then becomes

$$\oiint_{\text{sphere}} \frac{\partial \Phi}{\partial r} r^2 \sin \theta \, d\theta \, d\varphi = 0, \qquad \text{for a rigid body}$$

$$\neq 0, \qquad \text{but finite, for a deforming body} \quad (9.83)$$

It follows that $\partial \Phi / \partial r$ should behave at least as follows as r goes to infinity: as $r \to \infty$

$$\frac{\partial \Phi}{\partial r} \sim \frac{1}{r^3}, \qquad \text{if the body is rigid}$$

$$\frac{\partial \Phi}{\partial r} \sim \frac{1}{r^2}, \qquad \begin{array}{l} \text{if the body is undergoing} \\ \text{volume changes} \end{array} \qquad (9.84)$$

It can be shown, by the application of the curl theorem, Eq. (2.127), to the vector field grad Φ in the region R included between the surfaces S and Σ, that the derivatives $\partial \Phi / \partial \theta$ and $\partial \Phi / \partial \varphi$ should go to zero as $1/r^2$ as r goes to infinity:

$$\frac{\partial \Phi}{\partial \theta}, \qquad \frac{\partial \Phi}{\partial \varphi} \sim \frac{1}{r^2} \qquad \text{as } r \to \infty \qquad (9.85)$$

This is true for *rigid and deforming bodies*.

Recall that in the problem of the body moving through the fluid the conditions (9.84) and (9.85) apply on the actual velocity potential, while in the problem of flow past a fixed body they apply on the potential ϕ defined by the relation

$$\text{grad } \Phi = \text{grad } \phi + \mathbf{U}$$

(see Section 9.11).

Consider now the two dimensional flow in the region exterior to an infinite cylinder. Let $\mathscr{C}_0$ denote the boundary curve of the surface S of the cylinder, the curve being obtained by the intersection of a plane of motion with the cylinder (Fig. 9.21). Similarly, let $\mathscr{C}$ denote the boundary curve of the arbitrary surface Σ. Equation (9.82) then takes the form

$$\oint_{\mathscr{C}} \text{grad } \Phi \cdot \mathbf{n} \, ds = 0, \qquad \text{for a rigid cylinder}$$

$$\neq 0, \qquad \begin{array}{l} \text{but finite, for a} \\ \text{deforming cylinder} \end{array} \qquad (9.86)$$

where, as shown in the figure, **n** is the normal to $\mathscr{C}$ and ds is an element of the circuit $\mathscr{C}$. Now choose $\mathscr{C}$ to be a circle with center at some point close to the body, the center as the origin of coordinates and introduce cylindrical coordinates r, θ in the plane of motion. Equation (9.86) then becomes

$$\int_{\text{circle}} \frac{\partial \Phi}{\partial r} r\, d\theta = 0, \qquad \text{for a rigid cylinder}$$

$$\neq 0, \qquad \text{but finite, for a deforming body} \qquad (9.87)$$

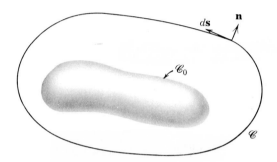

Fig. 9.21 Illustrating the determination of infinity conditions for two-dimensional flow exterior to an infinite cylinder.

It follows that $\partial \Phi / \partial r$ should behave at least as follows as r goes to infinity: as $r \to \infty$

$$\frac{\partial \Phi}{\partial r} \sim \frac{1}{r^2}, \qquad \text{if the cylinder is rigid}$$

$$\frac{\partial \Phi}{\partial r} \sim \frac{1}{r}, \qquad \begin{array}{l}\text{if the cylinder is undergoing}\\ \text{volume changes}\end{array} \qquad (9.88)$$

To examine the behavior of the derivative $\partial \Phi / \partial \theta$, we consider the circulation around the circuit $\mathscr{C}$. Since $\mathscr{C}$ is an irreducible circuit (the region exterior to $\mathscr{C}_0$ is doubly connected) the circulation around $\mathscr{C}$ may not vanish. We have*

$$\oint_{\mathscr{C}} \text{grad } \Phi \cdot \mathbf{ds} = \oint_{\mathscr{C}_0} \text{grad } \Phi \cdot \mathbf{ds} = \Gamma \qquad (9.89)$$

Now, as before, we choose a circle for $\mathscr{C}$ and obtain

$$\int_{\text{circle}} \frac{1}{r} \frac{\partial \Phi}{\partial \theta} r\, d\theta = \Gamma \qquad (9.90)$$

* Note that the application of the curl theorem to grad Φ in the region between $\mathscr{C}_0$ and $\mathscr{C}$ leads to the same result.

We conclude that $\partial\Phi/\partial\theta$ should behave as follows:
as $r \to \infty$

$$\frac{\partial\Phi}{\partial\theta} \sim \frac{1}{r}, \qquad \text{if } \Gamma = 0$$

$$\sim \text{finite}, \qquad \text{if } \Gamma \neq 0 \tag{9.91}$$

We note that in the problem of the cylinder moving through the fluid the conditions (9.88) and (9.91) apply on the actual velocity potential, while in the problem of flow past a fixed body they apply on the potential ϕ.

9.17 Velocity Components at Infinity

The behavior of the velocity components at infinity may now be stated. Consider first the motion arising out of a body moving through the fluid. For a finite body we have the following result:

$$u_r \sim \frac{1}{r^3}, \qquad \text{if body is rigid}$$

$$\sim \frac{1}{r^2}, \qquad \begin{array}{l}\text{if body undergoes} \\ \text{volume changes}\end{array} \tag{9.92}$$

and

$$u_\theta, u_\varphi \sim \frac{1}{r^3}$$

where u_r, u_θ, u_φ are components of the velocity. For an infinite cylinder the corresponding results are:

$$u_r \sim \frac{1}{r^2}, \qquad \text{if body is rigid}$$

$$\sim \frac{1}{r}, \qquad \text{if the body undergoes volume changes}$$

$$u_\theta \sim \frac{1}{r^2}, \qquad \text{if } \Gamma = 0 \tag{9.93}$$

$$\sim \frac{1}{r}, \qquad \text{if } \Gamma \neq 0$$

where u_r, u_θ are the components of the velocity. *We conclude that in the problem of a body moving through a fluid the fluid velocity at infinity is zero.* Consider next the steady flow past a fixed rigid body. For a finite body we have the following results:

$$(u_r - U_r), \qquad (u_\theta - U_\theta), \qquad (u_\varphi - U_\varphi) \sim \frac{1}{r^3} \qquad (9.94)$$

where U_r, U_θ, U_φ are the components of the uniform stream **U**, and u_r, u_θ, u_φ are the components of the actual velocity. For flow past an infinite cylinder the corresponding results are:

$$(u_r - U_r) \sim \frac{1}{r^2}$$

$$(u_\theta - U_\theta) \sim \frac{1}{r^2} \qquad \text{if } \Gamma = 0 \qquad (9.95)$$

$$\sim \frac{1}{r} \qquad \text{if } \Gamma \neq 0$$

where U_r and U_θ are the components of the uniform stream **U**, and u_r and u_θ are the components of the actual velocity. We conclude that *for the steady flow past a fixed rigid body the fluid velocity at infinity is equal to the original uniform velocity.*

9.18   Some Further Properties of Irrotational Motion

Simply Connected Region.   We shall now record some further properties of irrotational motion that are of particular interest to us. Consider first motion in a simply connected region.

(1) *The potential Φ can neither be a maximum nor minimum in the interior of the fluid.*

Consider a point P in the interior of the fluid. Let $\delta\tau$ be an infinitesimal volume element surrounding P. Let δS be the surface of the element. We then have

$$\oiint_{\delta S} \frac{\partial \Phi}{\partial n} \, dS = \oiint_{\delta S} \text{grad } \Phi \cdot \mathbf{n} \, dS$$

$$= \delta\tau \, \text{div} \, (\text{grad } \Phi) = 0$$

This shows that in the immediate neighborhood of P the spatial derivative $\partial\Phi/\partial n$ for all directions of the outward normal **n** can neither be wholly positive nor negative. Hence we conclude that Φ can neither be a minimum nor a maximum at P.

It follows that the maximum and minimum values of Φ will occur only on the boundary of the motion.

(2) *The spatial derivatives of Φ are also harmonic functions, that is they*

satisfy Laplace's equation. We have

$$\mathbf{V} = \text{grad } \Phi$$
$$\text{div } \mathbf{V} = \nabla^2\Phi = 0$$
$$\text{curl } \mathbf{V} = 0$$

From the last relation we obtain

$$\text{curl (curl } \mathbf{V}) = \text{grad (div } \mathbf{V}) - \nabla^2\mathbf{V}$$
$$= -\nabla^2\mathbf{V}$$

Hence it follows that

$$\nabla^2\mathbf{V} = 0$$

or,

$$\nabla^2 \text{ (grad } \Phi) = 0$$

In Cartesians, we have

$$\nabla^2\left(\frac{\partial\Phi}{\partial x}\right) = 0$$

$$\nabla^2\left(\frac{\partial\Phi}{\partial y}\right) = 0$$

$$\nabla^2\left(\frac{\partial\Phi}{\partial z}\right) = 0$$

Thus the first derivatives of Φ are also harmonic functions. In the same way, we can show that the higher derivatives of Φ are also harmonic.

(3) *The spatial derivatives of Φ can neither be maximum nor minimum in the interior of the fluid.* This follows from (1) if instead of Φ we consider its derivatives.

(4) *The velocity components can neither be a maximum nor a minimum in the interior of the fluid.* This follows from (3).

(5) *The magnitude of the velocity cannot be a maximum in the interior of the fluid.*

If Ψ and Φ are two scalar functions of position, Green's theorem (2.139) states that

$$\oiint_S \Psi \text{ grad } \Phi \cdot \mathbf{n} \, dS = \iiint_R (\Psi \nabla^2 \Phi + \text{grad } \Psi \cdot \text{grad } \Phi) \, d\tau$$

substitute Φ for Ψ. We then obtain, noting that $\nabla^2\Phi = 0$,

$$\oiint_S \Phi \text{ grad } \Phi \cdot \mathbf{n} \, dS = \iiint_R (\text{grad } \Phi)^2 \, d\tau$$

or

$$\frac{1}{2} \oint\!\!\!\oint_{S} \text{grad } \Phi^2 \cdot \mathbf{n} \, dS = \int\!\!\!\int\!\!\!\int_{R} (\text{grad } \Phi)^2 \, d\tau$$

Such a relation also holds for each of the derivatives $\partial\Phi/\partial x$, $\partial\Phi/\partial y$, $\partial\Phi/\partial z$ since they are also harmonic. Adding the corresponding relations for the derivatives, we obtain

$$\frac{1}{2} \oint\!\!\!\oint_{S} \text{grad } V^2 \cdot \mathbf{n} \, dS = \int\!\!\!\int\!\!\!\int_{R} [(\text{grad } u)^2 + (\text{grad } v)^2 + (\text{grad } w)^2] \, d\tau$$

$$> 0 \qquad\qquad (9.96)$$

where,

$$u = \frac{\partial\Phi}{\partial x}$$

$$v = \frac{\partial\Phi}{\partial y}$$

$$w = \frac{\partial\Phi}{\partial z}$$

$$V^2 = u^2 + v^2 + w^2$$

Now consider a point P in the interior of the fluid. Let δS be the surface of an infinitesimal volume element surrounding P. Equation (9.96) then takes the form

$$\oint\!\!\!\oint_{\delta S} \frac{\partial V^2}{\partial n} \, dS = \oint\!\!\!\oint_{\delta S} \text{grad } V^2 \cdot \mathbf{n} \, dS$$

$$> 0 \qquad\qquad (9.97)$$

This shows that in the immediate neighborhood of P the spatial derivative $\partial V^2/\partial n$ for all directions of the outward normal $\mathbf{n}$ can never be wholly negative. Hence we conclude that V^2 or equivalently the magnitude of the velocity can never be a maximum at a point in the interior of the fluid.

It follows that the maximum of V^2 will occur on the boundary of the motion.

It may be noted that no conclusion is drawn here regarding the possibility of the magnitude of the velocity attaining a minimum within the fluid.

(6) *The pressure attains its minimum at the boundary of the fluid.*

The pressure at any point is given by the Bernoulli equation

$$p = -\rho\left(\frac{\partial\Phi}{\partial t} + \frac{V^2}{2}\right) + f(t)$$

where the effect of the body forces is not included. Consider as before a point P in the interior of the fluid and let δS be the surface of an infinitesimal volume element $\delta\tau$ surrounding P. We then have

$$\oiint_{\delta S} \frac{\partial p}{\partial n}\, dS = \oiint_{\delta S} \operatorname{grad} p \cdot \mathbf{n}\, dS$$

$$= -\rho \frac{\partial}{\partial t} \oiint_{\delta S} \operatorname{grad} \Phi \cdot \mathbf{n}\, dS - \frac{\rho}{2} \oiint_{\delta S} \operatorname{grad} V^2 \cdot \mathbf{n}\, dS$$

$$= -\rho \frac{\partial}{\partial t} \iiint_{\delta\tau} \nabla^2\Phi\, d\tau - \frac{\rho}{2} \oiint_{\delta S} \frac{\partial V^2}{\partial n}\, dS$$

$$= -\frac{\rho}{2} \oiint_{\delta S} \frac{\partial V^2}{\partial n}\, dS$$

$$< 0$$

The last step follows from (9.97).

This shows that in the immediate neighborhood of P the spatial derivative $\partial p/\partial n$ for all directions of the outward normal $\mathbf{n}$ can never be wholly positive. Hence we conclude that the pressure can never be a minimum at a point in the interior of the fluid. It follows that the minimum pressure will occur on the boundary of the motion.

(7) *The solution of the Neumann exterior problem in a simply connected region is unique up to an additive constant.*

To recall, the problem is to solve the equation

$$\nabla^2\Phi = 0$$

in the region R exterior to the surface S with the condition

$$\frac{\partial \Phi}{\partial n} = f(\mathbf{r}, t) \quad \text{on } S$$

and certain conditions on the derivatives of Φ at infinity as given by Eqs. (9.84) and (9.85). Suppose that there are two solutions Φ_1 and Φ_2 satisfying the equation and the auxiliary conditions. Consider the difference $\Phi_1 - \Phi_2$. We have

$$\nabla^2(\Phi_1 - \Phi_2) = 0$$

$$\frac{\partial}{\partial n}(\Phi_1 - \Phi_2) = 0 \quad \text{on } S$$

Furthermore, the derivatives of $(\Phi_1 - \Phi_2)$ behave in a certain way at infinity.

This being the case, we can readily show that $\Phi_1 - \Phi_2$ must be a constant. Using Green's theorem (2.139) we obtain

$$\iiint_R [\text{grad}\,(\Phi_1 - \Phi_2)]^2 \, d\tau = \oiint_\Sigma (\Phi_1 - \Phi_2)\,\text{grad}\,(\Phi_1 - \Phi_2) \cdot \mathbf{n}\, dS$$

$$- \oiint_S (\Phi_1 - \Phi_2) \frac{\partial}{\partial n}(\Phi_1 - \Phi_2)\, dS$$

$$= \oiint_\Sigma (\Phi_1 - \Phi_2)\,\text{grad}\,(\Phi_1 - \Phi_2) \cdot \mathbf{n}\, dS$$

where Σ is an arbitrary surface enclosing S. The integral over S vanishes on account of the boundary condition. If we let the arbitrary surface Σ go to infinity, the integral over Σ also vanishes on account of the behavior of the integrand at infinity. Hence we obtain

$$\iiint_R [\text{grad}\,(\Phi_1 - \Phi_2)]^2 \, d\tau = 0$$

or

$$\text{grad}\,(\Phi_1 - \Phi_2) = 0$$

or

$$\Phi_2 = \Phi_1 + \text{a constant} \tag{9.98}$$

The constant may depend on time. This is the desired result.

Doubly Connected Region. The properties described in (1) to (6) above are also valid for irrotational motion in a doubly connected region. The uniqueness property given in (7) requires modification.

Consider the Neumann problem in the doubly connected region exterior to an infinite cylinder. Let $\mathscr{C}_0$ denote the boundary curve of the cylinder (Fig. 9.22). The problem is to determine Φ as the solution of

$$\nabla^2 \Phi = 0$$

in the region R exterior to $\mathscr{C}_0$ with the condition

$$\frac{\partial \Phi}{\partial n} = f(\mathbf{r}, t) \qquad \text{on } \mathscr{C}_0$$

The behavior of the derivatives of Φ as $r \to \infty$ is as given by Eqs. (9.88) and (9.91).

Suppose that there are two solutions Φ_1 and Φ_2. Then the difference

$$\Psi = \Phi_1 - \Phi_2$$

is a solution of

$$\nabla^2 \Psi = 0$$

in R and satisfies the condition

$$\frac{\partial \Psi'}{\partial n} = 0 \quad \text{on} \quad \mathscr{C}_0$$

Furthermore, as $r \to \infty$ the derivatives of Ψ' behave in the way required by Eqs. (9.88) and (9.91). To examine the nature of Ψ' in R, we wish to apply, as before, Green's theorem (2.139). This theorem applies to single-valued functions only. However, the region R is doubly connected, and

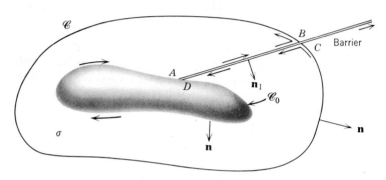

Fig. 9.22 Illustrating the uniqueness proof for the Neumann problem in a doubly connected region.

therefore Φ_1, Φ_2, and Ψ' may be many valued. We can make them single valued by inserting a barrier as shown in Fig. 9.22. Let $\mathscr{C}$ be an arbitrary circuit enclosing $\mathscr{C}_0$. Applying Green's theorem (2.139) to Ψ' in the region σ between $\mathscr{C}_0$ and $\mathscr{C}$, we obtain

$$\iint_{\sigma} (\text{grad } \Psi')^2 \, dS = \int_{\text{on } AB} \Psi' \frac{\partial \Psi'}{\partial n_1} \, ds + \int \Psi' \frac{\partial \Psi'}{\partial n} \, ds$$

$$- \int_{\text{on } CD} \Psi' \frac{\partial \Psi'}{\partial n_1} \, ds - \int_{\mathscr{C}_0} \Psi' \frac{\partial \Psi'}{\partial n} \, ds$$

where $\mathbf{n}_1$ and $\mathbf{n}$ respectively are the normals on the barrier and the circuits $\mathscr{C}_0$ and $\mathscr{C}$, and ds is an element of length on the boundary of σ. The integral over $\mathscr{C}_0$ vanishes since $\partial \Psi'/\partial n$ is zero on $\mathscr{C}_0$. We now let $\mathscr{C}$ go to infinity. In view of the behavior of Ψ' and $\partial \Psi'/\partial n$ as $r \to \infty$, we conclude that the integral over $\mathscr{C}$ vanishes as $r \to \infty$. Hence we obtain

$$\iint_{\sigma} (\text{grad } \Psi')^2 \, dS = -\int_{b_+} \Psi' \frac{\partial \Psi'}{\partial n_1} \, ds + \int_{b_-} \Psi' \frac{\partial \Psi'}{\partial n_1} \, ds$$

where b_+ denotes the CD side of the barrier and b_- the AB side of the barrier.

Now we have

$$\Psi'_{b_+} = (\Phi_1)_{b_+} - (\Phi_2)_{b_+}$$
$$\Psi'_{b_-} = (\Phi_1)_{b_-} - (\Phi_2)_{b_-}$$

Although Φ_1 and Φ_2 are multivalued, the derivatives $\partial\Phi_1/\partial n_1$ and $\partial\Phi_2/\partial n_1$ are single valued. This means that

$$\left(\frac{\partial\Psi'}{\partial n_1}\right)_{b_+} = \left(\frac{\partial\Psi'}{\partial n_1}\right)_{b_-}$$

It therefore follows that

$$\iint_\sigma (\text{grad }\psi)^2 \, dS = \int_{\text{barrier}} \{[(\Phi_1)_{b_-} - (\Phi_1)_{b_+}] - [(\Phi_2)_{b_-} + (\Phi_2)_{b_+}]\} \frac{\partial\psi}{\partial n_1} \, ds$$

The difference in the potential across the barrier is simply equal to the circulation around an irreducible circuit. Let the circulation corresponding to Φ_1 be denoted by Γ_1 and that corresponding to Φ_2 by Γ_2. We then have

$$\iint_\sigma (\text{grad }\Psi')^2 \, dS = (\Gamma_1 - \Gamma_2) \int_{\text{barrier}} \frac{\partial\Psi'}{\partial n_1} \, ds$$

or, in terms of Φ_1 and Φ_2,

$$\iint_\sigma [\text{grad }(\Phi_1 - \Phi_2)]^2 \, dS = (\Gamma_1 - \Gamma_2) \int_{\text{barrier}} \frac{\partial}{\partial n_1} (\Phi_1 - \Phi_2) \, ds \quad (9.99)$$

To proceed further we must specify the circulations Γ_1 and Γ_2. If they are equal, then

$$\Phi_2 = \Phi_1 + \text{a constant}$$

If Γ_1 and Γ_2 are not equal, then Φ_2 and Φ_1 are not equal.

We thus conclude that *the solution to the Neumann exterior problem in a doubly connected region is uniquely determined (up to an additive constant) only when the circulation is specified. For the same boundary and infinity conditions, different values of the circulation yield different solutions.*

The uniqueness theorems for the solution of the Neumann exterior problem are of considerable importance in aerodynamic theory. Without the theorems it would never be possible to assert that the flow pattern calculated for certain boundary conditions is the correct one. For a doubly connected region the theorem shows that the circulation must be specified in order to obtain a unique solution. It must be noted that *the value of the circulation cannot be specified on the basis of the considerations given so far.* The specification of the circulation must therefore rest on other considerations such as those derived from a physical understanding of the real features of the flow problem at hand.

9.19 Stream Functions and the Velocity Potential

We now consider the relation between the velocity potential and the stream functions, and the equations governing the stream functions in irrotational motion.

For incompressible motion the velocity field is related to the stream functions by

$$V = \text{grad}\,\Psi_1 \times \text{grad}\,\Psi_2$$

If the motion is also irrotational and

$$V = \text{grad}\,\Phi$$

we have

$$\text{grad}\,\Phi = \text{grad}\,\Psi_1 \times \text{grad}\,\Psi_2 \tag{9.100}$$

This then is the relation between the velocity potential and the stream functions.

An equation governing Ψ_1 and Ψ_2 in irrotational motion is given by the condition of irrotationality:

$$\text{curl}\,(\text{grad}\,\Psi_1 \times \text{grad}\,\Psi_2) = 0 \tag{9.101}$$

Two-Dimensional Motion. For such a motion we have in Cartesians (see Section 4.10)

$$\Psi_1 = \Psi(x, y)$$
$$\Psi_2 = z$$

It then follows that

$$\frac{\partial \Phi}{\partial x} = u = \frac{\partial \Psi}{\partial x}$$
$$\frac{\partial \Phi}{\partial y} = v = -\frac{\partial \Psi}{\partial y} \tag{9.102}$$

Similarly, in cylindrical coordinates we have

$$\Psi_1 = \Psi(r, \theta)$$
$$\Psi_2 = z$$

and

$$\frac{\partial \Phi}{\partial r} = u_r = \frac{1}{r}\frac{\partial \Psi}{\partial \theta}$$
$$\frac{1}{r}\frac{\partial \Phi}{\partial \theta} = u_\theta = -\frac{\partial \Psi}{\partial r} \tag{9.103}$$

The equation (9.101) becomes

$$\nabla^2 \Psi = \frac{\partial^2 \Psi}{\partial x^2} + \frac{\partial^2 \Psi}{\partial y^2} = 0 \tag{9.104}$$

in Cartesians, and

$$\nabla^2 \Psi' = \frac{1}{r} \frac{\partial}{\partial r}\left(r \frac{\partial \Psi'}{\partial r}\right) + \frac{1}{r^2} \frac{\partial^2 \Psi'}{\partial \theta^2} = 0 \qquad (9.105)$$

in cylindrical coordinates. Thus in two-dimensional irrotational motion of an incompressible fluid, the stream function obeys Laplace's equation.

Axisymmetric Motion. For such a motion we have in spherical coordinates r, θ, φ (see Section 4.11)

$$\Psi'_1 = \Psi'(r, \theta)$$

$$\Psi'_2 = \varphi$$

It then follows that

$$\frac{\partial \Phi}{\partial r} = u_r = \frac{1}{r^2 \sin \theta} \frac{\partial \Psi'}{\partial \theta}$$

$$\frac{1}{r} \frac{\partial \Phi}{\partial \theta} = u_\theta = -\frac{1}{r \sin \theta} \frac{\partial \Psi'}{\partial r} \qquad (9.106)$$

Similarly, in cylindrical coordinates r, φ, x we have

$$\Psi' = \Psi'(r, x)$$

$$\Psi'_2 = \varphi$$

and

$$\frac{\partial \Phi}{\partial r} = u_r = -\frac{1}{r} \frac{\partial \Psi'}{\partial x}$$

$$\frac{\partial \Phi}{\partial x} = u_x = \frac{1}{r} \frac{\partial \Psi'}{\partial r} \qquad (9.107)$$

Then equation (9.101) becomes

$$\frac{1}{\sin \theta} \frac{\partial^2 \Psi'}{\partial r^2} + \frac{1}{r^2} \frac{\partial}{\partial \theta}\left(\frac{1}{\sin \theta} \frac{\partial \Psi'}{\partial \theta}\right) = 0 \qquad (9.108)$$

in spherical coordinates r, θ, φ and

$$\frac{\partial^2 \Psi'}{\partial r^2} - \frac{1}{r} \frac{\partial \Psi'}{\partial r} + \frac{\partial^2 \Psi'}{\partial x^2} = 0 \qquad (9.109)$$

in cylindrical coordinates r, φ, x. We note that the equation for the stream function in axisymmetric irrotational incompressible motion is not Laplace's equation.

Chapter 10

Unsteady Acyclic Motion

We now begin the considerations for constructing the solutions of flow problems associated with the motion of a solid body through an ideal fluid. Our final aim is to determine the pressure distribution over the surface of the body and the force and moment acting on the body. In this chapter we shall concern ourselves with a few simple examples and some general results concerning the acyclic motion resulting from the motion of a rigid solid body translating through an infinitely extending ideal fluid.

10.1 Mathematical Problem

Our problem is to determine the velocity potential for the fluid motion resulting from the motion of a solid body through an infinitely extending fluid that was initially in a state of rest. We choose a space fixed reference frame and describe the surface of the body as

$$F(\mathbf{r}, t) = 0$$

and the velocity of an element of the surface of the body as $\mathbf{U}$. In general, the body may be translating, rotating and deforming. Consequently, $\mathbf{U}$ may in general be a function of position on the surface of the body and time. If the body is rigid and is in translatory motion only, then $\mathbf{U}$ is a function of time but uniform over the surface of the body.

For the problem under consideration we shall denote the velocity potential by ϕ rather than by Φ which has been employed generally so far for the velocity potential. The mathematical problem is to determine ϕ as the solution of the equation

$$\nabla^2 \phi = 0 \tag{10.1}$$

such that the solution satisfies the boundary condition

$$\frac{\partial F}{\partial t} + \operatorname{grad} \phi \cdot \operatorname{grad} F = 0 \quad \text{on } F(\mathbf{r}, t) = 0 \tag{10.2}$$

Furthermore, the components of the velocity

$$\mathbf{V} = \operatorname{grad} \phi$$

should vanish in a certain way with distance from the body as that distance tends to infinity. Henceforth, we shall refer to this requirement as the *infinity condition*.

Note that the boundary condition (10.2) may be expressed also in the form

$$\frac{\partial \phi}{\partial n} = \text{grad } \phi \cdot \mathbf{n} = \mathbf{U}(\mathbf{r}, t) \cdot \mathbf{n} \quad \text{on} \quad F(\mathbf{r}, t) = 0 \qquad (10.3)$$

or

$$\text{grad } \phi \cdot \text{grad } F = \mathbf{U} \cdot \text{grad } F \quad \text{on} \quad F(\mathbf{r}, t) = 0$$

This states that at every point on the surface of the body the component normal to the body of the fluid velocity is equal to the normal component of the body velocity. Once the velocity potential is determined, the pressure at any point is given by

$$p(\mathbf{r}, t) = -\rho \left[\frac{\partial \phi}{\partial t} + \frac{1}{2} (\text{grad } \phi)^2 \right] + f(t) \qquad (10.4)$$

The effect of the body forces on the pressure is not included in this relation. Such effect, as explained before, may be accounted for by calculating the hydrostatic pressure under the action of the body forces and adding it to the pressure calculated by (10.4).

Denote the pressure at infinity, which is a constant, by p_∞. Then the relation for the pressure may be written as

$$p(\mathbf{r}, t) = p_\infty - \rho \left[\frac{\partial \phi}{\partial t} + \frac{1}{2} (\text{grad } \phi)^2 \right] \qquad (10.5)$$

10.2 Expanding Sphere

We now consider a simple example of unsteady motion. A sphere immersed in the fluid expands uniformly in all directions. We wish to determine the pressure on the surface of the sphere.

We choose the center of the sphere as the origin of coordinates and introduce spherical coordinates r, θ, φ. If a denotes the radius of the sphere the surface of the sphere is described by

$$r = a(t) \qquad (10.6)$$

or by

$$F(r, t) = r - a(t) = 0 \qquad (10.7)$$

The boundary condition (10.2) then takes the form

$$\frac{\partial \phi}{\partial r} = \frac{da}{dt} \quad \text{on } r = a(t) \qquad (10.8)$$

In light of the boundary condition we realize that ϕ is a function of r and t only:

$$\phi = \phi(r, t)$$

Then the equation for ϕ becomes

$$\frac{\partial}{\partial r}\left(r^2 \frac{\partial \phi}{\partial r}\right) = 0$$

Integration yields

$$\frac{\partial \phi}{\partial r} = \frac{A(t)}{r^2} \tag{10.9}$$

and

$$\phi(r, t) = -\frac{A(t)}{r} + \text{const}$$

The constant, which may depend on time, may be set equal to zero. We determine $A(t)$ on the basis of the condition (10.8). We thus obtain

$$A(t) = a^2(t)\frac{da}{dt}$$

and

$$\phi(r, t) = -\frac{a^2(t)}{r}\frac{da}{dt} \tag{10.10}$$

The velocity is given by

$$\mathbf{V} = \text{grad } \phi = \frac{a^2(t)}{r^2}\frac{da}{dt}\, \mathbf{e}_r \tag{10.11}$$

For the pressure at any point we obtain

$$p(r, t) = p_\infty - \left[\frac{\partial \phi}{\partial t} + \frac{1}{2}(\text{grad } \phi)^2\right]$$

$$= p_\infty + \frac{\rho}{r}\left[\frac{d}{dt}\left(a^2\frac{da}{dt}\right) - \frac{1}{r^3}\left(a^2\frac{da}{dt}\right)^2\right] \tag{10.12}$$

The pressure on the surface of the sphere is given by

$$p[a(t), t] = p_\infty + \frac{\rho}{2}\left[3\left(\frac{da}{dt}\right)^2 + 2a\frac{d^2a}{dt^2}\right] \tag{10.13}$$

The pressure is uniform over the surface of the sphere and consequently the fluid exerts no resultant force or moment on the sphere. Work, however, has to be done in expanding the sphere against the pressure forces.

The expanding sphere is a simple example of a body undergoing volume changes. For such a body, we recall, the velocity components should vanish at least as $1/r^2$ as $r \to \infty$. This is borne out by the solution (10.9).

10.3 Problem for a Translating Body in Terms of a Body Fixed Reference Frame

Henceforth we shall be concerned with the problem of fluid motion resulting from the motion of a rigid body translating through the fluid. For such a situation the velocity of the body does not change with position on the surface of the body although the translatory velocity may be a function of time:

$$\mathbf{U} = \mathbf{U}(t) \text{ only} \tag{10.14}$$

For a rigid body the function specifying the surface of the body becomes independent of time if described from a reference frame fixed in the body. From a space fixed frame the function describing the body surface is time dependent even if the body is rigid and undergoing only translatory motion with a constant velocity. These observations combined with the fact that time does not appear explicitly in the governing equation for the velocity potential suggest that it will be advantageous to consider the analysis of the problem in terms of a reference frame fixed with respect to the body. We shall therefore use such a reference frame. Consequently we shall now express the problem, which has so far been expressed in terms of a space fixed reference frame, in terms of measurements made from a body fixed frame.

Denote by K_1 a reference frame fixed with respect to the moving body. We shall denote by the subscript 1 all measurements and operations made with respect to the frame K_1. The space fixed reference frame shall be denoted by K, and the measurements and operations made with respect to K shall be denoted without any subscript.

We now set up the connection between the descriptions in the two frames (Fig. 10.1). Since K_1 is translating with a velocity $\mathbf{U}(t)$ with respect to K we have, assuming that the two frames are coincident at time zero,

$$\mathbf{r}_1 = \mathbf{r}_1(\mathbf{r}, t) = \mathbf{r} - \int_0^t \mathbf{U}(\tau)\, d\tau$$
$$t_1 = t_1(\mathbf{r}, t) = t \tag{10.15}$$

For the potential ϕ_1 in K_1 corresponding to ϕ in K we have

$$\phi_1(\mathbf{r}_1, t_1) = \phi_1[\mathbf{r}_1(\mathbf{r}, t), t_1(\mathbf{r}_1, t)]$$
$$= \phi(\mathbf{r}, t) \tag{10.16}$$

Similarly for the function F_1 describing the body surface in K_1 and for the pressure p_1 in K_1 we have

$$F_1(\mathbf{r}_1, t_1) = F_1[\mathbf{r}_1(\mathbf{r}, t), t_1(\mathbf{r}, t)] = F(\mathbf{r}, t) \tag{10.17}$$
$$p_1(\mathbf{r}_1, t_1) = p_1[\mathbf{r}_1(\mathbf{r}, t), t_1(\mathbf{r}, t)] = p(\mathbf{r}, t) \tag{10.18}$$

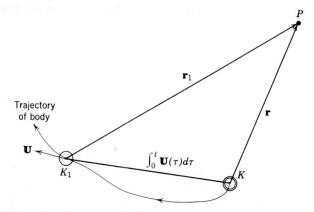

Fig. 10.1 Illustrating the relation between space-fixed and body-fixed descriptions.

The various differentiation operations in the two frames are related as follows:

$$\mathbf{\nabla} = \mathbf{\nabla}_1 \qquad (10.19)$$

$$\nabla^2 = \nabla_1{}^2 \qquad (10.20)$$

$$\frac{\partial}{\partial t} = \frac{\partial t_1}{\partial t}\frac{\partial}{\partial t_1} + \frac{\partial \mathbf{r}_1}{\partial t} \cdot \mathbf{\nabla}_1$$

Since from (10.15)

$$\frac{\partial t_1}{\partial t} = 1 \qquad (10.21)$$

and

$$\frac{\partial \mathbf{r}_1}{\partial t} = -\mathbf{U}(t)$$

we obtain

$$\frac{\partial}{\partial t} = \frac{\partial}{\partial t_1} - \mathbf{U}(t) \cdot \mathbf{\nabla}_1 \qquad (10.22)$$

In terms of the frame K_1 the equation for the potential becomes

$$\nabla_1{}^2 \phi_1 = 0$$

The boundary condition takes the form

$$\left(\frac{\partial F_1}{\partial t_1} - \mathbf{U} \cdot \mathbf{\nabla}_1 F_1\right) + \mathbf{\nabla}_1\phi_1 \cdot \mathbf{\nabla}_1 F_1 = 0 \quad \text{on} \quad F_1(\mathbf{r}_1, t_1) = 0$$

or

$$\mathbf{\nabla}_1\phi_1 \cdot \mathbf{n} = \mathbf{U}(t) \cdot \mathbf{n} \quad \text{on} \quad F_1(\mathbf{r}_1, t_1) = 0$$

or

$$\mathbf{\nabla}_1\phi_1 \cdot \mathbf{\nabla}_1 F_1 = \mathbf{U} \cdot \mathbf{\nabla}_1 F_1 \quad \text{on} \quad F_1(\mathbf{r}_1, t_1) = 0$$

It follows that

$$\frac{\partial F_1}{\partial t_1} = 0$$

or

$$F_1 = F_1(\mathbf{r}_1) \quad \text{only} \tag{10.23}$$

We conclude that *in terms of the frame K_1 the problem is to determine $\phi_1(\mathbf{r}_1, t_1)$ as the solution of the equation*

$$\nabla_1{}^2 \phi_1 = 0 \tag{10.24}$$

such that the solution satisfies boundary condition

$$\nabla_1 \phi_1 \cdot \mathbf{n} = \mathbf{U}(t) \cdot \mathbf{n} \quad \text{on} \quad F_1(\mathbf{r}_1) = 0 \tag{10.25}$$

Furthermore, the potential ϕ_1 and the components of the velocity

$$\mathbf{q}_1 = \nabla_1 \phi_1 \tag{10.26}$$

should satisfy the infinity conditions.

Once the potential ϕ_1 is determined the pressure at any point is obtained in terms of the measurements made in frame K_1 by the relation

$$p(\mathbf{r}, t) = p_1(\mathbf{r}_1, t_1) = p_\infty - \rho\left[\frac{\partial \phi_1}{\partial t_1} - \mathbf{U}(t) \cdot \nabla_1 \phi_1 + \frac{1}{2}(\nabla_1 \phi_1)^2\right] \tag{10.27}$$

This is obtained from (10.5).

We observe that in terms of the description in frame K_1 time enters the problem for the potential only through $\mathbf{U}$ in the boundary condition. We state that the time dependence of the potential comes through the time dependence of the velocity of the body. If the body is translating with a constant velocity, then the potential described in terms of a body fixed reference frame is time independent.

The problem for the potential in terms of the body fixed frame may be interpreted as a certain flow past a fixed rigid body. To see this, let us first write the boundary condition (10.25) in the form

$$[-\mathbf{U}(t) + \nabla_1 \phi_1] \cdot \mathbf{n} = 0$$

or

$$(-\mathbf{U}(t) + \mathbf{q}_1) \cdot \mathbf{n} = 0 \quad \text{on} \quad F_1(\mathbf{r}_1) = 0$$

Introduce now a velocity $\mathbf{V}_1$ and a potential Φ_1 such that

$$\mathbf{V}_1 = \nabla_1 \Phi_1 = -\mathbf{U}(t) + \mathbf{q}_1 = -\mathbf{U} + \nabla_1 \phi_1 \tag{10.28}$$

Equation (10.24) and the boundary condition (10.25) then take the form

$$\nabla^2 \Phi_1 = 0 \tag{10.29}$$

and

$$\nabla \Phi_1 \cdot \mathbf{n} = \mathbf{V}_1 \cdot \mathbf{n} = 0 \quad \text{on} \quad F(\mathbf{r}) = 0 \tag{10.30}$$

From (10.28) it follows that

$$(\nabla_1 \Phi_1 = \mathbf{V}_1) \to (-\mathbf{U}) \quad \text{as} \quad r \to \infty \tag{10.31}$$

The flow problem represented by (10.28, 29, 30) is that for flow past a fixed rigid body, the velocity at infinity being $-\mathbf{U}$ (see Fig. 10.2). It is seen that at each instant the flow field corresponds to that of steady flow past the fixed body, the uniform velocity at infinity being the negative of the velocity of the body at the instant under consideration. The only difference between a strictly steady flow past a fixed rigid body and that of the translating body as described in the body-fixed frame is that in the latter

Fig. 10.2   Flow in the body-fixed reference frame.

case the velocity at infinity changes from instant to instant. In this sense the solution for the flow problem for a translating body and that for the corresponding steady flow past the body is the same. These observations are in line with the considerations in Chapter 9 where it was shown that for an irrotational motion characterized by a single-valued potential the state of motion at one instant is independent of that at another instant.

When the problem for a translating rigid body is viewed in terms of the body-fixed frame as flow past a fixed rigid body, the potential ϕ_1 is referred to as the *disturbance potential*. This signifies that ϕ_1 arises due to the disturbance by the body of an originally uniform stream. The velocity

$$\mathbf{q}_1 = \nabla_1 \phi_1$$

is then known as the *disturbance velocity*. The potential Φ_1 and the velocity

$$\mathbf{V}_1 = -\mathbf{U} + \mathbf{q}_1$$

are called respectively the *total potential* and the *total velocity* or, simply the potential and the velocity. We note that the disturbance potential and velocity in the body-fixed frame correspond to the actual potential and velocity in the frame K.

10.4   Translating Sphere

A rigid sphere is translating through the fluid. We wish to determine the pressure distribution over the sphere and the force and moment on the sphere due to the reaction of the fluid. The sphere is of radius a. The velocity of the sphere is time dependent, that is the sphere is in accelerating motion.

We choose a body-fixed reference frame and introduce spherical coordinates r, θ, φ with origin at the center of the sphere (Fig. 10.3). The axis

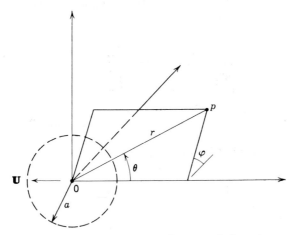

Fig. 10.3   Coordinate system for a translating sphere.

from which θ is measured is chosen opposite to the body velocity $\mathbf{U}$. The boundary condition,* from (10.25), is

$$\operatorname{grad} \phi \cdot \mathbf{e}_r = \mathbf{U} \cdot \mathbf{e}_r \qquad \text{on} \quad r = a$$

This reduces to

$$\frac{\partial \phi}{\partial r} = -U \cos \theta \quad \text{on} \quad r = a \qquad (10.32)$$

In light of this we expect the potential to be independent of the coordinate φ; that is, we expect the motion to be axisymmetric. Hence we have

$$\phi = \phi(r, \theta, t)$$

The equation for the potential then takes the form

$$\frac{\partial}{\partial r}\left(r^2 \sin \theta \frac{\partial \phi}{\partial r}\right) + \frac{\partial}{\partial \theta}\left(\sin \theta \frac{\partial \phi}{\partial \theta}\right) = 0 \qquad (10.33)$$

We construct the solution by separation of variables (consult for instance Hildebrand 1949 or, Sneddon 1957). We assume that

$$\phi(r, \theta) = R(r)\,\Theta(\theta) \qquad (10.34)$$

* We shall omit hereafter the subscript 1 to denote that the description is with respect to the body-fixed frame, for this is the only frame we shall use.

where R is a function of r only and Θ is a function of θ only. With (10.34), Eq. (10.33) may be separated into the form

$$\frac{1}{R}\frac{d}{dr}\left(r^2\frac{dR}{dr}\right) = -\frac{1}{\Theta \sin\theta}\frac{d}{d\theta}\left(\sin\theta\frac{d\Theta}{d\theta}\right) = k \qquad (10.35)$$

where k is a separation constant. This equation is equivalent to two separate equations for R and Θ:

$$\frac{d}{dr}\left(r^2\frac{dR}{dr}\right) - kR = 0 \qquad (10.36)$$

$$\frac{d}{d\theta}\left(\sin\theta\frac{d\Theta}{d\theta}\right) + k\Theta\sin\theta = 0 \qquad (10.37)$$

These are ordinary differential equations whose solutions are well known (see for instance Hildebrand).

We first consider (10.37). By putting

$$k = n(n+1) \qquad (10.38)$$

where n is non-negative, and by writing

$$\cos\theta = \mu \qquad (10.39)$$

(10.37) may be brought into the form

$$\frac{d}{d\mu}\left[(1-\mu^2)\frac{d\Theta}{d\mu}\right] + n(n+1)\Theta = 0 \qquad (10.40)$$

where Θ is expressed as a function of μ. This is *Legendre's differential equation*. The general solution of this equation is of the form

$$\Theta = A_nP_n(\mu) + B_nQ_n(\mu) \qquad (10.41)$$

where P_n and Q_n are *Legendre functions* and A_n and B_n are constant coefficients. The function $Q_n(\mu)$ becomes infinite on the axis $\theta = 0$ and $\theta = \pi$. Hence we set the coefficient B_n equal to zero. The function $P_n(\mu)$ becomes infinite on the axis if n is not an integer. Therefore *we restrict n to integral values* and write

$$\Theta = A_nP_n(\mu) \qquad (10.42)$$

The function $P_n(\mu)$ is a certain polynomial in μ of degree n. The first few polynomials are

$$P_0(\mu) = 1$$
$$P_1(\mu) = \mu = \cos\theta$$
$$P_2(\mu) = \tfrac{1}{2}(3\mu^2 - 1) = \tfrac{1}{2}(3\cos^2\theta - 1)$$
$$P_3(\mu) = \tfrac{1}{2}(5\mu^3 - 3\mu) = \tfrac{1}{3}(5\cos^2\theta - 1)\cos\theta$$

The general solution of the (10.36), replacing k by $n(n + 1)$, is of the form

$$R = C_n r^n + \frac{D_r}{r^{n+1}}$$

where C_n and D_n are constant coefficients. Since the solution for ϕ should at least be finite as $r \to \infty$, we set C_n equal to zero and write

$$R = \frac{D_n}{r^{n+1}} \qquad (10.43)$$

The solution for the potential is then of the form

$$\phi(r, \theta) = \sum_{n=0}^{\infty} \alpha_n \frac{P_n^{(\mu)}}{r^{n+1}} \qquad (10.44)$$

where α_n is a constant which in the present problem may be a function of time. Writing (10.44) explicitly we have

$$\phi(r, \theta) = \alpha_0 \frac{1}{r} + \alpha_1 \frac{\cos\theta}{r^2} + \alpha_2 \frac{P_2}{r^3} + \cdots \qquad (10.45)$$

We now determine the coefficients by application of the boundary condition (10.32). We have

$$\frac{\partial\phi}{\partial r} = -\frac{\alpha_0}{r^2} - 2\alpha_1 \frac{\cos\theta}{r^3} - \cdots$$

The boundary condition requires that the following relation be satisfied

$$-U\cos\theta = \frac{\partial\phi}{\partial r}(r = a) = -\frac{\alpha_0}{a^2} - 2\alpha_1 \frac{\cos\theta}{a^3} - \cdots \qquad (10.46)$$

This is satisfied by setting

$$\alpha_0, \alpha_2, \alpha_3, \ldots = 0$$

and

$$\alpha_1 = \frac{Ua^3}{2} \qquad (10.47)$$

We have now obtained the solution for ϕ which is

$$\phi(r, \theta, t) = \frac{U(t)a^3}{2} \cdot \frac{\cos\theta}{r^2} = -\frac{a^3}{2} \frac{U(t) \cdot \mathbf{r}}{r^3} \qquad (10.48)$$

We may now determine the streamline pattern for this motion. The motion is axisymmetric and the equation of a streamline in any axial plane is represented by

$$\psi(r, \theta) = \text{constant}$$

To determine this function ψ we have the relation that along a streamline

$$d\psi = -r \sin \theta u_\theta \, dr + r^2 \sin \theta u_r \, d\theta = 0 \qquad (10.49)$$

Now using (10.48) we obtain

$$u_\theta = \frac{1}{r} \frac{\partial \phi}{\partial \theta} = \frac{U a^3}{2} \frac{\sin \theta}{r^3}$$

$$u_r = \frac{\partial \phi}{\partial r} = -\frac{U a^3}{2} \frac{2 \cos \theta}{r^3}$$

With these relations (10.49) yields

$$\frac{U a^3}{2} \left(\frac{\sin^2 \theta}{r^2} \, dr - 2 \frac{\sin \theta \cos \theta}{r} \, d\theta \right) = 0$$

or

$$-\frac{U a^3}{2} d\left(\frac{\sin^2 \theta}{r} \right) = 0$$

Hence the stream function is given by

$$\psi(r, \theta) = -\frac{U a^3}{2} \frac{\sin^2 \theta}{r} + \text{a constant} \qquad (10.50)$$

The constant may be set equal to zero, so that $\psi = 0$ on $\theta = 0$. The streamlines are therefore described by the equation

$$\frac{\sin^2 \theta}{r} = \text{constant} \qquad (10.51)$$

The streamline picture given by (10.51) is shown in Fig. 10.4. Note that this picture describes the disturbance velocity field. It is the streamline pattern observed from a space-fixed reference frame that coincides with the body-fixed frame at the instant of time under consideration. In the preceding considerations we have omitted exhibiting explicitly the time dependence of ψ and U in the various relations.

The pressure over the surface of the sphere is given by

$$p(a, \theta, t) = p_\infty - \rho \left[\frac{\partial \phi}{\partial t} - \mathbf{U}(t) \cdot \nabla \phi + \frac{1}{2} (\nabla \phi)^2 \right]_{\text{at } r=a}$$

$$= p_\infty + \frac{\rho}{2} \left[\frac{U^2(t)}{4} (9 \cos^2 \theta - 5) - a \frac{dU}{dt} \cos \theta \right] \qquad (10.52)$$

If the sphere is moving with a constant velocity the pressure distribution over the sphere is given by

$$p(a, \theta) = p_\infty + \frac{\rho U^2}{8} (9 \cos^2 \theta - 5) \qquad (10.53)$$

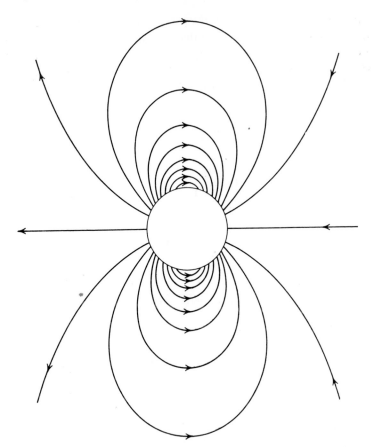

Fig. 10.4 Streamlines of the disturbance field resulting from a translating sphere.

We may express (10.52) in the form

$$\delta p \equiv (p - p_\infty) = \frac{\rho U^2(t)}{2}\left(1 - \frac{9}{4}\sin^2\theta\right) - \frac{\rho a}{2}\frac{dU}{dt}\cos\theta$$
$$= \delta p_1 + \delta p_2 \qquad (10.54)$$

where

$$\delta p_1 = \frac{\rho U^2(t)}{2}\left(1 - \frac{9}{4}\sin^2\theta\right) \qquad (10.55)$$

and

$$\delta p_2 = -\frac{\rho a}{2}\frac{dU}{dt}\cos\theta \qquad (10.56)$$

The variation of δp_1 and δp_2 over the sphere are shown in Fig. 10.5. It is seen that while δp_1 is symmetrical with respect to the plane $\theta = \pi/2$, δp_2 is not symmetrical. We may state that acceleration of the sphere leads to asymmetry in the pressure distribution and consequently to a force on the sphere. We further recognize that the force is entirely due to δp_2 which may be referred to as the pressure distribution due to acceleration.

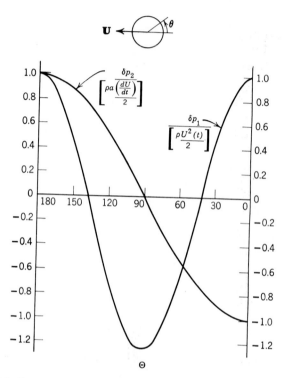

Fig. 10.5 Pressure distribution over the surface of a translating sphere.

The force on the sphere is given by

$$\mathbf{F} = -\oiint_{\text{sphere}} p\mathbf{n}\, dS = \frac{\rho a}{2}\frac{dU}{dt}\iint \mathbf{n}\cos\theta a^2 \sin\theta\, d\theta\, d\varphi$$

We note that $\mathbf{F}$ has a nonzero component only in the direction of $\mathbf{e}_x$, that is in the direction of the instantaneous velocity of the sphere. We find

$$\mathbf{F} = -\frac{2}{3}\pi a^3 \rho\frac{dU}{dt}\mathbf{e}_x \tag{10.57}$$

We conclude that a solid sphere accelerating through an ideal fluid experiences a resistance to its motion. The resisting force is known as the drag force or simply as the drag. If the sphere is moving with constant velocity the drag is zero! In other words, once the sphere has been set into a constant velocity motion no further application of an external force on the sphere is necessary to maintain its motion! The present theoretical result that the force on a sphere moving with constant velocity is zero is of course not substantiated by experiment. We shall take up further consideration of this result later; we note that there is no moment acting on the sphere.

Let us now ask about the force that must be applied to the sphere by an external agency so as to keep the sphere in motion through the fluid. Denoting by $\mathbf{F}_e$ such external force we have, according to Newton's second law of motion,

$$\frac{d}{dt}(m\mathbf{U}) = \mathbf{F}_e + \mathbf{F}$$

or

$$\mathbf{F}_e = \frac{d}{dt}(m\mathbf{U}) - \mathbf{F} \tag{10.58}$$

This states, as we know, that the external force should balance the inertial force of the body and the resisting force of the fluid. Substituting for $\mathbf{F}$ from (10.57) and noting that $\mathbf{U} = \mathbf{e}_x U$ we obtain

$$\mathbf{F}_e = \frac{d}{dt}\left[\left(m + \frac{2}{3}\pi a^3 \rho\right)\mathbf{U}\right] \tag{10.59}$$

on the basis of this relation we may state that the reaction of the fluid on a solid sphere accelerating (in translation) through the fluid is equivalent to increasing the mass of the sphere by the amount

$$m' = \tfrac{2}{3}\pi a^3 \rho \tag{10.60}$$

We refer to this mass as the *additional apparent* or *virtual mass*. Since the volume of the sphere is $(\tfrac{4}{3})\pi a^3$ it follows from (10.60) that the additional apparent mass is half of the mass of fluid displaced by the sphere.

10.5 Force on a Translating Body of Arbitrary Shape

The flow field resulting from the translatory motion of a rigid solid body of a more general shape than a sphere may be analyzed on similar lines as those followed in the preceding section. In analyzing such flow fields it is convenient to employ coordinates that are especially suited to the body shapes under consideration. For the solution of flow fields involving the motion of solid bodies of more general shape than a sphere reference may be made to Lamb (1932). We shall concern ourselves with some general results relating to the force and moment exerted by the fluid on a rigid body of arbitrary shape translating through the fluid.

We have seen that in the case of the sphere the fluid exerts no force on the sphere if the sphere is moving with a constant velocity. A reaction force by the fluid appears only if the sphere is accelerating. We now inquire whether such results hold for bodies of arbitrary shape also. It is possible to find the answer without having a detailed knowledge of the flow field under consideration.

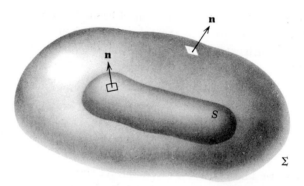

Fig. 10.6 Illustrating the determination of the force on a translating rigid body.

Consider a rigid body of arbitrary shape translating with a velocity $U(t)$ through the fluid. Let S denote the surface of the body (Fig. 10.6). The fluid force acting on the body is given by

$$F = -\oiint_S p\mathbf{n}\, dS \tag{10.61}$$

Our immediate aim is to express $\mathbf{F}$ in terms of the velocity potential. It is convenient, as already pointed out, to solve for the velocity potential by working in the body-fixed reference frame. If the space-fixed reference frame is chosen such that it coincides with the body-fixed frame at the time instant under consideration the pressure, according to Eq. (10.27), is given by

$$p(\mathbf{r}, t) = p_\infty - \rho\left[\frac{\partial \phi}{\partial t} - \mathbf{U} \cdot \mathbf{q} + \frac{q^2}{2}\right] \tag{10.62}$$

where

$$\mathbf{q} = \operatorname{grad} \phi \tag{10.63}$$

For convenience we shall employ often $\mathbf{q}$ for grad ϕ. The operations in the right member of (10.62) are with reference to the body-fixed frame.

With (10.62), the relation (10.61) for the force becomes

$$\mathbf{F} = \oiint\limits_{S} \rho \frac{\partial \phi}{\partial t}\, \mathbf{n}\, dS + \oiint\limits_{S} \rho \left[\frac{q^2}{2} - \mathbf{U} \cdot \mathbf{q}\right]\mathbf{n}\, dS \qquad (10.64)$$

Note that since the operations in the right member are with respect to the body-fixed frame and since with respect to that frame the surface S is time independent we can write

$$\oiint\limits_{S} \rho \frac{\partial \phi}{\partial t}\, \mathbf{n}\, dS = \frac{\partial}{\partial t} \oiint\limits_{S} \rho \phi \mathbf{n}\, dS. \qquad (10.65)$$

The other integral may be shown to vanish if the motion is acyclic or equivalently if the potential is single valued. First using the relation

$$\mathbf{U} \times (\mathbf{n} \times \mathbf{q}) = (\mathbf{U} \cdot \mathbf{q})\mathbf{n} - (\mathbf{U} \cdot \mathbf{n})\mathbf{q}$$

we write

$$\oiint\limits_{S} \left[\frac{q^2}{2} - \mathbf{U} \cdot \mathbf{q}\right]\mathbf{n}\, dS = \oiint\limits_{S} \left[\frac{q^2}{2}\mathbf{n} - (\mathbf{U} \cdot \mathbf{n})\mathbf{q}\right] dS - \mathbf{U} \times \oiint\limits_{S} (\mathbf{n} \times \mathbf{q})\, dS \qquad (10.66)$$

According to the boundary condition (10.25) we have

$$\mathbf{U} \cdot \mathbf{n} = \operatorname{grad} \phi \cdot \mathbf{n} = \mathbf{q} \cdot \mathbf{n} \quad \text{on} \quad S \qquad (10.67)$$

Hence we can write

$$\oiint\limits_{S} \left[\frac{q^2}{2}\mathbf{n} - (\mathbf{U} \cdot \mathbf{n})\mathbf{q}\right] dS = \oiint\limits_{S} \left[\frac{q^2}{2}\mathbf{n} - (\mathbf{q} \cdot \mathbf{n})\mathbf{q}\right] dS$$

Now, for any region R_0 enclosed by a fixed surface S_0 and containing fluid only, it can be shown that

$$\oiint\limits_{S_0} \left[\frac{q^2}{2}\mathbf{n} - (\mathbf{q} \cdot \mathbf{n})\mathbf{q}\right] dS = \iiint\limits_{R_0} [\mathbf{q} \cdot \nabla \mathbf{q} - \mathbf{q} \cdot \nabla \mathbf{q}]\, d\tau = 0$$

Consider the region of fluid enclosed between the surface S and another fixed surface Σ enclosing S (Fig. 10.6). The surface Σ is of arbitrary shape and located at an arbitrary distance from the body. We then have

$$\oiint\limits_{S} \left[\frac{q^2}{2}\mathbf{n} - (\mathbf{q} \cdot \mathbf{n})\mathbf{q}\right] dS = \oiint\limits_{\Sigma} \left[\frac{q^2}{2}\mathbf{n} - (\mathbf{q} \cdot \mathbf{n})\mathbf{q}\right] dS \qquad (10.68)$$

It follows that the value of this integral should be independent of the shape and location of the surface Σ. We have seen that for a finite rigid body, q vanishes as $1/r^3$ as $r \to \infty$ (see Section 9.17). The surface element

dS grows, as r^2 as $r \to \infty$. It therefore follows that for a finite rigid body

$$q^2 \, dS \sim \frac{1}{r^4} \quad \text{as} \quad r \to \infty$$

and

$$(\mathbf{q} \cdot \mathbf{n})q \, dS \sim \frac{1}{r^4} \quad \text{as} \quad r \to \infty$$

If the body is a rigid infinite cylinder we have seen that q vanishes as $1/r^2$ as $r \to \infty$ if the circulation is zero. If the circulation is not zero the r-component of $\mathbf{q}$ vanishes as $1/r^2$ but the θ-component of $\mathbf{q}$ vanishes as $1/r$ only (see Section 9.17). The surface element dS of Σ now grows as r as $r \to \infty$. Putting these results together, we state that for the infinite cylinder if the circulation is zero

$$q^2 \, dS \quad \text{and} \quad (\mathbf{q} \cdot \mathbf{n})q \, dS \sim \frac{1}{r^3} \quad \text{as} \quad r \to \infty$$

and if the circulation is not zero

$$q^2 \, dS \quad \text{and} \quad (\mathbf{q} \cdot \mathbf{n})q \, dS \sim \frac{1}{r} \quad \text{as} \quad r \to \infty$$

Hence it follows that the integral over Σ in Eq. (10.68) may be made arbitrarily small, both for a finite body and an infinite cylinder with circulation, by choosing the surface at larger and larger distances from the body. We conclude that the integral over Σ should be zero and that consequently the integral over S is zero: Hence we have

$$\oiint\limits_{S} \left[\frac{q^2}{2} \mathbf{n} - (\mathbf{U} \cdot \mathbf{n})\mathbf{q} \right] dS = 0 \tag{10.69}$$

It follows that the expression (10.64) for the force becomes

$$\mathbf{F} = \frac{\partial}{\partial t} \oiint\limits_{S} \rho \phi \mathbf{n} \, dS - \rho \mathbf{U}(t) \times \oiint\limits_{S} (\mathbf{n} \times \mathbf{q}) \, dS. \tag{10.70}$$

Consider next the integral $\mathbf{I} \equiv \oiint\limits_{S} \mathbf{n} \times \mathbf{q} \, dS$. This integral is related in a certain way to the circulation around the body and will vanish if the circulation is zero. To see this let us consider the component of the vector $\mathbf{I}$ in some fixed direction $\mathbf{e}$. We have

$$\mathbf{e} \cdot \mathbf{I} = \oiint\limits_{S} \mathbf{e} \cdot \mathbf{n} \times \mathbf{q} \, dS \tag{10.71}$$

By means of two planes normal to **e** cut out a small slice of the solid body (Fig. 10.7) and pick as shown an elemental area **n** dS on the surface of the slice. Note that the normal **n** in general need not be normal to **e**. Let $\mathbf{e}_1$ denote the direction along the curve of intersection between the body surface and a cutting plane. The unit vectors $\mathbf{e}_1$ and **e** are therefore

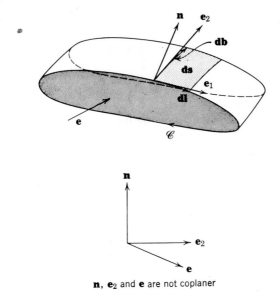

n, $\mathbf{e}_2$ and e are not coplaner

Fig. 10.7 Illustrating that a part of the force on the body is related to the circulation around the body.

normal. Also **n** is normal to $\mathbf{e}_1$. We denote the $\mathbf{e}_1$ edge of the surface element **n** dS by **dl**, and choose the other edge say **db** of **n** dS such that

$$\mathbf{n}\,dS = \mathbf{dl} \times \mathbf{db} \qquad (10.72)$$

Then we have

$$\mathbf{n}\,dS \times \mathbf{q} = (\mathbf{dl} \times \mathbf{db}) \times \mathbf{q} = (\mathbf{q} \cdot \mathbf{dl})\mathbf{db} - (\mathbf{q} \cdot \mathbf{db})\mathbf{dl}$$

It follows that

$$\mathbf{e} \cdot \mathbf{n} \times \mathbf{q}\,dS = (\mathbf{q} \cdot \mathbf{dl})\mathbf{e} \cdot \mathbf{db} \qquad (10.73)$$

since $\mathbf{e} \cdot \mathbf{dl}$ is zero. Let us write

$$dh = \mathbf{e} \cdot \mathbf{db} \qquad (10.74)$$

This is the normal distance between the cutting planes.

Using the relations (10.73) and (10.74) we obtain from (10.71)

$$\mathbf{e} \cdot \oiint_S \mathbf{n} \times \mathbf{q} \, dS = \int_{h_1}^{h_2} \left(\oint_{\mathscr{C}} \mathbf{q} \cdot \mathbf{dl} \right) dh = \int_{h_1}^{h_2} \Gamma_e(h) \, dh \qquad (10.75)$$

where h is distance measured along the fixed direction $\mathbf{e}$ and

$$\Gamma_e(h) = \oint_{\mathscr{C}} \mathbf{q} \cdot \mathbf{dl}$$

is the circulation around the curve of intersection between the body surface and the cutting plane situated at h. The limits h_1 and h_2 denote the extremities of the body measured along the direction $\mathbf{e}$.

From (10.75) it follows that if the circulation around any arbitrary circuit drawn on the surface of the body is zero, then the component of the integral $\mathbf{I}$ in any direction is zero, which means the integral itself is zero. We therefore conclude that *for motions without circulation or, equivalently for motions characterized by a single-valued potential the integral* $\mathbf{I}$ *vanishes*:

$$\mathbf{I} = \oiint_S \mathbf{n} \times \mathbf{q} \, dS = 0 \qquad (10.76)$$

We shall concern ourselves at present with acyclic motions only. We conclude that for acyclic motions the force on the body is given by

$$\mathbf{F} = \frac{\partial}{\partial t} \oiint_S \rho \phi \mathbf{n} \, dS \qquad (10.77)$$

The time dependence of the potential ϕ arises only through the time dependence of the velocity $\mathbf{U}$ of the body. In fact we have

$$\phi(\mathbf{r}, t) = \phi[\mathbf{r}; \mathbf{U}(t)]$$

and

$$\frac{\partial \phi}{\partial t} = \frac{\partial \phi}{\partial \mathbf{U}} \cdot \frac{d\mathbf{U}}{dt}$$

where $\partial \phi / \partial \mathbf{U}$ denotes the gradient of ϕ in the space of $\mathbf{U}$. On account of this we may immediately state that *if a finite rigid body is moving with a constant velocity through an infinitely extending ideal fluid the force on the body is zero*. This result is known as *d'Alembert's paradox*. We shall consider later the implications of this result.

10.6 Impulse

Let $\mathbf{F}_e$ denote the force applied externally to the body to translate it through the fluid. Then, according to Newton's second law of motion we have

$$\frac{d}{dt}(m\mathbf{U}) = \mathbf{F}_e + \mathbf{F} \qquad (10.78)$$

where m is the mass of the body. Rewrite the preceding equation as

$$\mathbf{F}_e = \frac{d}{dt}(m\mathbf{U}) - \mathbf{F} \qquad (10.79)$$

Substituting from (10.77) for $\mathbf{F}$ we obtain

$$\mathbf{F}_e = \frac{d}{dt}\left(m\mathbf{U} - \oiint_S \rho\phi\mathbf{n}\,dS\right) \qquad (10.80)$$

Note that the integral in (10.77) is actually a function of time only and consequently $\partial/\partial t$ in front of it actually means d/dt. Recall that for acyclic irrotational motion, that is for motion characterized by a single valued potential, $-\rho\phi$ is equal to the *impulsive pressure* ϖ (see Section 9.9). The integral of the impulsive pressures is called the *impulse*. Denoting it by $\mathscr{J}$ we have

$$\mathscr{J} = -\oiint_S \rho\phi\mathbf{n}\,dS = \oiint_S \varpi\mathbf{n}\,dS \qquad (10.81)$$

and

$$\mathbf{F} = -\frac{d\mathscr{J}}{dt} \qquad (10.82)$$

Equation (10.80) then takes the form

$$\mathbf{F}_e = \frac{d}{dt}(m\mathbf{U} + \mathscr{J}) \qquad (10.83)$$

This states that

the force applied externally on the body = the time rate of change of the momentum of the body + the time rate of change of the impulse applied on the fluid.

The impulse is the impulsive force required to establish impulsively the state of motion under consideration (see Section 9.9).

10.7 The Apparent Mass Tensor

Equation 10.83 suggests that the reaction of the fluid to a body translating through it is to change in some sense the inertial force of the body. The rate of change of the impulse vector, in general, is not in the direction of the acceleration of the body. This means *the external force* $\mathbf{F}_e$ *has to be applied, in general, in a direction different from that of the acceleration of the body through the fluid.* To illustrate these ideas and to develop computational formulas for $\mathscr{J}$ we now introduce the notion of *the apparent mass tensor.*

Consider, in terms of the body-fixed reference frame, the mathematical problem for determining ϕ. We have

$$\nabla^2 \phi = 0$$

$$\text{grad } \phi \cdot \mathbf{n} = \frac{\partial \phi}{\partial n} = \mathbf{U}(t) \cdot \mathbf{n} \quad \text{on} \quad S$$

and certain infinity conditions. Let us choose a Cartesian coordinate system x_1, x_2, x_3 and let $\mathbf{e}_1, \mathbf{e}_2, \mathbf{e}_3$ denote the reference unit vectors. We write

$$\mathbf{U} = (u_1, u_2, u_3) \tag{10.84}$$

and

$$\mathbf{n} = (n_1, n_2, n_3) \tag{10.85}$$

Since the equation and the boundary condition for ϕ are linear we seek a solution for ϕ in the form

$$\phi = \phi_1 + \phi_2 + \phi_3$$

where each of the functions ϕ_1, ϕ_2, ϕ_3 is a solution of the following problem*

$$\nabla^2 \phi_i = 0$$
$$\text{grad } \phi_i \cdot \mathbf{n} = \frac{\partial \phi_i}{\partial n} = u_i n_i \quad \text{on} \quad S \tag{10.86}$$

where i may be 1, 2 or 3. We note that it is convenient to set

$$\phi_i = u_i \varphi_i \qquad i = 1, 2, 3 \tag{10.87}$$

where φ_i unlike ϕ_i are scalar functions of position only. Time enters through u_i. The system (10.86) then takes the form

$$\nabla^2 \varphi_i = 0 \tag{10.88}$$

$$\text{grad } \varphi_i \cdot \mathbf{n} = \frac{\partial \varphi_i}{\partial n} = n_i \quad \text{on} \quad S, \qquad i = 1, 2, 3 \tag{10.89}$$

The derivatives of φ_i and, φ_i should satisfy the same type of infinity conditions as ϕ and its derivatives do. If φ_i are determined according to

---

* Those familiar with summation convention should note that such convention is not used here.

(10.88) and (10.89) and the infinity conditions, the potential ϕ is obtained from

$$\phi = u_1\varphi_1 + u_2\varphi_2 + u_3\varphi_3 = \mathbf{U}\cdot\boldsymbol{\varphi} \tag{10.90}$$

where $\boldsymbol{\varphi}$ is given by

$$\boldsymbol{\varphi} = (\varphi_1, \varphi_2, \varphi_3) \tag{10.91}$$

Now we express the impulse $\mathscr{J}$ in terms of $\boldsymbol{\varphi}$ and $\mathbf{n}$. We have

$$-\mathscr{J} = \oiint_S \rho\phi\mathbf{n}\, dS$$

$$= \oiint_S \rho(\mathbf{U}\cdot\boldsymbol{\varphi})\mathbf{n}\, dS$$

$$= \oiint_S \rho\left(\sum_k u_k\varphi_k\right)\mathbf{n}\, dS$$

$$= \sum_k\left(\oiint_S \rho\varphi_k\mathbf{n}\, dS\right)u_k \tag{10.92}$$

Let us denote by $\mathscr{J}_i$ the component of $\mathscr{J}$ in the $\mathbf{e}_i$ the direction. Then we have

$$\mathscr{J}_i = \mathbf{e}_i\cdot\mathscr{J} = \sum_k\left(-\oiint_S \rho\varphi_k n_i\, dS\right)u_k, \qquad i = 1, 2, 3 \tag{10.93}$$

We introduce the symbol m_{ki} with the meaning

$$m_{ki} = -\oiint_S \rho\varphi_k n_i\, dS. \tag{10.94}$$

Now, using (10.89) we may write the preceding equation in the form

$$m_{ki} = -\oiint_S \rho\varphi_k \frac{\partial\varphi_i}{\partial n}\, dS. \tag{10.95}$$

According to Green's theorem (2.141) if ψ_1 and ψ_2 are two harmonic functions the relation

$$\oiint_{S_0} \psi_1 \frac{\partial\psi_2}{\partial n}\, dS = \oiint_{S_0} \psi_2 \frac{\partial\psi_1}{\partial n}\, dS \tag{10.96}$$

holds for any surface S_0 enclosing a region in which $\nabla^2\psi_1$ and $\nabla^2\psi_2$ are zero. We consider the region enclosed between the surface S and an arbitrary surface Σ enclosing S, and apply (10.96) to the functions φ_i

and φ_k. Then we obtain

$$-\oiint_S \varphi_k \frac{\partial \varphi_i}{\partial n} \, dS + \oiint_S \varphi_k \frac{\partial \varphi_i}{\partial n} \, dS = -\oiint_S \varphi_i \frac{\partial \varphi_k}{\partial n} \, dS + \oiint_\Sigma \varphi_i \frac{\partial \varphi_k}{\partial n} \, dS$$

or

$$\oiint_S \varphi_i \frac{\partial \varphi_k}{\partial n} \, dS - \oiint_S \varphi_k \frac{\partial \varphi_i}{\partial n} \, dS = \oiint_\Sigma \left(\varphi_i \frac{\partial \varphi_k}{\partial n} - \varphi_k \frac{\partial \varphi_i}{\partial n} \right) dS.$$

We now let Σ go to infinity and conclude (on the basis of the behavior of the integrand as $r \to \infty$) that the integral over Σ must vanish. Hence we obtain

$$\oiint_S \varphi_i \frac{\partial \varphi_k}{\partial n} \, dS = \oiint_S \varphi_k \frac{\partial \varphi_i}{\partial n} \, dS. \tag{10.97}$$

It follows that

$$m_{ki} = m_{ik} \tag{10.98}$$

The components of the impulse $\mathcal{J}$ are therefore given by

$$\mathcal{J}_i = \mathbf{e}_i \cdot \mathcal{J} = \sum_k m_{ik} u_k \qquad i, k = 1, 2, 3 \tag{10.99}$$

where the m_{ik} are given by (10.94) or (10.95).

We now return to the expression (10.83) for the force applied externally to the body and write

$$\mathbf{F}_e = \frac{d}{dt} \left(m\mathbf{U} + \sum_i \left(\sum_k m_{ik} u_k \right) \mathbf{e}_i \right) \tag{10.100}$$

The component of $\mathbf{F}_e$ in the direction $\mathbf{e}_i$ is therefore given by

$$F_{e_i} = \mathbf{e}_i \cdot \mathbf{F}_e = \frac{d}{dt} \left(mu_i + \sum_k m_{ik} u_k \right) \tag{10.101}$$

Introduce the symbol δ_{ik} defined by

$$\begin{aligned} \delta_{ik} &= 0 \quad \text{if} \quad i \neq k \\ &= 1 \quad \text{if} \quad i = k \end{aligned} \tag{10.102}$$

where i and k may be 1, 2, or 3. Equation (10.101) may then be rewritten as

$$F_{e_i} = \sum_k (m\delta_{ik} + m_{ik}) \frac{du_k}{dt} \tag{10.103}$$

For example the $\mathbf{e}_1$ component is given by

$$F_{e_i} = (m + m_{11}) \frac{du_1}{dt} + m_{12} \frac{du_2}{dt} + m_{13} \frac{du_3}{dt}.$$

Equation 10.103 shows that in general the external force $\mathbf{F}_e$ is not in the direction of the acceleration of the body.

Equation 10.103 suggests that the coefficients m_{ik} may be thought of as additional apparent or virtual masses that need to be added in a suitable way to the mass of the body when determining the force that must be applied on the body so as to translate it through the fluid. The coefficients m_{ik} form a set of nine numbers which may be displayed as the array

$$
\begin{pmatrix}
m_{11} & m_{12} & m_{13} \\
m_{21} & m_{22} & m_{23} \\
m_{31} & m_{32} & m_{33}
\end{pmatrix}
$$

Since m_{ik} is equal to m_{ki} there are only six independent additional apparent masses. The set of m_{ik} is known as the *additional apparent* or *virtual mass tensor*. It depends on the shape of the body and, for a given choice of the body fixed axes, is a constant for any given body.

The expression $(m\delta_{ik} + m_{ik})$ may be referred to as the *apparent mass tensor*.

For any body it is possible to find three mutually perpendicular directions such that

$$
m_{ik} = 0 \quad \text{for} \quad i \neq k
$$

With respect to such axes 10.103 becomes

$$
F_{e_i} = (m + m_{ii}) \frac{du_i}{dt}, \qquad i = 1, 2, 3 \tag{10.104}
$$

When the body moves in the direction of one of such axes it would seem to have only an increased mass. The sum $m + m_{ii}$ is known as the *apparent mass for translation in the i-direction*, and the corresponding m_{ii} as the *additional apparent mass*. It is thus seen that for a translating rigid body there are at least three additional apparent masses. For bodies of revolution two of them are equal; for a sphere all the three are equal.

10.8 Kinetic Energy and Impulse

The additional apparent masses may be computed on the basis of (10.94) or (10.95). The computation involves first the determination of the vector $\boldsymbol{\varphi}$ and then integrations over the body surface of the products of the components of $\boldsymbol{\varphi}$ and those of $\mathbf{n}$. It is possible to calculate the m_{ik} by suitable differentiation of the kinetic energy of the fluid. We now show how this may be done.

Consider a region R_0 that is enclosed by a surface S_0 and that contains only fluid in motion. The kinetic energy T of the fluid in R_0 is given by

$$T = \iiint\limits_{R_0} \rho\, \frac{V^2}{2}\, d\tau = \frac{1}{2} \iiint\limits_{R_0} \rho(\text{grad } \phi)^2\, d\tau \qquad (10.105)$$

The volume integral may be related to a certain surface integral over S_0. For this purpose we use Green's theorem (2.139). From this theorem it follows that for a harmonic function such as ϕ we have

$$\iiint\limits_{R_0} (\text{grad } \phi)^2\, dt = \oiint\limits_{S_0} \phi \text{ grad } \phi \cdot \mathbf{n}\, dS$$

Hence, the kinetic energy of the fluid enclosed by S_0 is given by

$$T = \frac{1}{2} \oiint\limits_{S_0} \rho\phi \text{ grad } \phi \cdot \mathbf{n}\, dS = \frac{1}{2} \oiint\limits_{S_0} \rho\phi\, \frac{\partial \phi}{\partial n}\, dS \qquad (10.106)$$

Now consider the fluid in the region between the surface S of a moving body and an arbitrarily drawn surface Σ enclosing the body. For the kinetic energy of that fluid we obtain

$$T = -\oiint\limits_{S} \frac{\rho}{2}\, \phi\, \frac{\partial \phi}{\partial n}\, dS + \oiint\limits_{\Sigma} \frac{\rho}{2}\, \phi\, \frac{\partial \phi}{\partial n}\, dS$$

where $\mathbf{n}$ is directed as shown in Fig. 10.6. We now let Σ go to infinity. For acyclic motion involving a finite rigid body we have

$$\phi\, \frac{\partial \phi}{\partial n}\, dS \sim \frac{1}{r^3} \quad \text{as} \quad r \to \infty.$$

Hence we conclude that for such a situation

$$\oiint\limits_{\Sigma} \frac{\rho}{2}\, \phi\, \frac{d\phi}{\partial n}\, dS \to 0 \quad \text{as} \quad r \to \infty.$$

It follows that *for acyclic motion involving a finite rigid body the kinetic energy is given by*

$$T = -\frac{1}{2} \oiint\limits_{S} \rho\phi\, \frac{\partial \phi}{\partial n}\, dS = -\frac{1}{2} \oiint\limits_{s} \rho\phi \text{ grad } \phi \cdot \mathbf{n}\, dS \qquad (10.107)$$

where S is the surface of the body.

It is easily verified that *this result is true also for acyclic motion involving an infinite cylinder. It is not valid for cyclic motions, that is, for motions involving multivalued potentials.*

The kinetic energy as given by (10.107) may be readily related to the impulse $\mathscr{J}$. Using the boundary condition

$$\text{grad } \phi \cdot \mathbf{n} = \mathbf{U}(t) \cdot \mathbf{n} \quad \text{on} \quad S$$

we obtain

$$T = -\frac{1}{2} \oiint_S \rho \phi \mathbf{U} \cdot \mathbf{n} \, dS = -\frac{1}{2} \mathbf{U} \cdot \oiint_S \rho \phi \mathbf{n} \, dS = \frac{1}{2} \mathbf{U} \cdot \mathscr{J}. \quad (10.108)$$

This is similar to the general result that for a finite dynamical system twice the kinetic energy is equal to the scalar product of the momentum and the velocity.

Equation (10.108) may be expressed as

$$T = \frac{1}{2} \sum_i u_i \mathscr{J}_i = \frac{1}{2} \sum_i \left(\sum_k m_{ik} u_k \right) u_i$$

$$= \frac{1}{2} (m_{11} u_1{}^2 + m_{22} u_2{}^2 + m_{33} u_3{}^2 + 2m_{12} u_1 u_2 + 2m_{23} u_2 u_3$$

$$+ 2m_{31} u_3 u_1) \quad (10.109)$$

The kinetic energy of the fluid is thus a quadratic function of the components of the velocity of the body. The coefficients in this function are the additional apparent masses.

From (10.108) or (10.109) it readily follows that

$$\mathscr{J}_i = \frac{\partial T}{\partial u_i} \qquad (10.110)$$

and

$$m_{ik} = \frac{\partial}{\partial u_k}\left(\frac{\partial T}{\partial u_i}\right). \qquad (10.111)$$

As an example let us compute the coefficients m_{ik} for a translating sphere of radius a. According to (10.32) and (10.48) we have

$$\frac{\partial \phi}{\partial n} = -U(t) \cos \theta \quad \text{on} \quad S$$

and

$$\phi(a, \theta, t) = \frac{U(t)}{2} a \cos \theta.$$

Hence the kinetic energy is given by

$$T = -\frac{\rho}{2} \oiint_S \phi \frac{\partial \phi}{\partial n} \, dS = \frac{\pi a^3 \rho U^2}{2} \int_0^\pi \cos^2 \theta \sin \theta \, d\theta$$

$$= \frac{\pi a^3 \rho U^2}{3} = \frac{\pi a^3 \rho u_1{}^2}{3}. \qquad (10.112)$$

It follows that

$$m_{ik} = 0 \qquad \text{for all} \quad i \quad \text{and} \quad k \quad \text{not equal to} \quad 1$$

and

$$m_{11} = \tfrac{2}{3}\pi a^3 \rho \tag{10.113}$$

This is the result we had obtained before (see Eq. 10.60). In arriving at this result we had chosen the direction e_1 of the coordinate axes in a direction opposite to that of the velocity of the sphere. Instead, if we choose the axes such that the velocity of the sphere is not parallel to any of the axes we obtain the result

$$m_{11} = m_{22} = m_{33}$$

10.9 Moment on a Translating Body

We shall now obtain an expression for the moment exerted by the fluid on a finite rigid body translating through it. The *moment taken with respect to the origin of coordinates is given by*

$$\mathbf{M} = -\oiint_S \mathbf{r} \times p\mathbf{n} \, dS \tag{10.114}$$

where, as before, S denotes the surface of the body. To express this relation in terms of the velocity potential it is more convenient to proceed on the basis of the law of the angular momentum of a dynamical system than to follow the steps used for obtaining the expression for the force on the body.

We construct the required expression for $\mathbf{M}$ *first with respect to a space-fixed reference frame.* Later we shall relate that expression to the calculations in a body fixed reference frame. Consider the fluid that occupies at some instant the region R between the surface S of the moving body and an arbitrarily drawn *fixed* surface Σ enclosing S (Fig. 10.6). The angular momentum of that fluid with respect to the space-fixed origin is

$$\iiint_R \mathbf{r} \times \rho\mathbf{q} \, d\tau$$

where as before $\mathbf{q}$ denotes the fluid velocity and is used for convenience instead of grad ϕ. Then according to the considerations given in Section 6.9, the rate of change of angular momentum of fluid under consideration is given by

$$\frac{d}{dt}\iiint_R \mathbf{r} \times \rho\mathbf{q} \, d\tau + \oiint_\Sigma \mathbf{r} \times \rho(\mathbf{q} \cdot \mathbf{n})\mathbf{q} \, dS$$

$$+ \oiint_S \mathbf{r} \times \rho[(\mathbf{U} - \mathbf{q}) \cdot \mathbf{n}] \, dS \tag{10.115}$$

where R is to be considered fixed. Time differentiation with respect to the space-fixed reference frame is denoted by d/dt. Note that *in the term*

$$\frac{d}{dt} \iiint\limits_R \mathbf{r} \times \rho\mathbf{q} \, d\tau$$

it is not implied that R is a volume changing with time. On account of the boundary condition on S we have

$$(\mathbf{U} - \mathbf{q}) \cdot \mathbf{n} = 0 \quad \text{on} \quad S$$

Hence the expression (10.115) becomes

$$\frac{d}{dt} \iiint\limits_R \mathbf{r} \times \rho\mathbf{q} \, dt + \oiint\limits_\Sigma \mathbf{r} \times \rho(\mathbf{q} \cdot \mathbf{n})\mathbf{q} \, dS. \tag{10.116}$$

The moment exerted by the fluid on the body is denoted by $\mathbf{M}$, the moment being taken about the space-fixed origin. Hence the moment acting on the fluid in R is

$$-\mathbf{M} - \oiint\limits_\Sigma \mathbf{r} \times p\mathbf{n} \, dS \tag{10.117}$$

Since the rate of change of angular momentum of the fluid that is in R at some instant is equal to the moment on that fluid at that instant, we obtain, using (10.116) and (10.117), the following relation for $\mathbf{M}$:

$$-\mathbf{M} = \frac{d}{dt} \iiint\limits_R \mathbf{r} \times \rho\mathbf{q} \, d\tau + \oiint\limits_\Sigma \mathbf{r} \times \rho(\mathbf{q} \cdot \mathbf{n})\mathbf{q} \, dS + \oiint\limits_\Sigma \mathbf{r} \times p\mathbf{n} \, dS \tag{10.118}$$

Consider the integral over R. For any region R_0 containing only fluid enclosed by the surface S_0 we have

$$\iiint\limits_{R_0} \mathbf{r} \times \mathbf{q} \, d\tau = \iiint\limits_{R_0} \mathbf{r} \times \operatorname{grad} \phi \, d\tau = -\iiint\limits_{R_0} \operatorname{curl}(\phi\mathbf{r}) \, d\tau$$

$$= \oiint\limits_{S_0} \mathbf{r} \times \phi\mathbf{n} \, dS \tag{10.119}$$

It therefore follows that

$$\iiint\limits_R \mathbf{r} \times \rho\mathbf{q} \, d\tau = -\oiint\limits_S \mathbf{r} \times \rho\phi\mathbf{n} \, dS + \oiint\limits_\Sigma \mathbf{r} \times \rho\phi\mathbf{n} \, dS \tag{10.120}$$

When the calculations are done in a space-fixed reference frame the pressure is given by

$$p(\mathbf{r}, t) = p_\infty - \rho\left(\frac{\partial\phi}{\partial t} + \frac{q^2}{2}\right)$$

It therefore follows that

$$\oiint_{\Sigma} \mathbf{r} \times p\mathbf{n}\, dS = -\frac{d}{dt}\oiint_{\Sigma} \mathbf{r} \times \rho\phi\mathbf{n}\, dS - \oiint_{\Sigma} \mathbf{r} \times \rho\frac{q^2}{2}\mathbf{n}\, dS \quad (10.121)$$

From Eqs. (10.118, 120, 121) we obtain

$$\mathbf{M} = \frac{d}{dt}\oiint_{S} \mathbf{r} \times \rho\phi\mathbf{n}\, dS + \oiint_{\Sigma} \mathbf{r} \times \rho\left[\frac{q^2}{2}\mathbf{n} - (\mathbf{q}\cdot\mathbf{n})\mathbf{q}\right] dS \quad (10.122)$$

We now let Σ go to infinity. For acyclic motion involving a finite rigid body the integral over Σ vanishes as $1/r^3$ as $r \to \infty$. For acyclic motion involving an infinite rigid cylinder the integral vanishes as $1/r^2$ as $r \to \infty$. It therefore follows that the moment on the body is given by

$$\mathbf{M} = \frac{d}{dt}\oiint_{S} \mathbf{r} \times \rho\phi\mathbf{n}\, dS \quad (10.123)$$

In a similar way we can show that the fluid force on the body is given by

$$\mathbf{F} = \frac{d}{dt}\oiint_{S} \rho\phi\mathbf{n}\, dS \quad (10.124)$$

Let $\mathbf{F}_e$ and $\mathbf{M}_e$ denote the *external* force and moment applied on the body to translate it through the fluid. We then have

$$\frac{d}{dt}(m\mathbf{U}) = \mathbf{F}_e + \mathbf{F}$$

and

$$\frac{d}{dt}(\mathbf{L}) = \mathbf{M}_e + \mathbf{M}$$

where $m\mathbf{U}$ is the linear momentum of the body and $\mathbf{L}$ is its angular momentum. It follows that

$$\mathbf{F}_e = \frac{d}{dt}(m\mathbf{U}) - \mathbf{F} = \frac{d}{dt}\left(m\mathbf{U} - \oiint_{S}\rho\phi\mathbf{n}\, dS\right)$$

and

$$\mathbf{M}_e = \frac{d}{dt}\mathbf{L} - \mathbf{M} = \frac{d}{dt}\left(\mathbf{L} - \oiint_{S}\mathbf{r} \times \rho\phi\mathbf{n}\, dS\right)$$

We introduce the notation

$$\mathscr{I} = -\oiint_{S}\rho\phi\mathbf{n}\, dS = \oiint_{S}\varpi\mathbf{n}\, dS$$

and

$$\mathcal{J}_m = -\oiint_S \mathbf{r} \times \rho\phi\mathbf{n}\, dS = \oiint_S \mathbf{r} \times \varpi\mathbf{n}\, dS$$

where ϖ is the impulsive pressure (see Section 9.9). We call $\mathcal{J}$ the impulse, $\mathcal{J}_m$ the *moment impulse*. With this notation we write

$$\mathbf{F}_e = \frac{d}{dt}(m\mathbf{U} + \mathcal{J})$$

$$\mathbf{M}_e = \frac{d}{dt}(\mathbf{L} + \mathcal{J}_m)$$

$$\mathbf{F} = -\frac{d}{dt}\mathcal{J}$$

$$\mathbf{M} = -\frac{d}{dt}\mathcal{J}_m$$

Thus the force on the body is given by the negative of the rate of change of the impulse and the moment by the negative of the rate of change of the moment-impulse.

We shall now relate $d\mathcal{J}/dt$ and $d\mathcal{J}_m/dt$ which are time derivatives in a space-fixed reference frame to the corresponding time derivatives constructed in a body-fixed reference frame. Let us denote by K_1 the body-fixed frame and by $\mathcal{J}_1$ and $\mathcal{J}_{m_1}$ the impulse and the moment-impulse with respect to the K_1 frame; $\mathcal{J}_{m_1}$ being the moment with respect to O_1 the origin of coordinates in K_1. We denote the space-fixed frame by K and the origin in K by O. The frame K_1 translates with a velocity $\mathbf{U}(t)$ with respect to K.

At the instant under consideration let O_1 be at the position $\boldsymbol{\xi}$ with respect to O (Fig. 10.8). The impulse and moment-impulse computed in frame K

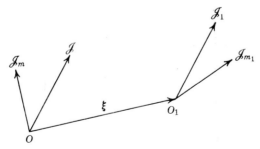

Fig. 10.8 Illustrating the relation between the space-fixed and body-fixed descriptions of impulse and moment-impulse.

are then related to those computed in frame K_1 by

$$\mathscr{I} = \mathscr{I}_1$$
$$\mathscr{I}_m = \mathscr{I}_{m_1} + \boldsymbol{\xi} \times \mathscr{I}_1$$

Furthermore, denoting the time derivatives in K and K_1 respectively by

$$\frac{^K d}{dt} \quad \text{and} \quad \frac{^{K_1} d}{dt}$$

we observe

$$\frac{^K d}{dt} \mathscr{I} = \frac{^{K_1} d}{dt} \mathscr{I}$$

and

$$\frac{^K d}{dt} \mathscr{I}_m = \frac{^{K_1} d}{dt} \mathscr{I}_{m_1} + \frac{^K d}{dt} \boldsymbol{\xi} \times \mathscr{I}_1 + \boldsymbol{\xi} \times \frac{^K d}{dt} \mathscr{I}$$

We choose the two frames to be coincident at the instant under consideration. We shall then have, noting that $^K d\boldsymbol{\xi}/dt$ is equal to **U**,

$$\frac{^K d}{dt} \mathscr{I}_m = \frac{^{K_1} d}{dt} \mathscr{I}_m + \mathbf{U} \times \mathscr{I}_1$$

It therefore follows that

$$\mathbf{F} = \frac{^K d}{dt} \iint_S \rho \phi \mathbf{n} \, dS = \frac{^{K_1} d}{dt} \iint_S \rho \phi \mathbf{n} \, dS \qquad (10.125)$$

and

$$\mathbf{M} = \frac{^K d}{dt} \iint_S \mathbf{r} \times \rho \phi \mathbf{n} \, dS = \frac{^{K_1} d}{dt} \iint_S \mathbf{r} \times \rho \phi \mathbf{n} \, dS + \mathbf{U}(t) \times \iint_S \rho \phi \mathbf{n} \, dS \qquad (10.126)$$

Henceforth we shall use only the body-fixed frame. Consequently we shall drop the superscript K_1 and also the subscript 1 on the impulse and moment-impulse computed in the body-fixed frame. With this understanding we write

$$\mathbf{F} = -\frac{d\mathscr{I}}{dt} = \frac{d}{dt} \iint_S \rho \phi \mathbf{n} \, dS \qquad (10.127)$$

and

$$\mathbf{M} = -\frac{d\mathscr{I}_m}{dt} - \mathbf{U} \times \mathscr{I}$$
$$= \frac{d}{dt} \iint_S \mathbf{r} \times \rho \phi \mathbf{n} \, dS + \mathbf{U} \times \iint_S \rho \phi \mathbf{n} \, dS. \qquad (10.128)$$

We now express $\mathcal{J}_m$ in a form similar to the expression (10.92) for $\mathcal{J}$. We have

$$\mathcal{J}_m = -\oiint_S \mathbf{r} \times \rho\phi\mathbf{n}\, dS$$

$$= -\oiint_S \mathbf{r} \times \rho(\mathbf{U} \cdot \varphi)\mathbf{n}\, dS$$

$$= \sum_k \left(-\oiint_S \rho\varphi_k \mathbf{r} \times \mathbf{n}\, dS\right) u_k$$

where use has been made of Eq. (10.90). The component of $\mathcal{J}_m$ in the direction $\mathbf{e}_i$ is then given by

$$\mathcal{J}_{m_i} = \sum_k I_{ki} u_k \tag{10.129}$$

where

$$I_{ki} = -\oiint_S \rho\varphi_k (\mathbf{r} \times \mathbf{n})_i\, dS \tag{10.130}$$

The coefficients I_{ki}, just like the mass coefficient m_{ik}, are constants that depend only on the body shape and the choice of the orientation of the coordinate axes in the body-fixed frame. It may be verified that

$$I_{ki} \neq I_{ik}$$

We note that to determine the force and moment on a rigid body translating through the fluid we need to compute eighteen coefficients, namely m_{ik} and I_{ik}, that are characteristic of the given body. Since m_{ik} is equal to m_{ki}, these are actually fifteen independent coefficients.

10.10   Uniform Translation

When a rigid body moves through the fluid with a constant velocity the force on the body is zero but the moment exerted by fluid on the body is not zero. We have

$$\mathbf{F} = 0$$

$$\mathbf{M} = -\mathbf{U} \times \mathcal{J}$$

$$= \mathbf{U} \times \oiint_S \rho\phi\mathbf{n}\, dS$$

$$\neq 0 \tag{10.131}$$

It follows that the body will tend to rotate; to maintain it in uniform translation through the fluid, an external moment must be applied to the body constantly.

As an example consider the case of a body of revolution moving with a constant velocity through the fluid. Let the velocity vector make an angle α with the axis of the body. We choose the body-fixed coordinates with the origin at the tip of the nose, the x_1-axis along the body axis, and the x_1–x_3-plane as the plane containing the body axis and the velocity vector (Fig. 10.9).

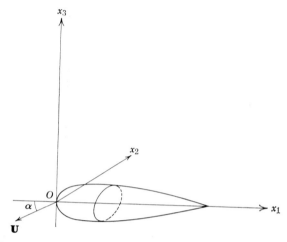

Fig. 10.9 A body of revolution translating uniformly through the fluid.

We have

$$\mathscr{I}_i = \sum_k m_{ik} u_k$$

It can be shown that

$$m_{ik} = 0 \quad \text{for} \quad i \neq k$$

Using this result and noting that

$$\mathbf{U} = (u_1, 0, u_3)$$

we obtain

$$\mathscr{I} = (m_{11}\, u_1, 0, m_{33}\, u_3)$$

Hence it follows that

$$\mathbf{M} = \mathbf{e}_2(m_{33} - m_{11})u_1 u_3 = \mathbf{e}_2 U^2(m_{33} - m_{11}) \cos \alpha \sin \alpha$$

It is seen that if m_{33} is larger than m_{11}, which usually is the case, the fluid-moment on the body tends to increase the angle α. In this sense the translatory motion under consideration is unstable. We note that there is no moment on the body if α is either zero or $\pi/2$.

10.11 Permanent Translation

From Eq. (10.131) we conclude that if the vectors U and $\mathcal{J}$ are parallel there is no moment on the body in steady translation. Hence in such a situation the body will remain permanently in steady translation. The axes along which permanent translation is possible are known as *axes of permanent translation*. For a body of revolution the axis of revolution and any transverse axis are such axes.

10.12 Remarks

In this chapter we were concerned mainly with the force and moment exerted by the fluid on a rigid body translating through it. The fluid motion is strictly acyclic. The results obtained are applicable to both a finite body and an infinite cylinder as long as the motion generated is without circulation. Modification of the results is necessary when motions with non-zero circulation are to be considered.

The extension of the present results to the case of acyclic fluid motion generated by a translating and rotating rigid body may be carried out by following a procedure similar to that used for the case of the translating body. It is found that for a translating and rotating body there are in general thirty-six additional apparent inertia coefficients such as the m_{ik} and I_{ik}. Of these there are actually twenty-one independent coefficients.

For the treatment of the subject of forces and moments acting on a body moving through an ideal fluid when the fluid motion is cyclic and when the body is both translating and rotating, and for examples of the application of the subject reference may be made to Lamb (1932) or to Kochin, Kibel, and Roze (1964).

Chapter 11

Steady Acyclic Motion

The aim of our investigations is to determine the velocity potential for the steady flow of an ideal fluid past certain solid bodies of aerodynamic interest. For this purpose we are to obtain the solution of Laplace's equation under certain prescribed boundary conditions. Because Laplace's equation forms the basis for a great deal of mathematical physics, its solutions have been investigated extensively, and, consequently, there exists a large body of mathematical theory about it. There are various methods of constructing its solutions. The task of constructing an exact solution directly by satisfying the given boundary conditions generally proves difficult. In view of this, it is fruitful to approach the solution to our problem from some simple known solutions of Laplace's equation and to discuss their significance in terms of fluid flow. Since the equation is linear we can find new solutions by superposing various known solutions. Simple flows may be used for building up the more complicated flow fields that must satisfy prescribed boundary conditions. In this chapter we shall see how the solution for the steady acyclic flow past a fixed body may be built up by superposing certain *singular solutions* of Laplace's equation. Having thus obtained the solutions for a few typical bodies we shall examine the theoretical results in light of observations. We start with a statement of the mathematical problem.

11.1 Statement of the Problem

Consider the steady flow past a fixed rigid body. We choose a body-fixed coordinate system and describe the body surface by

$$F(\mathbf{r}) = 0 \qquad (11.1)$$

Let the velocity and pressure of the undisturbed stream at infinity be denoted respectively by $\mathbf{U}$ and p_∞. Let the velocity and the velocity potential at any point of the flow field be denoted by $\mathbf{V}$ and Φ. We then have

$$\mathbf{V} = \operatorname{grad} \Phi = \nabla \Phi \qquad (11.2)$$

and the problem is to determine Φ as the solution of the equation

$$\nabla^2\Phi = 0 \tag{11.3}$$

in the region exterior to the body such that the solution satisfies the boundary conditions

$$\mathbf{V} \cdot \text{grad } F = \text{grad } \Phi \cdot \text{grad } F = 0 \quad \text{on} \quad F(\mathbf{r}) = 0 \tag{11.4}$$

and

$$\text{grad } \Phi = \mathbf{V} = \mathbf{U} \text{ at infinity} \tag{11.5}$$

If, as before, we denote the disturbance or perturbation velocity and potential by $\mathbf{q}$ and ϕ respectively we have

$$\mathbf{q} = \text{grad } \phi$$

The problem in terms of the perturbation potential is

$$\nabla^2\phi = 0$$

$$\text{grad } \phi \cdot \text{grad } F = -\mathbf{U} \cdot \text{grad } F \quad \text{on} \quad F(\mathbf{r}) = 0$$

$$\text{components of grad } \phi \to 0 \quad \text{as} \quad r \to \infty$$

This is precisely the problem for a rigid body moving with a velocity $-\mathbf{U}$ through an otherwise undisturbed fluid (see Section 10.3).

Once the velocity potential Φ is determined the pressure field is obtained from the steady-state Bernoulli equation

$$p(\mathbf{r}) = H - \frac{\rho}{2} (\text{grad } \Phi)^2 \tag{11.6}$$

where H is a constant that may be determined once the pressure and velocity at one reference point is given.

The solutions of (11.3) may be found by the *method of separation of variables* or by the *singularity method*. An example of the method of separation of variables is given in Section 10.4. Many other examples may be found in Lamb (1932). The singularity method consists of superposition of singular solutions. We shall use this method. With this in mind we shall investigate various simple functions that satisfy Laplace's equation to see what sort of boundary conditions they fulfill and how such functions could be effectively combined to build solutions that interest us.

11.2 Simple Polynomial Solutions

We choose Cartesian coordinates x, y, z and consider some simple functions of position. We shall first specialize them to satisfy the potential equation and then investigate the flow fields represented by them.

1. Let us then consider the expression

$$\Phi = Ax + By + Cz \tag{11.7}$$

where A, B, C are constants. This expression satisfies the potential equation, so it may be regarded as a velocity potential. The velocity components are given by

$$u = \frac{\partial \Phi}{\partial x} = A \quad \text{a constant}$$

$$v = \frac{\partial \Phi}{\partial y} = B \quad \text{a constant}$$

$$w = \frac{\partial \Phi}{\partial z} = C \quad \text{a constant}$$

Hence the velocity at every point has the same magnitude and direction. Thus the function (11.7) represents a flow that is uniform in space.

Conversely we can state that the potential for a fluid with a uniform velocity $\mathbf{V} = (U, V, W)$ is

$$\Phi \equiv \Phi(x, y, z) = Ux + Vy + Wz \tag{11.8}$$

The streamlines are all straight lines parallel to $\mathbf{V}$.

2. Consider now the function

$$\Phi = Ax^2 + By^2 + Cz^2 \tag{11.9}$$

where, as before, A, B, C are constants. Since (11.9) should satisfy the potential equation, we require that

$$A + B + C = 0 \tag{11.10}$$

This can be satisfied in many ways.

Suppose we desire that the potential be independent of the z-coordinate. Then we set

$$C = 0$$

and obtain

$$A + B = 0 \quad \text{or} \quad B = -A$$

In such a case the potential (11.9) becomes

$$\Phi = A(x^2 - y^2) \tag{11.11}$$

The corresponding velocity components are given by

$$u = 2Ax$$
$$v = -2Ay \tag{11.12}$$
$$w = 0$$

Equation (11.12), or equivalently the potential (11.11), represents a velocity field in which there is no velocity component in the z-direction and the flow appears the same in all planes perpendicular to the Z-axis; the motion is therefore two-dimensional.

The streamlines of the flow are given by the differential equation

$$\frac{dx}{u} = \frac{dy}{v}$$

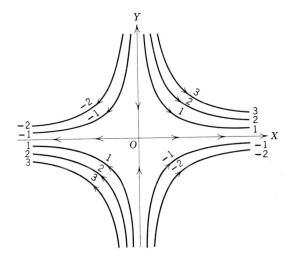

Fig. 11.1 Streamlines defined by xy = constant.

which, according to the relations (11.12), becomes

$$\frac{dx}{x} = -\frac{dy}{y} \tag{11.13}$$

Integrating this equation we obtain

$$xy = \text{const.} \tag{11.14}$$

as the equation for the streamlines, each streamline corresponding to a different value of the constant. The streamlines form a family of rectangular hyperbolas, as shown in Fig. 11.1. We notice that at the point $x = 0$, $y = 0$, the velocity is zero. Hence the origin is a *stagnation point*, and the streamline passing through the stagnation point is usually referred to as the *stagnation streamline*. In the present case it is given by the equation

$$xy = 0$$

It thus follows that the X- and Y-axes form the stagnation streamline. It may be noticed that at the stagnation point the directions of the four branches of the stagnation streamline are different from one another. In this sense the stagnation point exhibits a singular behavior and may be thought of as a singular point in the flow; it is a saddle.

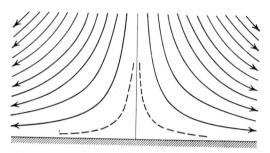

Fig. 11.2 Two-dimensional flow against a wall.

The flow field described by the potential (11.11) may be thought to represent certain local regions that occur as parts of more general flows. One example is the flow in the vicinity of the stagnation point for a two-dimensional flow against a wall, as shown in Fig. 11.2. Another example is the flow in the neighborhood of a rectangular corner, as shown in Fig. 11.3.

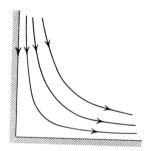

Fig. 11.3 Flow inside a rectangular corner.

It should be noted that the potential (11.11) and the velocity (11.12) exhibit singular behavior at infinity. In fact, the velocity at infinity is infinite.

The functions (11.7) and (11.9) are simple examples of polynomials in the coordinates x, y, z. We can proceed to consider polynomials of more general form, specialize them to satisfy the potential equation, and then investigate the particular flow fields represented by them. Finally, by a proper selection of such functions we may attempt to build a solution of Laplace's equation that satisfies prescribed boundary conditions. Instead of pursuing this tedious path, we shall now pass on to consider some specific functions that will quickly lead us to an analysis of the steady flow past a body.

In case of steady flow past a fixed body, we know that at infinity the velocity of the fluid reduces to that of a uniform stream. This suggests that the flow field may be considered to be built up of two parts—one that represents the uniform stream and the other the disturbance due to the presence of the body in the originally uniform stream. Since we have already determined the function that represents a uniform stream, we need to investigate further only functions that are likely to describe the disturbance field. From what we have said, these functions should be such that the disturbance velocity dies out with distance from the body. This means that such functions should be composed of terms involving inverse powers of the space coordinates. We shall now investigate a simple example of such a function.

11.3 The Source Potential

Let us consider the function

$$f(\mathbf{r}) = \frac{A}{r} \tag{11.15}$$

where A is a constant and r is the magnitude of the position vector $\mathbf{r}$ from a fixed reference point to a point in space. To use this function as a velocity potential (of some flow) we first check to see if it satisfies the potential equation. It turns out that*

$$\nabla^2 \left(\frac{A}{r} \right) \equiv \operatorname{div} \left(\operatorname{grad} \frac{A}{r} \right) = \frac{0}{r^3}$$

That is, the function (11.15) satisfies the potential equation at all points of space except at the point $\mathbf{r} = 0$, where it has the indeterminate form $0/0$.

To determine the value of $\nabla^2 (A/r)$ at the point $\mathbf{r} = 0$ we use the integral definition of divergence and write

$$\nabla^2 \left(\frac{A}{r} \right) = \operatorname{div} \left(\operatorname{grad} \frac{A}{r} \right) = \lim_{\Delta\tau \to 0} \frac{1}{\Delta\tau} \oiint_{\Delta S} \left(\operatorname{grad} \frac{A}{r} \right) \cdot \mathbf{n}\, dS$$

We choose for the arbitrary volume element $\Delta\tau$ a sphere of radius ε with center at $\mathbf{r} = 0$. If $\mathbf{e}_r$ is a unit vector in the direction of $\mathbf{r}$, we have

$$\operatorname{grad} \frac{A}{r} = -\frac{A}{r^2} \mathbf{e}_r$$

Consequently, we obtain

$$\left[\nabla^2 \frac{A}{r} \right]_{\text{at } r=0} = \lim_{\Delta\tau \to 0} \frac{1}{\Delta\tau} \oiint_{\substack{\text{over} \\ \text{sphere} \\ \varepsilon}} -\frac{A}{r^2} \mathbf{e}_r \cdot \mathbf{n}\, dS \tag{11.16}$$

Since *over the sphere*

$$\mathbf{n} = \mathbf{e}_r$$

* The result may easily be verified in Cartesians.

and

$$\frac{A}{r^2} = \frac{A}{\varepsilon^2}, \qquad \text{a constant}$$

(11.16) reduces to

$$\left[\nabla^2 \frac{A}{r}\right]_{\text{at } r=0} = -\infty \tag{11.17}$$

Thus the Laplacian of A/r is zero everywhere except at the point $\mathbf{r} = 0$, where it is infinite. It follows then that the function A/r may be regarded as the velocity potential of a certain flow field as long as the point $\mathbf{r} = 0$ is excluded from our considerations. The fact that $\mathbf{r} = 0$ is a peculiar or *singular point* should not discourage us from further consideration of A/r as a velocity potential. In fact, as we shall presently learn, this function will play a great part in building up the solution to the problem of flow over a body. In regard to the singular point, we try to arrange our considerations in such a way that it is outside the region of physical interest or, if this is not possible, we simply learn to recognize it as a peculiar point and expect such a point to indicate physically untenable results. During the course of our investigations we will meet other functions that are singular in some respect or other but we will soon learn to use them with advantage.

Now, let us look at the flow field represented by the potential

$$\Phi(\mathbf{r}) = \frac{A}{r} \tag{11.18}$$

The velocity at any point in space is given by

$$\mathbf{V}(\mathbf{r}) = \text{grad } \frac{A}{r} = -A \frac{\mathbf{e}_r}{r^2} = -A \frac{\mathbf{r}}{r^3} \tag{11.19}$$

We observe that the direction of the velocity is radial at all points of the field and that its magnitude is constant over the surface of any chosen sphere. The streamlines of the flow are straight rays through the point $\mathbf{r} = 0$. If the constant A is positive, the fluid flow is directed toward the point $\mathbf{r} = 0$. If instead of A we choose $-A$, that is, if we consider the potential

$$\Phi(\mathbf{r}) = -\frac{A}{r} \tag{11.20}$$

the fluid flow would be directed outward from the point $\mathbf{r} = 0$, since the corresponding velocity is given by

$$\mathbf{V}(\mathbf{r}) = A \frac{\mathbf{e}_r}{r^2} = A \frac{\mathbf{r}}{r^3} \tag{11.21}$$

In either case the magnitude of the velocity is inversely proportional to r^2 and as such increases as we approach the point $\mathbf{r} = 0$, where it actually becomes infinite. Conversely, the magnitude of the velocity decreases with distance from that point and completely vanishes at infinity. We notice again that the point $\mathbf{r} = 0$ is a singular point to be excluded from physical considerations.

From the velocity field (11.21) we see that fluid is being continuously created at the point $\mathbf{r} = 0$. Such a point is called a *source*, the corresponding potential (11.20) a *source potential* and the flow field represented by it a *source flow*. In the case of the velocity field (11.19), fluid is being continuously annihilated at the point $\mathbf{r} = 0$. Such a point is called a *sink*, the corresponding potential (11.18) a *sink potential*, and the flow field represented by it a *sink flow*. Point sources and sinks do not possess any physical reality as such but are important mathematical concepts useful in the analysis of fluid motion.

We now determine the quantity of fluid that is continuously appearing at a source or disappearing at a sink. The net outflow of volume* of fluid through any closed surface S drawn in the flow is given by the surface integral

$$\oiint_S \mathbf{V} \cdot \mathbf{n} \, dS = \oiint_S \operatorname{grad} \Phi \cdot \mathbf{n} \, dS$$

or, equivalently, by the volume integral

$$\iiint_R \operatorname{div} \mathbf{V} \, d\tau = \iiint_R \nabla^2 \Phi \, d\tau$$

where R is the region enclosed by the surface S. First consider a closed surface S_1 enclosing a region R_1 *that does not contain* the source. Then the net outflow of fluid through S_1 is

$$\oiint_{S_1} \mathbf{V} \cdot \mathbf{n} \, dS = \iiint_{R_1} \nabla^2 \Phi \, d\tau = 0$$

since $\nabla^2 \Phi = 0$ at all points except at the source. For the same reason, if S_0 is a surface enclosing the region R_0 that *contains* the source, we have

$$\oiint_{S_0} \mathbf{V} \cdot \mathbf{n} \, dS = \iiint_{R_0} \nabla^2 \Phi \, d\tau \neq 0$$

Before we compute the net outflow of fluid through S_0 we observe that if

* We talk about volume instead of mass because the density is constant for our fluid.

S_2 is any other surface enclosing S_0, and if R_2 is the region between S_0 and S_2,

$$\oiint_{S_2} \mathbf{V} \cdot \mathbf{n}\, dS - \oiint_{S_0} \mathbf{V} \cdot \mathbf{n}\, dS = \iiint_{R_2} \nabla^2 \Phi\, d\tau = 0$$

or

$$\oiint_{S_2} \mathbf{V} \cdot \mathbf{n}\, dS = \oiint_{S_0} \mathbf{V} \cdot \mathbf{n}\, dS$$

where $\mathbf{n}$ is an outward normal with respect to the region R_2. It follows, therefore, that the outflow of fluid through every surface enclosing the source is the same and is equal to the volume of fluid, say q, that is being created per unit time at the source. We thus set

$$q = \oiint_{S_0} \mathbf{V} \cdot \mathbf{n}\, dS \tag{11.22}$$

To evaluate this integral we choose for the arbitrary surface S_0 a sphere of radius r_0 with its center at the source; substitute for $\mathbf{V}$ from (11.21) and set $\mathbf{n} = \mathbf{e}_r$. Then (11.22) becomes

$$q = \oiint_{\substack{\text{sphere}\\ r_0}} \mathbf{V} \cdot \mathbf{n}\, dS = \oiint_{\substack{\text{sphere}\\ r_0}} \frac{A}{r^2} \mathbf{e}_r \cdot \mathbf{e}_r\, dS = 4\pi A \tag{11.23}$$

The quantity q is known as the *strength of the source* or simply the *source strength*. In terms of the source strength, the source potential may be expressed as

$$\Phi(\mathbf{r}) = -\frac{q}{4\pi}\frac{1}{r} \tag{11.24}$$

Similarly, in the case of a sink, the potential may be expressed as

$$\Phi(\mathbf{r}) = \frac{q}{4\pi}\frac{1}{r} \tag{11.25}$$

where q now represents the *strength* of the *sink* (or simply the *sink strength*) defined as the volume of fluid that disappears per unit time at the sink.

It should be noted that in the expressions (11.24) and (11.25) r is the magnitude of $\mathbf{r}$, the position vector drawn from the source, not from the origin of a chosen coordinate system, to a point in space (so-called field point). However, if the origin of the coordinate system is chosen to be at the source, then $\mathbf{r}$ is identical with the usual position vector. In such a case we speak of the expression (11.24) (or of 11.25) as the *potential due to a source* (or a sink) *situated at the origin*.

Consider now the case of a source q (i.e., of strength q) not situated at the origin of the chosen coordinate system. In Fig. 11.4 let point O represent the origin of the coordinate system, the point S the location of the source, and P a field point. Let

$\quad$ **s** denote the vector $\overrightarrow{OS}$

$\quad$ **r** denote the position vector $\overrightarrow{OP}$

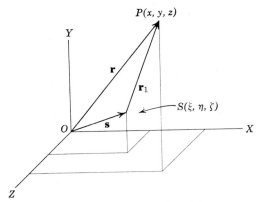

Fig. 11.4 Source not at the origin.

and

$\quad\quad$ $\mathbf{r}_1$ denote the vector $\overrightarrow{SP}$

Then the potential at P due to the source situated at S is

$$\Phi(\mathbf{r}) = -\frac{q}{4\pi}\frac{1}{r_1} = -\frac{q}{4\pi}\frac{1}{|\mathbf{r}_1|} = -\frac{q}{4\pi}\frac{1}{|\mathbf{r}-\mathbf{s}|} \qquad (11.26)$$

The fluid velocity at the point P is then given by (see 11.21)

$$\mathbf{V}(\mathbf{r}) = \frac{q}{4\pi}\frac{\mathbf{r}_1}{r_1^3} = \frac{q}{4\pi}\frac{(\mathbf{r}-\mathbf{s})}{|\mathbf{r}-\mathbf{s}|^3} \qquad (11.27)$$

Let us choose Cartesian coordinates and write

$$\mathbf{r} = (x, y, z)$$

and

$$\mathbf{s} = (\xi, \eta, \zeta)$$

Then the Cartesian form of the source potential (11.26) is

$$\Phi(x, y, z) = -\frac{q}{4\pi}\frac{1}{[(x-\xi)^2+(y-\eta)^2+(z-\zeta)^2]^{1/2}} \qquad (11.28)$$

If u, v, w denote the Cartesian components of the corresponding fluid velocity, then from Eq. (11.27) we obtain

$$u(x, y, z) = \frac{q}{4\pi} \frac{x - \xi}{[(x - \xi)^2 + (y - \eta)^2 + (z - \zeta)^2]^{3/2}}$$

$$v(x, y, z) = \frac{q}{4\pi} \frac{y - \eta}{[(x - \xi)^2 + (y - \eta)^2 + (z - \zeta)^2]^{3/2}} \qquad (11.29)$$

$$w(x, y, z) = \frac{q}{4\pi} \frac{z - \zeta}{[(x - \xi)^2 + (y - \eta)^2 + (z - \zeta)^2]^{3/2}}$$

We shall now look into the usefulness of the concept of source and sink.

11.4 Source in a Uniform Flow (Axisymmetric Flow over a Semi-infinite Body of Revolution)

A number of physically interesting flow fields can be obtained by combining the flow fields of suitable distributions of sources and sinks with that of a uniform stream. A simple example of such a combination is that of a single source with a uniform flow. Let the source strength be q and the velocity of the uniform stream be $\mathbf{U}$.

We choose Cartesian coordinates such that their origin is at the source and the X-axis points in the direction of the uniform stream $\mathbf{U}$ (Fig. 11.5). We thus set

$$\mathbf{U} = \mathbf{i}U$$

For the potential Φ of the combined flow field we have

$$\Phi(x, y, z) = \text{the potential due to the uniform stream}$$
$$+ \text{ the potential due to the source}$$

$$= Ux - \frac{A}{[x^2 + y^2 + z^2]^{1/2}} \qquad (11.30)$$

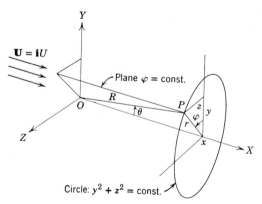

Fig. 11.5 Coordinate system for the problem of source in a uniform flow.

where

$$A = \frac{q}{4\pi} \tag{11.31}$$

For a given value of x, the potential and consequently the flow field is constant over a circle

$$y^2 + z^2 = \text{constant}$$

This means the flow field is symmetrical about the X-axis. It is an axisymmetric flow.

In axisymmetric flows it is convenient to use cylindrical coordinates and occasionally spherical coordinates. Sometimes in the analysis of a single problem it is profitable to go from one set of coordinates to another. In view of this possibility we denote, as shown in Fig. 11.5, by r, φ, x cylindrical coordinates and by R, θ, φ spherical coordinates. Their relation to Cartesians is then given by

Cylindrical

$$\begin{aligned} x &= x \\ y &= r \cos \varphi \\ z &= r \sin \varphi \end{aligned} \tag{11.32}$$

Spherical

$$\begin{aligned} x &= R \cos \theta \\ y &= R \sin \theta \cos \varphi \\ z &= R \sin \theta \sin \varphi \end{aligned} \tag{11.33}$$

It should be noted that in this notation R denotes the magnitude of the usual position vector, while

$$r = [y^2 + z^2]^{1/2} = R \sin \theta \tag{11.34}$$

We shall now look at our problem of a source in a uniform stream in terms of spherical coordinates. The potential (11.30) of the combined flow then takes the form

$$\Phi(R, \theta, \varphi) = UR \cos \theta - \frac{A}{R} \tag{11.35}$$

If u_R, u_θ, u_φ denote the components of the velocity $\mathbf{V}$ at any point R, θ, φ we have

$$\begin{aligned} u_R &= \frac{\partial \Phi}{\partial R} = U \cos \theta + \frac{A}{R^2} \\ u_\theta &= \frac{1}{R} \frac{\partial \Phi}{\partial \theta} = -U \sin \theta \\ u_\varphi &= \frac{1}{R \sin \theta} \frac{\partial \Phi}{\partial \varphi} = 0 \end{aligned} \tag{11.36}$$

Equations (11.35) and (11.36) show what we have already noted, that the potential and the flow are independent of the coordinate φ, that is, are axisymmetric. Thus the flow field looks the same in all planes defined by $\varphi = $ const. Therefore to analyze the flow field further we need only look at one of these planes. Such a plane is shown in Fig. 11.6.

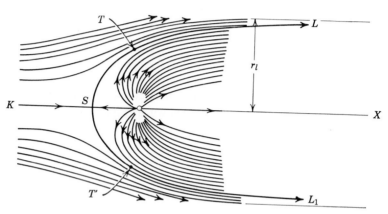

Fig. 11.6 Flow resulting from a source in uniform flow.

From the velocity field (11.36) we notice that both the velocity components vanish simultaneously at the point

$$\theta = \theta_s = \pi$$

and

$$R = R_s = \left(\frac{A}{U}\right)^{1/2} \tag{11.37}$$

The point (R_s, θ_s) is therefore a stagnation point in the flow and is denoted by S in the figure.

The streamlines in any axial plane are described by

$$\psi(R, \theta) = \text{constant}$$

where ψ is the streamfunction. Along a streamline we have

$$d\psi = \frac{\partial \psi}{\partial R}\, dR + \frac{\partial \psi}{\partial \theta}\, d\theta$$

$$= -\sin \theta u_\theta\, dR + R^2 \sin \theta\, u_R\, d\theta \tag{11.38}$$

See Eq. (5.49) or (9.106). With (11.36) and (11.38) we obtain

$$d\psi = d\left(\frac{UR^2 \sin^2 \theta}{2} - A \cos \theta\right)$$

Hence it follows that the streamlines are given by

$$\tfrac{1}{2}UR^2 \sin^2 \theta - A \cos \theta = \text{constant} = C \qquad (11.39)$$

By assigning different values for the constant, we obtain the various streamlines.

We now look at the stagnation streamline, that is, the streamline passing through the point R_s, θ_s. Substituting in (11.39) for R and θ from (11.37), we obtain

$$C = A$$

as the value of the constant for the stagnation streamline. Thus the *equation for the stagnation streamline* is

$$\tfrac{1}{2}UR^2 \sin^2 \theta - A(1 + \cos \theta) = 0$$

or

$$R = \left[\frac{2A}{U} \frac{1 + \cos \theta}{\sin^2 \theta} \right]^{\frac{1}{2}} \qquad (11.40)$$

$$= \sqrt{\frac{A}{U}} \, \text{cosec} \, \frac{\theta}{2}$$

We note that R is an even function of θ (as it should be, for the flow is axisymmetric), and thus for $\theta \neq \pi$ there are two branches of the stream-line. For $\theta = \pi$ all values of R are possible, that is, the whole negative X-axis is part of the stagnation streamline. This streamline appears as shown in Fig. 11.6.

In terms of the cylindrical coordinate r, the stagnation streamline is described by

$$r = R \sin \theta = \left[\frac{2A}{U} (1 + \cos \theta) \right]^{\frac{1}{2}} \qquad (11.41)$$

If T denotes the point where this streamline intersects the r-axis, then

$$r_T \equiv OT = \sqrt{\frac{2A}{U}} \qquad (11.42)$$

As θ goes to zero, the stagnation streamline reaches an asymptote for which, say, $r = r_l$. Then

$$r_l = r \quad \text{as} \quad \theta \to 0$$

$$= 2\left(\frac{A}{U}\right)^{\frac{1}{2}} \qquad (11.43)$$

The stagnation streamline may now be drawn in as shown in Fig. 11.6 by the line $KSTL - KST_1L_1 - OS$. Other streamlines, computed according to (11.39), appear as shown in the figure.

The flow field in all other axial planes (i.e., containing the X-axis) appears exactly the same as shown in the figure. We observe that the surface formed by the stagnation streamlines (Fig. 11.7) obtained by revolving the line *KSTL* about the X-axis, is a surface that divides the whole flow field into two distinct regions—one external and one internal to that surface.* Since the surface is a stream surface, no fluid flows across it. The internal region, therefore, consists entirely of the fluid emanating from the source. The source flow is pressed together (by the uniform

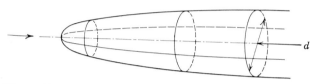

Fig. 11.7   The semi-infinite body of revolution generated by source in a uniform flow.

stream) and made to flow to the right. The external region consists only of the originally uniform stream now displaced. Thus the stream surface formed by the stagnation streamlines behaves just like that of a solid body placed in a uniform stream and may be regarded as such. The body appears as a semi-infinite body of revolution with a blunt nose, that is, a round nose.

From the preceding considerations we may now conclude that the *flow field due to the motion of an originally uniform stream past a semi-infinite body of revolution whose axis is parallel to the uniform stream can be represented by a simple superposition of a single source and the uniform stream.* The surface of the body cannot be prescribed arbitrarily but is to be described by an equation of the form (see 11.41)

$$r = \text{const.}\sqrt{1 + \cos \theta} \qquad (11.44)$$

At a large distance from the nose, the body is nearly cylindrical. If d denotes the diameter of this part of the body, then (11.44) becomes

$$r = \frac{d}{2\sqrt{2}} \sqrt{1 + \cos \theta} \qquad (11.45)$$

The strength of the source that represents the disturbance due to the body may be determined from (11.43) (since $d = 2r_l$), or directly as follows. Consider the flow within the surface of the body. The volume of fluid

---

* For this reason the stagnation streamline is sometimes referred to as the dividing streamline.

passing per unit time to the right through any cross section is equal to the strength q of the source. If the cross section is chosen in the cylindrical portion of the body, the fluid volume passing through it per unit time is simply $\pi(d^2/4)U$, since the flow velocity at a considerable distance from the source is equal to U, the velocity of the uniform stream. Therefore

$$q = \frac{\pi}{4} d^2 U$$

or

$$A = \frac{q}{4\pi} = \frac{d^2}{16} U \tag{11.46}$$

With the source strength known, the potential of the flow field is determined according to (11.35).

Finally, it may be noted that using the principle of superposition we have constructed a solution of the mathematical problem posed by (11.3) to (11.5), that is, a solution of Laplace's equation that satisfies prescribed boundary conditions at infinity and on the body.

11.5   Source and Sink in a Uniform Flow (Axisymmetric Flow over a Closed Body of Revolution)

The superposition of a single source and a uniform stream lead us to axisymmetric flow past a semi-infinite body of revolution. To obtain such a flow past a closed finite body of revolution, we may situate a sink on the axis of the semi-infinite body at some suitable distance downstream from the source and make the strength of the sink equal to that of the source. In this way the flow produced by the source is completely sucked in by the sink; the flow closes up behind the sink just as it opened out in front of the source; and we obtain the flow over an elongated body with rounded nose and tail.

Let us then consider the superposition of a uniform stream U, a source of strength q and a sink of strength q, the line joining the source and the sink being parallel to the uniform stream (Fig. 11.8). The source faces the uniform stream. We choose the origin O of the coordinate system midway between the source and sink and the X-axis parallel to U. The various coordinates are designated as shown in the figure. The velocity potential at the field point P is then given by

$$\Phi(P) = Ux - \frac{q}{4\pi} \frac{1}{R_2} + \frac{q}{4\pi} \frac{1}{R_1}$$

$$= Ux + A\left[\frac{1}{\sqrt{(x-d)^2 + r^2}} - \frac{1}{\sqrt{(x+a)^2 + r^2}}\right] \tag{11.47}$$

where

$$A = \frac{q}{4\pi}$$

and

$$a = \text{one half the distance between the source and sink}$$

The flow field, as before, is axisymmetric, and we look at the flow in any plane $\varphi = \text{constant}$.

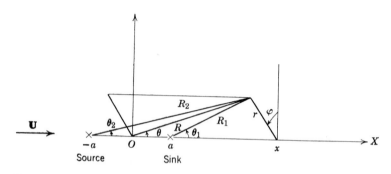

Fig. 11.8 Coordinates for the source-sink combination in a uniform stream.

If u_r, u_φ, u_x are the velocity components with respect to the cylindrical coordinates r, φ, x, we have

$$u_r = \frac{\partial \Phi}{\partial r} = -A\left\{\frac{r}{[(x-a)^2 + r^2]^{3/2}} - \frac{r}{[(x+a)^2 + r^2]^{3/2}}\right\}$$

$$u_\varphi = \frac{1}{r}\frac{\partial \Phi}{\partial \varphi} = 0 \qquad\qquad (11.48)$$

$$u_x = \frac{\partial \Phi}{\partial x} = U - A\left\{\frac{x-a}{[(x-a)^2 + r^2]^{3/2}} - \frac{x+a}{[(x+a)^2 + r^2]^{3/2}}\right\}$$

If x_s and r_s denote the coordinates of a stagnation point in the flow, they are determined by setting the right-hand members of (11.48) equal to zero. We thus obtain

$$r_s = 0 \qquad\qquad (11.49a)$$

and find that x_s is to be determined from the equation

$$U - A\left[\frac{x_s - a}{|x_s - a|^3} - \frac{x_s + a}{|x_s + a|^3}\right] = u_x \qquad (r = 0, x_s)$$

$$= 0 \qquad\qquad (11.49b)$$

We observe that

$$U - A\frac{4x_s a}{(x_s^2 - a^2)^2} = 0 \quad \text{for} \quad x_s > a$$

and

$$U + A\frac{4x_s a}{(x_3^2 - a^2)^2} = 0 \quad \text{for} \quad x_s < -a$$

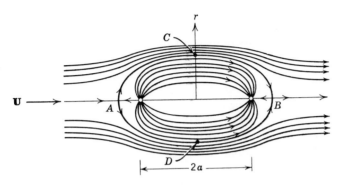

Fig. 11.9 Axial flow past a Rankine body.

Hence there are two stagnation points on the X-axis, one for which x_s is less than $-a$ and the other for which x_s is greater than a. They are shown as A and B in Fig. 11.9.

In any axial plane along a streamline we have

$$d\psi = \frac{\partial \psi}{\partial r}\,dr + \frac{\partial \psi}{\partial x}\,dx = ru_x\,dr - ru_r\,dx \qquad (11.50)$$

[see (5.50) or (9.107)]. With (11.48) and (11.50) we obtain after some simplification

$$d\psi = d\left(A\frac{x-a}{\sqrt{(x-a)^2 + r^2}} - A\frac{x+a}{\sqrt{(x+a)^2 + r^2}} + \frac{Ur^2}{2}\right)$$

Hence it follows that the streamlines are given by

$$A\left(\frac{x-a}{\sqrt{(x-a)^2 + r^2}} - \frac{x+a}{\sqrt{(x+a)^2 + r^2}}\right) + \frac{1}{2}Ur^2 = C, \quad \text{a constant}$$
$$(11.51)$$

Different streamlines correspond to different values of C.

We now look at the stagnation streamline. The value of the constant for this streamline is obtained by putting $r = r_s = 0$ and $x = x_s$. It is simply equal to zero. Thus the stagnation streamline is described by the equation

$$A\left[\frac{x-a}{\sqrt{(x-a)^2+r^2}} - \frac{x+a}{\sqrt{(x+a)^2+r^2}}\right] + \frac{1}{2}Ur^2 = 0 \qquad (11.52)$$

To trace the streamline it is convenient to express (11.52) in terms of R and θ as follows:

$$A(\cos\theta_1 - \cos\theta_2) + \frac{1}{2}UR^2\sin^2\theta = 0 \qquad (11.53)$$

where θ_1 and θ_2 are as designated in Fig. 11.8. From (11.53) it readily follows that the whole X-axis except the part between source and sink forms a part of the stagnation streamline.* The rest of this streamline is a closed curve shown by the line $ACBDA$ in the figure. We observe that, as before, this stagnation streamline is a dividing line between the originally uniform stream and the fluid flowing from the source to the sink. Thus a surface formed by the stagnation streamlines of the whole flow behaves just like a body of revolution obtained by revolving the line $ACBDA$ about the X-axis.

Consequently, we may conclude that the *particular* superposition we have considered of a source, sink, and uniform stream represents axisymmetric flow of an originally uniform stream past a closed body of revolution whose surface is described by an equation such as (11.52) or (11.53). The body is a symmetric ovoid whose maximum transverse radius r_c is given by the relation

$$r_c^2\sqrt{a^2+r_c^2} = 4a\frac{A}{U} \qquad (11.54)$$

which follows readily from (11.52) by setting $x = 0$ in it.

Bodies of this type are called *Rankine bodies* after Rankine who first suggested the idea of forming the type of flow under consideration (Fig. 11.9). Their nose and tail are blunt. The maximum speed on the body occurs at $x = 0$ and $r = r_c$. Denoting by V_m the maximum speed we obtain from (11.48)

$$u_r(r_c, x = 0) = 0$$

$$V_m = u_x(r_c, x = 0) = U + 2A\frac{a}{(a^2+r_c^2)^{3/2}} \qquad (11.55)$$

Eliminating A by using (11.54) we have

$$\frac{V_m}{U} = 1 + \frac{1}{2}\left(\frac{r_c}{a}\right)^2\frac{1}{1+(r_c/a)^2} \qquad (11.56)$$

* For points between the source and sink, $\theta_1 = \pi$ and $\theta_2 = 0$ and hence (11.52) is not satisfied; for $x < -a$ and $x > a$, $\theta_1 = \theta_2 = \pi$ or zero and hence (11.52) is satisfied.

We note that these Rankine bodies form a family of bodies for which only the ratio r_c/a can be prescribed for a given V_m/V.

11.6 Line Distribution of Sources and Sinks in a Uniform Flow: Axisymmetric Flow over Slender Bodies of Revolution

As an extension of the above simple example of axisymmetric flow over a certain closed body of revolution, we may next consider the superposition of a uniform stream and a suitable distribution of sources and sinks along

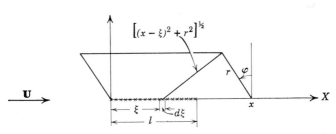

Fig. 11.10 Superposition of a uniform stream and a source distribution along a line parallel to the stream.

an axis parallel with the uniform stream. In this way, by selecting different distributions, it is possible to arrive at the axisymmetric flow over bodies of revolution of various shapes. *Since we are seeking closed bodies, the sum of the strengths of all the sources and sinks, no matter what their distribution, should be zero.* That is, for the body to be closed there must be enough sink strength to suck up all the fluid produced by the sources present. The sources and sinks are generally distributed continuously so as to obtain smooth bodies although single sources (and sinks) are also allowable. The method of analysis of the flow field in this case generally follows the lines of the preceding two sections.

We shall now consider briefly the superposition of a continuous line distribution of sources* and a uniform stream, the line containing the sources being parallel to the uniform stream. We choose, as before, cylindrical coordinates r, φ, x with the X-axis lying along the line of sources and pointing in the direction of **U**, the uniform stream (Fig. 11.10). Let

$$f = f(x)$$

denote the intensity of the source distribution per unit length. Positive values of $f(x)$ denote sources, negative values sinks. Let the source distribution extend from $x = 0$ to $x = l$. The condition that the total strength

* In this context the word "sources" means sources and sinks.

of the distribution is zero is expressed by

$$\int_0^l f(x)\,dx = 0$$

At $x = \xi$, consider an element $dx = d\xi$ of the distribution. Then, according to Eq. (11.26), the potential at a field point $P(x, r, \varphi)$ due to the elementary source $f(\xi)\,d\xi$ is

$$-\frac{1}{4\pi}\frac{f(\xi)\,d\xi}{R_\xi} = -\frac{1}{4\pi}\frac{f(\xi)\,d\xi}{\sqrt{(x-\xi)^2 + r^2}}$$

The potential at P due to all such elementary sources distributed along the X-axis from 0 to l is then given by

$$\Phi_1(x, r, \varphi) = -\frac{1}{4\pi}\int_0^l \frac{f(\xi)}{\sqrt{(x-\xi)^2 + r^2}}\,d\xi \qquad (11.57)$$

During this integration x, r, φ are kept constant, for they are the co-ordinates of the field point for which the potential is being computed.

Now, to obtain the total potential Φ of the flow field resulting from a superposition of the source distribution and the uniform stream U we add to the potential Φ_1 that due to the uniform stream. The latter is simply Ux. Thus the total potential is given by

$$\Phi(x, r, \varphi) = Ux + \Phi_1$$

$$= Ux - \frac{1}{4\pi}\int_0^l \frac{f(\xi)}{\sqrt{(x-\xi)^2 + r^2}}\,d\xi \qquad (11.58)$$

The velocity components u_x, u_r, u_φ are expressed by

$$u_x = \frac{\partial \Phi}{\partial x} = U + \frac{1}{4\pi}\int_0^l f(\xi)\frac{x - \xi}{[(x-\xi)^2 + r^2]^{3/2}}\,d\xi$$

$$u_r = \frac{\partial \Phi}{\partial r} = \frac{r}{4\pi}\int_0^l \frac{f(\xi)}{[(x-\xi)^2 + r^2]^{3/2}}\,d\xi \qquad (11.59)$$

$$u_\varphi = \frac{1}{r}\frac{\partial \Phi}{\partial \varphi} = 0$$

The flow field is axisymmetric.

Once the source distribution is specified, the velocity components can be evaluated. We can then proceed as before to determine the stagnation points, the streamlines, and, in particular, the stagnation streamlines of the flow. The practical calculation of the flow quantities is best carried out by numerical methods. We shall not enter into the details of these methods here but shall now pass on to some examples showing the results of such calculations.

In Fig. 11.11 we show four shapes of bodies of revolution that correspond to certain specified source distributions. These are taken from the calculations of Fuhrmann (1911) as reported by Prandtl (1925). In the figures the assumed source distributions are indicated on the axes. The upper halves of the figures show the streamlines of the disturbance flow field due to the source distribution, and the lower halves show the streamlines of the combined flows. We thus see that the superposition of a continuous line distribution of sources and a uniform stream parallel to that line represents the axisymmetric flow past a certain body of revolution.

The source potential (11.20) is the simplest singular solution of Laplace's equation. There are other singular solutions which, like the source, can be used to generate flow past a body. We shall now describe another such solution, the so-called *doublet singularity*.

11.7 The Doublet Potential

Consider the situation where a source of strength q and a sink of equal strength are situated very close together, say at a small distance l apart. Let us denote the sink by O, the source by S, and a field point by P (Fig. 11.12). Further, let

$$\mathbf{r} = \overrightarrow{OP}$$

$$\mathbf{r}_1 = \overrightarrow{SP}$$

$$\mathbf{l} = \overrightarrow{OS}$$

The potential at P due to the source and sink is given by

$$\Phi(P) = \frac{q}{4\pi}\left(\frac{1}{|\mathbf{r}|} - \frac{1}{|\mathbf{r}_1|}\right)$$

$$= \frac{q}{4\pi}\left(\frac{1}{r} - \frac{1}{r_1}\right)$$

$$= \frac{q}{4\pi}\left(\frac{r_1 - r}{rr_1}\right) \tag{11.60}$$

Now, if we bring the source and sink together by letting l go to zero, then the potential (11.60) will also go to zero. However, if at the same time as l goes to zero q is allowed to increase indefinitely in such a way that the product ql remains finite and equal to a constant μ, then the potential

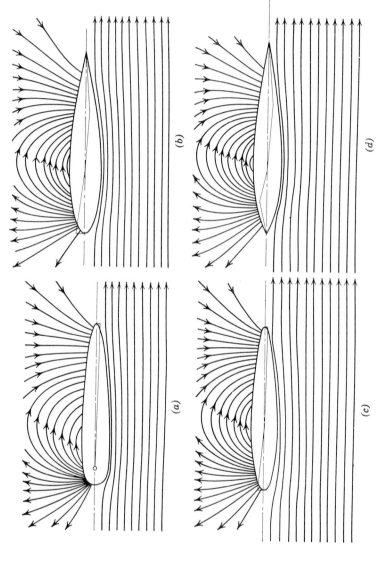

(a)

(b)

(c)

(d)

Fig. 11.11 Axial flow past bodies of revolution derived by combination of sources and uniform flow. Distribution of sources indicated on axis. Streamlines in upper half refer to the disturbance field; those in the lower half to the total flow field. (Fuhrmann 1911)

334

will not vanish. Instead it will assume the value given by

$$\Phi(P) = \lim_{\substack{l \to 0 \\ q \to \infty}} \frac{q}{4\pi}\left(\frac{r_1 - r}{rr_1}\right) \qquad (11.61)$$
$$\text{with}$$
$$ql = \mu$$

If θ is the angle between l and $\mathbf{r}$, we notice that as $l \to 0$,

$$(r - r_1) \to OT = l \cos \theta$$

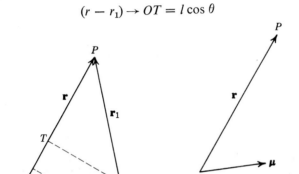

Fig. 11.12 Illustrating the derivation of the doublet potential.

and

$$rr_1 \to r^2$$

Hence the potential (11.61) becomes

$$\Phi(P) = -\frac{\mu}{4\pi}\frac{\cos \theta}{r^2} \qquad (11.62)$$

which is known as the *doublet potential*. The particular sink-source combination it represents is called a *doublet* or a *dipole* or a *double source*. The angle θ in (11.62) is measured with respect to a particular direction, namely the direction of the line segment extending from the sink to the source. The direction from the sink to the source of a doublet is known as the *axis of the doublet*. The constant μ is generally called the *strength of the doublet* or simply the *doublet strength*. Since μ originated as the product ql, it is also referred to as the *moment* of the doublet. We notice that to specify a doublet we need to give two quantities—a magnitude, namely, its strength, and a direction, namely, its axis. Thus we may describe a doublet as a vector

$$\boldsymbol{\mu} = \mu \mathbf{e}_{\mu} \qquad (11.63)$$

where $\mathbf{e}_{\mu}$ is a unit vector in the direction of the doublet axis.

Introducing the vector $\boldsymbol{\mu}$, we can express the doublet potential (11.62) as

$$\Phi(P) = -\frac{1}{4\pi}\frac{\boldsymbol{\mu}\cdot\mathbf{r}}{r^3} \qquad (11.64)$$

This equation is useful for conveniently obtaining the form of the doublet potential in terms of any desired coordinates.

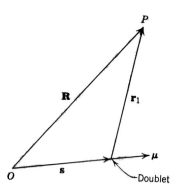

Fig. 11.13 Doublet not at the origin.

It should be noted that in the expression (11.64) $\mathbf{r}$ is the position vector drawn from the doublet—and not from the origin of a chosen coordinate system —to a field point. When the origin of coordinates is located at the doublet, then $\mathbf{r}$ becomes identical with the usual position vector. In such a case we speak of the expression (11.64) as the *potential due to a doublet situated at the origin.*

Consider now the case of a doublet $\boldsymbol{\mu}$ situated at $\mathbf{s}$ from the origin O of a coordinate system (Fig. 11.13). Let $\mathbf{R}$ denote the position vector from O to a field point P, and $\mathbf{r}_1$ the vector from the doublet to P. Then the potential at P due to the doublet at $\mathbf{s}$ is given, from (11.64), by

$$\Phi(\mathbf{R}) \equiv \Phi(P) = -\frac{1}{4\pi}\frac{\boldsymbol{\mu}\cdot\mathbf{r}_1}{|\mathbf{r}_1|^3}$$

$$= -\frac{1}{4\pi}\frac{\boldsymbol{\mu}\cdot(\mathbf{R}-\mathbf{s})}{|\mathbf{R}-\mathbf{s}|^3} \qquad (11.65)$$

We shall now write down the form of the doublet potential in each of the coordinate systems: Cartesian x, y, z; spherical R, θ, φ; and cylindrical r, φ, x. We choose the origin of the coordinates at the doublet, the X-axis in the direction of $\boldsymbol{\mu}$, and designate the various coordinates as shown in Fig. 11.14. Then from (11.64) or (11.65) we obtain

$$\Phi(x, y, z) = -\frac{\mu}{4\pi}\frac{x}{\{x^2 + y^2 + z^2\}^{3/2}} \qquad (11.66)$$

$$\Phi(R, \theta, \varphi) = -\frac{\mu}{4\pi}\frac{\cos\theta}{R^2} \qquad (11.67)$$

$$\Phi(r, \varphi, x) = -\frac{\mu}{4\pi}\frac{x}{\{x^2 + r^2\}^{3/2}} \qquad (11.68)$$

We notice that the *doublet potential*, and *consequently* the *flow field due to a doublet, is axisymmetric about the axis of the doublet.* This is in contrast to the source flow, which is spherically symmetric.

The velocity field of a doublet is readily obtained from the equation

$$\mathbf{V} = \text{grad } \Phi = -\frac{1}{4\pi} \text{grad } \frac{\boldsymbol{\mu} \cdot \mathbf{r}}{r^3}$$

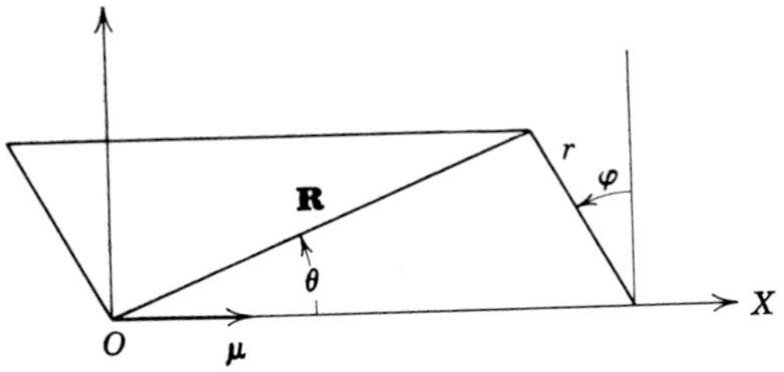

Fig. 11.14   Coordinates for doublet flow with doublet at origin.

If u_R, u_θ, u_φ are the velocity components with respect to the coordinates R, θ, φ, we have

$$u_R = \frac{\partial \Phi}{\partial R} = \frac{\mu}{2\pi} \frac{\cos \theta}{R^3}$$

$$u_\theta = \frac{1}{R} \frac{\partial \Phi}{\partial \theta} = \frac{\mu}{4\pi} \frac{\sin \theta}{R^3} \qquad (11.69)$$

$$u_\varphi = \frac{1}{R \sin \theta} \frac{\partial \Phi}{\partial \varphi} = 0$$

In any axial plane along a streamline we have

$$d\psi = -R \sin \theta u_\theta \, dR + R^2 \sin \theta u_R \, d\theta = d\left(\frac{\mu}{4\pi} \frac{\sin^2 \theta}{R}\right) = 0$$

Hence the streamlines are given by

$$\frac{\sin^2 \theta}{R} = C, \quad \text{a constant} \qquad (11.70)$$

Different values of the constant yield different streamlines. The streamlines appear as shown in Fig. 11.15. All the streamlines begin and end at the doublet. They proceed, as shown, from the source side of the doublet toward the sink side. Near the axis of the doublet, the streamline directions are the same as that of the axis. This is the motivation for defining the doublet axis as proceeding from the sink to the source.

The doublet, just like the source, is a singular solution of Laplace's equation. It does not satisfy that equation at the point where the doublet is situated. At that point both the potential and the velocity of the doublet become infinite. The doublet flow field, just like the source field, dies out with distance from the singularity and vanishes at infinity. In fact, the doublet dies out faster than the source.

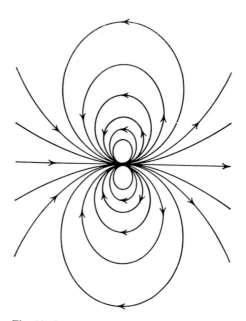

Fig. 11.15 Streamlines of a doublet flow.

There is a simple and significant relation between the doublet and source potentials. To see this we consider a doublet e_μ of unit strength and express its potential, following (11.64), by

$$\Phi_d = -\frac{1}{4\pi} e_\mu \cdot \frac{r}{r^3} \qquad (11.71)$$

Since

$$-\frac{r}{r^3} = \operatorname{grad}\left(\frac{1}{r}\right)$$

the potential (11.71) takes the form

$$\Phi_d = -e_\mu \cdot \operatorname{grad}\left(\frac{-1}{4\pi}\frac{1}{r}\right)$$

$$= -e_\mu \cdot \operatorname{grad}(\Phi_s)$$

$$= -\frac{\partial}{\partial n}(\Phi_s) \qquad (11.72)$$

where Φ_s is the potential due to a source of unit strength and

$\dfrac{\partial \Phi_s}{\partial n}$ is the partial derivative of Φ_s with respect to distance along the direction $\mathbf{e}_\mu$.

We thus see that the doublet potential can be generated by simply differentiating the source potential with respect to distance in a certain chosen direction. According to (11.72) the negative of this derivative would be the potential of the doublet and the chosen direction would be the axis of the doublet.*

In this way, by successive spatial differentiation of the source potential, we can build up a number of higher order singular solutions (referred to as sources or poles of higher order) of Laplace's equation. A suitable superposition of such higher order poles may then be used in the solution of specific problems. For our purposes the source and doublet singularities are adequate.

11.8   Doublet in a Uniform Stream: Flow over a Sphere

Let us now consider the superposition of a doublet μ and a uniform stream U whose direction is parallel to the doublet axis. Since the source side of the doublet should face the uniform stream, its axis opposes the uniform stream. We choose the origin of coordinates at the doublet, the X-axis in the direction of U, and designate the various coordinates as before (see, for instance, Fig. 11.14). We thus set

$$\mathbf{U} = \mathbf{i}U$$

and

$$\boldsymbol{\mu} = -\mathbf{i}\mu$$

In terms of spherical coordinates R, θ, φ the potential at a field point due to the uniform stream is

$$Ux = UR \cos \theta$$

and the potential due to the doublet, according to relation (11.65), is

$$\frac{1}{4\pi} \mu \frac{\cos \theta}{R^2}$$

Consequently, the potential due to the combined flow is given by

$$\Phi(R, \theta, \varphi) = UR \cos \theta + \frac{\mu}{4\pi} \frac{\cos \theta}{R^2}$$

$$= \left(UR + \frac{\mu}{4\pi} \frac{1}{R^2} \right) \cos \theta \qquad (11.73)$$

---

* The negative sign in (11.72) could have been avoided if we defined the axis as the direction from the source to the sink instead of from the sink to the source.

We notice that the potential, and consequently the flow field, is axisymmetric.

The velocity field is given by the components

$$u_R = \frac{\partial \Phi}{\partial R} = \left(U - \frac{\mu}{2\pi} \frac{1}{R^3} \right) \cos \theta$$

$$u_\theta = \frac{1}{R} \frac{\partial \Phi}{\partial \theta} = - \left(U + \frac{\mu}{4\pi} \frac{1}{R^3} \right) \sin \theta \qquad (11.74)$$

$$u_\varphi = \frac{1}{R \sin \theta} \frac{\partial \Phi}{\partial \varphi} = 0$$

The coordinates R_s, θ_s of the stagnation point are obtained by setting the right members of (11.74) equal to zero. We thus have

$$\theta_s = 0 \quad \text{or} \quad \pi$$

and

$$R_s = \left(\frac{\mu}{2\pi U} \right)^{1/3} \qquad (11.75)$$

We see, therefore, that there are two stagnation points shown by A and B in Fig. 11.16.

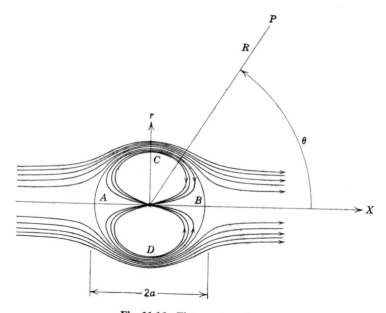

Fig. 11.16 Flow past a sphere.

In the axial plane along any streamline we have

$$d\psi = -R \sin \theta u_\theta \, dR + R^2 \sin \theta u_R \, d\theta$$

$$= d\left[\left(\frac{UR^2}{2} - \frac{\mu}{4\pi R}\right) \sin^2 \theta\right] = 0$$

Hence the streamlines are given by

$$\left(UR^2 - \frac{\mu}{2\pi R}\right) \sin^2 \theta = C, \quad \text{a constant} \tag{11.76}$$

By substituting the coordinates of the stagnation point, from (11.75), in the left-hand member of (11.76), we find that the value of C for the stagnation streamline is zero. Hence the stagnation streamline is described by the equation

$$\left(UR^2 - \frac{\mu}{2\pi} \frac{1}{R}\right) \sin^2 \theta = 0 \tag{11.77}$$

or, equivalently, by

$$\sin^2 \theta = 0 \quad \text{for any} \quad R \tag{11.77a}$$

and

$$UR^2 - \frac{\mu}{2\pi} \frac{1}{R} = 0 \quad \text{for any} \quad \theta \tag{11.77b}$$

Equation (11.77a) shows that the whole X-axis forms a part of this streamline. Equation (11.77b) shows that the circle

$$R = \left(\frac{\mu}{2\pi U}\right)^{\!1/3} = a, \quad \text{a constant} \tag{11.78}$$

describes the rest of that streamline (Fig. 11.16). Other streamlines of the flow appear as shown in the figure.

The surface containing all the stagnation streamlines of the whole field is obtained by rotating the circle of radius $R = a$ about the X-axis and is, therefore, a sphere of radius a. We may thus conclude that the flow resulting from a particular combination of a doublet and a uniform stream represents the flow due to a sphere in a uniform stream, the radius of the sphere being given by (11.78). Conversely, we may also state that the flow field due to the motion of an originally uniform stream of speed U past a sphere of radius a may be represented by the superposition of the uniform stream and a doublet whose axis opposes the stream and whose strength is given by

$$\mu = 2\pi U a^3 \tag{11.79}$$

Substituting (11.79) into Eq. (11.73) we obtain for the potential

$$\Phi(R, \theta) = U\left(1 + \frac{a^3}{2R^3}\right)R\cos\theta \qquad (11.80)$$

Consequently, the velocity components are given by

$$u_R(R, \theta) = U\left(1 - \frac{a^3}{R^3}\right)\cos\theta \qquad (11.81a)$$

$$u_\theta(R, \theta) = -U\left(1 + \frac{a^3}{2R^3}\right)\sin\theta \qquad (11.81b)$$

The disturbance field is that of a doublet [see also (10.48)].

11.9 Line Distribution of Doublets in a Uniform Stream: Lateral and Axisymmetric Flow Past a Body of Revolution

Since the sphere has no distinguished axis of revolution, the flow field due to the motion of a uniform stream past a sphere can be given another interesting and useful interpretation. Let us choose, as before, the origin

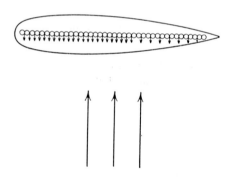

Fig. 11.17 Lateral flow past a body of revolution by combination of a doublet distribution and a uniform stream.

of coordinates at the center of the sphere and the X-axis parallel to the uniform stream. If we select the X-axis as the axis of revolution of the sphere, then the flow field is that due to a uniform stream parallel to the axis of revolution. However, if we consider any line in the plane normal to X as the axis of revolution of the sphere, then the flow field is that due to a uniform stream normal to the axis of revolution. Such a flow is generally referred to as *lateral* or *transverse flow* over a body of revolution. Whichever point of view we wish to take, the flow field is determined, as we have seen, by a superposition of the uniform stream and a doublet with its axis opposing the stream. This means that the lateral flow of a uniform stream past a certain body of revolution (namely, the sphere) can be represented by the superposition of the stream and a doublet on the axis of revolution of the body with the doublet axis normal to the body axis. One can generalize this idea and *represent the lateral flow over a nonspherical body of revolution by a suitable distribution of doublets along the axis of revolution with the axes of the doublets being normal to the body axis;* the doublet axes oppose the direction of the undisturbed stream at infinity (Fig. 11.17).

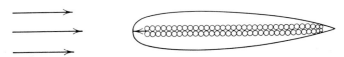

Fig. 11.18　Axial flow past a body of revolution by combination of a doublet distribution and a uniform stream.

A line distribution of doublets can also be used to generate a body of revolution in axisymmetric flow. Consider a distribution of doublets along a line such that the axes of the doublets point along the line (Fig. 11.18). Such a distribution is equivalent to a continuous one of sources together with a source at one end and a sink at the other.

The superposition of axial and lateral flow solutions leads to the flow past a body of revolution at an angle of yaw.

11.10　Flow Past Arbitrary Bodies of Revolution

We now turn to the practically important problem of determining the flow past a given body of revolution. If the surface of the body is *analytic*, that is, if it can be described analytically, the solution for the flow field can be found by using appropriate curvilinear coordinates and the method of separation of variables. Ellipsoids of revolution and the sphere are examples of such analytic bodies. Flow past such bodies has been extensively treated and for the details Lamb (1932) and Munk (1934) may be consulted. In this connection reference may be made also to Thwaites (1960) where further references will be found.

For the flow past a body of prescribed shape the powerful method of superposition of simple singular solutions is widely used. *The problem now is that of determining the distribution of singularities which represent the flow over a given body of revolution.* This direct problem is much more important than the indirect problem we have considered before where the flow past a body of revolution was generated by superposing a *specified* line distribution of sources and doublets on a uniform stream.

The direct problem appears to have been treated first by Karman (1927). For axisymmetric motion he used a continuous distribution of sources along the axis of the given body and gave a method for computing the distribution. For lateral flow past the given body he used a continuous distribution of doublets along the axis with the doublet axes opposing the undisturbed lateral flow. A method was given for computing the doublet

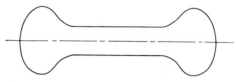

Fig. 11.19   Example of a body of revolution which cannot be represented by a distribution of singularities on its axis alone.

distribution. For the details consult Karman (1927). An approximate analytic solution for the flow past very *slender* bodies of revolution is given in Chapter 20.

The assumption that the flow about a given body of revolution can be represented exactly by a distribution of sources and doublets along the axis of the body is valid only if the shape of the body satisfies certain conditions. It is thus applicable to so-called *slender bodies* only. It is not applicable for a body with discontinuity in its surface slope, such as the body shown in Fig. 11.19. It can be shown, however, that *the acyclic flow past any arbitrary body may be represented by a distribution of sources or doublets on the surface of the body* (see next section). Hence the flow past arbitrary bodies of revolution may be found by using surface distributions of singularities. Such a treatment was originated by Flugge-Lotz (1931), who employed a surface distribution of source rings. Smith and Pierce (1958) gave a method of calculating, by means of an electronic computer, the axisymmetric flow past arbitrary bodies of revolution on the basis of surface distribution of sources. Hess (1962) gave a similar method for computing lateral flow past arbitrary bodies of revolution. For the details and specific problems their work should be consulted; for other references see Thwaites (1960).

11.11   Flow Past an Arbitrary Body

We now consider the problem of acyclic flow past an arbitrary body. Such a motion can be represented by a surface distribution of sources or doublets. This can be shown by the use of Green's theorem (2.141).

According to this theorem if ϕ_1 and ϕ_2 are two scalar functions of position we have

$$\oiint_{S_0} (\phi_1 \operatorname{grad} \phi_2 - \phi_2 \operatorname{grad} \phi_1) \cdot \mathbf{n} \, dS = \iiint_{R_0} (\phi_1 \nabla^2 \phi_2 - \phi_2 \nabla^2 \phi_1) \, d\tau$$

$$(11.82)$$

where R_0 is a region enclosed by the surface S_0. Let us set

$$\phi_1 = \frac{1}{r}$$

and

$$\phi_2 = \phi$$

where r is the distance from a fixed point P to another point and ϕ is a harmonic function. Then (11.82) becomes

$$\oiint_{S_0} \left(\frac{1}{r} \operatorname{grad} \phi - \phi \operatorname{grad} \frac{1}{r} \right) \cdot \mathbf{n} \, dS = -\iiint_{R_0} \phi \nabla^2 \left(\frac{1}{r} \right) d\tau \quad (11.83)$$

Suppose that the point P is external to S_0. Then $\nabla^2(1/r)$ vanishes at all points of R_0 and (11.83) reduces to

$$\oiint_{S_0} \left(\frac{1}{r} \operatorname{grad} \phi - \phi \operatorname{grad} \frac{1}{r} \right) \cdot \mathbf{n} \, dS = 0 \quad (11.84)$$

Suppose now that the point is in the region R_0. Since $\nabla^2(1/r)$ becomes infinite at r equal to zero, it is necessary to exclude this point from the region to which (11.83) applies. We draw a small sphere of radius ε with center at P and apply (11.83) to the region between ε and S_0 (Fig. 11.20). We obtain

$$-\oiint_{\text{sphere } \varepsilon} \left(\frac{1}{r} \frac{\partial \phi}{\partial r} + \frac{\phi}{r^2} \right) dS + \oiint_{S_0} \left(\frac{1}{r} \operatorname{grad} \phi - \phi \operatorname{grad} \frac{1}{r} \right) \cdot \mathbf{n} \, dS = 0$$

Now let ε go to zero. Noting that dS on the sphere is ε^2 times the solid angle we obtain

$$\phi(P) = \frac{1}{4\pi} \oiint_{S_0} \left(\frac{1}{r} \operatorname{grad} \phi - \phi \operatorname{grad} \frac{1}{r} \right) \cdot \mathbf{n} \, dS \quad (11.85)$$

where $\phi(P)$ is the value of ϕ at the point P. *This gives the value of ϕ at any point P in terms of the values of ϕ and $\partial\phi/\partial n$ on the boundary.*

Now, the source and doublet solutions of

$$\nabla^2 \phi = 0$$

are given respectively by

$$\phi_s = -\frac{1}{4\pi r}$$

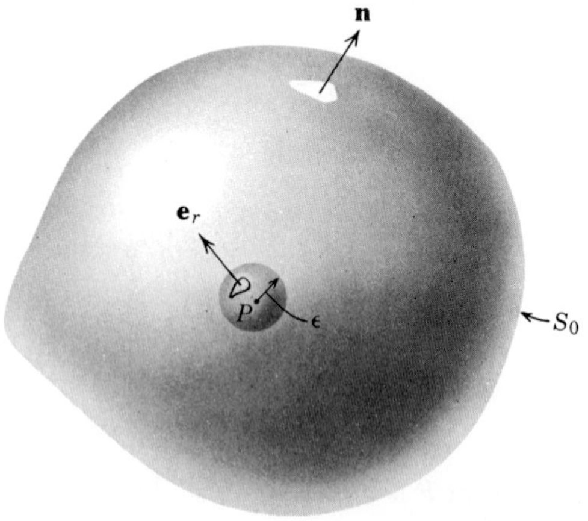

Fig. 11.20 Illustrating the application of Green's theorem to show that the flow past an arbitrary body may be represented by surface distribution of singularities.

and

$$\phi_d = -\mathbf{e}_\mu \cdot \mathrm{grad}\ \phi_s$$

where $\mathbf{e}_\mu$ denotes the axis of the doublet [see (11.72)]. With these relations Eq. (11.85) may be expressed as

$$\phi(P) = -\oiint_{S_0} \phi_s\ \mathrm{grad}\ \phi \cdot \mathbf{n}\ dS + \oiint_{S_0} \phi(\mathbf{n} \cdot \mathrm{grad}\ \phi_s)\ dS$$

This shows that ϕ at P is that due to a surface distribution of sources and doublets. The density of the sources is $-\mathrm{grad}\ \phi \cdot \mathbf{n}$ per unit area. The density of the doublets is ϕ per unit area. The doublet axes are along the inward normals to the surface.

The results (11.84) and (11.85) can be readily extended to the disturbance potential ϕ of the irrotational motion in the region R exterior to a solid body. For this purpose we first apply the results to the region between the surface S of the body and an arbitrary surface Σ enclosing S and then let Σ to to infinity. The integrals over Σ go to zero as Σ goes to infinity. In this way we obtain

$$\frac{1}{4\pi} \oiint_S \left(\frac{1}{r} \operatorname{grad} \phi - \phi \operatorname{grad} \frac{1}{r} \right) \cdot \mathbf{n} \, dS$$

$$= -\phi(P) \quad \text{if} \quad P \text{ is in } R$$
$$= 0 \qquad\quad \text{if} \quad P \text{ is outside } R$$

$$(11.86)$$

Note that $\mathbf{n}$ is the outward normal to S. The statements made in the preceding paragraph therefore apply also to this case.

The distribution expressed by (11.86) can, further, be replaced by a surface distribution of sources only or of doublets only. Let ϕ_1 be harmonic in the region R_1 enclosed by the surface S. Say ϕ is the disturbance potential under consideration. Now if P is a point in R we have

$$\phi(P) = -\frac{1}{4\pi} \oiint_S \left(\frac{1}{r} \operatorname{grad} \phi - \phi \operatorname{grad} \frac{1}{r} \right) \cdot \mathbf{n} \, dS$$

and

$$0 = \frac{1}{4\pi} \oiint_S \left(\frac{1}{r} \operatorname{grad} \phi_1 - \phi_1 \operatorname{grad} \frac{1}{r} \right) \cdot \mathbf{n} \, dS$$

Adding these two we obtain

$$\phi(P) = -\frac{1}{4\pi} \oiint_S \frac{1}{r} (\operatorname{grad} \phi - \operatorname{grad} \phi_1) \cdot \mathbf{n} \, dS$$

$$+ \frac{1}{4\pi} \oiint_S (\phi - \phi_1) \operatorname{grad} \frac{1}{r} \cdot \mathbf{n} \, dS$$

This again expresses that ϕ at P is that due to a surface distribution of sources and doublets. The density of the sources and doublets is however different from that in (11.86). This means the representation of $\phi(P)$ by a surface distribution of sources and doublets is not unique. This is in line with the uniqueness theorems. According to these theorems we know that ϕ is uniquely determined only when either ϕ or $\partial\phi/\partial n$ is prescribed on the boundary of the region in which ϕ is harmonic.

The function ϕ_1 is to be determined by prescribing either $\partial\phi_1/\partial n$ *or* ϕ_1 on the surface S. Suppose that we require $\phi_1 = \phi$ on S. Then the above equation reduces to

$$\phi(P) = -\frac{1}{4\pi} \oiint_S \frac{1}{r} (\operatorname{grad} \phi - \operatorname{grad} \phi_1) \cdot \mathbf{n} \, dS \qquad (11.87a)$$

This shows that ϕ at P is that due to a surface distribution of sources only of strength

$$(\text{grad } \phi - \text{grad } \phi_1) \cdot \mathbf{n}$$

per unit area.

Alternatively, if we require that

$$\text{grad } \phi \cdot \mathbf{n} = \text{grad } \phi_1 \cdot \mathbf{n}$$

on S we shall have

$$\phi(P) = \frac{1}{4\pi} \oiint (\phi - \phi_1) \text{ grad } \frac{1}{r} \cdot \mathbf{n} \, dS \qquad (11.87b)$$

This shows that $\phi(P)$ is that due to a surface distribution of doublets only. The representation (11.87) is unique.

We thus conclude that the acyclic flow past an arbitrary body may be represented by a distribution of sources alone or doublets alone on the surface of the body. Hess and Smith (1962) gave a method of computing with the aid of an electronic computer such a flow using a surface distribution of sources alone. For details and specific examples they may be consulted.

11.12   Pressures

We shall now describe the theoretically calculated pressure distribution over a sphere in a steady flow and over certain slender bodies of revolution in axial flow. Once the velocity potential has been determined, the pressure field is given by the Bernoulli equation

$$p + \tfrac{1}{2}\rho V^2 = H \quad \text{a constant}$$

where

$$V^2 = (\text{grad } \Phi)^2$$

The constant H can be evaluated by means of the *free stream* conditions, that is, the conditions in the undisturbed stream at infinity, or by means of the stagnation conditions, that is, those existing at a stagnation point. Thus if p_∞ and V_∞ denote the free stream pressure and velocity, and if p_s denotes the pressure at the stagnation point, we have

$$H = p_\infty + \tfrac{1}{2}\rho V_\infty^{\,2} = p_s$$

The pressure at any point is then given by

$$p = p_s - \tfrac{1}{2}\rho V^2$$

or, equivalently, by

$$p = p_\infty + \tfrac{1}{2}\rho V_\infty^{\,2}\left(1 - \frac{V^2}{V_\infty^{\,2}}\right) \qquad (11.88)$$

It is customary to express the pressure, particularly on the surface of a body, by means of a pressure coefficient defined as

$$C_p \equiv \frac{p - p_\infty}{\frac{1}{2}\rho V_\infty{}^2} \tag{11.89}$$

Hence, from Eq. (11.88), it follows that the pressure coefficient at any point is given by

$$C_p = 1 - \frac{V^2}{V_\infty{}^2} \tag{11.90}$$

Thus C_p at a stagnation point is 1.

   Sphere.   We consider first the case of a sphere of radius a immersed in an originally uniform stream **U**. The pressure distribution over the surface of the sphere, according to (11.90), is given by the coefficient

$$C_p(a, \theta, \varphi) = 1 - \frac{V^2(a, \theta, \varphi)}{U^2}$$

$$= 1 - \frac{u_R{}^2 + u_\theta{}^2}{U^2}$$

where a, θ, φ are the spherical coordinates of a point on the sphere. Over the sphere, from (11.81), we have

$$u_R = 0$$

and

$$u_\theta = -\tfrac{3}{2}U \sin \theta$$

Consequently, the pressure distribution over the sphere is given by

$$C_p(a, \theta, \varphi) = 1 - \tfrac{9}{4} \sin^2 \theta \tag{11.91}$$

This distribution, in any axial plane $\varphi = \text{const.}$, is shown in Fig. 11.21, where the angle β is measured from the forward stagnation point. Since the distribution is symmetrical with respect to β, the distribution shown in the figure applies equally to either the top or bottom half of the sphere. In the same figure some measured pressure distributions are also shown; we shall talk about them a little later.

   Since the theoretical pressure distribution over any meridian section of the sphere is symmetrical about the diameter AB and also about the diameter CD, it follows that the theoretical value for the force on the sphere is zero. Also, since all the pressure forces pass through the center

                                                              Ideal-Fluid Aerodynamics

of the sphere, the moment on the sphere is zero. This is in line with the considerations given in Chapter 10.

Slender Body of Revolution. We now consider the case of a slender body of revolution in axial flow obtained by combining a specified line distribution of sources and sinks with a uniform stream. The pressure distribution over the bodies shown in Fig. 11.11 were calculated by Fuhrmann, and his results are reproduced in Fig. 11.22. In the same

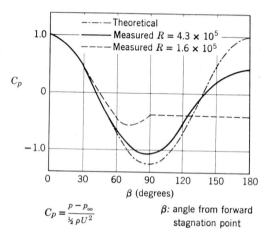

$$C_p = \frac{p - p_\infty}{\frac{1}{2}\rho U^2}$$

β: angle from forward stagnation point

Fig. 11.21 Pressure distribution over the sphere fixed in an originally uniform stream.

figure experimentally measured pressures are also shown; we shall return to these later. The force can be obtained by integration of the pressure forces over the body. Such calculations show that the force on each of the bodies considered is zero.

11.13 Discussion

Let us now examine the theoretical results we have obtained in light of experimental observations.

We first consider the case of the sphere. In Fig. 11.21 we have included the results of pressure measurements made at two values of the Reynolds number $\rho U d/\mu$, where d is the diameter of the sphere and μ is the viscosity of the fluid. We notice that the theoretical distribution differs from the measured one, which itself depends characteristically on the value of the Reynolds number. Over the front part of the sphere (i.e., in the neighborhood of the forward stagnation point A) there is a fair agreement, particularly at the larger Reynolds number, between the measured and theoretical results, but they differ drastically over the rear of the sphere.

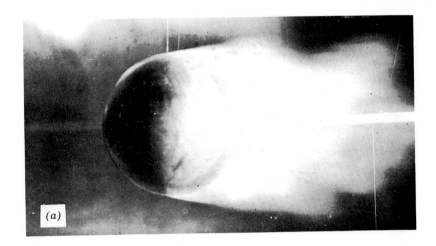

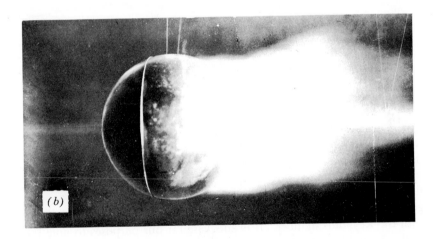

Plate 10 Flow past a sphere illustrating: (*a*) laminar separation; (*b*) turbulent separation. Courtesy of Professor O. G. Tietjens. Plate 14 of Prandtl and Tietjens (1934).

The theoretical pressure decreases from its stagnation value at A to a minimum value at C (where the velocity is a maximum) and rises again steeply to the stagnation value at the rear stagnation B. The measured pressure does not exhibit such recovery over the rear of the sphere. This means that the ideal fluid of our theory, in which effects of friction are completely absent, negotiates the rising pressure over the rear part of the

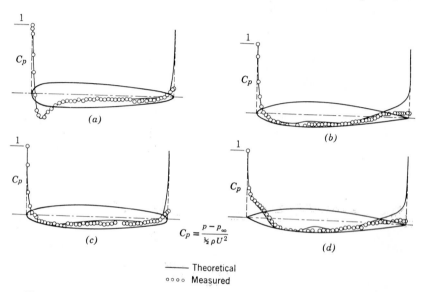

$$C_p = \frac{p - p_\infty}{\frac{1}{2}\rho U^2}$$

——— Theoretical
∘∘∘∘ Measured

Fig. 11.22 Pressure distribution over bodies described in Fig. 11.11.

sphere, whereas the actual fluid endowed with friction, no matter how small, fails to do so. The theoretical flow field looks as shown in Fig. 11-16 and does not depend on any parameter like the Reynolds number. On the other hand, the actual flow field of a real fluid depends on the role of fluid friction and thus on the Reynolds number. It is beyond our intent to go into the details of the variation of the actual flow field past a sphere with the Reynolds number. It is adequate for our purposes to note that at the Reynolds numbers that interest us, the actual flow past the sphere separates from the body (see Plate 10). The point of separation depends on the nature of the boundary layer on the forward part of the sphere. With a laminar layer separation occurs in the front part of the sphere, whereas with a turbulent layer it occurs in the rear. This explains why the measured pressure distributions differ between themselves. We thus see that the actual flow field past the sphere consists of separated flow in contrast to the unseparated flow predicted by the theory. Agreement between the two flow fields on the whole is better at the larger Reynolds number for which the boundary layer is turbulent and separation is removed to the rear end of the sphere.

From the measured pressures it follows that the sphere in the actual flow experiences considerable drag in contrast with the zero drag predicted by the theory.

On the basis of the foregoing discussion we conclude that the theoretical model we have employed for the flow past a sphere is indeed a poor approximation to the real flow. It should be replaced by a more realistic model based on a physical understanding of the actual flow. Further consideration of this matter is outside our present interest.

We now turn to the slender body of revolution in axial flow. Referring to Fig. 11.22, we see that in this case there is very good agreement between the theoretical and measured pressures over the whole body except right at the rear end. At the rear end the theoretical pressure steeply rises to the stagnation value, while actually such a rise does not take place. This deviation at the rear end is again attributable to the effects of friction. From the measured distribution we learn that there is a drag force acting on the body but that it is very small. This may be interpreted, to a certain extent, as a confirmation of the theoretical result that the drag is zero. It thus appears that the theoretical model we have adopted for the axial flow past a slender body of revolution is a good approximation to the actual flow.

From the point of view of the actual flow of a fluid past a body, the sphere and the body of revolution represent two different types of bodies. The sphere is an example of a bluff body for which separation of the flow is an important feature (see Section 1.9). Flow separation prevents the rise of pressure one expects, on the basis of ideal fluid theory, over the rear of the body and gives rise to a considerable drag force on the body. The drag depends on the position of separation and becomes smaller when the separation occurs nearer the rear of the body. In contrast to the sphere, the body of revolution in a uniform axial flow is an example of a streamlined body for which flow separation, if it occurs at all, does so very near the rear of the body (see Section 1.9). In this case the drag of the body is exceedingly small.

It should be borne in mind that the drag we have been talking about is that due to the normal pressures acting on the body. It is called *pressure* or *form drag* and is quite distinct from the skin friction drag arising out of frictional stresses acting tangentially on the body. Skin friction drag does not exist in the flow of an ideal fluid. For bluff bodies friction drag is small compared to the pressure drag, whereas for streamlined bodies the pressure drag is small compared to the friction drag.

11.14 Force on an Arbitrary Body: d'Alembert's Paradox

We consider, as we have been doing, steady acyclic irrotational motion past a fixed rigid body. On the basis of the considerations given in Chapter 10 (see Sections 10.5, 10.9, 10.10), the force and moment exerted by the fluid on the body are given by

$$\mathbf{F} = 0$$

$$\mathbf{M} = -\mathbf{U} \times \oiint_S \rho\phi\mathbf{n}\, dS$$

where ϕ is the disturbance potential. The result that the force on an arbitrary body is zero, as mentioned before, is known as *d'Alembert's Paradox*, after d'Alembert (1717–1783). The result implies that the so-called *drag force* on the body due to fluid resistance is zero. D'Alembert was the first man to attack the problem of fluid resistance by means of a rational theory and then meet with the unforeseen result of zero drag.* He wondered how one could explain by theory the resistance of fluids when a carefully laid out mathematical theory led him to such a paradoxical result.

Before the time of flying people were preoccupied with the problem of drag on a body moving uniformly through a fluid that is otherwise undisturbed. At that time d'Alembert's paradox generally meant the theoretical result of zero drag. What surprises us is not so much the zero drag force furnished by the ideal fluid theory developed so far but the zero *lift force* it predicts for *lifting bodies* such as the wings of an airplane. The absence of drag *for the type of motion we have considered* is understandable. If there is drag, work should be continually done by external forces acting on the body in order to maintain its motion. The work done should then rest in the fluid. It should appear as an increase in the internal energy of the fluid or as an increase in the kinetic energy of the fluid. In the latter case, energy should continually flow to infinity. Either of these possibilities is impossible in the motion of an ideal fluid *under the conditions we have envisaged so far*. We know that the internal energy of an ideal fluid is constant,† and we have seen that the disturbance velocity of the fluid due to the body dies out so rapidly with increasing distance from the body that there can be no flow of energy to infinity. With regard to the question of lift, it will be our main concern hereafter to understand the physical circumstances more clearly and develop the theory further so that it can furnish us with useful results.

---

* For additional interesting reading on this and related topics see Durand (1934).
† In fact, it does not appear in our considerations.

It is important to bear in mind the assumptions underlying d'Alembert's result. They are

1. The fluid is ideal, that is, inviscid and incompressible.
2. The fluid is unlimited, and the body is completely immersed in the fluid.
3. The body is moving uniformly, that is, with a constant velocity.
4. The motion is irrotational.
5. The circulation around every closed circuit is zero and consequently the velocity potential is single valued.

We shall now look at the significance of some of these assumptions. If the fluid is viscous, friction comes into play and the present theory is out of place. If the fluid is nonviscous but compressible, further investigation is necessary to decide about the applicability of d'Alembert's result. The theory of compressible inviscid fluid flow tells us that d'Alembert's paradox holds if the motion is completely subsonic and the assumptions (2) to (5) are made.

D'Alembert's result will not apply if the fluid is limited in extent with either a free surface or a solid boundary too near the uniformly moving body. It is known that a body submerged to an insufficient depth in a fluid with a free surface, such as water, and moving uniformly experiences a drag. This drag (known as *wave drag*) is a consequence of a system of waves that appear on the free surface and continually remove energy to infinity.

When a body moves nonuniformly (i.e., accelerates) through an ideal fluid, the fluid as we had seen exerts a force on the body.

D'Alembert's result, which is based on irrotational motion, has no significance for the rotational motion of an ideal fluid.

The assumption that the fluid motion is acyclic and the velocity potential is single valued has a very important consequence. *As we can surmise from the considerations of Section* 10.5 *it is this assumption that is responsible for the prediction of zero force on a body in a uniform stream.*

We emphasize that d'Alembert's result applies only to the force on the body. Even though the resultant of the pressure forces on the body turns out to be zero, the resultant of their moments (with respect so some reference point) generally need not vanish. Thus an arbitrary closed body executing a uniform motion under the conditions we are considering usually experiences a moment even though the force on it is zero.

11.15   Circulation as the Agency for Force

One of the principal aims of the present study of ideal fluid theory is to determine theoretically the fluid force acting on bodies of aerodynamic interest such as the wings of an airplane. But in light of d'Alembert's paradox it appears that any attempts to achieve such an aim may prove fruitless. To assure ourselves that the situation is not hopeless we shall now examine the steps that need to be taken to resolve d'Alembert's paradox and subsequently develop a satisfactory ideal fluid theory for the flow past a body such as an airplane wing.

Consider the steady flow past a fixed rigid body and *assume that the motion is cyclic. This means we assume that the circulation around circuits enclosing the body is not zero and that consequently the velocity potential is not singlevalued.* To fix ideas we may suppose that the region exterior to the body when it is finite is *somehow* made into a doubly connected region. In such a case, on the basis of the considerations given in Section 10.5, we conclude that the force on the body does not vanish and is in fact given by [see (10.70)]

$$\mathbf{F} = \rho \mathbf{U} \times \oiint_S \mathbf{n} \times \mathbf{q} \, dS \tag{11.92}$$

where $\mathbf{q}$ is the disturbance velocity and S, as usual, is the surface of the body. It was shown in Section 10.5 that if $\mathbf{e}$ is any fixed direction [see (10.75)], then

$$\mathbf{e} \cdot \oiint_S \mathbf{n} \times \mathbf{q} \, dS = \int_{h_1}^{h_2} \Gamma_e(h) \, dh$$

where Γ_e is the circulation around a certain circuit drawn on the body. Note that since

$$\oiint_S \mathbf{n} \times \mathbf{U} \, dS = 0$$

we have

$$\oiint_S \mathbf{n} \times \mathbf{q} \, dS = \oiint_S \mathbf{n} \times \mathbf{V} \, dS \tag{11.93}$$

where $\mathbf{V}$ is the total fluid velocity given by

$$\mathbf{V} = \mathbf{U} + \mathbf{q}$$

To illustrate explicitly the relation between the circulation and the force, let us consider the two-dimensional flow past an infinite cylinder with its generators normal to the free stream. A plane of the motion is shown in Fig. 11.23. The vectors $\mathbf{U}$, $\mathbf{n}$, and $\mathbf{V}$ all now lie in such a plane. Furthermore, since $\mathbf{V}$ must be tangential to the body surface, $\mathbf{n}$ and $\mathbf{V}$ are normal. It therefore follows that

$$\oint\!\!\!\oint_S \mathbf{n} \times \mathbf{q} \, dS = \oint\!\!\!\oint_S \mathbf{n} \times \mathbf{V} \, dS = \mathbf{e} \oint\!\!\!\oint_S V \, dS$$

where $\mathbf{e}$ is normal to the cross section of the cylinder in the direction of $\mathbf{n} \times \mathbf{V}$. Since $\mathbf{V}$ does not vary along the length of the cylinder we choose a *unit* length of the cylinder and write

$$dS = dl \times 1$$

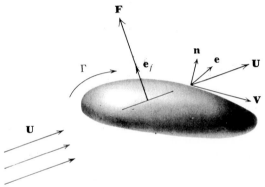

Fig. 11.23 Illustrating the relation between the circulation and the force for two-dimensional flow past an infinite cylinder.

where dl is an element of length along the contour of the cylinder. Denote by C this contour. Then, *per unit length* of the cylinder, we obtain

$$\mathbf{e} \oint\!\!\!\oint_S V \, dS = \mathbf{e} \oint_C V \, dl = \mathbf{e}\Gamma \tag{11.94}$$

where Γ denotes the circulation around C in the sense of right-hand rotation about $\mathbf{e}$. Substituting (11.94) in (11.92), we obtain the *force on the cylinder* as

$$\mathbf{F} = \rho \mathbf{U} \times \mathbf{e}\Gamma$$

per unit length. Since $|\mathbf{U} \times \mathbf{e}| = U$, and the direction of $\mathbf{U} \times \mathbf{e}$ is a unit vector $\mathbf{e}_f$ normal to $\mathbf{U}$ and $\mathbf{e}$ as shown in the figure, the *force becomes*

$$\mathbf{F} = \rho U\Gamma \mathbf{e}_f$$

per unit length. We speak of this force as the lift on the cylinder. There is no drag.

We thus see that if there is a circulation Γ around an infinite cylinder of arbitrary cross section placed in a uniform stream $\mathbf{U}$, the cylinder will experience a lift force

$$L = \rho U \Gamma \qquad (11.95)$$

per unit length (or span). This result is known as the *Kutta-Joukowski theorem*. We shall return to this later.

We have thus come to recognize that a nonzero circulation around a body in a uniform stream is essential if a force were to act on the body. This means that the velocity potential must be multivalued. From what we know so far, a multivalued potential is possible only if the region exterior to the body is multiply connected. Now the region outside a finite three-dimensional body is certainly not a multiply connected region. How can the potential in that region become multivalued, and how can we explain the lift force on certain finite lifting bodies such as wings? Our theory in its present state is of no help in resolving these difficulties. Any clues for straightening out the situation must come from a physical understanding of the nature of the flow over lifting bodies. To gain such understanding, a convenient place to start is with the infinitely long lifting body, the so-called infinite or two-dimensional wing, in a uniform stream.

In contrast with the three-dimensional situation, the region outside an infinite cylinder is doubly connected, and the potential, theoretically at least, can be multivalued. Then the circulation in any circuit enclosing the cylinder need not be zero. Now, however, the difficulty is that, given a cylinder in a certain stream, we have no way with our present theory to know whether or not there is circulation around the body; and if there is, how can one determine its magnitude? Again, clues to resolve these difficulties must come from a physical understanding of the flow over a lifting body.

Our task for the next few chapters is then clear. First we shall take up the problem of formulating the two-dimensional wing theory and analyzing some of the problems associated with it. We will then be in an advantageous position to take up the formulation of the three-dimensional wing theory and the analysis of some problems associated with it. As a first step in developing the analytical framework for the two-dimensional wing theory we shall consider in the next chapter steady acyclic two-dimensional motion.

Chapter 12

Steady Two-Dimensional Acyclic Motion

When the motion takes place in a series of planes parallel to a given plane and is the same in each of these planes, we speak of *plane* or *two-dimensional motion*. The velocity, the pressure, and other quantities related to the flow are equal at corresponding points of the planes. They are thus independent of a space coordinate measured along an axis normal to the planes. We shall denote this axis by Z and the corresponding coordinate by z. The velocity component w along Z is zero. The streamlines of the flow are plane curves and lie in parallel planes normal to the Z-axis.

Steady plane motion is possible only in the case of an *infinitely long cylindrical body* placed in a uniform stream with its generators normal to the stream. Such a body is known as a *two-dimensional body*. Physically there are no exact examples of two-dimensional flow, only situations where the motion can be considered a good approximation to two-dimensional flow. Studies of two-dimensional flows are, however, important. They contribute to our understanding of the nature of fluid motion. From the mathematical point of view, two-dimensional motion involves two independent variables in the governing equations, and this is a great help. It is particularly amenable, as we shall see, to mathematical analysis.

12.1 Recapitulation

We shall gather here for convenience some pertinent relations that have been introduced at several places in the preceding chapters. We choose the $z = 0$ plane as the representative plane of motion (Fig. 12.1). In analyzing two-dimensional flows we often use cylindrical polar coordinates besides Cartesians. Hence we shall record both the Cartesian and polar forms of the relevant terms and equations.

In analyzing two-dimensional motion we can work in terms of either the velocity potential, or the stream function, or both, as will be shown in Chapter 15. In this chapter we shall obtain the stream function and the potential for some simple flows.

First we use Cartesians x, y, z. The velocity $\mathbf{V}$, the potential Φ, the stream function ψ, and all other quantities are functions of x and y only.

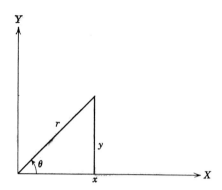

Fig. 12.1 Coordinates for two-dimensional flow.

We thus have

$$\mathbf{V} = [u(x, y), v(x, y), 0]$$

$$\Phi = \Phi(x, y)$$

$$\psi = \psi(x, y)$$

$$\frac{\partial \Phi}{\partial x} = u = \frac{\partial \psi}{\partial y}$$

$$\frac{\partial \Phi}{\partial y} = v = -\frac{\partial \psi}{\partial x}$$

$$\nabla^2 \Phi = \frac{\partial^2 \Phi}{\partial x^2} + \frac{\partial^2 \Phi}{\partial y^2} = 0$$

$$\nabla^2 \psi = \frac{\partial^2 \psi}{\partial x^2} + \frac{\partial^2 \psi}{\partial y^2} = 0$$

In terms of the cylindrical coordinates r, θ, z, we have

$$\mathbf{V} = [u_r(r, \theta), u_\theta(r, \theta), 0]$$

$$\Phi = \Phi(r, \theta)$$

$$\psi = \psi(r, \theta)$$

$$\frac{\partial \Phi}{\partial r} = u_r = \frac{1}{r} \frac{\partial \psi}{\partial \theta}$$

$$\frac{1}{r} \frac{\partial \Phi}{\partial \theta} = u_\theta = -\frac{\partial \psi}{\partial r}$$

$$\nabla^2 \Phi = \frac{1}{r} \frac{\partial}{\partial r}\left(r \frac{\partial \Phi}{\partial r}\right) + \frac{1}{r} \frac{\partial}{\partial \theta}\left(\frac{1}{r} \frac{\partial \Phi}{\partial \theta}\right) = 0$$

$$\nabla^2 \psi = \frac{1}{r} \frac{\partial}{\partial r}\left(r \frac{\partial \psi}{\partial r}\right) + \frac{1}{r} \frac{\partial}{\partial \theta}\left(\frac{1}{r} \frac{\partial \psi}{\partial \theta}\right) = 0$$

12.2 Further Considerations Relating to the Stream Function

We recall that along any streamline

$$\psi = C \quad \text{a constant} \tag{12.1}$$

or, equivalently

$$d\psi = \text{grad } \psi \cdot \mathbf{ds} = 0 \tag{12.2}$$

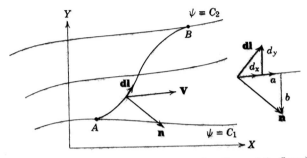

Fig. 12.2 Illustrating the relation between the stream function and the flow through an arbitrary curve.

In Cartesians this takes the form

$$\frac{\partial \psi}{\partial x} dx + \frac{\partial \psi}{\partial y} dy = -v\, dx + u\, dy = 0 \tag{12.3}$$

In polar coordinates we have

$$\frac{\partial \psi}{\partial r} dr + \frac{\partial \psi}{\partial \theta} d\theta = -u_\theta\, dr + ru_r\, d\theta = 0 \tag{12.4}$$

Consider any two streamlines $\psi = C_1$ and $\psi = C_2$ and an arbitrary curve AB joining them (Fig. 12.2). Denote by $\mathbf{dl}$ an element of length along the curve and by $\mathbf{n}$ the normal to that element. The net out flow of fluid mass through AB, measured per unit thickness normal to the plane of motion, is given by

$$m = \rho \int_{A \text{ via } AB}^{B} \mathbf{V} \cdot \mathbf{n}\, dl \tag{12.5}$$

Now, if in Cartesians,

$$\mathbf{dl} = (dx, dy)$$

and

$$\mathbf{n} = (a, b)$$

we then have

$$\mathbf{dl} \cdot \mathbf{n} = a\, dx + b\, dy = 0$$

and

$$|\mathbf{n}| = (a^2 + b^2)^{\frac{1}{2}} = 1$$

From these equations it follows that

$$\mathbf{n} = \left(\frac{dy}{dl}, \ -\frac{dx}{dl} \right) \tag{12.6}$$

and

$$\mathbf{V} \cdot \mathbf{n} \, dl = u \, dy - v \, dx \tag{12.7}$$

Substituting (12.7) into (12.5) we obtain

$$m = \rho \int_{A \text{ via } AB}^{B} (u \, dy - v \, dx)$$

In terms of the stream function this becomes

$$\frac{m}{\rho} = \int_{A}^{B} \left(\frac{\partial \psi}{\partial y} \, dy + \frac{\partial \psi}{\partial x} \, dx \right)$$

$$= \int_{A}^{B} d\psi$$

$$= \psi(B) - \psi(A)$$

$$= C_2 - C_1 \tag{12.8}$$

This shows that *the difference between the numerical values of the stream functions is equal to the volume rate of fluid flowing between them.*

Consider now a closed circuit $\mathscr{C}$ drawn in the region of the flow. The net outflow of fluid mass through $\mathscr{C}$ is then given by

$$m = \rho \int_{\mathscr{C}} \mathbf{V} \cdot \mathbf{n} \, dl$$

$$= \rho \int_{\mathscr{C}} d\psi$$

$$= \lim_{B \to A} [\psi(B) - \psi(A)]_{\text{via } \mathscr{C}}$$

where A and B are adjacent points on $\mathscr{C}$. If there are no sources of fluid in the region enclosed by the curve, the net outflow of mass through the curve is zero, and consequently the integral $\int d\psi$ around $\mathscr{C}$ is also zero. This means ψ *is a singlevalued function in any region in which there are no sources. It will be multivalued along circuits enclosing sources.*

In the plane of motion we can draw two sets of curves—one set described by the equation $\Phi = $ const., the so-called *equipotential lines,* and the other set described by the equation $\psi = $ const., that is, the streamlines. At any point of the flow field the gradient of the potential points in the direction of the velocity vector. Also, the direction of the streamline passing through that point is that of the velocity vector. Since grad Φ at any point is normal to the equipotential line passing through the point, it follows that the equipotential line and the streamline passing through any point are normal to each other at that point (Fig. 12.3). We thus see that *the equipotential lines* (Φ = const.) and *the streamlines* (ψ = const.) *form an orthogonal net.*

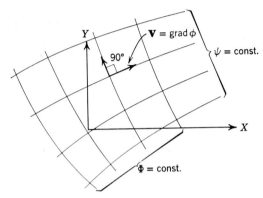

Fig. 12.3   Streamlines and equipotential lines are orthogonal.

12.3   Problem in Terms of the Stream Function

Consider steady flow past a cylinder. The problem of determining the flow field in terms of the stream function consists of solving the equation

$$\nabla^2 \psi = 0$$

in the region exterior to the cylinder such that ψ or its derivatives satisfy certain specified conditions on the contour of the body and at infinity. Since the contour of the body is a streamline, it follows that on that contour ψ must be a constant. Since the velocity at infinity must approach the undisturbed velocity, it follows that at infinity the spatial derivatives of ψ should assume the corresponding components of the undisturbed velocity. Thus, in Cartesians, the problem consists of determining $\psi(x, y)$ as the solution of the equation

$$\frac{\partial^2 \psi}{\partial x^2} + \frac{\partial^2 \psi}{\partial y^2} = 0$$

such that

$$\psi(x, y) = \text{a constant} \quad \text{on} \quad F(x, y) = 0$$

and that at infinity

$$\frac{\partial \psi}{\partial y} = \mathbf{U} \cdot \mathbf{i}$$

$$-\frac{\partial \psi}{\partial x} = \mathbf{U} \cdot \mathbf{j}$$

where $F(x, y) = 0$ describes the contour of the body and $\mathbf{U}$ is the free stream.*

As a first step in building up the two-dimensional flow fields past certain bodies, we shall obtain the stream function and the potential for three simple flow fields—uniform stream, a source, and a doublet. Our procedure will be to take the velocity field as given and then to find the stream function and the potential by integration of the velocity components.

12.4   Uniform Stream

Consider a uniform stream with velocity $\mathbf{Q}$. Let U and V denote its components along the X- and Y-axes (Fig. 12.4). Then

$$\frac{\partial \psi}{\partial y} = U = \frac{\partial \Phi}{\partial x}$$

and

$$-\frac{\partial \psi}{\partial x} = V = \frac{\partial \Phi}{\partial y}$$

It therefore follows that

$$\psi(x, y) = -Vx + Uy \tag{12.9}$$

and

$$\Phi(x, y) = Ux + Vy \tag{12.10}$$

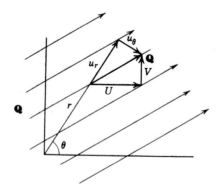

Fig. 12.4   Uniform flow.

* Such a problem is known as *Dirichlet's Problem*.

In terms of the polar coordinates r, θ, these equations take the form

$$\psi(r, \theta) = -Vr \cos \theta + Ur \sin \theta \qquad (12.11)$$

and

$$\Phi(r, \theta) = Ur \cos \theta + Vr \sin \theta \qquad (12.12)$$

12.5 Source Flow

In three-dimensional flow the source flow is such that the streamlines are radial lines in all directions from a point and the velocity is a function only of the distance from the source to a field point. A similar flow in two

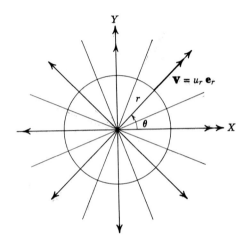

Fig. 12.5 Source flow.

dimensions is described when the streamlines in any plane of motion are all radial lines from a point in that plane and the velocity is a function only of the polar coordinate r measured from that point (Fig. 12.5). The velocity field, with respect to r, θ, and z of such a flow, is then represented by

$$\mathbf{V} = u_r \mathbf{e}_r \qquad (12.13)$$

where $u_r = u_r(r)$ only.

We first see whether this velocity field satisfies the incompressibility and irrotationality conditions. It is readily verified that the curl of $\mathbf{V}$ vanishes everywhere. The incompressibility condition requires that

$$\text{div } \mathbf{V} = 0$$

With Eq. (12.13) this becomes

$$\frac{\partial}{\partial r}(ru_r) = 0$$

or

$$u_r = \frac{B}{r} \quad \text{say}$$

where B is a constant. It therefore follows that the velocity

$$\mathbf{V} = u_r(r)\mathbf{e}_r = \frac{B}{r}\mathbf{e}_r \qquad (12.14)$$

is acceptable at all points except at the point $r = 0$, where it becomes infinite. This point is thus a singular point of the flow. It can further be seen that at this point div $\mathbf{V} \neq 0$ but is in fact infinite. This means the point is a *source*. It is often referred to as a *plane* or *two-dimensional source*.

Let q denote the strength of the source, that is, the volume measured per unit thickness normal to the plane of motion of fluid being created at the source per unit time. The strength q is then also equal to the volume of fluid flowing out of any curve enclosing the source. If, for simplicity, we choose a circle of radius ε with its center at the source we obtain

$$q = \int_{\text{circle }\varepsilon} \mathbf{V} \cdot \mathbf{e}_r \, dl$$

where dl is an element of length along the circle. From (12.14) $\mathbf{V} \cdot \mathbf{e}_r$ on the circle is equal to B/ε. Therefore

$$q = \frac{B}{\varepsilon} \int_{\text{circle }\varepsilon} dl = 2\pi B \qquad (12.15)$$

Now, in terms of the source strength we can rewrite (12.14) as

$$\mathbf{V} = \frac{q}{2\pi} \frac{\mathbf{e}_r}{r} \qquad (12.16)$$

The three-dimensional picture of a plane source is simply a doubly infinite *line source* obtained by a uniform distribution of point (i.e., three-dimensional) sources along a straight line. Hence q is the constant source strength per unit length. If we assume such a distribution along the Z-axis, we can obtain by integration over the distribution the flow field of a two-dimensional source.

To obtain the stream function we note that

$$\frac{\partial \psi}{\partial \theta} = ru_r = \frac{q}{2\pi}$$

and

$$\frac{\partial \psi}{\partial r} = -u_\theta = 0$$

Integration of these equations yields

$$\psi(r, \theta) = \frac{q}{2\pi} \theta + \text{const.} \tag{12.17}$$

To obtain the corresponding potential we note that

$$\frac{\partial \Phi}{\partial r} = u_r = \frac{q}{2\pi} \frac{1}{r}$$

and

$$\frac{\partial \Phi}{\partial \theta} = r u_\theta = 0$$

Integration of these equations yields

$$\Phi(r, \theta) = \frac{q}{2\pi} \log r + \text{const.} \tag{12.18}$$

In Cartesians, the stream and potential functions are

$$\psi(x, y) = \frac{q}{2\pi} \tan^{-1}\left(\frac{y}{x}\right) + \text{constant}$$

$$\Phi(x, y) = \frac{q}{4\pi} \log (x^2 + y^2) + \text{constant}$$

12.6 Combination of a Source and a Sink of Equal Strength

Consider a source and a sink each of strength q situated at the points A and B, respectively (Fig. 12.6). Let BC be an axis directed from the sink to the source. If P is any field point, let θ_1 denote the angle PAC and θ_2 the angle PBC.

The stream function at P due to the source at A is given by

$$\psi_1 = \frac{q}{2\pi} \theta_1$$

when the zero streamline is taken as the axis BAC.

The stream function at P due to the sink at B is given by

$$\psi_2 = -\frac{q}{2\pi} \theta_2$$

when the zero streamline is again taken as the axis BAC.

The stream function at P due to the combined flow is then given by

$$\psi = \psi_1 + \psi_2$$

$$= \frac{q}{2\pi}(\theta_1 - \theta_2) = \frac{q}{2\pi}\theta_3 \qquad (12.19)$$

where θ_3 is the angle APB.

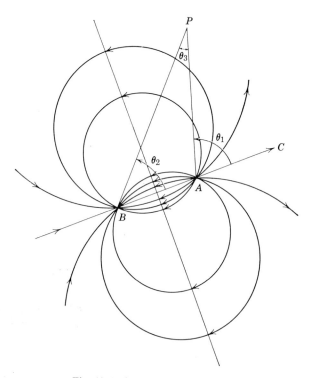

Fig. 12.6 Source-sink combination.

The streamlines are then described by the equation

$$\psi = \frac{q}{2\pi}\theta_3 = \text{constant}$$

or, equivalently, by

$$\theta_3 = \text{constant} \qquad (12.20)$$

The latter is the equation of a circle passing through the point P, the source at A, and the sink at B. Therefore the streamlines are circles all of which pass through the sink and the source (Fig. 12.6). The directions of the streamlines are as shown.

12.7 Doublet

The doublet in two dimensions is defined in the same way as the doublet in three dimensions. Thus if the distance l between a source and a sink of equal strength q is allowed to go to zero such that the product ql remains equal to a constant μ, we obtain a doublet of strength μ. The axis of the doublet is directed from the sink to the source.

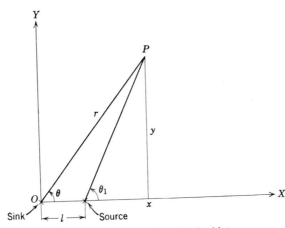

Fig. 12.7 Derivation of a doublet.

We obtain the stream function for the doublet as follows. Choose the sink point as the origin of coordinates and set up temporarily Cartesian axes X, Y such that the X-axis runs from the sink to the source (Fig. 12.7). Let P be any field point with the coordinates r, θ or, equivalently, x, y. Angles θ and θ_1 are measured as shown. The stream function at P due to a doublet situated at O with its axis in the direction of X is then given by

$$\psi(r, \theta) = \lim_{\substack{l \to 0 \\ \text{with } ql=\mu}} \frac{q}{2\pi}(\theta_1 - \theta) \qquad (12.21)$$

To evaluate the limit we observe that for small l we may write

$$\theta_1 - \theta \simeq \frac{l \sin \theta}{r - l \cos \theta}$$

Substituting this in (12.21) we obtain

$$\psi(r, \theta) = \frac{\mu}{2\pi} \frac{\sin \theta}{r}$$

$$\qquad (12.22)$$

$$\psi(x, y) = \frac{\mu}{2\pi} \frac{y}{x^2 + y^2}$$

The streamlines of the doublet flow are, therefore, described by the equation

$$\psi = \frac{\mu}{2\pi} \frac{\sin\theta}{r} = C, \quad \text{a constant}$$

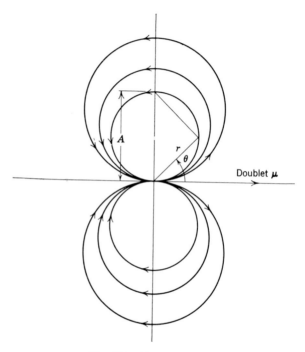

Fig. 12.8 Doublet flow.

This is the equation of a circle whose center is on the axis $\theta = \pi/2$ and whose diameter is given by

$$A = \frac{r}{\sin\theta} = \frac{\mu}{2\pi C}$$

The streamlines are as shown in Fig. 12.8. The potential for the doublet is given by

$$\Phi(r, \theta) = -\frac{2}{2\pi} \frac{\cos\theta}{r}$$

$$\Phi(x, y) = -\frac{\mu}{2\pi} \frac{x}{x^2 + y^2} \tag{12.23}$$

12.8 Source and Sink of Equal Strength in a Uniform Stream

Consider the superposition of a source, a sink each of strength q and a uniform stream U parallel to the direction from the source to the sink. We choose the origin of coordinates midway between the source and sink and the X-axis in the direction from the source to the sink (Fig. 12.9). Thus $U = iU$, the source is at $x = -a$, the sink is at $x = a$.

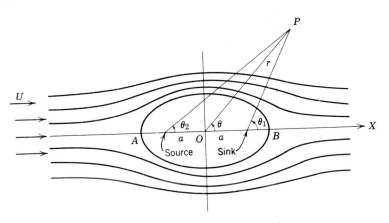

Fig. 12.9 Flow past a Rankine oval.

The stream function at any field point P with coordinates r, θ or, equivalently, x, y is given by

$$\psi = Ur \sin \theta + \frac{q}{2\pi}(\theta_2 - \theta_1)$$

It can be shown* that there are two stagnation points A and B such that

$$OA = OB = \sqrt{a^2 + \frac{qa}{\pi U}}$$

The streamlines are obtained from the equation $\psi = $ constant. The stagnation streamline is described by

$$UR \sin \theta + \frac{q}{2\pi}(\theta_2 - \theta_1) = 0$$

This shows that the whole X-axis except the segment between the source and the sink forms a part of the stagnation streamline. The rest of this line is a closed curve as shown in Fig. 12.9. The curve is symmetrical about X-axis. It is known as an oval.

* The algebraic details are left to the reader as an exercise.

We thus see (as expected) that the superposition we are considering represents the two-dimensional flow past an infinitely long symmetrical cylinder.

This method of source-sink superposition can be extended to represent the flow past symmetrical cylinders of arbitrary shape. Cylinders obtained in this way are called *Rankine ovals*.

12.9   Doublet in Uniform Stream: Flow over a Circular Cylinder

Let us consider now the combination of a doublet μ and a uniform stream U with the axis of the doublet opposing the stream. We choose the

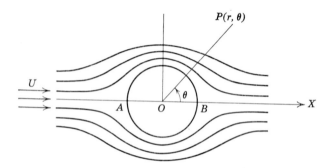

Fig. 12.10   Flow past a circular cylinder.

origin of coordinates at the doublet and the X-axis in the direction of the uniform stream (Fig. 12.10). Thus

$$\mathbf{U} = \mathbf{i}U$$

$$\boldsymbol{\mu} = -\mathbf{i}\mu$$

The stream function at any field point $P(r, \theta)$ due to the combined flow is then given by

$$\psi(r, \theta) = Ur \sin \theta - \frac{\mu}{2\pi} \frac{\sin \theta}{r}$$

$$= U\left(r - \frac{\mu}{2\pi U}\frac{1}{r}\right) \sin \theta \qquad (12.24)$$

The velocity components are

$$u_r(r, \theta) = \frac{1}{r}\frac{\partial \psi}{\partial \theta} = U\left(1 - \frac{\mu}{2\pi U}\frac{1}{r^2}\right) \cos \theta$$

and

$$u_\theta(r, \theta) = -\frac{\partial \psi}{\partial r} = -U\left(1 + \frac{\mu}{2\pi U}\frac{1}{r^2}\right) \sin \theta \qquad (12.25)$$

The stagnation points in the flow are obtained by setting these equations to zero. We thus find that there are two stagnation points A and B given by

$$A : r = \sqrt{\mu/2\pi U}, \quad \theta = \pi$$
$$B : r = \sqrt{\mu/2\pi U}, \quad \theta = 0 \tag{12.26}$$

The streamlines of the flow are described by the equation $\psi = $ constant. The constant for the stagnation streamline turns out to be zero. Hence this streamline is given by

$$\left(1 - \frac{\mu}{2\pi U}\frac{1}{r^2}\right) r \sin \theta = 0 \tag{12.27}$$

or by

$$\sin \theta = 0 \text{ for any } r \tag{12.28}$$

and

$$r = \sqrt{\mu/2\pi U}, \text{ a constant for any } \theta \tag{12.29}$$

Equations (12.28, 29) show that the stagnation streamline consists of the whose X-axis and a circle of radius $\sqrt{\mu/2\pi U}$ with its center at the doublet (see Fig. 12.10).

The surface formed by the stagnation streamlines lying in all the planes of motion is that of an infinitely long *circular* cylinder with its generators normal to those planes. Thus the superposition we are discussing represents the two-dimensional flow past a circular cylinder.*

It follows that if a circular cylinder of radius a is placed in a uniform stream **U**, the disturbance field due to the cylinder is represented by a doublet whose axis opposes the stream and whose strength is

$$\mu = 2\pi U a^2$$

The stream function and the velocity components are then given by

$$\psi(r, \theta) = U\left(1 - \frac{a^2}{r^2}\right) r \sin \theta \tag{12.30}$$

$$u_r(r, \theta) = U\left(1 - \frac{a^2}{r^2}\right)\cos \theta \tag{12.31}$$

$$u_\theta(r, \theta) = -U\left(1 + \frac{a^2}{r^2}\right)\sin \theta \tag{12.32}$$

* The flow is usually referred to as that over a circle. In two-dimensional flow similar nomenclature is generally used.

Let us look at the pressure distribution over the cylinder. In terms of the pressure coefficient, the pressure at any point on the cylinder is given by

$$C_p(a, \theta) = 1 - \frac{V^2(a, \theta)}{U^2} \qquad (12.33)$$

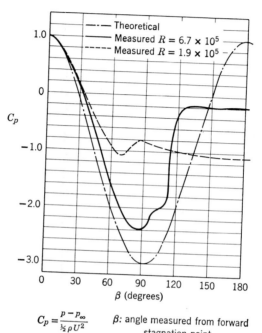

$$C_p = \frac{p - p_\infty}{\frac{1}{2}\rho U^2}$$ β: angle measured from forward stagnation point

Fig. 12.11   Pressure distribution over the circular cylinder.

On the cylinder, $u_r = 0$ and $u_\theta = -2U \sin \theta$ and therefore

$$V^2 = 4U^2 \sin^2 \theta$$

Hence Eq. (12.33) becomes

$$C_p(a, \theta) = 1 - 4 \sin^2 \theta \qquad (12.34)$$

This result is plotted in Fig. 12.11, which also contains measured pressure distributions for two values of the Reynolds number $\rho U d/\mu$, where d is the diameter of the cylinder. As expected, the theoretical distribution is symmetrical about the X- and Y- axes and gives rise to zero force on the cylinder.

As in the case of the sphere, the theoretical and experimental results show some agreement in the neighborhood of the forward stagnation point, but at the rear of the cylinder the discrepancies are considerable. Also notice that the experimental results vary with the Reynolds number. The cylinder, like the sphere, is a bluff body, and at the Reynolds numbers that interest us separation plays a major part in the actual flow over the cylinder (see Section 1.9). Experimentally, the cylinder experiences a considerable drag.

The theory so far developed, therefore, cannot be used to predict the actual flow of a fluid past the cylinder. In spite of this shortcoming the theoretical solution for the circular cylinder, as obtained above, is of considerable interest to us. It will play an important part, as we shall see, in our analysis of a two-dimensional lifting body, the so-called *infinite wing*.

In light of this consideration it is instructive to note that the theoretical result of zero force on the cylinder is again related to the absence of any circulation around it. This in turn is a consequence of the fact that the potential of the flow field is singlevalued.

12.10   Flow Past an Arbitrary Cylinder

We now turn to the motion of a uniform stream past a cylinder of arbitrary cross section. In this case we can attempt to represent the flow field by the superposition of a uniform stream and a certain distribution of plane sources alone or doublets alone along the cross-sectional contour of the cylinder.* The flow field thus obtained is characterized by a single-valued potential and hence zero circulation. Again, the force on the cylinder turns out to be zero.

Because our concern is to develop the theory so that it can predict useful results for the forces on a lifting body, we shall now have to look into flows in which the potential is multivalued and the circulation is not zero.

* See Smith and Pierce (1958).

Chapter 13

Circulation and Lift
for an Infinite Wing in Steady Flow

In this chapter we take up the formulation of the theory of lift of a two-dimensional lifting body, known as the *infinite wing*, fixed in a steady flow. For this purpose we first study flows with circulation. Such flows are called *circulatory flows*. As already pointed out, circulation is essential for lift. We wish to emphasize that the theory developed so far has no means of telling us whether or not there will be circulation around a body placed in a uniform stream.

It is interesting to note that around 1900 when mechanical flight was already realized as possible there was no rational theory to explain and predict the aerodynamic lift obtainable from certain bodies that we call *wings*. We owe to Lanchester, Kutta, and Joukowski the final recognition of the connection between lift of a wing and the circulatory flow around it. Kutta (1867–1944) and Joukowski (1847–1921) independently laid the foundations for a quantitative theory of lift of an infinite wing. For a finite (i.e., a three-dimensional) wing Lanchester (1878–1946) seems to have been the first to contemplate the connection between lift and circulation. However, he did not develop a practical mathematical theory. This was done by Prandtl (1875–1953).

13.1 Circulatory Flow with Constant Vorticity

Our aim is to seek some two-dimensional flows in which the circulation $\Gamma = \oint \mathbf{V} \cdot \mathbf{d}l$ is not zero. With this in mind let us suppose that there is some flow in which the streamlines are closed curves. Then the circulation taken around a streamline may not vanish, for the direction of the streamline at every point is the direction of the velocity, thus making $\mathbf{V} \cdot \mathbf{d}l$ positive all along that line. This suggests that as an initial step in searching for flows with circulation we should look for flows with closed streamlines. As a simple example let us consider a flow in which the *streamlines* are *concentric circles* (Fig. 13.1).

Let us choose the center of these circles as the origin of coordinates. Then with respect to cylindrical coordinates r, θ in the plane, the velocity field is described by

$$\mathbf{V} = u_\theta \mathbf{e}_\theta \tag{13.1}$$

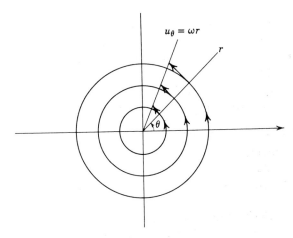

Fig. 13.1 Flow with circular streamlines: flow with constant vorticity.

only. In order for this to represent a physically possible flow field, it should satisfy the incompressibility condition (5.28), which now takes the form

$$\frac{\partial u_\theta}{\partial \theta} = 0$$

Therefore it follows that in (13.1) u_θ should be a function of r only, that is, we have

$$\mathbf{V} = u_\theta(r)\mathbf{e}_\theta \tag{13.2}$$

Let us take

$$u_\theta = Cr$$

or

$$\mathbf{V} = Cr\mathbf{e}_\theta \tag{13.3}$$

where C is a constant. It can be verified that (13.3) is compatible with the equation of motion (5.31).

The circulation around a streamline $r = $ const. is

$$\Gamma = \oint_{\text{circle } r} \mathbf{V} \cdot d\mathbf{l} = \oint Cr^2 \, d\theta = 2\pi Cr^2 \tag{13.4}$$

We thus see that (13.3) represents a physically possible circulatory flow.

We now inquire whether or not this flow is irrotational. The vorticity at any point in the flow is given by

$$\mathbf{\Omega}(r, \theta) = \mathbf{k}\,\frac{1}{r}\frac{dru_\theta}{dr} = \mathbf{k}\,\frac{1}{r}\frac{dCr^2}{dr}$$

$$= 2C\mathbf{k} \quad \text{a constant}$$

This means the velocity field (13.3) represents a rotational flow with uniform vorticity. Since vorticity is twice the angular velocity, it follows that the whole fluid is rotating like a rigid body with a constant angular velocity $\boldsymbol{\omega} = C\mathbf{k}$.

The flow represented by (13.3) is accordingly called *circulatory flow with constant vorticity* or *constant rotation*.*

13.2 Circulatory Flow without Rotation: Vortex Flow

Since our interest is in irrotational motion, we now ask whether or not there is a circulatory flow where the streamlines are circles but the flow field is irrotational. Such a flow is possible and can be determined as follows.

Now, the velocity field has to satisfy both the incompressibility and irrotationality conditions. From the condition of incompressibility we obtain, as before, that

$$\mathbf{V} = u_\theta(r)\mathbf{e}_\theta$$

The irrotationality condition then becomes

$$\mathbf{\Omega} = \mathbf{k}\,\frac{1}{r}\frac{dru_\theta}{dr} = 0 \tag{13.5}$$

It follows therefore that if we set

$$ru_\theta = K \quad \text{a constant}$$

or

$$\mathbf{V} = K\,\frac{\mathbf{e}_\theta}{r} \tag{13.6}$$

the motion would be irrotational except possibly at the point $r = 0$, where the vorticity, according to (13.5), becomes indeterminate and the velocity, according to (13.6), becomes infinite (Fig. 13.2a). In a moment we shall determine the value of the vorticity at that point.

The circulation along any streamline $r = $ const. is given by

$$\Gamma = \oint_{\text{circle } r} \mathbf{V} \cdot \mathbf{d}l = \oint K\,d\theta$$

$$= 2\pi K \quad \text{a constant} \tag{13.7}$$

* Note that because of the presence of vorticity everywhere, the circulation (13.4) varies from streamline to streamline.

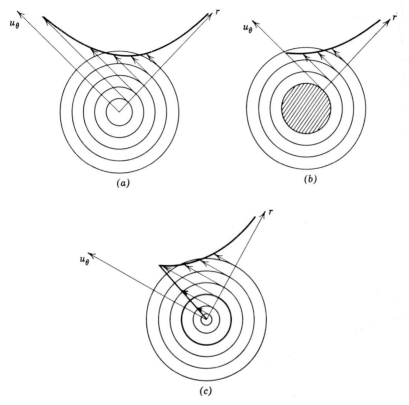

Fig. 13.2   Vortex flow: (a) point vortex; (b) circulating flow around a circular cylinder; (c) eddy.

We thus see that the velocity field given by (13.5) represents a circulatory flow that is irrotational everywhere except possibly at $r = 0$. We call this flow a *circulatory flow without rotation*.

To evaluate the value of the vorticity at the point $r = 0$ we use the relation between vorticity and circulation. According to this we have

$$(\mathbf{\Omega})_{r=0} \cdot \mathbf{k} \equiv (\text{curl } \mathbf{V})_{r=0} \cdot \mathbf{k}$$

$$= \lim_{\Delta S \to 0} \frac{1}{\Delta S} \oint_{C_k} \mathbf{V} \cdot d\mathbf{l} \tag{13.8}$$

where C_k is a small closed curve lying in the plane $\mathbf{k}$ and enclosing the point $r = 0$ and a small area ΔS. Choose C_k as a streamline $r = \varepsilon$. Then Eq. (13.8) becomes

$$(\mathbf{\Omega})_{r=0} = \lim_{\Delta S \to 0} \frac{\Gamma_\varepsilon}{\Delta S}$$

where Ω is the magnitude of the vorticity and Γ_ε is the circulation around the streamline $r = \varepsilon$. Now, from Eq. (13.7) we see that the circulation has the same constant value along all streamlines. Therefore

$$(\Omega)_{r=0} = \lim_{\Delta S \to 0} \frac{\text{constant}}{\Delta S} = \infty \qquad (13.9)$$

We thus see that the flow represented by (13.6) is irrotational everywhere except at the point $r = 0$, where the vorticity is infinite. It is a *circulatory flow with concentrated vorticity*. It is generally known as *vortex flow*. The center point of the vortex flow is called the *point vortex* or simply the *vortex*. It is customary to refer to the velocity at any point of the vortex flow as the *velocity induced by the vortex*. It must be understood that this is simply a matter of convenience and does not mean that the vortex is actually causing the flow, for they just coexist. Because of its significant properties, vortex flow plays an important part in aerodynamics.

Vortex flow can be used to represent approximately certain flows that can be realized physically. When doing this, however, we should remember that the flow near the vortex point cannot be physically true, for at that point both the velocity and the vorticity are infinite. To avoid this difficulty there are two possibilities. One possibility is to arrange the vortex point to be a fictitious one lying outside the fluid. For instance, the region enclosed by any of the circular streamlines of the flow can be considered as the cross section of a solid. In such a case the vortex lies within the body and thus is fictitious. The resulting flow represents *irrotational circulatory motion around a circular cylinder* (Fig. 13.2b).

Another possibility occurs when the center of the vortex lies in the fluid. In this case we assume (on the basis of experimental evidence and the theory of viscous flow) that there is a fluid core or nucleus surrounding the center of the flow and that the core rotates approximately like a solid body. Within the core we have circulatory flow with constant vorticity (see Section 13.1) and outside the core we have circulatory flow without rotation. Inside the core $u_\theta \sim r$ and outside $u_\theta \sim 1/r$ (Fig. 13.2c). Such a combination is known as an *eddy* or a *free vortex* or simply a *vortex*.* The central core is called the vortex core. The tornado and water spout are examples of such a flow. Also, we we shall see later, a lifting body trails behind it free vortices.

---

* If an eddy occurs in a fluid that is otherwise undisturbed, the spatial location of the eddy remains unaltered. However, if a uniform stream is superposed on it, it will move with the stream. Thus if a vortex is located at a point in the fluid where the velocity is $\mathbf{V}$, the vortex (i.e., the core and the associated outside circulatory flow field) will tend to move with the velocity $\mathbf{V}$. Such a vortex is therefore known as a *free vortex*.

The spatial picture of the point vortex is a doubly infinite straight line normal to the planes of motion. Such a line is called a *vortex line* or a *vortex filament*.

13.3   Circulation as the Strength of a Vortex Flow

Let us now look at some considerations related to the circulation in a vortex flow. Having already seen that the circulation around every streamline has the same value, we now seek the circulation around an arbitrary circuit.

Since the center of the vortex is a singular point, it is essential that the center be excluded from the rest of the flow field. We do this by surrounding the vortex center with an arbitrary closed curve that may be located as close as is necessary to the center and agree not to cross into the region enclosed by the curve. As a consequence the region exterior to the curve is doubly connected.*

According to the considerations of Section 9.14 we note the following:

Fig. 13.3   Symbol for vortex of strength Γ.

The circulation around any reducible circuit not enclosing the vortex center is zero. The circulation around any irreducible circuit enclosing the vortex center is not zero. The circulation around every irreducible circuit has the same value.

We thus see that the *circulation around any closed curve enclosing the center of vortex flow is a constant for the whole flow field*. It is therefore natural to choose the circulation Γ as the measure of a vortex flow. It is usual to call Γ the *strength* of the *vortex* flow or simply the strength of the vortex. We thus speak of a vortex of strength Γ and represent it diagrammatically as shown in Fig. 13.3.

From Eqs. (13.6) and (13.7) it follows that the velocity of a vortex flow can be expressed in terms of the circulation by

$$V(r, \theta) = \frac{\Gamma}{2\pi} \frac{\mathbf{e}_\theta}{r} \qquad (13.10)$$

It should be remembered that the sense of the circulation Γ in this equation is that of right-hand rotation about the Z-axis, that is, of a counterclockwise progression along a circuit in the XY-plane. The value of Γ needs to be specified.

---

* When singularities such as vortices and sources occur in the flow it is necessary to exclude them from the region of interest. In such a case the rest of the region will be multiply connected.

13.4 Stream and Potential Functions for a Vortex Flow

The stream and potential functions of a vortex flow are related to its velocity field as follows:

$$\frac{1}{r}\frac{\partial \psi}{\partial \theta} = \frac{\partial \Phi}{\partial r} = u_r = 0$$

and

$$-\frac{\partial \psi}{\partial r} = \frac{1}{r}\frac{\partial \Phi}{\partial \theta} = u_\theta = \frac{\Gamma}{2\pi r}$$

Integration of these equations yields

$$\psi(r, \theta) = -\frac{\Gamma}{2\pi}\log r + \text{a constant} \tag{13.11}$$

and

$$\Phi(r, \theta) = \frac{\Gamma}{2\pi}\theta + \text{a constant} \tag{13.12}$$

We note that the vortex potential is multivalued.

13.5 Uniform Flow Past a Circular Cylinder with Circulation

Let us consider the superposition of a uniform stream **U** past a circular cylinder of radius a and a circulatory flow around the cylinder. Such a combined flow plays a basic part, as we shall see, in the analysis of the flow past an infinite wing. It is not suggested here that when a circular cylinder is placed in an originally uniform stream the resulting steady flow will be such that there is a nonzero circulation around a circuit enclosing the cylinder. All that we are concerned with at present is to examine the consequences if the flow is one with circulation because mathematically such a flow is admissible.

Let us choose the origin of coordinates at the center of the cylinder and the X-axis in the direction of **U** (Fig. 13.4). We use polar coordinates r, θ. For the convenience of visualization of the flow field, we shall *take the sense of the circulatory flow clockwise and designate the circulation as* $-\Gamma$. *Note that the value of* Γ *is not specified.*

The stream function for the flow field due to the motion of the uniform stream past the cylinder is given by (12.30).

$$\psi_1(r, \theta) = U\left(1 - \frac{a^2}{r^2}\right)r \sin \theta$$

The stream function for the circulatory flow around the cylinder is, according to Eq. (13.11),

$$\psi_2(r, \theta) = \frac{\Gamma}{2\pi} \log\left(\frac{r}{a}\right)$$

The stream function for the combined flow is therefore given by

$$\psi(r, \theta) = \psi_1 + \psi_2 = U\left(1 - \frac{a^2}{r^2}\right) r \sin\theta + \frac{\Gamma}{2\pi} \log\left(\frac{r}{a}\right) \quad (13.13)$$

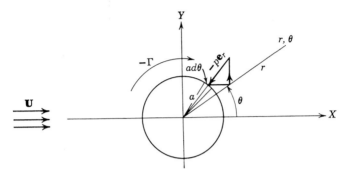

Fig. 13.4 Coordinates for flow with circulation past a circular cylinder.

The velocity components of the flow are furnished by

$$u_r(r, \theta) = \frac{1}{r}\frac{\partial \psi}{\partial \theta} = U\left(1 - \frac{a^2}{r^2}\right)\cos\theta \quad (13.14)$$

$$u_\theta(r, \theta) = -\frac{\partial \psi}{\partial r} = -U\left(1 + \frac{a^2}{r^2}\right)\sin\theta - \frac{\Gamma}{2\pi}\frac{1}{r} \quad (13.15)$$

At the stagnation points both u_r and u_θ are zero. They are therefore given by the equations

$$\left(1 - \frac{a^2}{r^2}\right)\cos\theta = 0$$

$$U\left(1 + \frac{a^2}{r^2}\right)\sin\theta = -\frac{\Gamma}{2\pi}\frac{1}{r} \quad (13.16)$$

from which we deduce the following results.

1. If $\Gamma = 0$, there are two stagnation points with the coordinates $r = a$, $\theta = \pi$ and $r = a$, $\theta = 0$. This result we had obtained previously (Fig. 13.5a).

2. If $\Gamma \neq 0$, the stagnation points should be located such that $\sin\theta$ is negative, that is, θ lies between π and 2π.

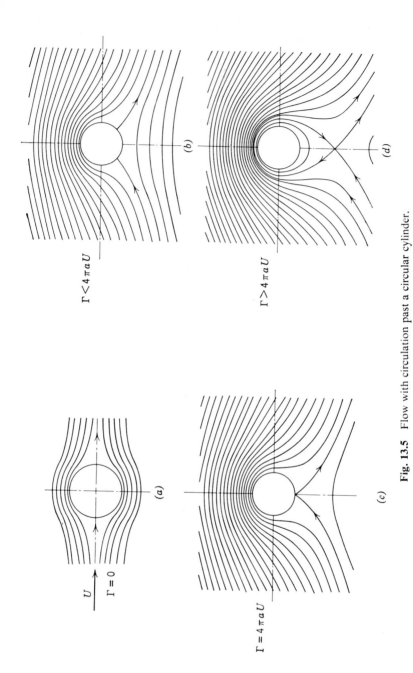

$\Gamma < 4\pi a U$

(b)

$\Gamma > 4\pi a U$

(d)

U

$\Gamma = 0$

(a)

$\Gamma = 4\pi a U$

(c)

Fig. 13.5 Flow with circulation past a circular cylinder.

384

3. One solution of the equations is

$$r = a$$

and

$$\theta = \sin^{-1}\left(\frac{-\Gamma}{4\pi U a}\right)$$

This shows that in this case $\Gamma \leq 4\pi U a$.

Thus if $\Gamma < 4\pi U a$, there are two stagnation points on the surface of the cylinder, as shown in Fig. 13.5b.

If $\Gamma = 4\pi U a$, the two points coincide and only one stagnation point occurs on the cylinder at $\theta = 3\pi/2$ (Fig. 13.5c).

4. An alternate solution of the equations is given by

$$\theta = \frac{3\pi}{2}$$

$$r = \frac{1}{4\pi U}\left[\Gamma \pm \sqrt{\Gamma^2 - (4\pi U a)^2}\right]$$

This means r is real only if $\Gamma \geq 4\pi U a$. We thus see that if $\Gamma > 4\pi U a$, there is a stagnation point outside the cylinder such that $\theta = 3\pi/2$ and $r > a$ (Fig. 13.5d).* This case will not concern us in our work.

We emphasize that *the location of the stagnation points on the body depends crucially on the value of the circulation.* In the *present case* they move downward as the circulation increases. The flow field, which is symmetrical with respect to both X- and Y-axes when the circulation is zero, becomes more and more unsymmetrical with respect to the X-axis as the value of the circulation is increased. We are concerned here with the case of $\Gamma < 4\pi U a$.

The type of flow we are discussing here can be realized with a real viscous imcompressible fluid by rotating a circular cylinder placed in an otherwise uniform stream.† Pictures obtained experimentally of the flow are shown in Plate 11. There is striking similarity between the streamlines determined experimentally and theoretically.

Let us now look at the pressure distribution over the cylinder. It is given by the Bernoulli equation

$$p(a, \theta) = H - \tfrac{1}{2}\rho V^2(a, \theta)$$

---

* There is a similar stagnation point inside the cylinder, but this is irrelevant in our present context. If r_1 and r_2 (corresponding to $-$ and $+$ in the above equation) are the r coordinates of the inside and outside points, $r_1 r_2 = a^2$.

† Viscosity is the agency for generating the circulatory motion around the rotating cylinder. Rotation of a circular cylinder in an inviscid stream has no effect.

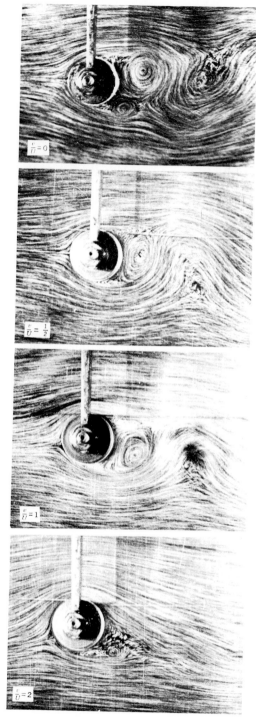

$\frac{v}{U}=0$

$\frac{v}{U}=\frac{1}{2}$

$\frac{v}{U}=1$

$\frac{v}{U}=2$

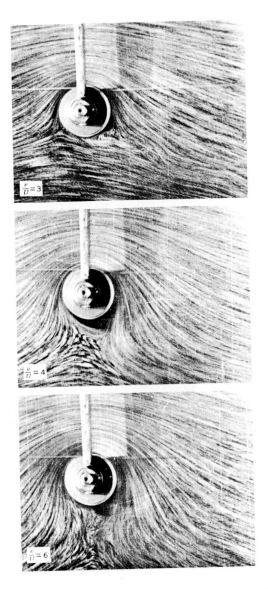

Plate 11   Flow past a rotating cylinder; U is the speed of the undisturbed uniform stream, v is the peripheral speed of the cylinder. Courtesy of Professor O. G. Tietjens. Plates 7 to 9 of Prandtl and Tietjens (1934). Last two pictures appear also as Fig. V of O. G. Tietjens: Strömungslehre, Vol. I, Springer-Verlag, 1960.

On the surface of the cylinder $u_r = 0$ and, therefore, using Eq. (13.15), we have

$$V^2(a, \theta) = u_\theta^2(a, \theta)$$

$$= 4U^2 \sin^2 \theta + \frac{\Gamma^2}{4\pi^2} \frac{1}{a^2} + \frac{2U\Gamma}{\pi a} \sin \theta \qquad (13.17)$$

The pressure on the cylinder is then obtained by substituting (13.17) into the Bernoulli equation.

The pressure distribution is symmetrical with respect to the Y-axis but unsymmetrical with respect to the X-axis. This means that the cylinder experiences a nonzero force in the Y-direction but a zero force in the X-direction, that is, there is a lift but no drag on the cylinder. The *lift L per unit length* of the cylinder is obtained readily by integrating the Y-components of the pressure forces over a unit length of the cylinder. Thus

$$L = -\int_0^{2\pi} p \sin \theta \, a \, d\theta$$

$$= \frac{\rho}{2} a \int_0^{2\pi} u_\theta^2(a, \theta) \sin \theta \, d\theta$$

Substituting in this for u_θ^2 from Eq. (13.17) and integrating we obtain

$$L = \rho U \Gamma \qquad (13.18)$$

This is a significant result in that for the first time we have a nonzero force on a body. Further, it emphasizes the crucial importance of circulation around a body as an agency of lift on the body.

The occurrence of lift on the cylinder can be explained in a simple manner. With no circulation, the flow over the cylinder in a uniform stream is symmetrical and the velocities above and below are equal. If a clockwise circulatory flow is now superposed (Fig. 13.5b) the velocity above the cylinder increases, whereas the velocity below it decreases. Consequently, according to Bernoulli's theorem, above the cylinder there is a low pressure (so-called "suction"), whereas below it there is a high pressure. This results in a lift on the cylinder. This simple physical picture of the connection between circulation and lift in the case of the cylinder was first given by Rayleigh. It is interesting to know that Rayleigh gave this explanation to account for the irregular flight of a spinning tennis ball. He did not, however, explain the origin of the circulation. The lift experienced by rotating bodies (e.g., spheres and cylinders) is usually known as the *Magnus effect*.

The use of the rotating cylinder as a lifting device and as a device for preventing separation and eddy formation, that is, as a device for the so-called boundary layer control, has been the object of several experimental investigations. For some details consult, for example, Goldstein (1938) where other references may be found.

13.6   Flow with Circulation Past an Arbitrary Cylinder

The concept of steady flow with circulation can be extended to the flow past a cylinder of arbitrary cross-sectional shape. The extension is based on the definition of circulation. If for the flow past an arbitrary cylinder we require that the circulation around a circuit enclosing the cylinder be not zero we speak of the flow as one with circulation. *It is of course not implied that if a cylinder of arbitrary cross-sectional shape is placed in an originally uniform stream the resulting steady flow will be one with circulation.* Whether it is so or not is a matter for independent consideration. We recall that mathematically the region exterior to a cylinder is a doubly connected region, and consequently the solution for the flow past the cylinder, given the shape of the cylinder contour and the free stream velocity at infinity, is not unique until the circulation is specified (see Section 9.18).

Since complex irrotational flows satisfying certain prescribed conditions can be built up by superposing several simpler flows, we may regard the flow with circulation past an arbitrary cylinder as the result of super-position of two flows: (1) a flow with zero circulation past the cylinder and, (2) a purely circulatory motion about the given cylinder. The latter flow may be thought of as resulting from a continuous distribution of vortices along the contour of the cylinder. It can be verified that the fluid velocity in such a flow at any point of the body contour is tangential to the contour and that the velocity will vanish at infinity. Furthermore, it can be shown that the velocity is finite on the contour if there are no discontinuities in the slope of the contour. At points where the slope is discontinuous the velocity will become infinite. Some of these features are considered again in Sections 17.7 and 17.8 where the disturbance field due to an airfoil in an originally uniform stream is represented as resulting from a continuous distribution of vortices on the airfoil surface.

13.7   Kutta-Joukowski Theorem and the Problem
of the Circulation Theory of Lift

We consider again the steady flow with circulation past an arbitrary cylinder. Choose Cartesians with the X-axis in the direction of the free stream U. The circulation around the cylinder, as shown, is taken in the clockwise direction. As shown in Section 11.15 the force on the cylinder, per unit span, is given by

$$\mathbf{F} = \rho U \Gamma \mathbf{j} \qquad (13.19)$$

or

$$L = \rho U \Gamma$$

where Γ is the value of the circulation, **j** the direction of the Y-axis, and L the *lift*. This result, known as the *Kutta-Joukowski theorem, states that if there is a circulation of magnitude Γ around the cylinder and if the undisturbed velocity at infinity is of magnitude U, then a lift exists, the magnitude of which is $\rho U \Gamma$ per unit span.* Kutta (1902) and Joukowski (1906), independently of each other, arrived at this result.

Since the circulation Γ is not known, this theorem does not permit an immediate determination of the lift on the cylinder. It, however, furnishes the foundation for the *theory of lift.* It shows that *such a theory must rest on the possibility of a finite circulation existing around a lifting cylinder and on the possibility of predicting theoretically the value of that circulation, given the shape of the contour of the cylinder and the free-stream velocity.* To develop the theory of lift we need to consider:

1. the circumstances under which the steady flow past an infinite cylinder is a flow with nonzero circulation,

2. the criterion that determines the value of the circulation when it exists,

3. the physical basis for the existence of a steady flow with circulation past a cylinder when such a flow exists.

13.8 Airfoils, Circulation, and the Kutta Condition

From experimental observations it had been known, from the very beginning of flight, that *only certain bodies that have a profile with a sharp* (i.e., pointed) *trailing edge are suitable as lifting bodies or wings.* Only wings with a sharp trailing edge appear to have well-defined lift on them. We can describe a wing roughly as a flat or slightly cambered plate, symmetric with respect to a median plane. The thickness of the wing is much smaller than its other dimensions. The cross sections of the wing in planes parallel to the median plane are called *airfoil profiles* or simply profiles. A picture of such a profile was shown already in Chapter 1. For sometime our concern will be solely with a wing that is an infinite cylinder of invariable profile. The geometry of such an infinite wing is completely determined by the shape of the airfoil profile. For this reason *we shall refer to the infinite wing as the airfoil* (Fig. 1.10).

On the basis of the early remarks in the preceding paragraph it is natural to expect that only the steady flow past an airfoil will be a flow with well-defined circulation. This expectation is well born out by experiment.

Now we face the question of how to determine the circulation around the airfoil, given its shape and the free-stream velocity. Experiment shows that the flow past the airfoil and the lift on it are uniquely determined once the airfoil shape and the free stream are given. Theoretically, on the basis of the mathematical considerations given so far, a solution for the flow field past the airfoil can be obtained for any value of the circulation and thus the solution is not unique until a definite value of the circulation is specified. The theoretical flow pattern for three different values of the circulation is shown in Fig. 13.6.

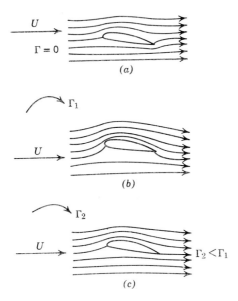

Fig. 13.6   Theoretical flow pattern for flow with circulation past an airfoil.

The theoretical considerations given so far, we emphasize, do not help specify the particular value of the well-defined circulation that must exist in the steady flow past an airfoil. To be able to specify the circulation we bring in additional considerations. Consider the flow of an ideal fluid past a corner (see Fig. 13.7). It can be verified (by solving for the flow field by the method of separation of variables) that the velocity of the flow in the neighborhood of the corner may be described by

$$u_\theta \sim r^{(\pi-\alpha)/\alpha} \sin \frac{\pi\theta}{\alpha}$$

$$u_r \sim r^{(\pi-\alpha)/\alpha} \cos \frac{\pi\theta}{\alpha}$$

where r and θ are polar coordinates with the origin at the corner (see also

Section 15.2). *We see that at the corner*

$$u_r \to 0 \quad \text{if } \alpha < \pi$$

that is when flow is inside a corner, and

$$u_r \to \infty \quad \text{if } \alpha > \pi$$

that is when flow is outside a corner.

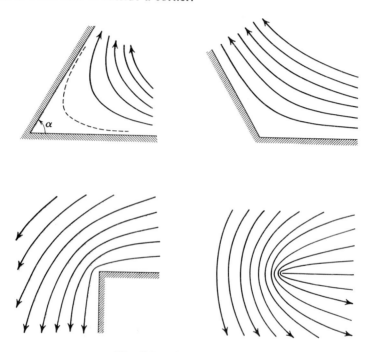

Fig. 13.7   Flow past a corner.

Now we return to the flow past the airfoil. *For arbitrary values of the circulation* Γ *there will be in general flow around the trailing edge from one side to the other and in all such cases the velocity at the edge (which is a sharp corner with* $\alpha > \pi$*) will be infinite. For one particular value of* Γ, *however, there will be no flow around the trailing edge and hence no infinite velocity at that edge; in such a case the flow will leave the trailing edge smoothly.*

This observation may be made the basis for determining a unique value for the circulation in the flow past an airfoil and consequently for determining a unique solution for that flow. We require *that the circulation set up around an airfoil be of a strength just sufficient to make the flow leave the airfoil smoothly at the trailing edge.* This condition was put forth by

Kutta (1902) and independently by Joukowski (1906). It is usually known as the *Kutta condition*. This theoretical condition has been found to agree completely with experimental observations (Section 13.9).

It follows that when an airfoil is set into uniform motion through a fluid such as air or water, a circulatory flow around the airfoil must somehow come about. We shall describe briefly this aspect of the generation of circulation in the next section.

The addition of the Kutta condition to our considerations completes the framework necessary for an adequate ideal-fluid theory of the lift on an airfoil in steady flow. The resulting theory is known as *the circulation theory of lift*. The recognition of the crucial role of circulation in the generation of lift and the determination of a unique value of the circulation by means of the Kutta condition are land marks in the development of modern aerodynamics. Recall the opening remarks of this chapter.

13.9   The Generation of Circulation

It remains to describe the physical basis of the preceding considerations. For this purpose we examine *the sequence of events that are observed experimentally* when an airfoil is set into uniform motion from rest through a real fluid such as air or water. The flow pattern at the first instant of motion, as it appears from a body-fixed reference frame, is as shown in Fig. 13.8*a*. The flow is actually like the flow without circulation of an ideal fluid. Thus at the first instant of motion the real fluid goes around the sharp trailing edge with a very high velocity. Owing to the action of the viscosity of the fluid (no matter how small the viscosity) such a motion cannot, however, continue; soon a surface of discontinuity (i.e., a vortex sheet) emanates from the edge and a vortex begins to form near the edge. Such a vortex is known as *the starting vortex*. As the airfoil proceeds in its motion the starting vortex grows rapidly in intensity while the extent of the vortex sheet increases (Fig. 13.8*b*).

As a reaction to the generation and development of the starting vortex, which is a rotation of a part of the fluid, a rotation in a sense opposite to that of the starting vortex is created in the rest of the fluid. In particular this reaction appears as a circulatory flow of the fluid around the airfoil. The growth of the circulatory flow follows that of the vortex. The growing circulatory flow modifies continuously, as shown in Fig. 13.8, the flow pattern around the airfoil.

As the airfoil proceeds, the strength of the starting vortex and that of the circulation around the airfoil grow simultaneously until the flow field around the airfoil is such that the fluid flows off smoothly from the trailing edge as shown in Fig. 13.8*c*. The full development of the starting vortex and of the circulation around the airfoil takes place quite quickly (usually in about the time it takes for the airfoil to move its own chord length). When such a condition has been reached and the airfoil has been in motion for a sufficiently long time the starting vortex is left way behind the airfoil. It has then practically no influence on the flow around the airfoil. Whenever the condition of smooth flow at the trailing edge is disturbed, say by a

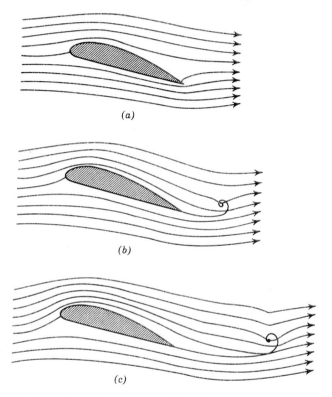

Fig. 13.8 Development of flow pattern, as shown by experiment, around an airfoil: (a) Flow at the first instant of motion; (b) Flow a little later showing the growth of the starting vortex; (c) Flow afterwards when the growth of the starting vortex and the associated circulation around the airfoil is complete so that the flow leaves the trailing edge smoothly.

change in the speed of the airfoil or in its angle of attack, a new starting vortex is formed, and a new value of the circulation is established such as to restore smooth flow at the trailing edge.

Photographs of the consecutive stages of flow around an airfoil starting from rest are shown in Plate 12. Plate 13 shows the flow as viewed from a space-fixed reference frame. Picture 13a refers to the situation soon after the formation of the starting vortex. Picture 13b refers to the situation that results when the airfoil is first accelerated from rest and then immediately stopped.

We thus see that the experimental observations amply support the essential features of the circulation theory of lift described in the preceding section. Furthermore, as we shall see in the later chapters, the lift calculated by means of this theory is in fair agreement with experimental observations.

We can now see, in light of the considerations given in this section, why the existence of circulation around a body or the lack of it could not be decided solely on the concept of an ideal fluid. The generation of circulation depends entirely on the nature of the viscous flow of a real fluid past certain bodies. Once the role of viscosity in generating the circulation has been recognized, we return to the idea of an inviscid fluid and take into account the role of viscosity by assuming that a well-defined circulation exists around the airfoil and that it is determined by the Kutta condition.

13.10   Mathematical Problem

We conclude this chapter with a statement of the mathematical problem for the circulation theory of lift. In terms of the velocity potential Φ the problem is to determine Φ as the solution of the equation

$$\nabla^2 \Phi = 0$$

in the region exterior to the airfoil such that

$$\text{grad } \Phi \cdot \mathbf{n} = 0 \quad \text{on the surface of the airfoil}$$
$$\text{grad } \Phi \to \mathbf{U} \quad \text{at infinity}$$

and the circulation Γ satisfies the Kutta condition. In terms of the stream function ψ the problem is to determine ψ as the solution of the equation

$$\nabla^2 \psi = 0$$

in the region exterior to the airfoil such that $\psi = $ a constant on the airfoil surface, the spatial derivatives of ψ approach the corresponding components of $\mathbf{U}$ at infinity, and Γ satisfies the Kutta condition.

The Kutta condition may be expressed in a more explicit form. *If the trailing edge angle is finite, as shown in Fig. 13.9a, the velocity at the trailing edge must be zero; otherwise a velocity discontinuity, which cannot be permitted, will result at the edge.* That a velocity discontinuity cannot be allowed is seen as follows. A velocity discontinuity, which can be only a tangential discontinuity, requires that the component of the velocity normal to the discontinuity must be continuous through the discontinuity

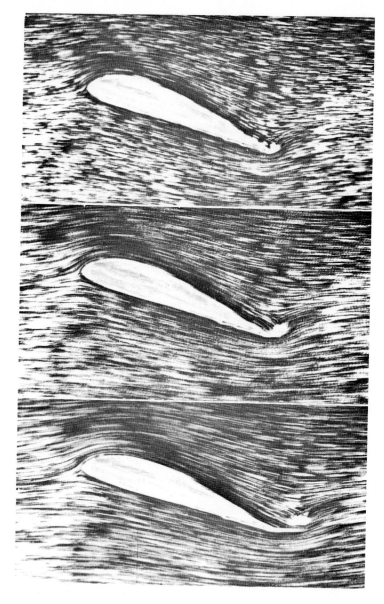

Plate 12 Consecutive stages of flow past an airfoil starting from rest. Courtesy of Professor O. G. Tietjens. Plates 17 and 18 of Prandtl and Tietjens (1934).

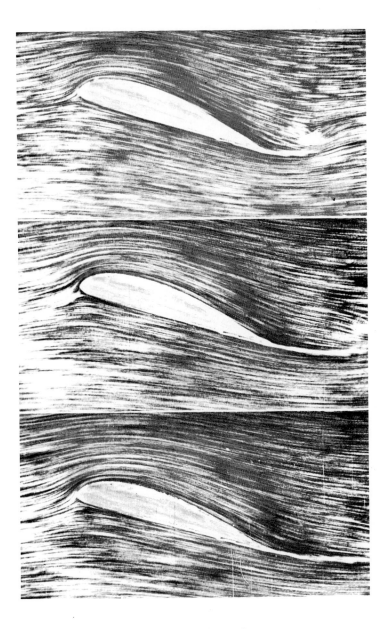

Plate 12 (*Continued*)

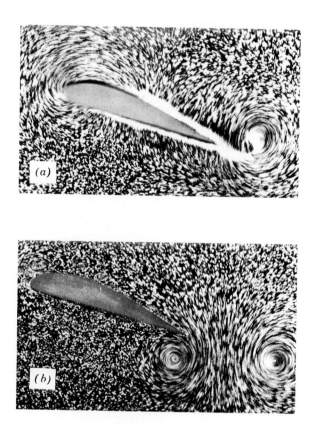

Plate 13 Flow past an airfoil as viewed from a space-fixed reference frame. (a) Immediately after starting the airfoil. Note the starting vortex and the formation of the circulatory flow over the airfoil. (b) Situation when the airfoil was stopped after the formation of the starting vortex. Note the "stopping" vortex and the decay of the circulatory flow. The starting vortex had been detached. Courtesy of Professor O. G. Tietjens. Plate 22 of Prandtl and Tietjens (1934).

(see Chapter 7). Now at the trailing edge the velocity on the upper side of the airfoil is parallel to that side whereas the velocity on the lower side is parallel to the lower side (Fig. 13.9a). This means, if the velocities are finite, the components normal to the discontinuity that must emanate from the trailing edge in that case will *not* be continuous. We conclude that the velocities on either side of the airfoil at the trailing edge must be zero; in other words *a trailing edge with a finite angle must be a stagnation point.*

Consider now a cusped trailing edge as shown in Fig. 13.9b. Now the velocities on either side of the airfoil at the edge are in the same direction,

$$(a) \qquad\qquad (b)$$

Fig. 13.9 Shape of trailing edge: (a) trailing edge with a finite angle; (b) cusped trailing edge.

and according to the Kutta condition we may expect that all that is necessary is for these velocities to be finite. As can be readily shown the magnitudes of the velocities must be the same. If their magnitudes are not the same a discontinuity must be assumed to emanate from the edge. Then, because the pressure must be continuous across the discontinuity, we should have, according to Bernoulli's relation, at the trailing edge

$$p_+ + \frac{\rho}{2} u_+{}^2 = p_0 = p_- + \frac{\rho}{2} u_-{}^2$$

where p denotes the pressure, u the velocity, p_0 the stagnation pressure, and the plus and minus subscripts denote respectively the upper and lower sides of the airfoil. Since p_+ is equal to p_-, it follows that the velocities on either side must have the same magnitude. We conclude that *a cusped trailing edge need not be a stagnation point, but at such an edge the velocity must be finite and have the same magnitude on either side of the airfoil.*

From the preceding considerations it follows that *there cannot be a surface of discontinuity* (i.e., *a vortex sheet*) *emanating from the trailing edge of an airfoil in steady flow whether the trailing edge angle is finite or zero.* Note that this result does not hold for an airfoil in unsteady motion or for a finite wing in steady flow.

We return to consider the mathematical problem for determining the steady flow past an airfoil. As has been pointed out before, it is in general not possible to obtain in a direct manner an exact solution of the problem for an arbitrary airfoil. Approximate closed form solutions may, however, be constructed for the thin airfoil by the method of superposition of simple singular solutions. This method is described in Chapter 17.

In constructing the solution for the airfoil problem we are concerned with solving Laplace's equation in a plane, given some auxiliary conditions on an arbitrary contour in the plane. Difficulties arise when the contour is not a simple one such as a circle or an ellipse. In view of this we may introduce in the plane a new set of coordinates in terms of which the contour may be represented as a simple form and attempt to solve the problem in terms of the new coordinates. In general, however, when a transformation of coordinates is employed the form of the governing equation for the problem will also change unless the transformation satisfies certain conditions. Let us consider this matter in a little more detail.

In Cartesians we are to seek the solution of the equation

$$\frac{\partial^2 f}{\partial x^2} + \frac{\partial^2 f}{\partial y^2} = 0 \tag{13.20}$$

where $f(x, y)$ may be either the velocity potential or the stream function. For the reasons outlined above, let us introduce a new set of coordinates q_1 and q_2 such that

$$\begin{aligned} q_1 &= q_1(x, y) \\ q_2 &= q_2(x, y) \end{aligned} \tag{13.21}$$

Let $g(q_1, q_2)$ correspond to $f(x, y)$. It may be verified that (13.20) then takes the form

$$\left[\left(\frac{\partial q_1}{\partial x} \right)^2 + \left(\frac{\partial q_1}{\partial y} \right)^2 \right] \frac{\partial^2 g}{\partial q_1{}^2} + \left[\left(\frac{\partial q_2}{\partial x} \right)^2 + \left(\frac{\partial q_2}{\partial y} \right)^2 \right] \frac{\partial^2 g}{\partial q_2{}^2}$$

$$+ \left(\frac{\partial^2 q_1}{\partial x^2} + \frac{\partial^2 q_1}{\partial y^2} \right) \frac{\partial g}{\partial q_1} + \left(\frac{\partial^2 q_2}{\partial x^2} + \frac{\partial^2 q_2}{\partial y^2} \right) \frac{\partial g}{\partial q_2}$$

$$+ 2 \left(\frac{\partial q_1}{\partial x} \frac{\partial q_2}{\partial x} + \frac{\partial q_1}{\partial y} \frac{\partial q_2}{\partial y} \right) \frac{\partial^2 g}{\partial q_1 \partial q_2} = 0 \tag{13.22}$$

If we desire that this equation reduce to Laplace's equation in q_1 and q_2, that is,

$$\frac{\partial^2 g}{\partial q_1{}^2} + \frac{\partial^2 g}{\partial q_2{}^2} = 0 \tag{13.23}$$

we must require that

$$\left(\frac{\partial q_1}{\partial x}\right)^2 + \left(\frac{\partial q_1}{\partial y}\right)^2 = \left(\frac{\partial q^2}{\partial x}\right)^2 + \left(\frac{\partial q_2}{\partial y}\right)^2 \neq 0$$

$$\frac{\partial^2 q_1}{\partial x^2} + \frac{\partial^2 q_1}{\partial y^2} = 0$$

$$\frac{\partial^2 q_2}{\partial x^2} + \frac{\partial^2 q_2}{\partial y^2} = 0$$

$$\frac{\partial q_1}{\partial x}\frac{\partial q_2}{\partial x} + \frac{\partial q_1}{\partial y}\frac{\partial q}{\partial y} = 0$$

It can be shown that the transformation (13.21) which satisfies the above four conditions is given by a relation of the form

$$(q_1 + iq_2) = h(x + iy) \tag{13.24}$$

where $i = \sqrt{-1}$, and h denotes an arbitrary function of the definite function $(x + iy)$. As is known, $(x + iy)$ and $(q_1 + iq_2)$ are complex variables. We thus conclude that the transformation of coordinates we are seeking must be affected by means of the functions of a complex variable. We are thus naturally led to the study of the functions of a complex variable and of their use in analyzing the two-dimensional irrotational motion of an ideal fluid. We now turn to this study.

Chapter 14

Elements of the Theory of Functions

of a Complex Variable

In this chapter we shall acquaint ourselves with the elements of the theory of the functions of a complex variable. We introduce the complex variable through the general solution of Laplace's equation in two dimensions. Finding the solutions of Laplace's equation is the essence of our problem no matter how we wish to attack it. Following the introduction of the complex variable, we study the differential and integral calculus of an *analytic function* of a complex variable. A differentiable function is an analytic function. We shall find that the properties exhibited by the so-called real and imaginary parts of an analytic function are identical with those exhibited by the potential and stream functions of a two-dimensional potential motion of an ideal fluid. Furthermore, analytic functions provide us with the type of transformation of independent variables we were inquiring about in the last chapter. In the next chapter we shall apply the results we obtain here to the problem of two-dimensional potential motion.

The theory of functions of a complex variable is an elegant and rigorous study. Our main concern here is with the role of this theory in two-dimensional potential motion, and as such we must content ourselves in presenting the elements of the theory with only the necessary rigor. The reader should bear this in mind and should refer to the works cited at the end of this book to appreciate the full beauty and scope of the theory.

14.1 General Solution of Laplace's Equation in Two Dimensions: Introduction of the Complex Variable

Let $g(x, y)$ be any function governed by the equation

$$\nabla^2 g = \frac{\partial^2 g}{\partial x^2} + \frac{\partial^2 g}{\partial y^2} = 0 \tag{14.1}$$

With reference to two-dimensional irrotational motion, $g(x, y)$ may represent any one of the quantities: the velocity potential Φ, the stream

function ψ, the velocity components u and v, any new independent coordinates q_1, q_2 which may be employed to relate the potential problem
connected with a complex geometry to that connected with a simple
geometry. To investigate the general solution of (14.1) we introduce new
independent variables $\xi(x, y)$ and $\eta(x, y)$, which will reduce the equation to
a form that may be integrated readily. It is found, as may be verified, that
if we choose

$$\xi(x, y) = x + \sqrt{-1}y$$
$$\eta(x, y) = x - \sqrt{-1}y \tag{14.2}$$

Eq. (14.1) reduces to

$$\frac{\partial^2 \tilde{g}}{\partial \xi \, \partial \eta} = 0 \tag{14.3}$$

where

$$\tilde{g}(\xi, \eta) \equiv g[x(\xi, \eta), y(\xi, \eta)]$$

The functions $x(\xi, \eta)$ and $y(\xi, \eta)$ are simply the inverses of the transformation relations (14.2). Equation (14.3) readily integrates to

$$\tilde{g}(\xi, \eta) = \tilde{f}_1(\xi) + \tilde{f}_2(\eta)$$

where $\tilde{f}_1(\xi)$ and $\tilde{f}_2(\eta)$ are any arbitrary functions of the variables ξ and η.
It therefore follows that

$$g(x, y) = f_1(x + \sqrt{-1}y) + f_2(x - \sqrt{-1}y) \tag{14.4}$$

where f_1 and f_2 are arbitrary functions of the single variables $(x + \sqrt{-1}y)$
and $(x - \sqrt{-1}y)$, respectively. These variables are, however, *definite
independent functions* of x and y. The functions f_1 and f_2 are entirely
arbitrary* except for the obvious requirement that f_1 should possess the
first and second derivatives with respect to its *own* argument $(x + \sqrt{-1}y)$
and that f_2 should possess the first and second derivatives with respect to
its *own* argument $(x - \sqrt{-1}y)$.

If a significant meaning can be attached to the entity $\sqrt{-1}$, the single
variables $(x + \sqrt{-1}y)$ and $(x - \sqrt{-1}y)$ may be interpreted as simple
combinations of a pair of real variables or numbers. Then a whole class
of functions of the variables $(x + \sqrt{-1}y)$ and $(x - \sqrt{-1}y)$ become
available for constructing solutions of Laplace's equation. Square roots
of negative numbers made their appearance early in mathematical pursuits,
and a system of analysis involving such quantities has been developed
extensively.† The entity $\sqrt{-1}$ is called the *imaginary unit* and is usually

* f_2 and f_1 should be such that together they yield a function $g(x, y)$, which, being a
physical quantity, should not involve $\sqrt{-1}$.
† They arose as early as the Middle Ages when mathematicians sought a general
solution of quadratic equations.

denoted by the letter i. A combination of the type $x + iy$ is known as a complex or imaginary number. If x and y are variable, the single variable $x + iy$ is said to be a *complex variable*. A function of the variable $x + iy$ is known as *a function of a complex variable* or simply a *complex function*. The other independent combination $x - iy$ is known as the *complex conjugate* of $x + iy$. In the same way, $x + iy$ is the complex conjugate of $x - iy$. We thus see that in seeking the solutions of Laplace's equation in two dimensions we are naturally led to a study of the theory of functions of the type $f(x + iy)$ and $F(x - iy)$. We shall now go into some of the details of this theory. The theory is developed around the variable $x + iy$. The role of the conjugate $x - iy$ and its functions will become apparent during the study of the functions of the variable $x + iy$.

14.2 Nomenclature and Algebra of Complex Numbers

We introduce the imaginary unit i through the relation

$$i^2 = -1 \tag{14.5}$$

and define a complex number as any combination of the form

$$a + ib$$

where a and b are real numbers. A *complex number thus represents a pair of real numbers.*

We denote a complex number by a single letter, say z, and write

$$z = a + ib \tag{14.6}$$

The real numbers a and b are known, respectively, as the *real* and *imaginary parts* of a complex number z and are denoted by

$$a = \text{Re } z$$
$$b = \text{Im } z \tag{14.7}$$

The complex number $a - ib$ is called the conjugate of the number $z = a + ib$ and is denoted by the symbol $\bar{z}$.

The algebraic operations of addition, multiplication, and division for complex numbers are defined in the same manner as for real numbers. Thus, recalling (14.5), we have

Addition and subtraction

$$(a + ib) \pm (c + id) = (a \pm c) + i(b \pm d) \tag{14.8}$$

Multiplication

$$(a + ib)(c + id) = (ac - bd) + i(ad + bc) \tag{14.9}$$

Division

$$\frac{(a + ib)}{(c + id)} = \frac{(a + ib)(c - id)}{(c + id)(c - id)}$$

$$= \frac{(ac + bd)}{c^2 + d^2} + i\,\frac{bc - ad}{c^2 + d^2} \tag{14.10}$$

where $c + id \neq 0$.
Two complex numbers $a + ib$ and $c + id$ are equal only when

$$a = c \quad \text{and} \quad b = d$$

14.3   Geometric Interpretation

Every complex number $a + ib$ represents an *ordered* pair of real numbers (a, b).* The order of a number in the pair denotes whether it is the real *or* imaginary part of the complex number. Now, every ordered pair of real numbers represents a point in a plane. This means *complex numbers may be represented geometrically by points in a plane.* For each complex number there corresponds only a single point in the plane, and, conversely, for each point there corresponds a single complex number.

Let us set up a Cartesian system of axes to mark out points in a plane or, equivalently, to represent a system of complex numbers. Complex numbers that have only real parts are represented by number pairs of the form $(a, 0)$ and lie on one of the axes. Such an axis is called the *real axis*. Complex numbers that have only imaginary parts are represented by number pairs of the form $(0, b)$ and lie on the other axis, which is called the *imaginary axis*. An arbitrary complex number z represented by a number pair (a, b) is then given by a point whose coordinates with respect to the real and imaginary axes are, respectively, a and b.

A plane, the points of which represent a system of complex numbers, is known as a plane of complex numbers or simply a complex plane.† It is usual to denote the real axis by X, the imaginary axis by Y, and an arbitrary complex number z by the number pair (x, y). Since a number pair (x, y) uniquely specifies a vector in a plane, *a complex number may also be interpreted as a vector in a plane.* Thus a complex number $z = (x + iy)$, an ordered pair of real numbers (x, y), a point in a plane, a vector in a plane are all equivalent. We may thus alternately denote a complex number by

$$z = (x, y) \tag{14.11}$$

---

* Instead of introducing the imaginary unit i we may define a complex number as an ordered pair of real numbers and develop the algebra of complex numbers as the algebra of ordered pairs of real numbers. See for instance, Konrad Knopp (1952).
† The names Argand plane and Gaussian number plane are also used.

The magnitude of the vector that represents a complex number z is known as the modulus or magnitude or absolute value of the number z and is denoted by $|z|$. Since z is the number $x + iy$, we have

$$|z| = \sqrt{x^2 + y^2} \tag{14.12}$$

The angle between the real axis and the vector z is known as the argument of z and denoted by *

$$\arg z = \tan^{-1} \frac{y}{x} \tag{14.13}$$

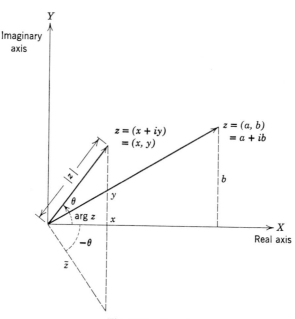

Fig. 14.1 Complex plane.

This shows that the argument of any given complex number is only determined to within an added multiple of 2π. In other words, the arg z is multivalued, with any two successive values differing by 2π. The value of the argument corresponding to the range 0 and 2π (or π and $-\pi$) is usually known as its *principal value*.

The conjugate $\bar{z} = x - iy$ has the following properties (see Fig. 14.1)

$$|\bar{z}| = |z| \tag{14.14}$$

and

$$\arg \bar{z} = -\arg z \tag{14.15}$$

Thus $\bar{z}$ is the reflection of the point z in the real axis (see Fig. 14.1).

* Sometimes also known as the amplitude of z.

The algebraic operations on complex numbers have simple geometrical interpretations. It is seen that equality and addition of complex numbers corresponds to equality and addition of vectors in a plane. Multiplication of complex numbers may also be given geometric meaning, but this we shall do later.

Just as for vectors, the question of whether one complex number is greater or less than another does not arise. All we can say is that the complex numbers are either equal or not equal. Of course, we can always compare the magnitudes of the complex numbers.

14.4 Polar and Exponential Forms of a Complex Number

In polar coordinates, a point or a position vector in the plane is indicated by the number pair (r, θ), where r is the length of the position vector while

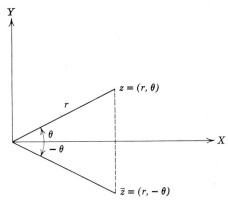

Fig. 14.2   Polar forms of a complex number and its conjugate.

θ gives its direction (see Fig. 14.2). Thus in polar coordinates a complex number z may be represented by the number pair (r, θ) and denoted as

$$z = (r, \theta) \tag{14.16}$$

We then have

$$|z| = r \tag{14.17}$$

$$\arg z = \theta \tag{14.18}$$

For any given z, r has a definite single positive value, while θ, as pointed out earlier, has infinitely many values that differ by multiples of 2π.

If we consider the complex number z in its *Cartesian form* $x + iy$ and substitute

$$x = r \cos \theta, \qquad y = r \sin \theta$$

we obtain

$$z = r(\cos \theta + i \sin \theta) \tag{14.19}$$

This is known as the *polar form* of the complex number z. The expression $(\cos\theta + i\sin\theta)$ is a complex number whose magnitude is unity and argument is θ. It is simply a unit vector in the direction of the complex number z.*

Differentiating the expression $(\cos\theta + i\sin\theta)$ with respect to θ, we obtain

$$\frac{d(\cos\theta + i\sin\theta)}{\cos\theta + i\sin\theta} = i\,d\theta$$

This integrates to

$$(\cos\theta + i\sin\theta) = e^{i\theta} \qquad (14.20)$$

This relation is generally referred to as *Euler's formula.*

Using Eqs. (14.19) and (14.20) we may express a complex number z as

$$z = re^{i\theta} \qquad (14.21)$$

This is known as the *exponential form* of a complex number.

Putting together the several forms of z we have

$$z = re^{i\theta} = r(\cos\theta + i\sin\theta) = x + iy \qquad (14.22)$$

and

$$|z| = r = \sqrt{x^2 + y^2} \qquad (14.23)$$

$$\arg z = \theta = \tan^{-1}\frac{y}{x} \qquad (14.24)$$

All the different forms of the complex number are found useful in applications. The form that is more suitable depends entirely on the problem on hand.

Using the exponential form, we see that the product of two complex numbers $z_1 = r_1 e^{i\theta}$ and $z_2 = r_2 e^{i\theta_2}$ is given by

$$z_1 z_2 = r_1 r_2 e^{i(\theta_1 + \theta_2)}$$

that is,

$$\left.\begin{array}{l} |z_1 z_2| = r_1 r_2 = |z_1|\,|z_2| \\ \arg|z_1 z_2| = \theta_1 + \theta_2 = \arg z_1 + \arg z_2 \end{array}\right\} \qquad (14.25)$$

and

We notice that *multiplication of any complex number z by a number of the form $e^{i\varphi}$, where φ is real, is equivalent to rotating the vector representing z through an angle φ in the complex plane* (Fig. 14.3). This lends geometrical meaning to the product of complex numbers.

Using the exponential form we obtain the result that

$$(\cos\theta + i\sin\theta)^n = (\cos n\theta + i\sin n\theta) \qquad (14.26)$$

This is known as *De Moivre's theorem.*

---

* Recall that any vector may be represented as a magnitude times a unit vector.

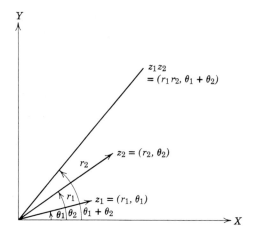

Fig. 14.3   Illustrating the product of complex numbers.

14.5   Function of a Complex Variable

If for each value of a complex variable z a new complex number ζ is generated by some rule, we say that ζ is a *function* of the complex variable z and write

$$\zeta = \zeta(z)$$

(Other notations such as $\zeta = f(z)$ or $F(z)$ may also be employed.) Thus the notion of a complex function is exactly the same as that of a real function. If a single value of ζ corresponds to each value of z, we say $\zeta = \zeta(z)$ is a *single-valued* function; if more than one value of ζ is produced for each z, we say $\zeta(z)$ is a *multiple-valued* function.

Expressing the complex numbers ζ and z in terms of their real and imaginary parts

$$\zeta = (\xi, \eta) = (\xi + i\eta)$$
$$z = (x, y) = (x + iy)$$

we interpret the function $\zeta = \zeta(z)$ as equivalent to defining two real numbers (ξ, η) for each value of a pair of real variables (x, y). We thus write

$$\xi = \xi(x, y)$$
$$\eta = \eta(x, y)$$

and

$$\zeta(z) = \xi(x, y) + i\eta(x, y)$$

where $\xi(x, y)$ and $\eta(x, y)$ are called the real and imaginary parts of the complex function ζ. In this form the complex function $\zeta = \zeta(z)$ is simply a combination of a pair of real functions of two real variables. We will find this interpretation of a complex function sometimes useful. Generally, however, important information results by using the unseparated form where a function $\zeta(z)$ is exhibited in terms of the single complex variable z.

14.6 Analytic Function

Of all the functions of a complex variable, the ones that are of interest in our applications (as in all physical sciences) are those that are *differentiable*. Such functions are called *analytic functions*. For a complex function the concepts of calculus, such as differentiability, continuity, and limit, are formally the same as for a function of a real variable. Differentiability of a complex function implies its continuity and guarantees, as shall be seen later, that the function may be repeatedly differentiated or integrated any number of times. We shall concern ourselves with the details of only the *condition of differentiability*.

A complex function $\zeta = \zeta(z)$ is said to be differentiable at a point z if as z_1 approaches z the limit

$$\lim_{z_1 \to z} \frac{\zeta(z_1) - \zeta(z)}{z_1 - z}$$

exists. If it is possible to construct this limit, we call it the *derivative* of the function $\zeta(z)$ at the point z and denote it by $d\zeta/dz$ or $\zeta'(z)$. Thus we write

$$\frac{d\zeta}{dz} \equiv \zeta'(z) = \lim_{z_1 \to z} \frac{\zeta(z_1) - \zeta(z)}{z_1 - z}$$

or, putting $\zeta(z_1) - \zeta(z) = \Delta\zeta$ and $z_1 - z = \Delta z$,

$$\frac{d\zeta}{dz} = \lim_{\Delta z \to 0} \frac{\Delta\zeta}{\Delta z} \qquad (14.28)$$

This idea of differentiability and derivative for a complex function is the same as that for a function of a real variable. *There is, however, an important distinction that we should bear in mind.* Since z_1 and z are complex numbers or, equivalently, points of a plane, there are infinitely many directions or paths along which Δz, that is, the point z_1, may be chosen. Under such circumstances we should state that *for a complex function $\zeta = \zeta(z)$ to be differentiable at a point z, the limit:* $\lim_{\Delta z \to 0} \Delta\zeta/\Delta z$, *or, equivalently, the derivative $d\zeta/dz$ should assume the same value no matter from what direction Δz approaches zero.* This, then, is precisely the *condition* for a complex function to be *analytic*. Such a requirement does not arise explicitly in case of a function of a real variable, for all values of a real variable lie on a single straight line.*

---

* For a more rigorous discussion consult books on the theory of functions of a complex variable.

Points of the complex domain z, where the function $\zeta(z)$ is analytic, are called *regular points* of the function. Points where the function is not analytic are called *singular points* of the function.

The rules for differentiation of analytic functions are formally the same as for real functions. The proof may be carried out in exact analogy. Thus we have the following simple results:

$$\frac{d}{dz}[f(z) + g(z)] = f'(z) + g'(z)$$

$$\frac{d}{dz}\left[f(z)\, g(z)\right] = f'(z)\, g(z) + f(z)\, g'(z)$$

$$\frac{d}{dz}\left[\frac{f(z)}{g(z)}\right] = \frac{f'(z)\, g(z) - f(z)\, g'(z)}{g(z)^2}$$

if $g(z) \neq 0$.

If $\zeta(z)$ is an analytic function, and if $w(\zeta)$ is another analytic function, the function given by

$$g(z) = w[\zeta(z)]$$

is also analytic, for we have

$$\frac{dg(z)}{dz} = \frac{dw}{d\zeta}\frac{d\zeta}{dz}$$

Thus *an analytic function of an analytic function is analytic.* From these various rules one may be able to obtain a wide class of analytic functions from a few basic analytic functions. Later we shall consider some examples of such functions.

14.7 Cauchy-Riemann Conditions

The condition for the differentiability of a complex function implies important conditions on its real and imaginary parts. Since a complex function $\zeta(z) = \xi + i\eta$ is no more than a pair of real functions, $\xi(x, y)$ and $\eta(x, y)$, it is natural to suppose that the complex function is differentiable if ξ and η are differentiable. This, however, is not true. The *differentiability of $\xi(x, y)$ and $\eta(x, y)$ does not in itself imply the differentiability of* the function $\zeta(z) = \xi + i\eta$. For example, as can be readily verified, the simple function $\zeta(z) = 2x + 6yi$ is not differentiable although ξ and η are. Besides differentiability, the real and imaginary parts of an analytic function must fulfill other important conditions, which we shall now obtain.

Consider the function

$$\zeta(z) = \xi(x, y) + i\eta(x, y)$$

Writing

$$\Delta z = \Delta x + i\,\Delta y$$

$$\Delta \zeta = \Delta \xi + i\,\Delta \eta$$

we express the derivative of ζ as

$$\frac{d\zeta}{dz} = \lim_{\Delta z \to 0} \frac{\Delta \zeta}{\Delta z} = \lim_{(\Delta x + i \Delta y) \to 0} \frac{\Delta \xi + i\,\Delta \eta}{\Delta x + i\,\Delta y} \qquad (14.29)$$

Of the infinitely many directions along which $\Delta z = \Delta x + i\,\Delta y$ may be taken, we choose specifically the two directions corresponding to $\Delta z_1 = \Delta x$ and $\Delta z_2 = i\,\Delta y$, that is, parallel to the x- and y-axes, respectively (Fig. 14.4). For the derivative along Δz_1 we have

$$\left(\frac{d\zeta}{dz}\right) = \lim_{\Delta x \to 0} \frac{\Delta \xi + i\,\Delta \eta}{\Delta x} = \frac{\partial \xi}{\partial x} + i\frac{\partial \eta}{\partial x} \qquad (14.30)$$

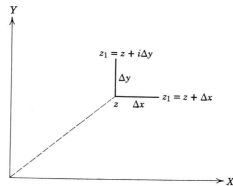

Fig. 14.4 Illustrating the computation of the derivative of a complex function.

For the derivative along Δz_2 we have

$$\left(\frac{d\zeta}{dz}\right) = \lim_{\Delta y \to 0} \frac{\Delta \xi + i\,\Delta \eta}{i\,\Delta y} = \frac{1}{i}\frac{\partial \xi}{\partial y} + \frac{\partial \eta}{\partial y} \qquad (14.31)$$

It follows, therefore, that if the derivative of ζ has to assume the same value from both these directions, we should require that

$$\frac{d\zeta}{dz} = \frac{\partial \xi}{\partial x} + i\frac{\partial \eta}{\partial x} = \frac{\partial \eta}{\partial y} + \frac{1}{i}\frac{\partial \xi}{\partial y}$$

This equation reduces to the following pair of partial differential equations, which are known as *Cauchy-Riemann* conditions:

$$\frac{\partial \xi}{\partial x} = \frac{\partial \eta}{\partial y}$$

$$\frac{\partial \xi}{\partial y} = -\frac{\partial \eta}{\partial x} \qquad (14.32)$$

Now, it is natural to expect that similar additional conditions might be required to establish that the derivative $d\zeta/dz$ will actually assume a single definite value irrespective of the direction from which Δz approaches zero. This, however, is not the case, for it can be shown that the Cauchy-Riemann equations are *sufficient* to ensure this independence of the derivative. We thus conclude that *the real and imaginary parts of an analytic function are a pair of real functions that, besides being differentiable, satisfy the Cauchy-Riemann conditions.*

14.8   Some Consequences of Cauchy-Riemann Equations

We shall now consider some significant results that follow directly from the Cauchy-Riemann conditions.

1. If the real and imaginary parts of a function of a complex variable satisfy the Cauchy-Riemann equations, the complex function is an analytic function.

2. From Eqs. (14.32) we obtain

$$\frac{\partial^2 \xi}{\partial x^2} = -\frac{\partial^2 \eta}{\partial x \, \partial y} = -\frac{\partial^2 \xi}{\partial y^2}$$

or

$$\nabla^2 \xi = \frac{\partial^2 \xi}{\partial x^2} + \frac{\partial^2 \xi}{\partial y^2} = 0 \qquad (14.33)$$

Similarly, we have

$$\nabla^2 \eta = \frac{\partial^2 \eta}{\partial x^2} + \frac{\partial^2 \eta}{\partial y^2} = 0 \qquad (14.34)$$

This shows that both the real and imaginary parts of an analytic function satisfy Laplace's equation. They both are thus *harmonic functions*.

3. The real and imaginary parts are, however, *not* independent harmonic functions. For, if the real part $\xi(x, y)$ is specified, the Cauchy-Riemann equations determine the imaginary part $\eta(x, y)$ to within an arbitrary additive constant. In this sense the real and imaginary parts of an analytic function are said to be *conjugate harmonic functions*.

4. Consider the two families of curves in the x-y plane described by the functions

$$\xi(x, y) = \text{const.}$$

$$\eta(x, y) = \text{const.}$$

where ξ and η are the real and imaginary parts of an analytic function. The angle of intersection between the two families is given by

$$\cos^{-1} \frac{\text{grad } \xi \cdot \text{grad } \eta}{|\text{grad } \xi| \, |\text{grad } \eta|}$$

where

$$\text{grad } \xi \cdot \text{grad } \eta = \frac{\partial \xi}{\partial x}\frac{\partial \eta}{\partial x} + \frac{\partial \xi}{\partial y}\frac{\partial \eta}{\partial y}$$

However, by the Cauchy-Riemann conditions this dot product is zero. This means the two families of curves intersect each other at right angles. We therefore state that *the two families of curves defined by the real and imaginary parts of an analytic function are orthogonal to each other.*

5. If the function

$$\zeta = \zeta(z)$$

is analytic and if the derivative

$$\zeta' \equiv \frac{d\zeta}{dz} \neq 0$$

then the inverse function

$$z = z(\zeta)$$

exists and is analytic with its derivative given by

$$z' \equiv \frac{dz}{d\zeta} = \frac{1}{\zeta'} = \frac{1}{d\zeta/dz}$$

This result is of vital significance in the theory and application of conformal transformation. The proof of the result is left as a problem for the reader.

14.9 Remarks

It is appropriate at this stage to recognize the intimate connection between the analytic functions of a complex variable and the solutions for the problem of two-dimensional irrotational motion of an ideal fluid. We realize that *the properties exhibited by the real and imaginary parts of an analytic function are identical with the properties exhibited by the velocity potential and stream functions of a two-dimensional irrotational motion of an ideal fluid and vice versa.* The problem of such a motion, consequently reduces to that of finding an analytic function whose real and imaginary parts satisfy certain prescribed boundary conditions.

Consider now the question of relating the solution for the two-dimensional potential motion for one geometry with that for another geometry. For this purpose, as we have seen, we must introduce coordinate transformations that obey certain rules (see Section 13.10). These rules, as we now realize, are identical with those governing the real and imaginary parts of an analytic function of a complex variable.

We may thus conclude that the theory of two-dimensional potential motion and the theory of an analytic function of a complex variable are identical.

14.10   Some Analytic Functions

We now consider some examples of analytic functions. Such functions will be used repeatedly in our later considerations.

One of the simplest functions is a *power function* z^n, where n is a positive integer. It can be readily seen that it is analytic at all points of the complex plane. It follows, at once, that a so-called *polynomial function* $a_0 + a_1 z + a_2 z^2 + \cdots + a_n z^n$, where n is a positive integer and the a's are constants (complex, in general), is also analytic in the entire complex plane. We further conclude that a *rational function*

$$\frac{a_0 + a_1 z + a_2 z^2 + \cdots + a_n z^n}{b_0 + b_1 z + b_2 z^2 + \cdots + b_n z^n}$$

is analytic at all points where the denominator does not vanish.

From these considerations we may conclude that a power series

$$\sum_0^\infty a_n z^n = a_0 + a_1 z + a_2 z^2 + \cdots + a_n z^n + \cdots$$

where n is an integer and the a's are constants is an analytic function within the region of convergence of the series.* Conversely, it is possible, as we will indicate later, to express an analytic function—within the region it is analytic—in terms of a power series. This procedure is very helpful in discussing the theory and applications of complex functions.

Now, the *exponential function* e^z, the *trigonometric functions* $\sin z$ and $\cos z$, and the *hyperbolic functions* $\sinh z$ and $\cosh z$ can all be defined, just as in the case of the corresponding real functions, by means of power series that are convergent everywhere. Thus all these functions are analytic in the entire complex plane.

Another important function is the *logarithmic function* $\log z$. Let us write

$$z = re^{i\theta}$$

where $r = z$ and $\theta = \arg z$. As pointed out earlier, although r is single valued, θ has infinitely many values that differ by multiples of 2π. To show this explicitly we set

$$\theta = \theta_p + 2\pi k$$

where $k = 0, \pm 1, \pm 2, \ldots$, and θ_p is the principal value of $\arg z$, that is, the value of θ in the range 0 to 2π. We then have

$$z = re^{i(\theta_p + 2\pi k)}$$

and

$$\log z = \log r + i(\theta_p + 2\pi k) \tag{14.35}$$

This shows that the logarithm has infinitely many values that differ by multiples of $2\pi i$. Thus $\log z$ is a multivalued function. The various

---

* For the convergence of series, and so forth, reference may be made to Knopp's or any other suitable book.

values of such a function are called *branches**. The principal value branch of $\log z$ is $(\log r + i\theta_p)$.

Let the point z describe a closed curve from an initial point z_0 in the z-plane (Fig. 14.5). Considering the curve C_1, which does not enclose the

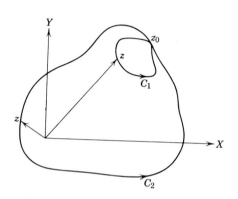

Fig. 14.5 Illustrating the multivaluedness of $\log z$.

point $z = 0$, we observe that $\log z$ returns to its initial value as z returns to its initial point on the curve. Considering the curve C_2 which does enclose the point $z = 0$, we see that $\log z$ changes by $\pm 2\pi i$ according to whether the curve is described in the positive or negative sense of θ. We thus conclude that to go from one branch to another of $\log z$, we must make a circuit enclosing the point $z = 0$; otherwise we remain in the same branch of the function. The point $z = 0$ is called a *branch point*.

For any particular branch of $\log z$, the derivative

$$\frac{d}{dz} \log z = \frac{1}{z}$$

is defined for all points except the point $z = 0$. Thus each branch of the logarithmic function is analytic at all points except $z = 0$.

14.11 Geometrical Significance of a Complex Function: Notion of Mapping

Consider the function $\zeta = \zeta(z)$. According to this function, for every given value of z a value of ζ is generated. Each value of z may be represented by a point in the complex plane of the variable z and each

---

* Note that although $\arg z$ is multivalued, z itself is single valued. The multiplicity of $\log z$ arises out of the multiplicity of $\arg z$.

corresponding value of ζ may be represented by a point in the complex plane of the variable ζ. Henceforth we shall refer to these planes respectively as the z- and ζ-planes. We thus see that *geometrically the function* $\zeta(z)$ *maps or transforms points of the z-plane into points of the ζ-plane.* In this sense a function $\zeta(z)$ may be considered as a *transformation* or a *mapping* function. It transforms curves and regions in the z-plane into curves and regions in the ζ-plane.

Generally it is not convenient to represent the two complex planes of z and ζ superposed on each other. Therefore to represent transformations

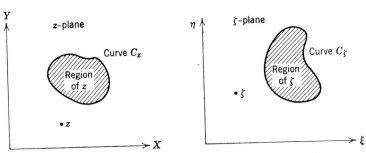

Fig. 14.6   Notation for mapping.

or mappings we utilize separate representations of the two planes (see, for example, Fig. 14.6). Writing $z = x + iy$ and $\zeta = \xi + i\eta$, the points in the z- and ζ-planes are given by the number pairs (x, y) and (ξ, η), respectively.

The meaning of the function $\zeta(z)$ as a transformation is also clear from the point of view of its real and imaginary parts, which are given by

$$\xi = \xi(x, y)$$
$$\eta = \eta(x, y)$$

This pair of real functions produce a pair of real numbers (ξ, η) for every given pair of real numbers (x, y).

14.12   Some Simple Transformations

We now consider the over-all features of some simple transformations.
1. Consider the transformation

$$\zeta(z) = z + b$$

where b is a *complex constant*. This represents a simple *translation* of the entire z-plane by the vector b (see Fig. 14.7). It follows that under this transformation the shape of any geometrical figure remains unchanged.

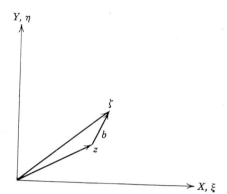

Fig. 14.7 Mapping giving rise to translation.

The mapping $\zeta = z$ is known as *identity* transformation, for it leaves every point unchanged. The points that map into themselves are called the *fixed points* of a transformation.

2. Consider next the transformation

$$\zeta = Az$$

where A, in general, may be a complex constant. To discuss the properties of this transformation it is convenient to use the polar form of a complex number. Let us therefore write

$$z = re^{i\theta}$$

$$A = ae^{i\beta}$$

We then have

$$\zeta = ar\, e^{i(\theta+\beta)}$$

or

$$|\zeta| = a\,|z|$$

$$\arg \zeta = \theta + \beta$$

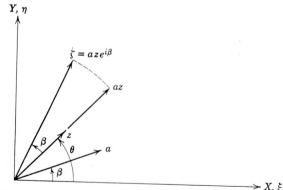

Fig. 14.8 Mapping giving rise to rotation and magnification.

It is thus seen that the vector z is *stretched* by the factor $a = |A|$ and then *rotated* through the angle $\beta = \arg A$ (see Fig. 14.8). Thus the transformation $\zeta = Az$ produces a *rotation* and *magnification*. Under such a transformation the shape (not the size and orientation) of any geometrical figure remains unchanged. The points ∞ and 0 are both fixed points of the mapping.

The transformation $\zeta = az$, where a is real, denotes a pure stretching, whereas the transformation $\zeta = ze^{i\beta}$ denotes a pure rotation by the angle β.

3. It follows that the transformation

$$\zeta(z) = Az + b$$

where A and b are complex constants, represents a translation, a rotation, and a magnification. Under such a transformation geometrical figures remain similar.

4. Consider next the transformation

$$\zeta(z) = \frac{1}{z}$$

The function $1/z$ is well defined except for the point $z = 0$ for which it becomes infinite. If we agree to speak of a point at infinity, the function $1/z$ may be considered defined throughout the z-plane. We use polar coordinates as before and write the transformation as

$$\zeta = \frac{1}{z} = \frac{1}{r}e^{-i\theta}$$

We thus have

$$|\zeta| = \frac{1}{|z|} = \frac{1}{r}$$

and

$$\arg \zeta = -\arg z = -\theta$$

The transformation may also be exhibited as

$$\zeta = \left(\frac{1}{r}e^{i\theta}\right)e^{-2i\theta} = z_1 e^{-2i\theta}$$

where z_1 is a point that lies on the same line as z and has a magnitude given by the relation

$$|z_1| = \frac{1}{|z|}$$

The point z_1 is said to be the *inversion* of z with respect to the unit circle

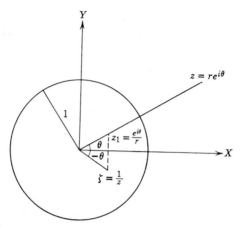

Fig. 14.9 Mapping giving rise to inversion and reflection.

(see Fig. 14.9). The vector ζ is obtained from z_1 by *reflection* in the real axis. Thus the transformation $\zeta = 1/z$ denotes an inversion with respect to the unit circle and then a reflection in the real axis. As may be verified, generally the shape of geometrical figures is not preserved by this transformation.

14.13 Conformal Transformation: Transformation by Analytic Functions

We shall now go into the nature of the transformation brought about by an analytic function. For this purpose we study the local changes in a transformation.

Let $\zeta = \zeta(z)$ be an analytic function. With this as a transformation, points and curves in the z-plane are mapped into points and curves in the ζ-plane. Let ζ be the *image* point in the ζ-plane of the point z in the z-plane. Also let curve C_ζ be the *image* in the ζ-plane of curve C_z in the z-plane (see Fig. 14.10). The direction along C_ζ corresponds to the direction along C_z. At z, let δz be an infinitesimal element of the curve C_z drawn in the direction of the curve. Then, at ζ, the infinitesimal element $\delta \zeta$ of the curve C_ζ is the image of δz.

The elements $\delta \zeta$ and δz are then related by

$$\delta \zeta = \frac{d\zeta(z)}{dz} \delta z \qquad (14.36)$$

where $d\zeta/dz$ is the derivative of $\zeta(z)$ at the point z.* The quantities $\delta \zeta$,

* Recall that $\zeta(z)$ is analytic, and therefore $d\zeta/dz$ does not depend on the direction of curve C_z.

$d\zeta/dz$, and δz are all complex numbers. To proceed, we assume that at the point z the derivative $d\zeta/dz$ does not vanish:

$$\frac{d\zeta}{dz} \neq 0$$

We shall later consider the case where $d\zeta/dz$ does vanish. From (14.36) we obtain

$$|\delta\zeta| = \left|\left(\frac{d\zeta}{dz}\right)_z\right| |\delta z| \qquad (14.37)$$

and

$$\arg \delta\zeta = \arg \left(\frac{d\zeta}{dz}\right)_z + \arg \delta z \qquad (14.38)$$

This shows that an element δz through a point z under the transformation $\zeta = \zeta(z)$ has its magnitude magnified by the factor $|d\zeta/dz|$ and has its

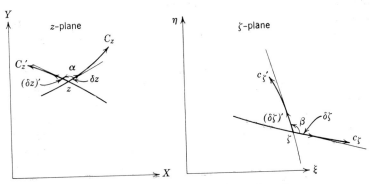

Fig. 14.10 Illustrating conformal transformation.

argument increased by the amount $\arg(d\zeta/dz)$, where the derivative $d\zeta/dz$ is evaluated at the point z. In this sense $d\zeta/dz$ is the *local scale factor* (*complex naturally*) *of the transformation*. Since $\zeta(z)$ is analytic, $d\zeta/dz$ is independent of the direction δz. Therefore *all infinitesimal elements passing through the point z are scaled by the same factor $d\zeta/dz$.*

Now, consider another curve C_z' passing through the same point z and let C_ζ' be its image passing through the point ζ. Let α be the angle of intersection between the curves C_z and C_z' measured from C_z to C_z', and let β be the angle of intersection between the curves C_ζ and C_ζ' measured from C_ζ to C_ζ' (see Fig. 14.10). Denote by $(\delta z)'$ an element of curve C_z' through z and by $(\delta\zeta)'$ its image. We then have the following relations

$$(\delta\zeta)' = \left(\frac{d\zeta}{dz}\right)(\delta z)' \qquad (14.39)$$

and

$$\arg{(\delta\zeta)}' = \arg\left(\frac{d\zeta}{dz}\right) + \arg{(\delta z)}' \tag{14.40}$$

Subtracting (14.38) from (14.40) we obtain

$$\left.\begin{array}{c}\arg{(\delta\zeta')} - \arg{(\delta\zeta)} = \arg{(\delta z)}' - \arg{(\delta z)}\\[4pt]\text{or}\\[4pt]\beta = \alpha\end{array}\right\} \tag{14.41}$$

We have thus proved that *under the transformation* $\zeta = \zeta(z)$, *which is analytic, the angle between any two curves passing through any point* z *at which* $d\zeta/dz \neq 0$ *is preserved and also its sense remains unchanged.* Such a transformation is known as a *conformal transformation.* Furthermore, all infinitesimal elements of length passing through z receive the same magnification of their magnitudes, in fact, the (complex) scale factor of the mapping is the same for all the elements passing through a given point.

Summing up the preceding considerations we may state that the mapping affected by an analytic function is conformal in a neighborhood of every point at which the derivative of the mapping function does not vanish. Elemental figures at any point z go into elemental figures at the image point ζ. Since the transformation is conformal, the figures remain similar. Another way of saying it is that when regions are transformed by an analytic function, geometrical similarity is preserved for the minute structure of the regions. The scale of mapping at any point is given by $d\zeta(z)/dz$, a complex number, and generally this scale factor changes from point to point of the z-plane. Because of this the regions involved in a transformation may not look alike on the aggregate although their elemental structures are similar.

In concluding this section we point out the significance of the requirement that the transformation be an analytic function. If it is not an analytic function, the derivative $d\zeta/dz$ may have different values in different directions. In such a case the angle between any two curves passing through a point will not be preserved, that is, the mapping will no longer be conformal. In light of these considerations, one could say that the definition of an analytic function could have been based on the requirement that a transformation be conformal. Recalling our inquiry with regard to a transformation that would relate a potential problem with respect to one geometry with that of another geometry, we now see that what we are after is a conformal transformation.

14.14 Critical Point of a Transformation

We now consider the nature of the transformation at a point where the derivative of the mapping function is zero. Such a point is referred to as the *critical point of the transformation*. Consider first the example of a transformation $\zeta = z^n$, where n is a positive integer. The function is *analytic* in the entire z-plane with the derivative

$$\zeta' \equiv \frac{d\zeta}{dz} = nz^{n-1}$$

This is zero at the origin, whereas everywhere else it is different from zero. Therefore the transformation is conformal everywhere except *perhaps for the one point $z = 0$, which is a critical point.*

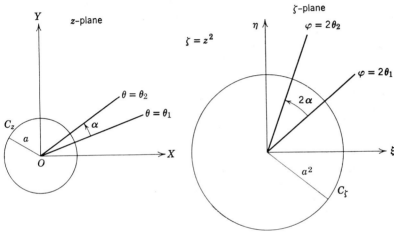

Fig. 14.11 Illustrating the nonconformal nature at a critical point of a transformation.

Writing $z = re^{i\theta}$ and $\zeta = \rho e^{i\varphi}$, we express the transformation as

$$\rho e^{i\varphi} = r^n e^{in\theta}$$

or

$$\rho = r^n$$

$$\varphi = n\theta$$

From this we conclude that circles about the origin in the z-plane (i.e., curves defined by $r = |z| = $ const.) transform into circles about the origin in the ζ-plane and rays emanating from the origin in the z-plane (i.e., curves defined by $\theta = \arg z = $ const.) transform into rays emanating from the origin in the ζ-plane. See Fig. 14.11, where the transformation $\zeta = z^2$ is used. If $\theta = \theta_1$ and $\theta = \theta_2$ are any two rays in the z-plane, their images in the ζ-plane are given by $\varphi = n\theta_1$ and $\varphi = n\theta_2$. Therefore the angle between the image rays is n times the angle between the rays in the z-plane. In other words, the mapping $\zeta = z^n$ is not conformal at the critical point $z = 0$. At this point the transformation has the property of multiplying angles by n.

If instead of the transformation $\zeta = z^n$ we consider the transformation $\zeta(z) = (z - z_0)^n$ we conclude, on the same lines as above, that the transformation is conformal everywhere except at the critical point $z = z_0$, where it has the property of multiplying angles by n.

Now let us consider a transformation in the general form $\zeta = \zeta(z)$ and let us assume that the point $z = z_0$ is a critical point. We wish to know the nature of the transformation at this point. The function $\zeta(z)$ is analytic and the point z_0 and its neighborhood are regular points. Let ζ_0 be the image of z_0. For points ζ in the neighborhood of ζ_0, using Taylor's expansion, we may write

$$\zeta(z) = \zeta(z_0) + a_1(z - z_0) + a_2(z - z_0)^2 + \cdots$$
$$+ a_{n-1}(z - z_0)^{n-1} + a_n(z - z_0)^n + a_{n+1}(z - z_0)^{n+1} + \cdots \quad (14.42)$$

where the coefficients a_k are given by

$$a_k = \frac{1}{k!}\left(\frac{d\zeta}{dz^k}\right)_{z=z_0}$$

Since z_0 is a critical point, $d\zeta/dz$ is zero at that point. Consequently, the coefficient a_1 is zero. Not knowing the function $\zeta(z)$ explicitly, we have no knowledge of its higher derivatives at z_0. *Let us suppose that the first $(n - 1)$ derivatives are zero at the point $z = z_0$.* Then the expansion (14.42) reduces to

$$\zeta = \zeta_0 + a_n(z - z_0)^n \left[1 + \frac{a_{n+1}}{a_n}(z - z_0) + \frac{a_{n+2}}{a_n}(z - z_0)^2 + \cdots \right]$$

This may further be rewritten as

$$\zeta - \zeta_0 = (z - z_0)^n f(z) \quad (14.43)$$

where $f(z)$ is the function given by a_n times the expression in the square brackets. We thus see that *the transformation $\zeta = \zeta(z)$ near a critical point may be represented in the form of the equation* (14.43). We observe that the derivative df/dz at the point $z = z_0$ does not vanish. In other words, the point z_0 is not a critical point of the function $f(z)$ and this function is a conformal transformation at the point z_0. With this understanding we realize that the point $z = z_0$ is a critical point of the transformation $\zeta = \zeta(z)$ only through the function $(z - z_0)^n$. From the previous discussion about the properties of the transformation $(z - z_0)^n$, we conclude the following.

The transformation $\zeta = \zeta(z)$ is not conformal at a critical point. At such a point it has the property of multiplying angles by n. The factor n is the order of the derivative $d^n\zeta/dz^n$, which first becomes nonzero at the critical point.

Let us apply this result to the transformation $\zeta = z^n$. At the critical point $z = 0$, not only the first derivative of ζ is zero but all the first $n - 1$ derivatives are also zero. This means the transformation $\zeta = z^n$ is non-conformal at $z = 0$, where it multiplies all angles by the factor n.

14.15 Complex Integrals

We now pass on to the integration of complex functions. In the next sections we shall develop several important results relating to complex integrals. As we shall see, many of the considerations underlying these

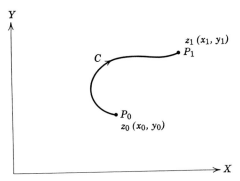

Fig. 14.12   Defining a line integral.

results are similar to those introduced in Sections 9.13 and 9.14, which deal with circulation and its relation to the convectivity of space. The notions of simply and multiply connected regions and of reducible and irreducible circuits will again be employed.

Consider that a complex function $\zeta = \zeta(z)$, not necessarily analytic, is defined in a certain region of the complex plane z. In that region let C be any path (or curve) joining two arbitrary points P_0 and P_1 (see Fig. 14.12). The complex numbers denoting P_0 and P_1 are z_0 and z_1, respectively. The expression

$$\int_{z_0}^{z_1} \zeta(z)\, dz$$

is known as the (line) integral of $\zeta(z)$ from z_0 to z_1 along the path C. The actual value of this integral, in general, depends on the path C and the endpoints z_0 and z_1.

The foregoing definite integral of a function of complex variable $z = x + iy$ may be expressed in terms of line integrals of the real and imaginary parts of the function. For this purpose we set

$$\zeta(z) = \xi(x, y) + i\eta(x, y)$$

and obtain

$$\int_{z_0}^{z_1} \zeta(z)\, dz = \int_{x_0,y_0}^{x_1,y_1} (\xi\, dx - \eta\, dy) + i\int_{x_0,y_0}^{x_1,y_1} (\eta\, dx + \xi\, dy) \quad (14.44)$$

where (x_0, y_0) and (x_1, y_1) are, respectively, the points z_0 and z_1 and the integrations are performed along the given curve C. This equation is the basis of further discussion.

14.16 The Cauchy Integral Theorem

We are most interested in complex integrals whose values depend only on the endpoints and not on the particular path that joins them. To investigate such integrals we think of the complex integrals in terms of the line integrals of real and imaginary parts (as expressed by 14.44) and ask for the conditions under which the values of these integrals depend only on the endpoints and not on the path that joins them.

From calculus* we recall the important theorem: *If a region is simply connected and in that region $u(x, y)$, $v(x, y)$ are differentiable functions, then the line integral*

$$\int_{P_0(x_0,y_0)}^{P_1(x_1,y_1)} [u(x, y)\, dx + v(x, y)\, dy]$$

will be independent of the path joining P_0 and P_1 if and only if

$$\frac{\partial u}{\partial y} = \frac{\partial v}{\partial x}$$

This condition assures us that the differential $u\, dx + v\, dy$ is an exact differential, say of the function $\phi(x, y)$. We then have

$$\int_{P_0}^{P_1} (u\, dx + v\, dy) = \int_{P_0}^{P_1} d\phi = \phi(P_1) - \phi(P_0)$$

This by itself does not imply that $\phi(P_1) - \phi(P_0)$ is the same no matter what path joins the endpoints P_0 and P_1. For $\phi(P_1) - \phi(P_0)$ to be independent of the path, we must require that $\phi(x, y)$ be single valued. The single-valuedness of ϕ is assured by the requirement that the region under consideration be simply connected.

We now apply the above theorem to the line integrals

$$\int_{x_0,y_0}^{x_1,y_1} (\xi\, dx - \eta\, dy)$$

* See for instance Courant (1934).

and

$$\int_{x_0,y_0}^{x_1,y_1} (\eta \, dx + \xi \, dy)$$

which constitute the real and imaginary parts of the complex integral

$$\int_{z_0}^{z_1} \zeta(z) \, dz$$

For a simply connected region and differentiable ξ and η, these integrals will be independent of the path if the following conditions are satisfied:

$$\frac{\partial \xi}{\partial y} = -\frac{\partial \eta}{\partial x}$$

$$\frac{\partial \xi}{\partial x} = \frac{\partial \eta}{\partial y}$$

But these are precisely the Cauchy-Riemann conditions. If these are satisfied, the function $\zeta(z)$ is analytic.

We thus arrive at the important theorem: *In a simply connected region the integral*

$$\int_{z_0}^{z_1} \zeta(z) \, dz$$

is independent of the path of integration if $\zeta(z)$ is analytic in that region. In such a case $\zeta(z) \, dz$ will be given an exact differential of some function, say $F(z)$. We will then have

$$\int_{z_0}^{z_1} \zeta(z) \, dz = \int_{z_0}^{z_1} d[F(z)] = F(z_1) - F(z_0)$$

Consider now the integration of a complex function $\zeta(z)$ around a circuit (i.e., a closed curve) C (Fig. 14.13). *For the direction of integration along a closed curve we adopt the convention that the positive sense is that for which the region enclosed by the curve lies on the left.* The integral of $\zeta(z)$ around C is then denoted by

$$\oint_C \zeta(z) \, dz$$

Let $P_0 = z_0$ and $P_1 = z_1$ be any two arbitrary points on the circuit and let C_1 and C_2 denote its two parts that connect z_0 to z_1 (see Fig. 14.13). Positive orientation of curves C_1 and C_2 is from z_0 to z_1. We then have

$$\oint_C \zeta(z) \, dz = \int_{z_0}^{z_1} \quad \text{along} \quad C_1 + \int_{z_1}^{z_0} \quad \text{along} \quad -C_2$$

If the region is simply connected and $\zeta(z)$ is analytic, it follows from the previous theorem that

$$\int_{z_0}^{z_1} \text{ along } C_1 = \int_{z_0}^{z_1} \text{ along } C_2 = -\int_{z_1}^{z_0} \text{ along } -C_2$$

or

$$\oint_C \zeta(z)\, dz = 0$$

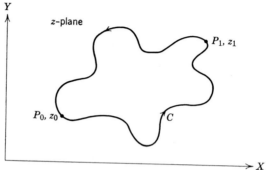

Fig. 14.13   Illustrating the line integral around a circuit.

We thus arrive at the famous *Cauchy Integral theorem: If $\zeta(z)$ is analytic in a simply connected region, then*

$$\oint_C \zeta(z)\, dz = 0 \tag{14.45}$$

for all closed curves C in that region. The contents of this and the previous theorem are equivalent.*

14.17   Integration in Multiply Connected Regions

On the basis of the Cauchy theorem several significant results may be derived for complex integrals in multiply connected regions. Such regions may arise, as we know, because of physical considerations in a problem, or they may arise because of our desire to exclude from a region of consideration points at which a complex function is singular. For instance, suppose that in a certain region bounded by the closed curve C the function $f(z)$ is analytic everywhere except at certain points of the region, say at the points $z_1, z_2, \ldots z_n$ (see Fig. 14.14). If we are to concern ourselves only with

* The Cauchy theorem may be proved for less restrictive conditions than assumed here. The interested reader should refer to the works cited at the end of this book.

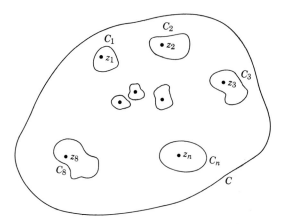

Fig. 14.14 Multiply connected region resulting from nonanalyticity of a complex function.

the region of analyticity of the function $f(z)$, we surround the singular points by closed curves $C_1 \ldots C_n$ and exclude from consideration the regions included by these curves. This artifice makes the region of interest multiply connected.

We consider for simplicity a doubly connected region such as that exterior to a closed curve $\mathscr{C}$ (see Fig. 14.15) and a function $\zeta(z)$ that is analytic in that region. Consider first a *reducible circuit* such as C_1 (see figure). The region enclosed by C_1 is simply connected and by the Cauchy

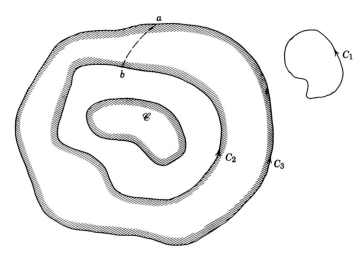

Fig. 14.15 Illustrating integration in a doubly connected region.

theorem we obtain

$$\oint_{C_1} \zeta(z)\, dz = 0$$

Consider next an *irreducible circuit* such as C_2. Now we cannot use the Cauchy theorem and assert that the integral

$$\oint_{C_2} \zeta(z)\, dz$$

vanishes. It may or may not be zero.

Let C_3 be another *irreducible circuit* but *reconcilable* with C_2 (see figure). The region bounded by the curves C_2 and C_3 is not simply connected but is, in fact, doubly connected. We render this region simply connected by introducing the *barrier* such as ab from a point a on C_3 to a point b on C_2 (see figure). We then apply the Cauchy theorem to the integral over the new boundary curve of this region. The new boundary is the closed curve $C_3 + ab - C_2 + ba$. The positive sense of the curve is determined according to the convention that the enclosed region should lie on the left. We then obtain

$$\oint_{C_3} \zeta(z)\, dz - \oint_{C_2} \zeta(z)\, dz = 0$$

or

$$\oint_{C_3} \zeta(z)\, dz = \oint_{C_2} \zeta(z)\, dz$$

These results for an *analytic function* $\zeta(z)$ in a *doubly connected region* may be summed up as follows:

1. The integral $\oint \zeta(z)\, dz$ over any reducible circuit is zero.
2. The integral over any irreducible circuit may not vanish in general. At this stage its value cannot be evaluated.*
3. The integral has the same value for all irreducible circuits.

These results may be readily extended to multiply connected regions. Consider a n-ply connected region that is bounded by the curves $C, C_1, C_2, \ldots, C_n$, where the interiors of $C_1 \ldots C_n$ are contained in the interior of C (see Fig. 14.16). This region may be rendered simply connected by introducing nonintersecting barriers as shown by dotted lines in the figure. If $\zeta = \zeta(z)$ is analytic in that region, we obtain the following results:

1. The value of the integral $\oint \zeta(z)\, dz$ over any reducible curve is zero.
2. The value of the integral over the circuit C is equal to the sum of the integrals over the circuits $C_1 \ldots C_n$, where all the integrations are carried

* See Section 14.23 for this evaluation.

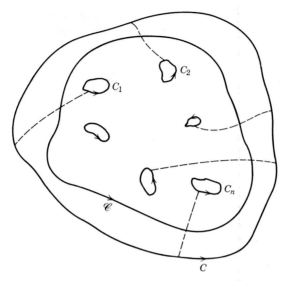

Fig. 14.16   Illustrating integration in a multiply connected region.

out in the same direction. We write

$$\oint_C = \oint_{C_1} + \oint_{C_2} + \cdots + \oint_{C_n} \tag{14.46}$$

The value of the integral over C is the same as over any other irreducible curve that encloses all the curves $C_1 \ldots C_n$ and lies in the region under consideration:

$$\oint_C = \oint_{\mathscr{C}}$$

3. The value of the integral over any two reconcilable but irreducible circuits $\mathscr{C}_1$ and $\mathscr{C}_2$, both of which enclose only the curves $C_1 \ldots$ to C_k and lie in the region under consideration, is the same and is equal to the sum of the integrals over the curves C_1 to C_k:

$$\oint_{\mathscr{C}_1} = \oint_{\mathscr{C}_2} = \oint_{C_1} + \oint_{C_2} + \cdots \oint_{C_k} \tag{14.47}$$

14.18   Some Simple Integrals

To illustrate the preceding considerations we seek the integral of the power function*

$$\zeta(z) = z^n$$

* The following considerations apply equally well to the function Az^n, where A is a complex constant.

where n is a positive or negative integer. The integral of this function is of vital significance in the applications and, in fact, in the integration of more general functions. We treat four separate cases.

1. *Case when $n = 0$.* Then the function is simply a constant equal to unity, and we obtain

$$\oint \zeta(z)\, dz = \oint dz = 0$$

for any curve whatsoever.

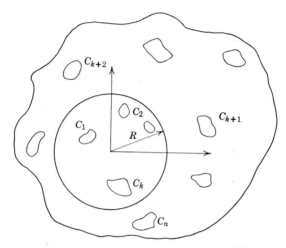

Fig. 14.17 Illustrating the integration of z^n in a multiply connected region.

2. *Case when $n > 0$.* Now the derivative of the function is given by

$$\frac{d\zeta}{dz} = nz^{n-1}$$

Since $n > 1$, the derivative exists everywhere and the function is analytic everywhere. We may immediately state on the basis of Cauchy's theorem that in a simply connected region

$$\oint z^n\, dz = 0 \quad \text{with} \quad n > 0$$

for any circuit whatsoever.

Consider a multiply connected region such as is shown in Fig. 14.17. For such a region, on the basis of the results given in the previous section, it is sufficient only to consider the value of the integral over *irreconcilable irreducible* circuits. Furthermore, the shape of the circuit is indeed immaterial. Consider then a circle $z = R$ and say it is an irreducible circuit that includes in its interior the excluded interiors of the curves $C_1 \ldots C_k$ (see Figs. 14.16 and 14.17). We now evaluate the integral around the circle. Using the polar form we obtain

$$
\oint_{\text{circle } R} z^n \, dz = \oint_{\text{circle } R} (Re^{i\theta})^n i Re^{i\theta} \, d\theta
$$

$$
= iR^{n+1} \oint e^{i(n+1)\theta} \, d\theta
$$

$$
= iR^{n+1} \frac{1}{n+1} e^{i(n+1)\theta} \Big|_0^{2\pi}
$$

$$
= 0
$$

From this and the results given in the previous section it follows that the integral of z^n, $n > 0$ over any circuit whatsoever lying even in a multiply connected region vanishes.

3. *Case when $n < -1$.* Let us write $n = -m$, where m is a positive integer greater than 1. We then have

$$
\zeta(n) = \frac{1}{z^m}, \qquad m > 1
$$

and

$$
\frac{d\zeta}{dz} = -\frac{m}{z^{m+1}}
$$

It follows that the function is analytic everywhere except at the point $z = 0$. We exclude this point by enclosing it in the interior of a circuit C. If we assume that there are no other excluded regions (which may arise due to physical considerations in a problem), the region external to C is a doubly connected region. Then from the results of the previous section we observe that for any circuit not enclosing the point $z = 0$,

$$
\oint \frac{1}{z^m} \, dz = 0, \qquad m > 1
$$

and that for all circuits enclosing the point $z = 0$ the integral has the same value. We evaluate this value by choosing a circle $|z| = R$. We shall then find that its value is zero.

Suppose that in addition to the singularity at the point $z = 0$, there are other excluded regions. The region then is n-ply connected instead of being double connected. Even in such a situation we shall find, just in the same manner as was done above for $n > 0$, that the integral over any circuit whatsoever is zero.

We thus conclude that for any circuit whatsoever we obtain the result

$$\oint \frac{1}{z^m}\, dz = 0, \qquad m > 1$$

whether the region under consideration is doubly connected or n-ply connected.

4. *Case when $n = -1$.* We then have

$$\zeta(z) = \frac{1}{z}$$

and

$$\frac{d\zeta}{dz} = -\frac{1}{z^2}$$

The function $1/z$ is therefore analytic everywhere except at the point $z = 0$. We exclude this point by enclosing it in the interior of circuit C. We assume that there are no other excluded regions. Then for any circuit not enclosing the point $z = 0$,

$$\oint \frac{1}{z}\, dz = 0$$

and for all circuits enclosing the point $z = 0$, the integral has the same value. This value is readily found. We have

$$\oint_C \frac{dz}{z} = \log z \Big|_{z_1}^{z_2}$$

where z_1 and z_2 are coincident points on the circuit C. We know that $\log z$ is multivalued and that its value changes by $2\pi i$ over any positively directed circuit that encloses the point $z = 0$ once (see Section 14.10). We thus obtain

$$\oint_C \frac{dz}{z} = 2\pi i$$

for every circuit C that contains in its interior the point $z = 0$.

The same conclusions are valid, as may be verified, if in addition to the singularity at $z = 0$ there are other excluded regions and the region is thus n-ply connected.

The above results concerning the integral of the function

$$\zeta(z) = z^n$$

where n is a positive or negative integer, may be summed up as follows: In any simply or multiply connected region

1. $$\oint_C z^n\, dz = 0, \qquad n \gtrless 0 \tag{14.48}$$

for every circuit C.

2. $\oint_C \dfrac{dz}{z^m} = 0, \qquad m > 1$ 　　　　　　　　　　　(14.49)

for every circuit C.

3. $\oint_C \dfrac{dz}{z} = 0$ 　　(for every circuit not enclosing $z = 0$)　　(14.50)

$\qquad\qquad = 2\pi i$ 　　(for every circuit enclosing $z = 0$ once).

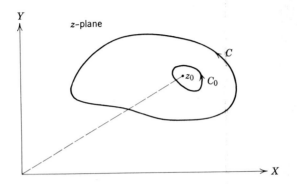

Fig. 14.18　Illustrating the derivation of the Cauchy integral formula.

For the function $\zeta(z) = (z - z_0)^n$ we readily obtain the results:

1. $\oint_C (z - z_0)^n \, dz = 0, \qquad n \gtrless 0 \quad \text{or} \quad n < -1$ 　　　(14.51)

for every circuit C.

2. $\oint \dfrac{dz}{(z - z_0)^n} = 0$ 　　(for every circuit not enclosing $z = z_0$) (14.52)

$\qquad\qquad = 2\pi i$ 　　(for every circuit enclosing $z = z_0$ once).

14.19　The Cauchy Integral Formula

Let us consider a simply connected region and seek the integral of the function

$$\frac{\zeta(z)}{z - z_0}$$ 　　　　　　　　　　　　　　　　　(14.53)

where $\zeta(z)$ is analytic in that region and z_0 is any point in it. This function (14.53) is analytic everywhere except at $z = z_0$. We exclude this point by a closed circuit C_0, as shown in Fig. 14.18. From the results of Section 14.17

we know that

$$\oint_C \frac{\zeta(z)}{z - z_0}\, dz = 0$$

where C is any circuit in the region and does *not* contain z_0 in its interior, and that

$$\oint_C \frac{\zeta(z)}{z - z_0}\, dz = \oint_{C_0} \frac{\zeta(z)}{z - z_0}\, dz \qquad (14.54)$$

where C is any circuit in the region and contains z_0 in its interior (see Fig. 14.18).

We express the right-hand side of (14.54) as

$$\oint_{C_0} \frac{\zeta(z)}{z - z_0}\, dz = \zeta(z_0)\oint_{C_0} \frac{dz}{z - z_0} + \oint_{C_0} \frac{\zeta(z) - \zeta(z_0)}{z - z_0}\, dz$$

$$= 2\pi i\zeta(z_0) + \oint_{C_0} \frac{\zeta(z) - \zeta(z_0)}{z - z_0}\, dz \qquad (14.55)$$

Recall that all these integrals are independent of the choice of the curve C_0 (or, equivalently, of the curve C) as long as C_0 lies in the region of interest and contains in its interior z_0. Let us choose C_0 as the circle $|z - z_0| = \rho$ and write

$$\oint_{C_0} \frac{\zeta(z) - \zeta(z_0)}{z - z_0}\, dz = \oint_{\text{circle } \rho} \frac{\zeta(z) - \zeta(z_0)}{z - z_0}\, dz \qquad (14.56)$$

Now, if $|\zeta(z) - \zeta(z_0)|$ does not become infinite on the circle (which is what we assume), there must be some number ε such that on the circle

$$|\zeta(z) - \zeta(z_0)| \lessgtr \varepsilon$$

Then the absolute value of the integral (14.56) is equal to or less than the product: length of the circle times the absolute value of the integrand. Therefore we have

$$\left| \oint_{\text{circle } \rho} \frac{\zeta(z) - \zeta(z_0)}{z - z_0}\, dz \right| \lessgtr 2\pi\rho\, \frac{\varepsilon}{\rho} \quad \text{or} \quad 2\pi\varepsilon$$

The number ε, and consequently the absolute value of the integral, can be made smaller and smaller by choosing the radius of the circle smaller and smaller. But the integral does not depend on ρ. Hence the integral should vanish:

$$\oint_{\text{circle } \rho} \frac{\zeta(z) - \zeta(z_0)}{z - z_0}\, dz = 0$$

It therefore follows that (14.54) reduces to

$$\oint_C \frac{\zeta(z)}{z - z_0}\, dz = 2\pi i \zeta(z_0) \tag{14.57}$$

This result constitutes the *Cauchy Integral Formula: In a simply con-nected region in which $\zeta(z)$ is an analytic function*

$$\zeta(z_0) = \frac{1}{2\pi i} \oint_C \frac{\zeta(z)}{z - z_0}\, dz \tag{14.58}$$

where C is any circuit in the region and z_0 is any point in its interior. If we use the letter z instead of z_0 for the general point, we express the formula as

$$\zeta(z) = \frac{1}{2\pi i} \oint_C \frac{\zeta(\alpha)}{\alpha - z}\, d\alpha \tag{14.59}$$

where now α is the (dummy) variable of integration.

Cauchy's integral formula is a very remarkable statement. It shows the strong interrelation among the various values of an analytic function. The values of the function at every point of a region are determined by its values on the boundary of the region. To know the values in the interior we need specify only the values of the boundary. The connection with the properties of the solution of Laplace's equation is readily seen.

14.20 Unlimited Differentiability of an Analytic Function

Using Cauchy's integral formula, we show that an analytic function $\zeta = \zeta(z)$ may be repeatedly differentiated any number of times. By definition we have

$$\left(\frac{d\zeta}{dz}\right)_{\text{at } z} = \lim_{z_0 \to z} \frac{\zeta(z_0) - \zeta(z)}{z_0 - z}$$

where z and z_0 are any two neighboring points. If the region is simply connected and $\zeta(z)$ is analytic in it, and if C is any curve that lies in the region and contains z and z_0, using (14.59) we obtain

$$\frac{d\zeta}{dz} = \lim_{z_0 \to z} \frac{1}{z_0 - z} \frac{1}{2\pi i} \oint_C \left(\frac{1}{\alpha - z_0} - \frac{1}{\alpha - z} \right) \zeta(\alpha)\, d\alpha$$

$$= \frac{1}{2\pi i} \lim_{z_0 \to z} \oint_C \frac{\zeta(\alpha)}{(\alpha - z_0)(\alpha - z)}\, d\alpha$$

$$= \frac{1}{2\pi i} \oint \frac{\zeta(\alpha)}{(\alpha - z)^2}\, d\alpha$$

Repeating this process, we find

$$\frac{d^2\zeta}{dz^2} = \frac{2!}{2\pi i} \oint_C \frac{\zeta(\alpha)}{(\alpha - z)^3} \, d\alpha$$

$$\frac{d^n\zeta}{dz^n} = \frac{n!}{2\pi i} \oint_C \frac{\zeta(\alpha)}{(\alpha - z)^n} \, d\alpha \qquad (14.60)$$

We thus obtain an *integral representation for the derivatives of an analytic function*. The integrals involved may be shown to be finite. Therefore we conclude that a complex function that is once differentiable may be differentiated any number of times. The function and all its derivatives are analytic. We repeat this important result: *An analytic function has derivatives of all orders in the region in which it is analytic.*

14.21 Taylor Series

Since a power function z^n, where n is a positive integer, is an analytic function, a power series $\Sigma A_n z^n$, where A_n is a complex coefficient, represents in its circle of convergence an analytic function. We now show

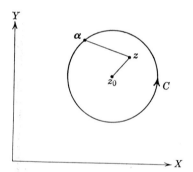

Fig. 14.19   Illustrating the derivation of Taylor series.

that if $\zeta(z)$ is an analytic function, then it may be expressed as a power series in a circle about any point of the region of its analyticity.

Consider a simply connected region in which the function $\zeta(z)$ is an analytic function. Let z and z_0 be any two points in that region and let C be any circle about z_0 such that z is contained in the interior of the circle (see Fig. 14.19). According to the Cauchy formula we have

$$\zeta(z) = \frac{1}{2\pi i} \oint_C \frac{\zeta(\alpha)}{\alpha - z} \, d\alpha$$

$$= \frac{1}{2\pi i} \oint_C \frac{\zeta(\alpha)}{(\alpha - z_0) - (z - z_0)} \, d\alpha \qquad (14.61)$$

where the number $\alpha - z$ is expressed in terms of the numbers $\alpha - z_0$ and $z - z_0$.

Now we write

$$\frac{1}{(\alpha - z_0) - (z - z_0)} = \frac{1}{\alpha - z_0} \frac{1}{1 - (z - z_0)/(\alpha - z_0)}$$

If z is in the interior of the circle C,

$$|z - z_0| < |\alpha - z_0|$$

and we may use the series representation

$$\left(1 - \frac{z - z_0}{\alpha - z_0}\right)^{-1} = \sum_{n=0}^{\infty} \left(\frac{z - z_0}{\alpha - z_0}\right)^n$$

which converges. Introducing this series we rewrite Eq. (14.61) as

$$\begin{aligned}
\zeta(z) &= \frac{1}{2\pi i} \oint_C \frac{\zeta(\alpha)}{\alpha - z} \, d\alpha \\
&= \frac{1}{2\pi i} \oint_C \zeta(\alpha) \left[\sum_{n=0}^{\infty} \frac{(z - z_0)^n}{(\alpha - z_0)^{n+1}}\right] d\alpha \\
&= \sum_{n=0}^{\infty} \left[\frac{1}{2\pi i} \oint_C \frac{\zeta(\alpha)}{(\alpha - z_0)^{n+1}} \, d\alpha\right] (z - z_0)^n \\
&= \sum_{n=0}^{\infty} A_n (z - z_0)^n
\end{aligned} \tag{14.62}$$

where

$$A_n = \frac{1}{2\pi i} \oint_C \frac{\zeta(\alpha)}{(\alpha - z_0)^{n+1}} \, d\alpha \tag{14.63}$$

Using the equation (14.60) of the previous section we observe that

$$A_n = \frac{1}{n!} \left(\frac{d^n \zeta}{dz^n}\right)_{z=z} \tag{14.64}$$

Equation (14.62) then becomes

$$\begin{aligned}
\zeta(z) &= \sum_{n=0}^{\infty} \frac{1}{n!} \left(\frac{d^n \zeta}{dz^n}\right)_{z=z_0} (z - z_0)^n \\
&= \zeta(z_0) + \left(\frac{d\zeta}{dz}\right)_{z_0} (z - z_0) + \left(\frac{d^2 \zeta}{dz^2}\right)_{z_0} \frac{(z - z_0)^2}{2!} + \cdots
\end{aligned} \tag{14.65}$$

This is the Taylor series for $\zeta(z)$ about the point z_0. The series converges in any circle contained in the circle C. It is interesting to observe that in the interior of the circle C the value of the function $\zeta(z)$ at every point is completely determined by the values of the function and its derivatives at the point z_0.

14.22   Laurent Series

The Taylor series is an expansion for an analytic function about a regular point lying in a simply connected region in which the function is analtyic. We now consider a doubly connected region and inquire into the possibility of expanding an analytic function about a point lying in the excluded region. If the excluded region contains singular points of the function, our inquiry is with regard to the expansion of the function about a singular point.

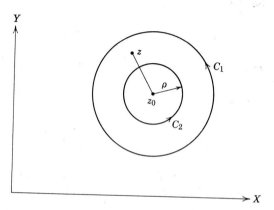

Fig. 14.20   Illustrating the derivation of Laurent series.

Consider an annular region bounded by two circles C_1 and C_2 concentric with respect to the point z_0 (see Fig. 14.20). In that region let the function $\zeta = \zeta(z)$ be analytic. If z is any point in that region, then according to the Cauchy integral formula we have

$$\zeta(z) = \frac{1}{2\pi i}\oint_C \frac{\zeta(\alpha)}{\alpha - z}\,d\alpha$$

$$= \frac{1}{2\pi i}\oint_{C_1} \frac{\zeta(\alpha)}{\alpha - z}\,d\alpha - \frac{1}{2\pi i}\oint_{C_2} \frac{\zeta(\alpha)}{\alpha - z}\,d\alpha \qquad (14.66)$$

As in the previous section we can express

$$\frac{1}{2\pi i}\oint_{C_1} \frac{\zeta(\alpha)}{\alpha - z}\,d\alpha = \sum_{n=0}^{\infty} A_n(z - z_0)^n \qquad (14.67)$$

where

$$A_n = \frac{1}{2\pi i}\oint_{C_1} \frac{\zeta(\alpha)}{(\alpha - z_0)^{n+1}}\,d\alpha \qquad (14.68)$$

To handle the integral around C_2 we proceed in a similar manner. Since for α on C_2

$$|\alpha - z_0| < |z - z_0|$$

we write

$$\frac{1}{z - \alpha} = \frac{1}{(z - z_0) - (\alpha - z_0)} = \frac{1}{z - z_0}\left(1 - \frac{\alpha - z_0}{z - z_0}\right)^{-1}$$

$$= \frac{1}{z - z_0} \sum_{n=0}^{\infty} \left(\frac{\alpha - z_0}{z - z_0}\right)^n$$

Then the integral over C_2 in Eq. (14.66) may be written as

$$-\frac{1}{2\pi i} \oint_{C_2} \frac{\zeta(\alpha)}{\alpha - z} \, d\alpha = \sum_{n=1}^{\infty} A_{-n}(z - z_0)^{-n} \qquad (14.69)$$

where

$$A_{-n} = \frac{1}{2\pi i} \oint_{C_2} \zeta(\alpha)(\alpha - z_0)^{n-1} \, d\alpha \qquad (14.70)$$

Combining the results we obtain

$$\zeta(z) = \sum_{n=0}^{\infty} A_n(z - z_0)^n + \sum_{n=1}^{\infty} A_{-n}(z - z_0)^{-n} \qquad (14.71)$$

Introducing a single notation, this may be rewritten as

$$\zeta(z) = \sum_{n=-\infty}^{\infty} A_n(z - z_0)^n$$

$$= A_0 + A_1(z - z_0) + A_2(z - z_0)^2 + \cdots$$

$$+ \frac{A_{-1}}{z - z_0} + \frac{A_{-2}}{(z - z_0)^2} + \cdots \qquad (14.72)$$

where

$$A_n = \frac{1}{2\pi i} \oint_C \frac{\zeta(\alpha)}{(\alpha - z_0)^{n+1}} \, d\alpha \quad (n \text{ an integer}) \qquad (14.73)$$

The curve C may be taken as any circle between C_1 and C_2 or any equivalent circuit. This series representation, Eq. (14.71) or (14.72), for an analytic function in an annular region is known as the *Laurent expansion*. The series converges for z in the annular region, that is, for every z such that $\rho_2 < |z - z_0| < \rho_1$, where ρ_1 and ρ_2 are, respectively, the radii of the outer and inner circles.

14.23   Integration of a Function with Singularities: the Residue Theorem

Consider a region interior to a circuit C. Let $\zeta(z)$ be a function that is analytic everywhere in the region except at a point z_0. From the results in

Section 14.16 we know that the integral

$$\oint \zeta(z)\, dz$$

vanishes for all circuits lying in the region and *not* enclosing the point z_0 and that it has the same value for all circuits in the region and enclosing the point z_0. The value of the integral could not, however, be determined at that time. Now, using the Laurent expansion, we can immediately find its value.

Let C_0 be a circuit in the interior of C such that the singular point z_0 is contained in the interior of C_0. Then, according to the Laurent expansion, about the point z_0 (14.72), the value of $\zeta(z)$ at any point on C_0 is given by

$$\zeta(z) = \sum_{-\infty}^{\infty} A_n (z - z_0)^n$$

We therefore obtain

$$\oint_{C_0} \zeta(z)\, dz = \sum_{-\infty}^{\infty} A_n \oint (z - z_0)^n$$
$$= 2\pi i A_{-1} \tag{14.74}$$

since all the integrals except that of $(z - z_0)^{-1}$ vanish, and the value of $\oint (z - z_0)^{-1}\, dz$ is simply equal to $2\pi i$ (see Section 14.18). The coefficient A_{-1} is assumed known from the Laurent expansion. Hence the value of the integral is found.

The coefficient of $(z - z_0)^{-1}$ term in the Laurent expansion of a function $\zeta(z)$ about z_0, which is a singular point of $\zeta(z)$, is called the residue of $\zeta(z)$ at z_0. Equivalently, the residue of $\zeta(z)$ at a singular point z_0 is equal to

$$\frac{1}{2\pi i} \oint_{C_0} \zeta(z)\, dz$$

where C_0 is a circuit that contains only z_0 in its interior. Equation (14.74) states *the value of the integral is equal to $2\pi i$ times the residue.*

Consider now an arbitrary region in which $\zeta(z)$ is analytic except for a finite number of singularities at the points $z_1, z_2, \ldots z_n$ (see Fig. 14.21). Let $C_1, C_2, \ldots C_n$ be sufficiently small circles about the respective centers $z_1 \ldots z_n$. If C is any circuit that contains in its interior only the singularities $z_1 \ldots z_k$, say, we have

$$\oint_C \zeta(z)\, dz = \oint_{C_1} \zeta(z)\, dz + \oint_{C_2} \zeta(z)\, dz + \cdots + \oint_{C_k} \zeta(z)\, dz$$
$$= 2\pi i [A_{-1}(z_1) + A_{-1}(z_2) + \cdots + A_{-1}(z_k)]$$
$$= 2\pi i \sum_{j=1}^{k} A_{-1}(z_j) \tag{14.75}$$

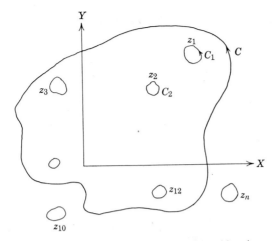

Fig. 14.21   Illustrating the derivation of the residue theorem.

where
$$A_{-1}(z_j) \equiv \text{the residue at the singular point} \quad z = z_j$$

Equation (14.75) constitutes the so-called *residue theorem: the value of the integral of a function $\zeta(z)$ over an arbitrary closed curve C is equal to the sum of the residues of $\zeta(z)$ at the singular points enclosed by C.* This theorem naturally has many applications. In applications, the residue will be known from the Laurent expansion, and hence one may find the value of the integral.

Chapter 15

Two-Dimensional Motion
and the Complex Variable

In this chapter we shall learn how to represent and to analyze the two-dimensional potential motion of an ideal fluid in terms of the functions of a complex variable. We begin with the notions of what are called *complex potential* and *complex velocity*, which, as we shall see, are the combinations $\Phi + i\Psi$ and $u - iv$ respectively. The problem reduces to that of finding an analytic function that must take certain values on the prescribed boundaries cf the motion. We shall see how this may be done by using the method of conformal mapping. In illustrating the application of the various ideas we consider mainly the problem of steady flow past a fixed arbitrary cylinder.

15.1 Complex Potential and Complex Velocity

Let us recall the relationship between the potential and the stream function and that between the velocity components in two-dimensional motion. Denoting, as before, the potential by Φ, the stream function by Ψ, and the velocity components by u and v, we have

$$\Phi = \Phi(x, y), \qquad \Psi = \Psi(x, y)$$

$$u = u(x, y), \qquad v = v(x, y)$$

The functions Φ and Ψ are related through the differential equations

$$\frac{\partial \Phi}{\partial x} = u = \frac{\partial \Psi}{\partial y}$$

$$\frac{\partial \Phi}{\partial y} = v = -\frac{\partial \Psi}{\partial x}$$

They both obey Laplace's equation

$$\nabla^2 \Phi = 0, \qquad \nabla^2 \Psi = 0$$

444

The equipotential lines described by $\Phi(x, y) = $ const. and the streamlines described by $\Psi(x, y) = $ const. form an orthogonal set of curves.

Similar properties are exhibited by the velocity components, for we have

$$\frac{\partial u}{\partial x} = -\frac{\partial v}{\partial y} \qquad (condition\ of\ incompressibility)$$

$$\frac{\partial u}{\partial y} = \frac{\partial v}{\partial x} \qquad (condition\ of\ irrotation)$$

and

$$\nabla^2 u = 0, \qquad \nabla^2 v = 0$$

Of course, the curves $u(x, y) = $ const. and $v(x, y) = $ const. form an orthogonal net.

We thus see that the interrelations between the potential and the stream function and those between the u- and v-components are exactly the same as those existing between the real and imaginary parts of an analytic function. *We, may, therefore, naturally combine the potential and the stream function into an analytic function of a complex variable or, conversely, assert that the real and imaginary parts of any analytic function represent, respectively, the potential and the stream function of a certain two-dimensional potential flow.* Equivalently, the u- and v-components may be interpreted as real and imaginary components of an analytic function and vice versa.*

It is customary to combine Φ and Ψ into an analytic function and call it the *complex potential.* Denoting it by $F(z)$ we write

$$F(z) = \Phi(x, y) + i\Psi(x, y) \qquad (15.1)$$

The x, y-plane in this context becomes the plane of the complex variable z. The derivative of the complex potential is given by

$$F'(z) \equiv \frac{dF}{dz} = \frac{\partial \Phi}{\partial x} + i\frac{\partial \Psi}{\partial x}$$

$$= \frac{\partial \Psi}{\partial y} - i\frac{\partial \Phi}{\partial y} = u(x, y) - iv(x, y) \qquad (15.2)$$

The right-hand side of this equation is the complex conjugate of $u + iv$, which is the velocity vector. It has, however, become customary to call the conjugate $u - iv$ the *complex velocity.* It is usually denoted by $W(z)$.

---

* In fact, in two dimensions the components of any vector field whose divergence and curl are both zero may be combined into an analytic function.

We thus have

$$W(z) \equiv u(x, y) - iv(x, y) = \frac{dF}{dz} \tag{15.3}$$

15.2 Flows Represented by Some Simple Functions

For the flows introduced in Chapters 12 and 13 we may readily find the complex potentials and the complex velocities. Here we shall consider some simple analytic functions and see what flow fields are represented by them. We investigate the function Az^n, *where n is an integer* and A is a complex constant.

Let the complex velocity be represented by

$$W(z) = Az^n$$

Then the complex potential is given by

$$F(z) = \frac{A}{n+1} z^{n+1}$$

1. *Case when n = 0.* We have

$$u - iv = W = A = U - iV \text{ where } U \text{ and } V \text{ are constants.} \tag{15.4}$$
$$\Phi + i\Psi = F = Az = (Ux + Vy) + i(Uy - Vx) \tag{15.5}$$

The function

$$F(z) = Az$$

therefore represents a *uniform flow* in an arbitrary direction.

2. *Case when n = 1.* *If A is real*, we obtain

$$u - iv = W = Az = A(x + iy)$$
$$\Phi + i\Psi = F = \frac{Az^2}{2} = \frac{A}{2}(x^2 - y^2) + iAxy$$

The equipotential lines and the streamlines form an orthogonal net of rectangular hyperbolas (see Fig. 15.1 and compare Section 11.2).

3. *Case when n > 1.* Express the potential $F(z)$ as Az^n and *assume that A is real.* Using the polar form,

$$\Phi + i\Psi = F(z) = Az^n$$
$$= Ar^n(\cos n\theta + i \sin n\theta)$$

Consider the stream function given by

$$\Psi = Ar^n \sin n\theta$$

We see that Ψ becomes zero on the rays $\theta = 0$ and $\theta = \alpha$, where α is given by

$$n\alpha = \pi$$

The angle α is restricted to the range 0 to 2π. This means that the function Az^n, where A is real and n is greater than one, represents the flow around a

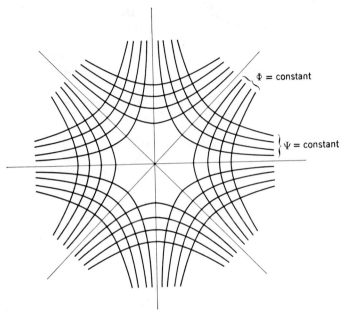

Fig. 15.1 Equipotential lines and streamlines defined by the function $Az^2/2$, where A is a real constant.

corner of angle $\alpha = \pi/n$, or, conversely, the flow in a corner of angle α is expressed by the complex potential

$$F(z) = Az^{\pi/\alpha} \qquad (15.6)$$

Then for the complex velocity we have

$$W(z) = A\,\frac{\pi}{\alpha}\,z^{(\pi-\alpha)/\alpha} = A\,\frac{\pi}{\alpha}\,r^{(\pi-\alpha)/\alpha}e^{i(\pi-\alpha)/\alpha} \qquad (15.7)$$

4. *Case when* $n = -1$. First we assume that A is real and positive. We then have

$$u - iv = W(z) = \frac{A}{z} = Ar(\cos\theta - i\sin\theta) \qquad (15.8)$$

$$\Phi + i\Psi = F(z) = A\log z = A\log r + iA\theta \qquad (15.9)$$

We conclude that the flow is that due to a *source* at the origin. The strength of the source is given by

$$q = 2\pi A \tag{15.10}$$

Consider next that A is purely imaginary, equal, say, to ib with $b > 0$. We then have

$$u - iv = W = i\frac{b}{z} = ibr(\cos\theta - i\sin\theta) \tag{15.11}$$

$$\Phi + i\Psi = F = ib\log z = -b\theta + ib\log r \tag{15.12}$$

We conclude that the flow is that due to a *vortex* at the origin. The circulation Γ of the vortex is given by

$$\Gamma = -2\pi b \tag{15.13}$$

Recall that according to our convention the positive direction of Γ is counterclockwise. We thus observe that the complex potentials for the source and vortex flows are of the form

$$F(z) = A\log z$$

where A, in general, is a complex coefficient. If A is purely real we obtain a source flow, whereas if A is purely imaginary we have a vortex flow. In this sense the source and vortex belong to the same class of singular flows.

5. *Case when $n = -2$.* We assume that A is real and positive. We then have

$$u - iv = W = \frac{A}{z^2} = A\frac{\cos 2\theta}{r^2} - iA\frac{\sin 2\theta}{r^2} \tag{15.14}$$

$$\Phi + i\Psi = F = -\frac{A}{z} = -A\frac{\cos\theta}{r} + iA\frac{\sin\theta}{r} \tag{15.15}$$

We conclude that the flow is that due to a *doublet* situated at the origin. The axis of the doublet is in the direction of x and the strength of the doublet is given by

$$\mu = 2\pi A \tag{15.16}$$

Proceeding in this manner we may obtain the corresponding flows for $n < -2$. All these flows, as we know, are singular flows with the singularity at the origin. For convenience we list on the following page the important flows of the function Az^n, where n is an integer.

$F(z)$	$W(z)$	Flow
A complex constant	0	No Flow
Az A complex	A	Uniform flow in an arbitrary direction.
Az^n $n > 1$ A real	Anz^{n-1}	Flow around a corner of angle $\alpha = \pi/n$.
$A \log z$ A real and > 0	$\dfrac{A}{z}$	Source at origin. Strength $q = 2\pi A$.
$ib \log z$ $b > 0$	$i\dfrac{b}{z}$	Vortex at origin. Circulation $\Gamma = -2\pi b$ (positive circulation counterclockwise).
$-\dfrac{A}{z}$ A real and > 0	$\dfrac{A}{z^2}$	Doublet at origin. Axis in x-direction. Strength $\mu = 2\pi A$.

By the superposition of the complex potentials for elementary flows, such as those described above, we may arrive at the complex potential for more general flow fields. For example, consider the combination of a uniform stream and doublet with its axis opposing the stream. Choosing the x-axis in the direction of the stream, we have for the combined potential

$$F(z) = Uz + \frac{A}{z}$$
$$= \left(Ux + \frac{A \cos \theta}{r} \right) + i\left(Uy - \frac{A \sin \theta}{r} \right) \tag{15.17}$$

Hence the stream function of the combined flow is given by

$$\Psi = Uy - \frac{A \sin \theta}{r} = Ur \sin \theta - \frac{A \sin \theta}{r}$$

We observe that $\Psi = 0$ on the circle

$$r = \sqrt{\frac{A}{U}}$$

Thus the potential (15.17) represents uniform flow past a circular cylinder whose radius is equal to $\sqrt{A/U}$. There is no circulation around the cylinder.

To obtain the uniform flow past a circular cylinder with circulation around it we add to the potential (15.17) the potential

$$-ib \log \frac{z}{a}$$

for a clockwise circulation, where a is a real constant. Addition of the constant has no effect on the flow field. We then have

$$F(z) = Uz + \frac{A}{z} - ib \log \frac{z}{a} \qquad (15.18)$$

The stream function of this flow is given by

$$\Psi = Ur \sin \theta - \frac{A \sin \theta}{r} - b \log \frac{r}{a}$$

This becomes zero on the circle

$$r = a = \sqrt{\frac{A}{U}}$$

Thus the potential (15.18) represents flow past a circular cylinder with circulation. The radius of the cylinder is again $\sqrt{A/U}$.

15.3 Circulation and Source Strength

We shall now begin our considerations with respect to the steady flow past any fixed arbitrary cylinder. As a preliminary in this regard we show that the circulation around any closed curve and the outflow of fluid through that circuit may be expressed as the real and imaginary parts of a complex function. By definition we have

$$\Gamma = \oint_C \mathbf{V} \cdot \mathbf{ds} = \oint_C (u \, dx + v \, dy) = \oint_C d\Phi$$

According to (12.10a), the volume outflow of fluid through the circuit is given by

$$Q = \oint_C (u \, dy - v \, dx) = \oint_C d\Psi$$

Therefore it follows that

$$\Gamma + iQ = \oint_C W(z) \, dz = F(z_2) - F(z_1) \qquad (15.19)$$

where z_2 and z_1 are coincident points on the curve C. Thus $\Gamma + iQ$ may be expressed as the jump in the complex potential around the circuit. If either Γ or Q is nonzero, the complex potential must be multivalued. The outflow Q is the total strength of all the sources contained in the interior of the circuit C, whereas Γ is the total circulation of all the vortices contained in the circuit.

15.4   Flow Past an Arbitrary Cylinder

We now consider the steady flow past a fixed cylinder of arbitrary cross section. Far from the body the velocity is uniform. *A priori* we do not know whether or not there is a circulation around the body. We assume, however, that there are no sources or vortices or other singularities in the flow field. Our goal, as before, is to obtain a solution for the flow field given the uniform velocity at infinity and the shape of the surface of the body.

We choose the origin of the coordinates in the interior of the body. Then in the region exterior to the body the complex velocity may be represented by a Laurent expansion about the origin.* Since at infinity the complex velocity should represent the uniform stream and near the cylinder a perturbation, we write

$$W(z) = A_0 + \frac{A_1}{z} + \frac{A_2}{z^2} + \frac{A_3}{z^3} + \cdots = \sum_{n=0}^{\infty} \frac{A_n}{z^n} \qquad (15.20)$$

where the coefficients A_n are complex in general. The corresponding complex potential is given as the integral of (15.20), which is

$$F(z) = A_0 z + A_1 \log z - \sum_{n=2}^{\infty} A_n \frac{1}{n-1} \frac{1}{z^{n-1}} + \text{a constant} \quad (15.21)$$

Our problem is simply to determine these coefficients so as to satisfy the prescribed conditions, such as the body condition, and perhaps the Kutta condition, if the body is of the appropriate shape. The problem, although it could be stated so simply, is not usually as simple to solve! However, we can derive, on the basis of the theory of the functions of a complex variable, several important results without actually computing all the coefficients in the expansion (15.20).

We can readily remark about the coefficients A_0 and A_1. If *we choose the X-axis in the direction of the uniform stream*, we should have

$$A_0 = U \qquad (15.22)$$

Now, for any closed circuit C we have

$$\Gamma + iQ = \oint_C W(z)\, dz$$

$$= 0 \quad \text{if } C \text{ does not enclose the body}$$

$$= 2\pi i A_1 \quad \text{for all circuits enclosing the body} \qquad (15.23)$$

* As there are no singularities on the surface of the body, the expansion would include the surface.

Expressing A_1 as

$$A_1 = a_1 + ib_1$$

we have

$$\Gamma = -2\pi b_1 \qquad (15.24)$$

$$Q = 2\pi a_1 \qquad (15.25)$$

Since the body is closed, the total source strength should be zero. Hence we should set

$$a_1 = 0 \qquad (15.26)$$

As regards b_1, we cannot set it equal to zero, for we are dealing with a multiply connected region and the circulation [or equivalently $\oint W(z)\, dz$] may or may not be zero (see Sections 9.14 and 14.17). Thus b_1 remains as an undetermined coefficient. On the basis of Eq. (15.26) the series expansion (15.20) for the complex velocity may be rewritten as

$$W(z) = A_0 + \frac{ib_1}{z} + \sum_{n=2}^{\infty} \frac{A_n}{z^n} \qquad (15.27)$$

The complex potential then takes the form

$$F(z) = A_0 z + ib_1 \log z - \sum_{n=2}^{\infty} A_n \frac{1}{n-1} \frac{1}{z^{n-1}} + \text{a constant} \quad (15.28)$$

15.5 Flow Past a Circular Cylinder

To illustrate the direct determination of the coefficients in the expansion (15.27), let us determine the flow field due to a cylinder of radius a placed in an originally uniform stream $\mathbf{U}$. We choose the direction of the X-axis as that of $\mathbf{U}$. The coefficients are to be determined according to the surface condition that

$$\Psi(r = a, \theta) = 0 \qquad (15.29)$$

In view of this boundary condition it is convenient to use the polar form and write the complex potential given by (15.28) in the form

$$\Phi + i\Psi = F(z) = Ure^{i\theta} + ib_1(\log r + i\theta) + (\alpha + i\beta)$$

$$- \sum_{n=2}^{\infty} \frac{a_n + ib_n}{n-1} \frac{e^{-i(n-1)\theta}}{r^{n-1}} \qquad (15.30)$$

where we have set

$$A_0 = U$$

$$\text{constant} = \alpha + i\beta$$

$$A_1 = a_n + ib_n$$

and

$$z = re^{i\theta}$$

From (15.30) it follows that

$$\Psi(a, \theta) = Ua \sin \theta + b_1 \log a + \beta$$
$$- \sum_{n=2}^{\infty} \frac{1}{n-1} \frac{1}{a^{n-1}} \left\{ \frac{b_n \cos(n-1)\theta - a_n}{\sin(n-1)\theta} \right\}$$

Using (15.29) we have

$$\beta = -b_1 \log a$$
$$b_1 \text{ is undetermined}$$
$$b_n = 0 \text{ for } n = 2, 3, \ldots$$
$$a_2 = -Ua^2$$
$$a_n = 0 \text{ for } n = 3, 4, \ldots$$

Substituting these values in (15.28), we obtain the complex potential for the flow past a circular cylinder as

$$F(z) = U\left(z + \frac{a^2}{z}\right) + ib_1 \log \frac{z}{a} \qquad (15.31)$$

If the circulation is clockwise and its magnitude is Γ, this equation becomes

$$F(z) = U\left(z + \frac{a^2}{z}\right) + i\frac{\Gamma}{2\pi} \log \frac{z}{a}$$

The complex velocity is given by

$$W(z) = U\left(1 - \frac{a^2}{z^2}\right) + i\frac{\Gamma}{2\pi}\frac{1}{z}$$

The circulation naturally remains undetermined.

15.6   Complex Representation of Forces and Moments Acting on an Arbitrary Body: Blasius' Relations

One of the important results that can be established on the basis of the series expansion (15.27) is the famous *Kutta-Joukowski theorem* about the lift on an arbitrary cylinder. To do this, and for the sake of further applications, we express in complex notation the forces and moments acting on an arbitrary cylinder. Using the principles of conservation of linear and angular momentum we can set up, for steady conditions, formulas for the forces and moments acting on a fixed arbitrary cylinder immersed in a two-dimensional flow field (see Appendix A). If C_0 is any arbitrary circuit that contains the body in its interior, and if X and Y denote, respectively, the x- and y-components of the force on the body, we obtain, as shown in Appendix A,

$$X = \oint_{C_0} \frac{\rho}{2} V^2 \, dy - \oint_{C_0} \rho u(u \, dy - v \, dx)$$

$$Y = -\oint_{C_0} \frac{\rho}{2} V^2 \, dx - \oint_{C_0} \rho v(u \, dy - v \, dx)$$

The moment M on the body with respect to the origin is given by

$$M = -\left[\frac{\rho}{2} \oint_{C_0} V^2(x\,dx + y\,dx) + \rho \oint_{C_0} (vx - uy)(u\,dy - v\,dx)\right]$$

with M taken positive for rotation about the z-axis according to the right-hand rule.

These formulas, when expressed in terms of the complex velocity $W(z) = u - iv$ and the complex variable $z = x + iy$, lead to very simple relations for the forces and moment. One combines the equations for the force components to define a complex force $X - iY$, in analogy to the complex potential and complex velocity. We thus obtain*

$$X - iY = i\frac{\rho}{2} \oint_{C_0} [W(z)]^2\,dz \qquad (15.32)$$

$$M = -\frac{\rho}{2} \operatorname{Re}\left\{\oint_{C_0} [W(z)]^2 z\,dz\right\} \qquad (15.33)$$

These equations are generally known as *Blasius' relations*. These relations may be readily extended to situations where more than one cylinder is immersed in the flow field. The simplicity of the Blasius relations is indeed striking.

15.7 Force and Moment on an Arbitrary Cylinder: Kutta-Joukowski Theorem

Consider the steady flow past a fixed arbitrary cylinder (see Fig. 15.2). The uniform velocity at infinity is **U**. We choose the X-axis in the direction of **U**. According to (15.27), the complex velocity for this flow may be expressed as

$$W(z) = A_0 + \frac{A_1}{z} + \frac{A_2}{z^2} + \sum_{n=3}^{\infty} \frac{A_n}{z_n}$$

where

$$A_0 = U$$

$$A_1 = ib_1 = -i\frac{\Gamma}{2\pi} \quad \text{(for positive } \Gamma\text{)}$$

We then obtain

$$W^2(z) = B_0 + \frac{B_1}{z} + \frac{B_2}{z^2} + \frac{()}{z^3} + \frac{()}{z^4} + \cdots \qquad (15.34)$$

and

$$W^2(z) \cdot z = B_0 z + B_1 + \frac{B_2}{z} + \frac{()}{z^2} + \frac{()}{z^3} + \cdots \qquad (15.35)$$

* The reader should do this as a problem.

where

$$B_0 = A_0{}^2$$
$$B_1 = 2A_0A_1$$
$$B_2 = A_1{}^2 + 2A_0A_2 \quad \text{etc.} \tag{15.36}$$

Since we wish to integrate $W^2(z)$ and $zW^2(z)$, we really are interested only in the coefficients of $1/z$ in (15.34) and (15.35).

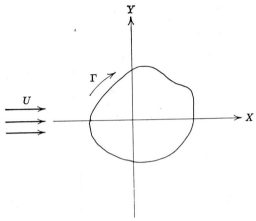

Fig. 15.2 Flow with circulation past an arbitrary cylinder.

The force and moment on the body are therefore given by

$$X - iY = i\,\frac{\rho}{2} \int_{C_0} W^2\,dz$$

$$= i\,\frac{\rho}{2}\,2\pi i B_1$$

$$= -\rho\pi(2A_0A_1)$$

$$= i\rho U\Gamma \quad (\Gamma \text{ counterclockwise})$$

$$= -i\rho U\Gamma \quad (\Gamma \text{ clockwise as shown in Fig. 15.2}) \tag{15.37}$$

or $X = 0,\ Y = \rho U\Gamma$.

The moment on the body is given by

$$M = -\frac{\rho}{2}\,\text{Re}\left\{\oint_{C_0} zW^2\,dz\right\}$$

$$= -\frac{\rho}{2}\,\text{Re}\,(2\pi i B_2)$$

$$= -\pi\rho\,\text{Re}\,(iB_2)$$

$$= -2\pi\rho U\,\text{Re}\,(iA_2) \tag{15.38}$$

We thus see, as anticipated, that the force on a cylinder of any cross section whatsoever is equal to only a lift force of magnitude $\rho U \Gamma$ per unit span of the cylinder. Indeed, we do not yet know what the value of the circulation is, if it exists. In general, there exists a moment on the cylinder. It is interesting to observe that if we are concerned only with the over-all force and moment, all we need to know are the coefficients A_1 (i.e., circulation) and A_2 in the series expansion for $W(z)$.

For the circular cylinder the reader may verify that the moment is zero.

15.8 Mapping of Flows

To proceed with the analysis of flow problems involving complicated boundaries we employ the technique of mapping. In this section we give some general results relating to such mapping.

Consider the xy-plane, which we refer to as the z-plane, and a certain flow field within a given region. The flow field is a solution of Laplace's equation and satisfies certain prescribed conditions on the boundaries of the region. It is characterized by the potential $\Phi(x, y)$, or the stream function $\Psi(x, y)$, or by the complex potential $F(z) = \Phi + i\Psi$, which is an analytic function.

We now employ a transformation $\xi = \xi(x, y)$, $\eta = \eta(x, y)$ which will map the region and its boundaries in the z-plane into another desired configuration in the ξ,η-plane, which we refer to as the ζ-plane. The differential equation and the boundary conditions are also transformed correspondingly, and the solutions in the two planes are images of each other except perhaps at certain peculiar points. Let us denote by Φ_ζ and Ψ_ζ, respectively, the images of Φ and Ψ. We have

$$\Phi_\zeta = \Phi_\zeta(\xi, \eta) = \Phi[x(\xi, \eta), y(\xi, \eta)]$$
$$\Psi_\zeta = \Psi_\zeta(\xi, \eta) = \Psi[x(\xi, \eta), y(\xi, \eta)]$$

The complex potential in the ζ-plane which is an image of $F(z)$ is given by

$$\tilde{F}(\zeta) = \Phi_\zeta + i\Psi_\zeta$$

We know that if the transformation is such that $\xi(x, y) + i\eta(x, y)$ forms an analytic function, say $\zeta(z)$, then for all points where

$$\left|\frac{d\zeta}{dz}\right|^2 = \left(\frac{\partial \xi}{\partial x}\right)^2 + \left(\frac{\partial \xi}{\partial y}\right)^2 = \left(\frac{\partial \eta}{\partial x}\right)^2 + \left(\frac{\partial \eta}{\partial y}\right)^2 \neq 0$$

the complex potential $\tilde{F}(\zeta)$ is analytic. In other words, $\tilde{F}(\zeta)$ is analytic at all points where the transformation $\zeta(z) = \xi + i\eta$ is *not* critical.

We could also have reached this conclusion from the following considerations. Under the transformation $\zeta(z)$ the flow field in the z-plane characterized by $F(z)$ is mapped into a flow field in the ζ-plane characterized by the complex potential $\tilde{F}(z)$. We then have

$$F(z) = \tilde{F}(\zeta) = \tilde{F}[\zeta(z)]$$

Therefore it follows that

$$\frac{dF}{dz} = \frac{d\tilde{F}}{d\zeta}\frac{d\zeta}{dz}$$

or,

$$\frac{d\tilde{F}}{d\zeta} = \frac{dF}{dz}\frac{1}{d\zeta/dz}$$

It immediately follows that *the derivative of $\tilde{F}(\zeta)$ exists at all points where $d\zeta/dz$ does not vanish. At the critical points of the transformation $\tilde{F}(\zeta)$ ceases to be analytic.*

The two flow fields are related as follows. We express the flow in the ζ-plane in terms of the flow in the z-plane. The inverse of the transformation

$$\zeta(z) = \xi(x, y) + i\eta(x, y) \tag{15.39}$$

is expressed as

$$z = z(\zeta) = x(\xi, \eta) + iy(\xi, \eta) \tag{15.40}$$

From the results of Section 14.8 we know that if the derivative $(d\zeta/dz) \neq 0$, then the inverse function exists with its derivative given by

$$\frac{dz}{d\zeta} = \frac{1}{d\zeta/dz} \tag{15.41}$$

The complex potential $\tilde{F}(\zeta)$ is then obtained from $F(z)$ by replacing z by $z(\zeta)$:

$$\tilde{F}(\zeta) = F[z(\zeta)] \tag{15.42}$$

Denoting the complex velocity in the ζ-plane by $\tilde{W}(\zeta)$ we have

$$\tilde{W}(\zeta) = \frac{d\tilde{F}}{d\zeta}$$

$$= \frac{dF}{dz}\frac{1}{(d\zeta/dz)} = W(z)\frac{1}{d\zeta/dz}$$

$$= W[z(\zeta)]\frac{dz}{d\zeta} \tag{15.43}$$

We see that *the complex velocity in the ζ-plane will become infinite at the critical points of the transformation. Therefore in applications one usually chooses the transformation such that the critical points do not lie in the region of interest. However, if for one reason or another they must be allowed in the* *region of interest (as in the case of an airfoil, as we shall see) one requires that, in order to keep $\tilde{W}(\zeta)$ from becoming infinite, the complex velocity $W(z)$ should be zero at the critical points.*

In applications the known flow in the z-plane is usually that of uniform motion or that of flow past a circular cylinder or the flow for some other simple geometry. One then seeks the flow in the ζ-plane over a comparatively complex geometry. The problem then is simply to determine the mapping function $\zeta(z)$ given a few requirements with regard to the correspondence between the two planes.

15.9 Transformation of Circulation and Source Strength

When mapping flow fields we shall be interested in knowing how singular flows, such as sources and vortices, transform. To investigate this aspect consider C_z in the z-plane. Let Γ and Q represent, respectively, the circulation around C_z and the total strength of the sources included in the interior of C_z. The circulation P is the total circulation of the vortices included in C_z. Under the mapping $\zeta = \zeta(z)$, let the contour C_z go into the contour C_ζ in the ζ-plane and, correspondingly, the interior of C_z into the interior of C_ζ. Then for the circulation Γ_ζ around C_ζ and the total source strength Q_ζ enclosed by C_ζ we have

$$
\Gamma_\zeta + iQ_\zeta = \oint_{C_\zeta} \tilde{W}(\zeta)\, d\zeta
$$

$$
= \oint_{C_\zeta} W(z)\, \frac{1}{d\zeta/dz} \frac{d\zeta}{dz}\, dz
$$

$$
= \oint_{C_z} W(z)\, dz = \Gamma + iQ \tag{15.44}
$$

We thus see that the circulation and the total source strength, with respect to the contour C_ζ, are equal to those with respect to its image C_z.

It may be verified that if a source or a vortex is situated at a critical point of the transformation, the circulation and the source strength are not preserved under the transformation.

15.10 Transformation of Flow Past an Arbitrary Cylinder into that Past a Circular Cylinder

The Transformation. We now proceed directly to the problem of determining by means of mapping the flow past an arbitrary cylinder. For this purpose we take the flow past a circular cylinder, for which the solution is known, as the basic flow field for the transformation. Following the usual custom, we shall denote the plane of motion for the arbitrary cylinder as the

z-plane and that for the circular cylinder as the ζ-plane.* We first inquire whether there is a unique transformation $\zeta = \zeta(z)$ or, equivalently, $z = z(\zeta)$ that will map an arbitrary closed curve in the z-plane into a circle in the ζ-plane and the region exterior to the curve into the region exterior to the circle. In answer to this inquiry it can be shown that there exists such a transformation and that it is uniquely determined if a given point of the z-plane (in the region considered) and a given direction through that point are required to go into a given point of the ζ-plane and a given direction through the point.†

Consider the z-plane, that is, the plane of the arbitrary cylinder (Fig. 15.3). We denote the undisturbed stream by the velocity vector $\mathbf{V}_\infty$ whose

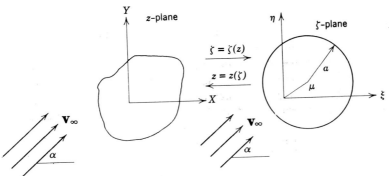

Fig. 15.3 Transformation of flow past an arbitrary cylinder into that past a circular cylinder.

magnitude is V_∞ and whose direction is given by the angle α measured from the X-axis. The complex velocity at infinity in the z-plane is therefore given by $V_\infty e^{-i\alpha}$. In transforming the flow past the arbitrary cylinder into the flow past the circle‡ in the ζ-plane we require that the velocity at

---

* Previously we regarded the z-plane as the plane of the known flow field and the ζ-plane as that of the required flow field. The present interchange of the roles of the symbols z and ζ should cause no confusion.

† This result is a simple extension of Riemann's theorem: Given in the z-plane a region A bounded by a single curve (or simple boundary), and in the ζ-plane a circle C, there exists an analytic function $\zeta = \zeta(z)$, analytic in A, such that to each point of A there is a corresponding point in the interior of C and, conversely, a point in the circle corresponds to one and only one point of A. The function $\zeta(z)$ becomes unique if a given point of A and a direction through that point are chosen to correspond to a given point in the interior of C and a given direction through that point.

The boundary of A is transformed uniquely and continuously into the circumference of the circle. This theorem is readily extended to exterior regions by using the notion of *inversion*. We shall not go into the proof of the theorem or its extension.

‡ We use circle and circular cylinder synonymously, also arbitrary curve and arbitrary cylinder.

infinity in the ζ-plane should also be $V_\infty e^{-i\alpha}$. This is an obvious requirement, for the perturbation field due to a body must vanish at infinity no matter what the shape of the body.

Now, if $z = z(\zeta)$ is the mapping function, we have

$$\tilde{W}(z) = W(\zeta) \frac{1}{dz/d\zeta}$$

Since we require that at infinity the velocities be equal, it follows that we should have

$$\frac{dz}{d\zeta} = 1 \quad \text{for} \quad \zeta = \infty \tag{15.45}$$

or

$$z = \zeta \quad \text{for} \quad \zeta = \infty$$

This means that the transformation function should be such that the point at infinity in the z-plane goes into the point at infinity in the ζ-plane. Since the directions of the velocities at infinity in the two planes are the same, we have required that in the transformation the direction α through the point $z = \infty$ go into the direction α through the point $\zeta = \infty$. These requirements are met by setting the scale factor $dz/d\zeta$ equal to unity at $\zeta = \infty$.

We may then express the transformation in the series*

$$z(\zeta) = \zeta + \frac{C_1}{\zeta} + \frac{C_2}{\zeta^2} + \frac{C_3}{\zeta^3} + \cdots$$

$$= \zeta + \sum_{n=1}^{\infty} \frac{C_n}{\zeta^n} \tag{15.46}$$

The derivative of $z(\zeta)$ is then given by the series

$$\frac{dz}{d\zeta} = 1 - \frac{C_1}{\zeta^2} - 2\frac{C_2}{\zeta^3} - 3\frac{C_3}{\zeta^4} - \cdots$$

$$= 1 - \sum_{n=1}^{\infty} n \frac{C_n}{\zeta^{n+1}} \tag{15.47}$$

Note that there is no constant term such as C_0 in the series (15.46). This, however, is no restriction, for the location of the circle is arbitrary and we can choose the coordinates of the circle in such a way that the constant term in the series expansion vanishes. Presence or absence of a constant term in the series (15.46) has no effect on the series (15.47).

Our problem in the mapping of the flow past an arbitrary cylinder into that past a circle is then to determine the coefficients $C_1, C_2, \ldots,$ of the

* The series in (15.46) and (15.47) are Laurent expansions about the point at infinity in the ζ-plane.

series (15.46) as function of the shape of the boundary curve of the arbitrary cylinder. We shall return to this problem in the next chapter, where we shall consider the mapping of an airfoil into a circle. At present we proceed, on the basis of the series representation (15.46), to an evaluation of the flow field in the z-plane (i.e., in the plane of the arbitrary cylinder) in terms of the flow field in the ζ-plane (i.e., in the plane of the circle).

Complex Potential and Complex Velocity for the Flow Past the Circle. Let $\zeta = \mu$ denote the center of the circle and a its radius (Fig. 15.3). The complex function $\zeta = \mu + ae^{i\theta}$ describes the circle. Recall that the complex potential for the flow past a circular cylinder where the origin of

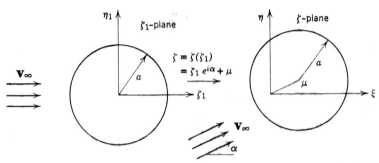

Fig. 15.4 Illustrating the derivation of the complex potential and velocity for the flow past the circle.

the coordinates is situated at the center of the circle and the real axis is directed along the direction of the undisturbed stream is given by

$$F(\zeta_1) = V_\infty \zeta_1 + i\,\frac{\Gamma}{2\pi}\,\log\frac{\zeta_1}{a} + V_\infty\,\frac{a^2}{\zeta_1} \qquad (15.48)$$

where ζ_1 is the complex variable whose plane is as shown in Fig. 15.4 and Γ is the magnitude of the clockwise circulation around the circle. To obtain the complex potential for the flow past the circle in the ζ-plane where the complex velocity at infinity is $V_\infty e^{-i\alpha}$ and the circle is located at $\zeta = \mu$ we use the transformation

$$\zeta = \zeta(\zeta_1) = \zeta_1 e^{i\alpha} + \mu \qquad (15.49)$$

The complex potential in the ζ-plane is then given by

$$F(\zeta) = V_\infty(\zeta - \mu)e^{-i\alpha} + i\,\frac{\Gamma}{2\pi}\,\log\frac{\zeta - \mu}{ae^{i\alpha}} + \frac{V_\infty a^2}{\zeta - \mu}\,e^{i\alpha} \qquad (15.50)$$

The circulation, the value of which we do not know *a priori*, remains unaltered during the transformation.

For the complex velocity in the ζ-plane we then obtain

$$W(\zeta) = \frac{dF(\zeta)}{d\zeta}$$

$$= V_\infty e^{-i\alpha} + i\frac{\Gamma}{2\pi}\frac{1}{\zeta - \mu} - V_\infty a^2 e^{i\alpha}\frac{1}{(\zeta - \mu)^2} \qquad (15.51)$$

It is convenient to have $W(\zeta)$ expressed as a series in $1/\zeta$. For this purpose we use the following expansions:

$$(\zeta - \mu)^{-1} = \frac{1}{\zeta}\left(1 - \frac{\mu}{\zeta}\right)^{-1}$$

$$= \frac{1}{\zeta} + \frac{\mu}{\zeta^2} + \frac{\mu^2}{\zeta^3} + \cdots$$

$$(\zeta - \mu)^{-2} = \frac{1}{\zeta^2} + 2\frac{\mu}{\zeta^3} + 3\frac{\mu}{\zeta^4} + \cdots$$

We may then express the complex velocity as

$$W(\zeta) = A_0 + \frac{A_1}{\zeta} + \frac{A_2}{\zeta^2} + \cdots = \sum_{n=0}^{\infty}\frac{A_n}{\zeta^n} \qquad (15.52)$$

where

$$A_0 = V_\infty e^{-i\alpha}$$

$$A_1 = \frac{i\Gamma}{2\pi}$$

$$A_2 = \frac{i\Gamma}{2\pi}\mu - V_\infty a^2 e^{i\alpha} \text{ etc.}$$

Velocity Field in the Plane of the Arbitrary Cylinder. The complex velocity in the z-plane is given by

$$W(z) = \frac{dF(z)}{dz} = \frac{d}{d\zeta}F(\zeta)\frac{1}{dz/d\zeta}$$

$$= W(\zeta)\frac{1}{dz/d\zeta} \qquad (15.53)$$

where $W(\zeta)$ and $dz/d\zeta$ are given, respectively, by Eqs. (15.52) and (15.47).

The Pressure Field in the z-plane. According to Bernoulli's equation the pressure at any point x, y in the z-plane is given by

$$p(z, y) = p_\infty + \tfrac{1}{2}\rho V_\infty^2\left(1 - \frac{V^2}{V_\infty^2}\right)$$

where p_∞ is the pressure at infinity. For the magnitude of the velocity at any point $z = (x, y)$ we have

$$V^2(x, y) = |W(z)|^2 = |W(\zeta)|^2 \frac{1}{|dz/d\zeta|^2}$$

To obtain the pressure distribution over the surface of the cylinder we set $\zeta = \mu + ae^{i\theta}$ in the functions $W(\zeta)$ and $dz/d\zeta$.

Force and Moment on the Arbitrary Cylinder. We had seen that the force on any arbitrary cylinder placed in an originally uniform stream of speed V_∞ is simply a *lift force* whose magnitude is $\rho V_\infty \Gamma$ per unit length of the cylinder (see Section 15.7). Since under conformal transformation the value of the circulation is preserved, we conclude that the lift force on the arbitrary cylinder in the z-plane is equal to that on the circular cylinder in the ζ-plane.

The moment on the arbitrary cylinder is given by

$$M_0 = - \frac{\rho}{2} \operatorname{Re} \oint_{C_0} W^2(z) z \, dz \qquad (15.54)$$

where C_0 is a control circuit enclosing the cylinder and M_0 is the moment with respect to the origin of coordinates in the z-plane and is positive when anticlockwise. The right-hand side of (15.54) is readily expressed as an integral in the plane of the circle, for we have

$$W^2(z) \, dz = W^2(\zeta) \frac{1}{(dz/d\zeta)^2} \frac{dz}{d\zeta} d\zeta$$

$$= W^2(\zeta) \frac{1}{(dz/d\zeta)} d\zeta \qquad (15.55)$$

Equation (15.54) then becomes

$$M_0 = - \frac{\rho}{2} \oint_{C_\zeta} W^2(\zeta) \frac{z(\zeta)}{dz/d\zeta} d\zeta \qquad (15.56)$$

where C_ζ is any circuit enclosing the circular cylinder.*

We immediately proceed to express the moment M_0 in terms of the coefficients of the series expansions for the complex velocity $W(\zeta)$ and the mapping function $z(\zeta)$. Using the expansions (15.46), (15.47), and (15.52), we have

$$B_0 = A_0{}^2 = V_\infty{}^2 e^{-2i\alpha}$$

$$B_2 = (2A_0A_1 + A_1{}^2) \qquad (15.57)$$

$$= i \frac{V_\infty \Gamma}{\pi} \mu e^{-i\alpha} - 2V_\infty{}^2 a^2 - \frac{\Gamma^2}{4\pi^2} \qquad (15.58)$$

---

* C_ζ need not be the image of C_0, for all circuits enclosing the circle are reconcilable with one another.

and

$$\mathrm{Re}\,2\pi i(2B_0 C_1 + B_2) = \mathrm{Re}\,(i4\pi C_1 V_\infty^2 e^{-2i\alpha} - 2V_\infty \Gamma \mu e^{-i\alpha})$$

Hence the moment on the cylinder is given by

$$M_0 = \mathrm{Re}\,(\rho V_\infty \Gamma \mu e^{-i\alpha} - i2\pi\rho V_\infty^2 C_1 e^{-2i\alpha})$$
$$= \mathrm{Re}\,(L\mu e^{-i\alpha} - i2\pi\rho V_\infty^2 C_1 e^{-2i\alpha}) \tag{15.59}$$

where L = lift on the cylinder = $\rho V_\infty \Gamma$

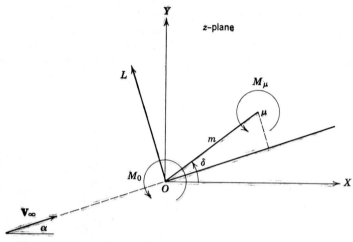

Fig. 15.5 Illustrating the force and moment on the arbitrary cylinder.

To reduce (15.59) further let us write the complex numbers μ and C_1 as

$$\mu \equiv me^{i\delta} \tag{15.60}$$
$$C_1 \equiv \mathscr{C}^2 e^{2i\gamma} \tag{15.61}$$

with these, Eq. (15.59) takes the form

$$M_0 = Lm\cos(\delta - \alpha) + 2\pi\rho V_\infty^2 \mathscr{C}^2 \sin 2(\gamma - \alpha) \tag{15.62}$$

The force system on the arbitrary cylinder is shown in Fig. 15.5. There is a force equal to the lift L acting at the origin O and a moment M_0 about O. If $\mu = me^{i\delta}$ is a point in the plane of the arbitrary cylinder, the moment about such a point is given by

$$M_\mu = M_0 - Lm\cos(\delta - \alpha) \tag{15.63}$$

Using (15.62) we obtain

$$M_\mu = 2\pi\rho V_\infty^2 \mathscr{C}^2 \sin 2(\gamma - \alpha) \tag{15.64}$$

In view of this, Eq. (15.62) may be rewritten as

$$M_0 = Lm \cos(\delta - \alpha) + M_\mu \qquad (15.65)$$

Remarks. We have thus seen that the velocity and pressure fields in the flow past an arbitrary cylinder may be completely determined by the method of conformal transformation. The problem reduces to that of finding the coefficients in the series expansion for the transformation function and of specifying the circulation around the given cylinder. We know that well-defined circulation appears only on certain types of cylinders that have airfoil-like cross sections with sharp trailing edges and then the circulation may be fixed by means of the Kutta condition. In the next chapter we shall consider airfoils and see how to determine analytically the circulation around them. Also, we shall extend the results of this chapter and obtain the aerodynamic properties of a certain type of airfoil

Chapter 16

The Problem of the Airfoil

The problem of the airfoil may concern one or the other of the following aspects—the determination of the aerodynamic characteristics, such as circulation, lift, moment, and pressure distribution for a given airfoil; the design of an airfoil to exhibit certain specified characteristics; the determination of the effects on the airfoil characteristics of changes in the profile and in the flow configuration. In a direct approach to the problem, one attempts to express the characteristics of the airfoil in terms of the shape of the airfoil surface, the speed of the uniform stream, and the attitude of the airfoil with respect to the uniform stream. For this purpose Laplace's equation is to be solved directly for prescribed boundary conditions. If the method of conformal transformation is employed, the problem reduces to that of determining the mapping function that transforms an arbitrary airfoil into a circle. This is rather difficult. A procedure for finding such a mapping function has been given by Theodorsen, and we shall describe his method in the last section of this chapter. The solution according to this method is not available in a convenient closed form and generally involves considerable computation. The problem of the arbitrary airfoil may, however, be solved approximately by assuming that the airfoil is sufficiently thin so that the boundary conditions may be simplified. We shall present the simplified theory of the arbitrary airfoil in the next chapter. In this chapter we shall concern ourselves mainly with the indirect problem, namely, that of transforming a circle into a airfoil-like profile and of determining the properties of an airfoil thus generated.

We begin with some nomenclature associated with an airfoil.

16.1 Nomenclature

There are two points L and T on an airfoil such that the straight line LT is greater in length than any other straight line joining two points on the airfoil. These points are called the leading and trailing edges, respectively. The leading edge is the one that faces the oncoming stream. The line joining the leading and trailing edges is known as the chord line (or axis

466

of the airfoil) or simply the chord (see Fig. 16.1). We denote the length of the chord by l although usually it is denoted by c.

We choose the X-axis along the chord from the leading edge to the trailing edge and the origin of coordinates at some point on the chord, usually at the leading edge or at the center of the chord. Then the surface of the airfoil above the X-axis is known as the upper surface and that below the X-axis is known as the lower surface. We denote them as

$$y = \eta_u(x) \quad \text{for} \quad y > 0$$
$$= \eta_l(x) \quad \text{for} \quad y < 0$$

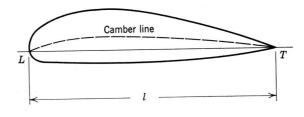

Fig. 16.1 Illustrating the airfoil nomenclature.

The mean line of the airfoil is known as the camber line of the airfoil (see Fig. 16.1). It is defined by

$$\eta_c = \tfrac{1}{2}(\eta_n + \eta_l)$$

The thickness of the airfoil is defined by

$$\eta_t = \tfrac{1}{2}(\eta_u - \eta_l)$$

The actual thickness is given by $2\eta_t$. Thus one speaks of the airfoil as composed of a *thickness envelope wrapped around a camber line*.

If $\eta_c \equiv 0$, the *airfoil is symmetrical;* otherwise it is cambered. In theory, we talk of airfoils for which $\eta_t = 0$, that is, of an airfoil of zero thickness.

The angle between the direction of the undisturbed stream at infinity and the direction of chord line from the leading edge to the trailing edge is called the *angle of attack or the angle of incidence.*

16.2 Mapping at the Trailing Edge

As the preliminary step in specifying the circulation around an airfoil, we consider the properties that should be exhibited by the mapping at the trailing edge of an airfoil. Let $z = z_T$ denote the trailing edge, and let the point on the circle corresponding to z_T be denoted by ζ_T. We refer to both z_T and ζ_T as the trailing-edge points. Consider the elements dz_1 and dz_2

of the airfoil curve passing through the points z_T (see Fig. 16.2). Let $d\zeta_1$ and $d\zeta_2$ be the corresponding image elements at the image point ζ_T. The angle ψ_z between dz_1 and dz_2 is required to be different from π, as shown in the figure, since the trailing edge is to be a sharp edge. The angle between the image elements $d\zeta_1$ and $d\zeta_2$ is, however, equal to π. This means that the mapping should not preserve angles at the trailing edge, that is, should not be conformal. This, in turn, means that the point $\zeta = \zeta_T$ should be a critical point of the transformation $z = z(\zeta)$.

The transformation will, in general, possess certain numbers of critical points. These points are given by the solution of the equation $(dz/d\zeta) = 0$.

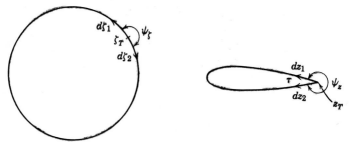

Fig. 16.2   Mapping conditions at the trailing edge.

Thus what we require of the mapping is that one of its critical points should lie on the circle while all the rest of them are included in the interior of the circle so as to leave the regions exterior to the circle and the airfoil conformal to each other. The critical point on the circle then goes into the sharp trailing edge of the airfoil.

The angle between the elements dz_1 and dz_2 of the airfoil curve at the trailing edge z_T is readily determined. If the first $(n-1)$ derivatives of the transformation function are zero at the point ζ_T, then we have

$$\psi_z = n\pi$$

The angle

$$\tau = 2\pi - n\pi = \pi(2 - n)$$

is defined as the *trailing-edge angle*.

16.3   Kutta Condition and the Value of Circulation

We immediately proceed to determine the circulation around the airfoil by application of the Kutta condition. According to this condition we require that the velocity at the trailing edge be finite and continuous. The complex velocity at the trailing edge is given by

$$W(z = z_T) = W(\zeta = \zeta_T) \frac{1}{(dz/d\zeta)_{\zeta = \zeta_T}}$$

Since at the trailing edge $dz/d\zeta$ is zero, $W(z = z_T)$ will be infinite unless $W(\zeta = \zeta_T)$ is also zero. Hence we must set

$$W(\zeta = \zeta_T) = 0 \qquad (16.1)$$

Thus the *Kutta condition requires that on the circle the complex velocity be zero at the point* $\zeta = \zeta_T$, *which corresponds to the trailing edge of the airfoil.* In other words, the point ζ_T is a stagnation point.

In the ζ-plane let us choose the real axis along the line from the origin O to the point ζ_T. The positive direction of the axis is that from O to ζ_T.

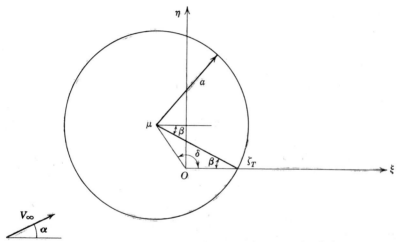

Fig. 16.3   Defining some symbols for the flow past the circle.

Denote the center of the circle by the point $\zeta = \mu$ and the radius of the circle by a (see Fig. 16.3). It follows that

$$a = |\zeta_T - \mu|$$

If the argument of the vector $\zeta_T - \mu$ is denoted by $-\beta$, we then have

$$\zeta_T - \mu = ae^{-i\beta} \qquad (16.2)$$

Using (15.51) and (16.2) we obtain for the complex velocity at the point ζ_T

$$W(\zeta_T) = V_\infty e^{-i\alpha} + i\frac{\Gamma}{2\pi}\frac{1}{\zeta_T - \mu} - V_\infty a^2 e^{i\alpha}\frac{1}{(\zeta_T - \mu)^2}$$

$$= V_\infty e^{-i\alpha} + i\frac{\Gamma}{2\pi}\frac{1}{ae^{-i\beta}} - V_\infty\frac{e^{i\alpha}}{e^{-i2\beta}}$$

With this, (16.1) takes the form

$$2\pi a V_\infty[1 - e^{2i(\alpha+\beta)}] + i\Gamma e^{i(\alpha+\beta)} = 0$$

From this we obtain

$$\Gamma = 4\pi a V_\infty \sin(\alpha + \beta) \tag{16.3}$$

This equation shows that the circulation depends, in general, on the airfoil shape represented by the parameters a and β, the speed V_∞, and the angle of attack α.

16.4   Lift on the Airfoil

The lift on the airfoil is given by

$$L = \rho V_\infty \Gamma = 4\pi \rho V_\infty^2 a \sin(\alpha + \beta) \tag{16.4}$$

For the lift coefficient we obtain

$$C_L = \frac{L}{\frac{1}{2}\rho V_\infty^2 l} = 8\pi\left(\frac{a}{l}\right)\sin(\alpha + \beta) \tag{16.5}$$

For small angles of attack (strictly for small $\alpha + \beta$), the lift coefficient may be expressed as

$$C_L \approx 8\pi\left(\frac{a}{l}\right)(\alpha + \beta) \tag{16.6}$$

The chord of the airfoil, as we shall see, is about $4a$, being equal to it in the limiting case of a so-called flat-plate airfoil. Thus we may say that *the lift coefficient is about $2\pi \sin(\alpha + \beta)$ or $2\pi(\alpha + \beta)$ for small angles of attack*.

The lift on the airfoil is zero when the angle of attack is equal to $-\beta$. We call this angle of attack the *no-lift angle* (or the *zero-lift angle*) and denote it by α_0. We thus have

$$\alpha_0 = -\beta \tag{16.7}$$

Introducing α_0, it is customary to designate $\alpha - \alpha_0$ as the *effective angle of attack* and denote it by α_e:

$$\alpha_e = \alpha - \alpha_0 = \alpha + \beta \tag{16.8}$$

The formulas (16.3) to (16.5) and the consequent realization that the lift coefficient depends linearly on the angle of attack (for small angles of attack) are fundamental results and agree satisfactorily with experimental observations (see Fig. 16.4). They are landmarks in the theory of lift. Their discovery was of vital importance in the eventual achievement of winged flight. Before their discovery, that is, until about the year 1900, the outlook for a rational and practically useful theory of lift was extremely pessimistic.*

---

* For interesting and illuminating historical surveys, reference may be made to Durand (1934), Pritchard (1957), and Kármán (1954).

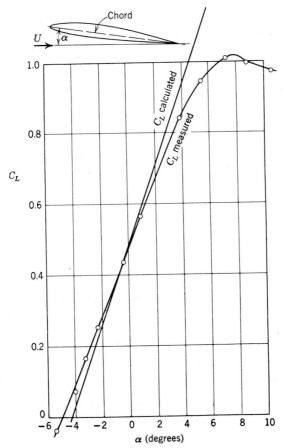

Fig. 16.4 Comparison of theoretical and experimental results for the lift of an airfoil.

16.5 Moment on the Airfoil

It is customary to define the positive direction of the moment on an airfoil as that which tends to increase the angle of attack (as the so-called nose-up moment). Henceforth we shall use this convention. Then, according to Eq. (15.59), the moment on the airfoil is given by

$$M_0 = -2\pi\rho V_\infty^2 \mathscr{C}^2 \sin 2(\gamma - \alpha) - Lm \cos (\delta - \alpha)$$
$$= M_\mu - Lm \cos (\delta - \alpha) \tag{16.9}$$

where m, δ, $\mathscr{C}^2$, and γ are defined by the relations (15.60) and (15.61):

$$\mu = me^{i\delta}$$
$$C_1 = \mathscr{C}^2 e^{i2\gamma}$$

Let us compute the moment, denoted by M, about any point z in the plane of the airfoil. Introducing the vector (see Fig. 16.5)

$$z - \mu = he^{i\varphi}$$

we obtain

$$M = M_\mu + Lh \cos(\varphi - \alpha)$$

$$= -2\pi\rho V_\infty^2 \mathscr{C}^2 \left[\sin 2(\gamma - \alpha) + 2\,\frac{ah}{2}\, \sin(\alpha + \beta) \cos(\varphi - \alpha) \right] \quad (16.10)$$

where use has been made of the expressions for M_μ and L. This equation shows that M, like L, depends on the airfoil shape, the speed V_∞, and the

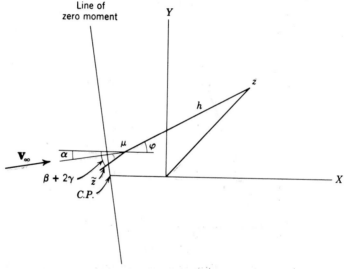

Fig. 16.5 Defining the symbols used in relation to the moment on the airfoil.

angle of attack α. Furthermore, unlike the lift, the moment explicitly involves a coefficient, $C_1 = \mathscr{C}^2 e^{i2\gamma}$, of the transformation function.

Moment at No-lift. · From (16.9) we see that even when the lift on the airfoil is zero, which happens for $\alpha = -\beta$, there is a moment. In such a situation the moment about any one point is the same as the moment about any other point. We denote this *moment at no-lift* by $\tilde{M}$. Then from Eq. (16.9) we obtain

$$\tilde{M} = -2\pi\rho V_\infty^2 \mathscr{C}^2 \sin 2(\gamma + \beta) \quad (16.11)$$

Aerodynamic Center and the Moment about the Aerodynamic Center. We inquire whether there is a point in the plane of the airfoil about which the moment does not change when the angle of attack is varied. Such an inquiry is natural in view of the importance of the role of moments in the dynamics of an airfoil or of a winged vehicle. We answer the inquiry as follows.

Equation 16.10, which gives the moment about any point, may be re-written as*

$$M = 2\pi\rho V_\infty^2 \mathscr{C}^2 \left\{ \sin 2(\alpha - \gamma) + \frac{ah}{\mathscr{C}^2} [\sin (\beta + \varphi) + \sin (2\alpha + \beta - \varphi)] \right\}$$

(16.12)

From this equation we see that if we choose the point $z = \tilde{z}$ such that

$$\tilde{z} - \mu = \frac{\mathscr{C}^2}{a} e^{i(\beta + 2\gamma + \pi)}$$

$$= -\frac{\mathscr{C}^2}{a} e^{i(\beta + 2\gamma)}$$

(16.13)

then the moment about $\tilde{z}$ becomes independent of the angle of attack.† We thus conclude that there is a point about which the moment remains unaltered as the angle of attack is varied. Such a point we call the *aerodynamic center*. Its coordinates are given by (16.13). See also Fig. 16.5.

The moment about the aerodynamic center obtained by substituting (16.13) into (16.12) turns out to be

$$-2\pi\rho V_\infty^2 \mathscr{C}^2 \sin 2(\beta + \gamma)$$

which indeed is independent of α. Furthermore, this is exactly the moment at no-lift. We thus write

$$\tilde{M} = \text{moment at no-lift}$$

$$= \text{moment about the aerodynamic center}$$

$$= -2\pi\rho V_\infty^2 \mathscr{C}^2 \sin 2(\beta + \gamma)$$

(16.14)

This shows that, for given values of ρ and V_∞, $\tilde{M}$ depends only on the shape of the airfoil.

The moment about any point z may now be expressed as

$$M = \tilde{M} + L |z - \tilde{z}| \cos [\arg (z - \tilde{z}) - \alpha]$$

From this it follows that *the moment is zero about any point lying on a line*

* Use $\sin -\psi = -\sin \psi$ and $2 \sin A \cos B = \sin (A + B) + \sin (A - B)$.
† The reader should verify this.

parallel to the direction of the lift and at a perpendicular distance

$$k = \frac{\tilde{M}}{L} \tag{16.15}$$

from the aerodynamic center $\tilde{z}$ (see Fig. 16.5).

The intersection of this line with the chord line of the airfoil is usually referred to as the *center of pressure*, denoted by the symbol C.P. This should not be misunderstood as the point of application of the resultant of the pressure forces acting on the airfoil, for in general there is no such point; there is only a line of action of the resultant force. The change in the center of pressure with variation of the angle of attack is an important characteristic of airfoils, particularly for considerations of stability.

The moment about any point may also be expressed in nondimensional form

$$C_M = \frac{M}{\frac{1}{2}\rho V_\infty^2 l^2}$$

Thus the *moment coefficient* for the moment about the aerodynamic center is given by

$$\tilde{C}_M = \frac{\tilde{M}}{\frac{1}{2}\rho_\infty V_\infty l^2} = -4\pi \left(\frac{\mathscr{C}}{l}\right)^2 \sin 2(\gamma + \beta) \tag{16.16}$$

Introducing the coefficients C_L and $\tilde{C}_M$ we may rewrite Eq. (16.15) as

$$\frac{k}{l} = \frac{\tilde{C}_M}{C_L} \tag{16.17}$$

16.6 Velocity and Pressure Distributions on the Airfoil Surface

We may readily obtain expressions also for the distributions of velocity and pressure on the airfoil. The complex velocity on the airfoil surface is given by

$$W(z) \text{ on the surface} = [W(\zeta) \text{ on the circle}]\left[\frac{1}{(dz/d\zeta)} \text{ on the circle}\right]$$

Denoting the points on the circle by

$$\zeta = \mu + ae^{i\theta}$$

we obtain, using (15.51) and (16.3),

$$W(\zeta) \text{ on the circle} = V_\infty e^{-i\alpha} - V_\infty e^{i(\alpha - 2\theta)} + i2V_\infty \sin(\alpha + \beta)e^{-i\theta}$$
$$= i2V_\infty[\sin(\alpha + \beta) - \sin(\alpha - \theta)]e^{-i\theta} \tag{16.18}$$

The velocity distribution on the airfoil is given by multiplying the right-hand side of (16.18) by $(dz/d\zeta)^{-1}$ on the circle. The pressure distribution on the airfoil is given by the Bernoulli equation

$$p \text{ on the airfoil} = p_\infty - \frac{\rho}{2} V_\infty^2 \left[1 - \frac{|W(z)|^2 \text{ on airfoil}}{V_\infty^2} \right]$$

The magnitude of the velocity on the airfoil surface is obtained from (16.18) as

$$|W(z)| \text{ on the airfoil} = 2V_\infty [\sin (\alpha + \beta)$$
$$- \sin (\alpha - \theta)] \frac{1}{|dz/d\zeta| \text{ on the circle}}$$

This completes the essential results in the theory of lift of an airfoil. It is remarkable that we were able to obtain these results without constructing explicitly the solution of the flow field. This was possible because of the use of the simple but powerful methods afforded by the theory of functions of a complex variable. In the next chapter we shall see how the so-called simplified theory of an arbitrary airfoil may be handled without employing the complex-variable representation. Now, we proceed to consider some details of the mapping between a circle and an airfoil.

16.7 Transformation of a Circle into an Airfoil

We now consider the problem of generating airfoil-like profiles by choosing a circle and a transformation function. The transformation from a circle to an airfoil may be expressed as

$$z = z(\zeta) = \zeta + \sum_1^\infty \frac{C_n}{\zeta^n} \tag{16.19}$$

Such a function, as we had seen (see Section 15.10), meets the requirements at infinity: that the point at infinity in the z-plane goes into the point at infinity in the ζ-plane and that the complex velocity at infinity be mapped into itself.

Consider the derivative of the transformation (16.19)

$$\frac{dz}{d\zeta} = 1 - \frac{C_1}{\zeta^2} - 2\frac{C_2}{\zeta^3} - 3\frac{C_3}{\zeta^4} \tag{16.20}$$

This becomes infinite at the point $\zeta = 0$, that is, at the origin of the coordinates. It may become zero at a number of points. Let us say that there are k such points and let us denote them by $\zeta_1, \zeta_2, \ldots, \zeta_k$. These are then the solutions of the equation

$$\left(1 - \frac{\zeta_1}{\zeta}\right)\left(1 - \frac{\zeta_2}{\zeta}\right) \cdots \left(1 - \frac{\zeta_k}{\zeta}\right) = \frac{dz}{d\zeta} = 0$$

In the expansion of the left-hand side of this equation, the coefficient of the $1/\zeta$ term is

$$\sum_{i=1}^{k} \zeta_i$$

This must equal the coefficient of $1/\zeta$ in the expansion given in (16.20). This latter coefficient is, however, zero. Therefore we should have

$$\sum_{i=1}^{k} \zeta_i = 0 \tag{16.21}$$

This means that the centroid of the critical points of the transformation is the origin of coordinates in the ζ-plane.

We now have sufficient information to choose the circle and construct the transformation. The procedure is as follows:

1. Choose k points in the ζ-plane to be the zeros of $dz/d\zeta$. We denote these zeros, as done above, by the points $\zeta_1, \zeta_2, \ldots, \zeta_k$.

2. The transformation is then given by

$$\frac{dz}{d\zeta} = \left(1 - \frac{\zeta_1}{\zeta}\right)\left(1 - \frac{\zeta_2}{\zeta}\right) \cdots \left(1 - \frac{\zeta_k}{\zeta}\right) \tag{16.22}$$

The integral of this equation may be expressed as

$$z = z(\zeta) = \zeta + \frac{C_1}{\zeta} + \frac{C_2}{\zeta^2} + \cdots + \frac{C_k}{\zeta^k} \tag{16.23}$$

where the complex coefficients $C_1, C_2, \ldots, C_k$ are determined by $\zeta_1, \ldots, \zeta_k$.

3. Choose the centroid of the zeros of $dz/d\zeta$ as the origin O of the coordinates.

4. Choose one of these zero points to be the trailing-edge point. As before, denote it by ζ_T.

5. Draw a circle with center at any point but such that it encloses all the zeros of $dz/d\zeta$ except the point ζ_T and passes through ζ_T. Denote, as before, the center of the circle by the point μ. The radius of the circle is then given by

$$a = |\zeta_T - \mu|$$

6. Choose the line O to ζ_T as the real axis ξ. The imaginary axis η is then automatically fixed. Denoting, as before, by $-\beta$ the argument of $\zeta_T - \mu$ we have

$$ae^{-i\beta} = \zeta_T - \mu$$

7. Choosing different sets of zeros of $dz/d\zeta$ and different circles, we may obtain an infinity of different airfoil shapes

16.8   The Joukowski Transformation

We now apply the preceding ideas to obtain the simplest transformation. For this purpose we look for a function $z(\zeta)$ whose derivative has only two zeros. Denoting them by ζ_1 and ζ_2 we should have, according to (16.21),

$$\zeta_1 + \zeta_2 = 0$$

or

$$\zeta_2 = -\zeta_1$$

Now, one of these points should be the trailing-edge point. Let us set

$$\zeta_1 = \zeta_T \tag{16.24}$$

Then it follows that

$$\zeta_2 = -\zeta_T \tag{16.25}$$

Equation (16.22) now takes the form

$$\frac{dz}{d\zeta} = \left(1 - \frac{\zeta_T}{\zeta}\right)\left(1 + \frac{\zeta_T}{\zeta}\right)$$

$$= 1 - \frac{\zeta_T^{\,2}}{\zeta^2} \tag{16.26}$$

The transformation function then becomes

$$z(\zeta) = \zeta + \frac{\zeta_T^{\,2}}{\zeta} \tag{16.27}$$

Choosing the real axis along the vector ζ_T we set

$$\zeta_T = C \tag{16.28}$$

where *C is a real positive number.* The transformation (16.27) then takes the form

$$z(\zeta) = \zeta + \frac{C^2}{\zeta} \tag{16.29}$$

This simple transformation is known as the *Joukowski transformation.* It not only leads to the *Joukowski airfoils* but also plays a basic role in many approximate theories of the airfoil. We shall return to a consideration of the Joukowski profiles in the next two sections. Now let us examine the basic properties of the transformation (16.29).

The points $\zeta = C$ and $\zeta = -C$ map, respectively, into the points $z = 2C$ and $z = -2C$ (see Fig. 16.6).

Consider the *circle* described by

$$\zeta = Ce^{i\theta}$$

Points on the circumference of this circle go into points on the x-axis (in the z-plane) along the strip $-2C \leq x \leq 2C$. As we traverse along the top half of the circle from A to B to D, we cover the top of the strip from A to B to D. As we traverse along the bottom half of the circle from D to E to A, we cover the bottom of the strip from D to E to A.

Denote the points in the ζ-plane by

$$\zeta = re^{i\theta}$$

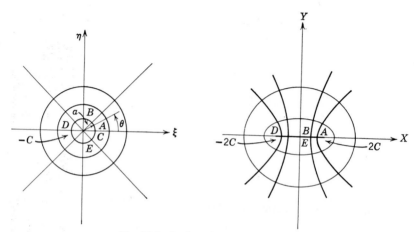

Fig. 16.6 Joukowski transformation.

Then the transformation (16.29) may be expressed as

$$x = \left(r + \frac{C^2}{r}\right) \cos\theta$$

$$y = \left(r - \frac{C^2}{r}\right) \sin\theta \qquad (16.30)$$

For r equal to a constant r_0 we obtain, eliminating θ,

$$\frac{x^2}{\left(r_0 + \dfrac{C^2}{r_0}\right)^2} + \frac{y^2}{\left(r_0 - \dfrac{C^2}{r_0}\right)^2} = 1 \qquad (16.31)$$

This equation represents an ellipse in the z-plane having foci at $x = \pm 2C$ (see Fig. 16.6). For θ equal to a *constant* θ_0 we obtain from (16.30) on eliminating r,

$$\frac{x^2}{(2C \cos\theta_0)^2} - \frac{y^2}{(2C \sin\theta_0)^2} = 1 \qquad (16.32)$$

This equation represents confocal hyperbolas having foci at $x = \pm 2C$.

We thus conclude that the transformation (16.29) maps circles about the origin in the ζ-plane into a family of confocal ellipses in the z-plane and rays through the origin in the ζ-plane into a family of confocal hyperbolas in the z-plane. The foci of the ellipses and hyperbolas are the same and are located at $x = \pm 2C$. The ellipses and the hyperbolas form an orthogonal net.

Consider the mapping of the circles exterior to the circle $r = C$. As the radius r_0 of these circles goes to C, the corresponding ellipses tend to the strip $-2C \leq x \leq 2C$. Thus we see that the exterior to the circle C is mapped into the exterior to the strip. What about the mapping of the interior to the circle $r = c$? The interior to the circle is again mapped into the region exterior to the strip. To see this let us consider the mapping of the two circles

$$|\zeta| = r_0$$

and

$$|\zeta| = \frac{C^2}{r_0}$$

where r_0 is greater than C. Thus the circle $|\zeta| = r_0$ is exterior to the circle C, and the circle $|\zeta| = (C^2/r_0)$ is interior to the circle C. From (16.31) we conclude that both the circles are mapped into the same ellipse. We thus conclude that the mapping represented by (16.29) is double valued. To avoid this double valuedness, we introduce two sheets of the z-plane, one over the other and suitably joined across the strip, and consider that the exterior of the circle C is mapped into one of the sheets and the interior into the other sheet. The points $x = \pm 2C$ are then known as branch points, and the strip $-2C \leq x \leq 2C$ is known as a branch cut. The sheets are known as Riemann sheets.

16.9 Joukowski Airfoils

The Circle. We now apply the Joukowski transformation to a circle. We choose the circle such that it passes through $\zeta = \zeta_T = C$ and encloses the point $\zeta = -\zeta_T = -C$. If, as before, μ denotes the center of the circle and a its radius, we have

$$me^{i\delta} = \mu = ae^{-i\beta} - C \qquad (16.33)$$

The quantities m and β may be taken as parameters defining the circle. Generally m and β are chosen small.

The Trailing-Edge Angle. This is given by

$$\tau = \pi(2 - n)$$

where n is the order of the derivative of $z(\zeta)$ which first becomes nonzero at the trailing edge (see Section 16.2). For the Joukowski transformation, it may be readily verified, n is 2. Therefore it follows that

$$\tau = 0$$

We thus conclude that *the airfoils generated by the Joukowski transformation have cusped trailing edges.*

The Flat-Plate Airfoil. Choose the circle with center at the origin and radius equal to C. We thus have

$$\mu = 0, \quad a = C, \quad \beta = 0$$

We had seen that such a circle transforms into a strip on the X-axis extending from $x = -2C$ to $x = +2C$. We talk of this as the flat-plate airfoil. Its chord is equal to $4C$.

The Circular-Arc Airfoil. Choose $\mu = m e^{i\pi/2} = im$. Then β is not zero and a is equal to $C \sec \beta$ (see Fig. 16.7). To interpret the airfoil

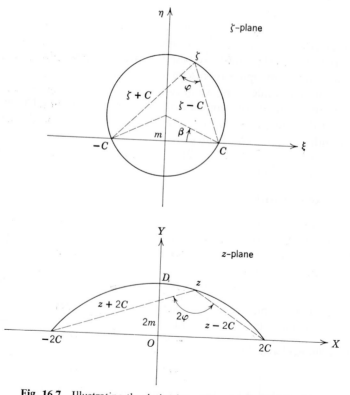

Fig. 16.7 Illustrating the derivation of the circular-arc airfoil.

generated from this circle we first observe that the Joukowski transformation may be expressed as*

$$\frac{z - 2C}{z + 2C} = \frac{(\zeta - C)^2}{(\zeta + C)^2} \tag{16.34}$$

Denoting the argument of $(\zeta - C)/(\zeta + C)$ by φ (see Figure 16.7) we obtain

$$\arg(z - 2C) - \arg(z - 2C) = 2\varphi \tag{16.35}$$

Now, for points on the circle φ is a constant. This means the corresponding image points should describe a curve such that for those points the left-hand side of (16.35) remains a constant equal to 2φ. Such a curve is simply a circular arc passing through the points $z = -2C$ and $z = 2C$ and the point D (see Figure 16.7). This point, as may be verified, is given by

$$OD = 2im$$

We thus conclude that the present choice for the circle generates a circular arc airfoil. Naturally it has no thickness, just like the flat plate.

The chord of the airfoil is $4C$. The maximum camber is given by the ratio of the maximum ordinate to the chord:

$$\frac{2m}{4C} = \tfrac{1}{2} \tan \beta$$

Thus the angle β is a measure of the camber of the airfoil

The Symmetrical Airfoil. Choose $\mu = -m$. Then β is equal to zero. Setting

$$m = \varepsilon C \quad \text{where } \varepsilon < 1 \tag{16.36}$$

we have

$$a = m + C = C(1 + \varepsilon)$$

(see Fig. 16.8.) A point on the circle is given by

$$\zeta = \mu + ae^{i\theta} = ae^{i\theta} - m \tag{16.37}$$

The corresponding point on the airfoil is given by

$$z = (ae^{i\theta} - m) + \frac{C^2}{(ae^{i\theta} - m)} \tag{16.38}$$

* From $z = \zeta + (C^2/\zeta)$ we have

$$z - 2C = \frac{\zeta^2 + C^2 - 2C\zeta}{\zeta} = \frac{(\zeta - C)^2}{\zeta}$$

and, similarly,

$$z + 2C = \frac{(\zeta + C)^2}{\zeta}.$$

Equation 16.34 immediately follows.

If on the right-hand side of this equation we replace θ by $-\theta$, the left-hand side will become $\bar{z}$. This means that the airfoil curve below the real axis is a reflection of the portion of the curve above that axis. In other words, the airfoil is symmetrical with respect to the real axis. It has thus thickness *but no camber*.

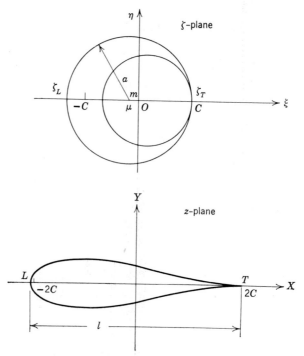

Fig. 16.8 Illustrating the derivation of the symmetric Joukowski airfoil.

The chord of the airfoil is given by

$$l = 2C + |z_L|$$

where z_L is the leading-edge point of the airfoil. We have*

$$z_L = \zeta_L + \frac{C^2}{\zeta_L} = -C(1 + 2\varepsilon) - \frac{C}{1 + 2\varepsilon}$$

$$= -2C(1 + 2\varepsilon^2 + \cdots)$$

Therefore we obtain

$$l = 4C(1 + \varepsilon^2 + \cdots) \tag{16.39}$$

* Note that $\zeta_L = -(a + m) = -C(1 + 2\varepsilon)$.

The thickness of the airfoil at the center is given by

$$t_c = 2y\left(\theta = \frac{\pi}{2}\right) = 4C\varepsilon$$

The maximum thickness occurs at the point where $dy/d\theta$ is zero. This occurs at $\theta = (2\pi/3)$, which corresponds to *quarter-chord point from the leading edge*. We thus obtain

$$t_{max} = 4C\,\frac{3\sqrt{3}}{4}\,\varepsilon$$

The thickness ratio is defined by

$$\tau = \frac{t_{max}}{\text{chord}} \tag{16.40}$$

Hence we have

$$\tau = \frac{3\sqrt{3}}{4}\,\varepsilon = 1.299\varepsilon$$

We *thus see that ε is a measure of the thickness of the airfoil.* It is referred to as the eccentricity.

Airfoil with Camber and Thickness. Choose the center of the circle at any arbitrary point $\mu = me^{i\delta}$, where $\delta \neq (\pi/2)$ or π as in the preceding two cases. Let μ_0 denote the intersection of the vector $ae^{-i\beta}$ with y-axis (see Fig. 16.9). A circle with center at μ_0 and radius equal to $|\mu_0 - \zeta_T|$ transforms into a circular arc. The arbitrary circle, the points of which are described by $\zeta = ae^{-i\theta} + \mu$, goes into an airfoil that includes in its interior

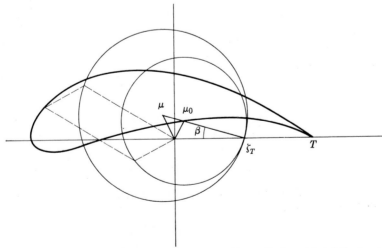

Fig. 16.9 Illustrating the derivation of an arbitrary Joukowski airfoil.

the circular arc. We say that the circular arc forms the skeleton of the airfoil. The angle β determines the mean curvature of the profile. The distance $|\mu - \mu_0|$ determines the thickness of the profile.

A simple graphical construction may be given for the arbitrary Joukowski airfoil. This follows directly from the geometrical meaning of the simple transformations $w(\zeta) = \zeta$ and $w(\zeta) = (C^2/\zeta)$ (see Section 14.12). Such a construction is known as the Trafftz graphical construction.

16.10   Properties of Joukowski Airfoils

These properties, like the properties of any airfoil, are conveniently expressed in nondimensional form. To obtain these properties we special-ize the results of Section 16.4 and 16.5 by using the geometrical properties of the Joukowski airfoils. First we observe that the coefficient $C_1 = \mathscr{C}^2 e^{i2\gamma}$ (16.19) in the general transformation

$$z = z(\zeta) = \zeta + \frac{C_1}{\zeta} + \frac{C_2}{\zeta^2} + \cdots$$

turns out to be

$$C_1 = \mathscr{C}^2 e^{i2\gamma} = C^2$$

for the Joukowski transformation. Hence we have

$$\mathscr{C} = C \quad \text{and} \quad \gamma = 0 \tag{16.41}$$

For the general Joukowski airfoil with thickness and camber we then obtain the following results:*

$$C_L = 8\pi\left(\frac{a}{l}\right)\sin(\alpha + \beta) \tag{16.42a}$$

$$\alpha_0 = -\beta \tag{16.42b}$$

$$\frac{\tilde{z}}{l} = \frac{\mu}{l} - \frac{C^2}{al}e^{i\beta} \tag{16.42c}$$

$$\tilde{C}_M = -4\pi\left(\frac{C}{l}\right)^2\sin 2\beta \tag{16.42d}$$

$$\frac{k}{l} = \frac{\tilde{C}_M}{C_L} \tag{16.46e}$$

Recall that the center μ of the circle, its radius a, and the angle β are connected by (16.33), which is

$$me^{i\delta} = \mu = ae^{-i\beta} - C$$

* See Sections 16.4 and 16.5 for the definitions of the quantities involved.

The corresponding properties for the flat plate, circular arc, and symmetric airfoils are given in the following table. For the circular arc the results are expressed in terms of the maximum camber defined non-dimensionally by $2m/l$ (see Fig. 16.7). Similarly, for the symmetric profile they are expressed in terms of the maximum thickness defined by the parameter τ (see 16.40).

We see that the angle of no-lift is directly related to the camber of the airfoil. The lift coefficient is affected by both camber and thickness. It may be verified that a 10 per cent change in $4m/l$ causes only a 1 per cent change in C_L. Thus we may say that *the circular-arc camber has no significant effect on the lift; it only shifts the no-lift angle.* On the other hand, thickness affects the lift more significantly than the camber. Increase in thickness increases lift. Because of viscous effects, however, this increase is not realized practically.

Airfoil

Property	*Flat Plate*	*Circular Arc*	*Symmetric*
$-\alpha_0 = \beta$	0	$\tan^{-1}\dfrac{4m}{l}$	0
C_L	$2\pi \sin \alpha$	$2\pi\left[1 + \left(\dfrac{4m}{l}\right)^2\right]^{-\frac{1}{2}} \sin(\alpha + \beta)$	$\simeq 2\pi(1 + 0.77\tau)\sin \alpha$
$\tilde{z}/l$	$-\dfrac{1}{4}$	$i\dfrac{m}{l} - \dfrac{1}{4}\left[1 + \left(\dfrac{4m}{l}\right)^2\right]^{-\frac{1}{2}} e^{i\beta}$	$\simeq -\dfrac{1}{4}$
$\tilde{C}_M$	0	$-\dfrac{\pi}{4}\sin 2\beta$	0
k/l	0	$-\dfrac{1}{8}\left[1 + \left(\dfrac{4m}{l}\right)^2\right]^{-\frac{1}{2}} \dfrac{\sin 2\beta}{\sin(\alpha + \beta)}$	0

The aerodynamic center is at the quarter-chord point for the flat plate and practically at that point for the symmetric airfoil. For these the line of action of the lift passes through the aerodynamic center. We may therefore say that for these airfoils the center of pressure and the aerodynamic center coincide.

Joukowski profiles have been investigated extensively both theoretically and experimentally. Examples of calculated and measured pressure distributions over a Joukowski profile are shown in Fig. 16.10. The agreement between them is satisfactory. The differences can be explained as due to the influence of the viscosity of the fluid which, of course, is neglected in the theory. A comparison of the calculated and measured lift on a

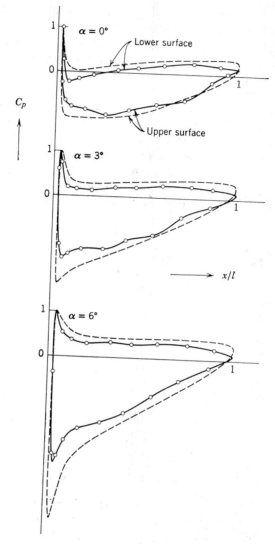

Fig. 16.10 Comparison of theoretical and experimental results for the pressures over a Joukowski airfoil.

Joukowski profile is shown in Fig. 16.11. The agreement is again satisfactory, particularly in the region of small angles of attack. The calculated lift is somewhat larger than the measured value, indicating that the calculated value of the circulation is slightly larger than the actual value. This again may be explained by the influence of viscosity.

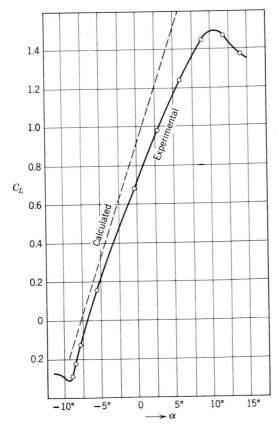

Fig. 16.11 Comparison of theoretical and experimental results for the lift on a Joukowski airfoil.

16.11 Other Airfoils

With the Joukowski airfoils of moderate thickness there is no freedom of varying the camberline appreciably from a circular arc. We have seen that an increase in the camber of the circular arc does not improve the lift considerably. Also, the maximum camber lies rather close to the center of the chord, whereas a position within the forward portion of the chord is usually preferred. To avoid these disadvantages one may generate airfoil profiles from a circle by means of a transformation that includes more than two zeros of the derivative of the mapping function. Airfoils derived in this way are known as *Mises Airfoils*. The procedure for actually constructing them is similar to that for the Joukowski family of airfoils. Their properties may be readily determined. In this connection reference may be made to Mises (1959).

The Kármán-Trefftz Airfoils. The Joukowski family and the Mises family of airfoils all have cusped trailing edges. Naturally it is not possible to construct such edges in practice. Hence one seeks a transformation that will generate from a circle an airfoil profile whose trailing-edge angle τ is not zero. A simple transformation of this kind is set up as follows.

We had seen that if $\zeta = \zeta_0$ is a critical point (specifically a zero of the first derivative) of the transformation $z = z(\zeta)$, then we may express $z(\zeta)$ in a series expansion about the point ζ_0 and write

$$z(\zeta) - z(\zeta_0) = (\zeta - \zeta_0)^n f(\zeta) \qquad (16.43)$$

where n is the order of the first nonzero derivative of $z(\zeta)$ at ζ_0, and $f(\zeta)$ and its first derivative does not become zero or infinite at ζ_0 (see Section 14.14). At ζ_0 angles are not preserved but are multiplied by the factor n.

In application to the airfoil problem, the trailing-edge point $\zeta = \zeta_T$ is a zero of $dz/d\zeta$. Hence the transformation may be expressed in the form

$$z(\zeta) - z(\zeta_T) = (\zeta - \zeta_T)^n f(\zeta) \qquad (16.44)$$

Now, if we choose only the zeros for $dz/d\zeta$, as done in the case of the Joukowski transformation, the other zero is at the point

$$\zeta = -\zeta_T$$

because the origin of coordinates is to be at the centroid of the zeros. Then the transformation may also be expressed as

$$z(\zeta) - z(-\zeta_T) = (\zeta + \zeta_T)^n f(\zeta) \qquad (16.45)$$

From the relations (16.44) and (16.45) we obtain

$$\frac{z(\zeta) - z(\zeta_T)}{z(\zeta) - z(l\mu_T)} = \frac{(\zeta - \zeta_T)^n}{(\zeta + \zeta_T)^n}$$

Setting $z(\zeta_T) = z_T$ and $z(-\zeta_T) = -z_T$, we have

$$\frac{z - z_T}{z + z_T} = \frac{(\zeta - \zeta_T)^n}{(\zeta + \zeta_T)^n} \qquad (16.46)$$

We may also express (16.46) in the series form:

$$z = \frac{z_T}{n\zeta_T} \zeta + \frac{n^2 - 1}{3} \frac{\zeta_T^2}{\zeta} + \cdots \qquad (16.47)$$

We require that the transformation should be such that for $\zeta \to \infty$, $dz/d\zeta = 1$. To satisfy this requirement, on the basis of (16.47), we should set

$$z_T = n\zeta_T \tag{16.48}$$

Then the transformation takes the form

$$z = \zeta + \frac{n^2 - 1}{3} \frac{\zeta_T^2}{\zeta} + \cdots \tag{16.49}$$

or the form

$$\frac{z - n\zeta_T}{z + n\zeta_T} = \frac{(\zeta - \zeta_T)^n}{(\zeta + \zeta_T)^n} \tag{16.50}$$

Finally, if we set $\zeta_T = C$, where C is a real positive number, we may write (16.50) as

$$\frac{z - nC}{z + nC} = \frac{(\zeta - C)^n}{(\zeta + C)^n} \tag{16.51}$$

This is the simplest transformation that will produce a sharp trailing edge with a finite angle. The trailing-edge angle is given by

$$\tau = \pi(2 - n)$$

In terms of this angle the transformation may be written as

$$\frac{z - \dfrac{2\pi - \tau}{\pi} C}{z + \dfrac{2\pi - \tau}{\pi} C} = \frac{(\zeta - C)^{(2\pi - \tau)/\pi}}{(\zeta + C)^{(2\pi - \tau)/\pi}} \tag{16.52}$$

If the trailing edge is required to be a cusp, we set $\tau = 0$, whence (16.52) reduces to

$$\frac{z - 2C}{z + 2C} = \frac{(\zeta - C)^2}{(\zeta + C)^2}$$

which is simply the Joukowski transformation as given previously by (16.34).

The transformation (16.52) is known as the Kármán-Trefftz transformation. It is a simple extension of the Joukowski transformation so as to obtain a finite trailing edge. The airfoils generated by it are known as the Kármán-Trefftz family. A typical airfoil of this family is shown in Fig. 16.16. The skeleton of the airfoil is made up of two circular arcs. The

Fig. 16.12 A Kármán-Trefftz airfoil.

properties of such airfoils may be readily determined by suitably specializing the general relations previously developed. In this connection reference may be made to Durand (1934).

16.12 Theodorsen's Method for the Arbitrary Airfoil

To deal with the problem of an airfoil of arbitrary shape we need to determine the mapping function that transforms the given airfoil profile into a circle. A method suggested by Theodorsen (1932) is here briefly described; full details can be found in Theodorsen.

The mapping function is built up in two stages; in the first the airfoil profile in the z plane is mapped into a contour in the ζ' plane by employing the Joukowski transformation

$$z = \zeta' + \frac{C^2}{\zeta'} \tag{16.53}$$

If the airfoil is an ellipse, this transformation will map the airfoil into an exact circle. Hence, to have the image contour in the ζ' plane as close to a circle as possible, it is necessary to position the coordinate axes in the airfoil plane (i.e., the z plane) such that the airfoil profile is as nearly elliptical as possible with respect to those axes. The contour in the ζ' plane is referred to as the pseudo-circle.

The second stage consists of finding the mapping function that transforms the pseudo-circle into an exact circle in the ζ plane. Thoedorsen suggests the mapping

$$\zeta' = \zeta \exp \left(\sum_1^\infty \frac{C_n}{\zeta^n} \right) \tag{16.54}$$

where the coefficients C_n, complex in general, have to be determined.

Denote a point *on the pseudo-circle* by

$$\zeta' = C e^{\psi(\theta)} e^{i\theta} \tag{16.55}$$

Note that $Ce^{i\theta}$ describes a circle in the ζ' plane which goes into a Joukowski airfoil in the z plane. Thus $e^{\psi(\theta)}$ denotes the deviation of the image contour in the ζ' plane from the Joukowski circle.

The corresponding points on the airfoil and pseudo-circle are related as follows, from (16.53), by

$$x = 2C \cosh \psi \cos \theta$$
$$y = 2C \sinh \psi \sin \theta$$

from which we find

$$2 \sin^2 \theta = P + \left[P^2 + \left(\frac{y}{C} \right)^2 \right]^{1/2} \tag{16.56}$$

$$2 \sin^2 \psi(\theta) = -P + \left[P^2 + \left(\frac{y}{C} \right)^2 \right]^{1/2} \tag{16.57}$$

where

$$P = 1 - \left(\frac{x}{2C}\right)^2 - \left(\frac{y}{2C}\right)^2$$

This establishes the function $\psi(\theta)$.

Denote a point on *the image circle* in the ζ plane by

$$\zeta = Ce^{\psi_0}e^{i\phi} = Re^{i\phi} \tag{16.58}$$

where ψ_0 is a constant that is yet to be determined. The corresponding points on the pseudo-circle and the exact circle are related according to (16.54). Thus we have

$$\frac{Ce^{\psi(\theta)}e^{i\theta}}{Ce^{\psi_0}e^{i\phi}} = \frac{\zeta'}{\zeta} = \exp\left(\sum_1^\infty \frac{C_n}{\zeta^n}\right)$$

setting

$$\frac{C_n}{\zeta^n} = \frac{A_n + iB_n}{\zeta^n} = \frac{A_n + iB_n}{R^n}e^{-in\phi} \tag{16.59}$$

we obtain

$$\psi - \psi_0 = \sum_1^\infty \frac{1}{R^n}(A_n \cos n\phi + B_n \sin n\phi) \tag{16.60}$$

$$\theta - \phi = \sum_1^\infty \frac{1}{R^n}(B_n \cos n\phi - A_n \sin n\phi) \tag{16.61}$$

where $R = Ce^{\psi_0}$, a constant. We further have

$$\psi_0 = \frac{1}{2\pi}\int_0^{2\pi} \psi(\phi)\, d\phi \tag{16.62}$$

$$\frac{A_n}{R^n} = \frac{1}{\pi}\int_0^{2\pi} \psi(\phi) \cos n\phi\, d\phi \tag{16.63}$$

$$\frac{B_n}{R^n} = \frac{1}{\pi}\int_0^{2\pi} \psi(\phi) \sin n\phi\, d\phi \tag{16.64}$$

These equations determine the three unknowns ψ_0, A_n, and B_n. To find them, however, we need to know $\psi(\phi)$ or, equivalently, $\theta(\phi)$, which is what we wish to find. It follows that they need to be solved by iteration (see Theodorsen) or by numerical methods. For the details reference should be made to the literature.

Chapter 17

Elements of Thin Airfoil Theory

The problem of calculating the flow field and the aerodynamic properties of any given arbitrary airfoil with no restrictions as to its thickness, camber, or angle of attack is complex in practice. The calculation could be done by using Theodorsen's method. This method, however, is not convenient either for a rapid estimation of the velocity or pressure distribution over the airfoil or for designing an airfoil profile that will have a prescribed surface distribution of velocity or pressure. It thus appears profitable to have an alternative approach that will in some way simplify the mathematical conditions of the problem and enable the construction of a closed form solution (approximate naturally) for the problem. Such an alternative approach is provided by the so-called thin airfoil theory. The elements of this theory are presented in this chapter. Thin airfoil theory had its beginnings in the early days of aerodynamics, some of the original attempts being those of Munk (1922), Birnbaum (1923), and Glauert (1926).

17.1 Formulation of the Problem in Terms of the Perturbation Field

Consider the steady flow past a fixed airfoil of arbitrary shape (Fig. 17.1). The flow field is the solution of the following mathematical problem:

Differential equation

$$\nabla^2 \Phi = \frac{\partial^2 \Phi}{\partial x^2} + \frac{\partial^2 \Phi}{\partial y^2} = 0 \tag{17.1}$$

Boundary condition

$$\text{grad } \Phi \cdot \text{grad } F = 0 \quad \text{or} \quad F(x, y) = 0 \tag{17.2}$$

Infinity condition

$$\text{grad } \Phi = \mathbf{V}_\infty \text{ at infinity} \tag{17.3}$$

Kutta condition

The circulation around the airfoil is such that

the velocity is finite and continuous at the trailing edge $\qquad$ (17.4)

492

In these relations $\Phi = \Phi(x, y)$ is the velocity potential and $F(x, y) = 0$ describes the airfoil profile.

Denoting by $\mathbf{q} = \mathbf{q}(x, y)$ the *perturbation or disturbance velocity* due to the airfoil, we write
$$\mathbf{V}(x, y) = \mathbf{V}_\infty + \mathbf{q}(x, y) \tag{17.5}$$

where $\mathbf{V}_\infty$ as usual is the free stream velocity. Denoting the components of $\mathbf{q}$ by u and v, and the components of $\mathbf{V}$ by u_t and v_t, we have

$$u_t(x, y) = V_\infty \cos \alpha + u(x, y) \tag{17.6}$$
$$v_t(x, y) = V_\infty \sin \alpha + v(x, y) \tag{17.7}$$

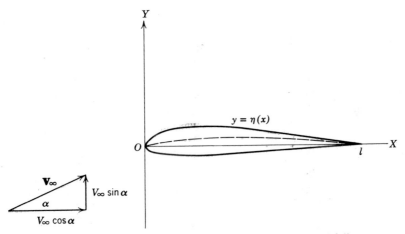

Fig. 17.1 Coordinates and symbols for the flow past an airfoil.

where α is the angle of attack (see Fig. 17.1). Introduce $\phi = \phi(x, y)$ as the potential for the perturbation velocity. We then have

$$u(x, y) = \frac{\partial \phi}{\partial x} \tag{17.8}$$

$$v(x, y) = \frac{\partial \phi}{\partial y} \tag{17.9}$$

and
$$\operatorname{grad} \Phi = \mathbf{V}_\infty + \operatorname{grad} \phi$$
$$\Phi(x, y) = (xV_\infty \cos \alpha + yV_\infty \sin \alpha) + \phi \tag{17.10}$$

In terms of the perturbation field, the above mathematical problem takes the following form:

Differential equation
$$\nabla^2 \phi = \frac{\partial^2 \phi}{\partial x^2} + \frac{\partial^2 \phi}{\partial y^2} = 0 \tag{17.11}$$

Boundary condition

$$(\mathbf{V}_\infty + \text{grad } \phi) \cdot \text{grad } F = 0 \quad \text{or} \quad F(x, y) = 0 \qquad (17.12)$$

Infinity condition

$$\text{components of grad } \phi \to 0 \text{ at infinity} \qquad (17.13)$$

Kutta condition

The circulation around the airfoil should be such that the perturbation velocity $\mathbf{q} = \text{grad } \phi$ should be finite and continuous at the trailing edge $\qquad (17.14)$

As we know, it is rather difficult to solve this direct problem. To make some progress with the construction of the solution we simplify the problem by introducing the approximation of small perturbations. With such an approximation the boundary condition may be reduced to a simpler form than (17.12) and transferred to the axis (chord line) of the airfoil. Then the problem itself may be represented as a superposition of three other simpler problems. We now proceed with the development of such a scheme.

17.2 Simplification of the Boundary Condition

Consider the boundary condition as expressed by (17.12). For convenience let us temporarily work in terms of u and v instead of ϕ. In component form (17.12) appears as

$$(V_\infty \cos \alpha + u)\frac{\partial F}{\partial x} + (V_\infty \sin \alpha + v)\frac{\partial F}{\partial y} = 0 \quad \text{on} \quad F(x, y) = 0 \quad (17.15)$$

Suppose the airfoil curve is described by

$$y = \eta(x) = \eta_u(x) \text{ for the upper curve}$$
$$= \eta_l(x) \text{ for the lower curve} \qquad (17.16)$$

Then we have

$$F(x, y) = \eta(x) - y = 0 \qquad (17.17)$$

$$\frac{\partial F}{\partial x} = \frac{d\eta}{dx} \qquad (17.18)$$

$$\frac{\partial F}{\partial y} = -1 \qquad (17.19)$$

on substituting (17.18) and (17.19) into (17.15), we obtain

$$v = (V_\infty \cos \alpha + u)\frac{d\eta}{dx} - V_\infty \sin \alpha \quad \text{on} \quad y = \eta(x), 0 \leq x \leq l \quad (17.20)$$

We observe that v would be a known linear function of dy/dx but for the presence of u in the first term of the right-hand side of (17.20). In this sense this equation is said to express the boundary condition in a non-linear form.

To simplify the equation (17.20), we introduce certain assumptions. We *assume that the airfoil is sufficiently thin and elongated* (*that it is only a small deviation from its chord line*) *and that the angle of attack is sufficiently small so that the perturbation velocities u and v are small compared to V_∞:*

$$u \ll V_\infty, \qquad v \ll V_\infty$$

The shape of the airfoil is such that

$$\frac{\partial F}{\partial x} < \frac{\partial F}{\partial y}$$

In other words, the normal to the airfoil curve is practically normal to the chord line. With these assumptions we set

$$V_\infty \cos \alpha = V_\infty$$
$$V_\infty \sin \alpha = V_\infty \alpha$$

and neglect $u\, \partial F/\partial x = u\, d\eta/dx$ in comparison with other terms appearing in the boundary condition. Equation 17.20 then becomes

$$v[x, \eta(x)] = V_\infty \frac{d\eta}{dx} - V_\infty \alpha, \qquad 0 \leq x \leq l \qquad (17.21)$$

This simplified relation is sometimes known as the *linearized form of the boundary condition.*

We immediately observe that the assumption of small perturbations is not valid at and near the stagnation points occurring on the airfoil for in such regions the perturbation velocity is of the same order of magnitude as the undisturbed velocity. We further note that if the airfoil has a rounded leading edge, the assumption that $\partial F/\partial x$ is much less than $\partial F/\partial y$ is violated at and near that edge. On the basis of these considerations, we conclude that the simplification we made gives rise inherently to certain regions where the solution of the resulting problem is not valid. Fortunately for applications such regions are quite small and the solution of the simpler problem is remarkably useful.

17.3 Transfer of the Boundary Condition

Although (17.21) is linear, its application is rather inconvenient for, the value of the velocity component has to be computed at points on the airfoil. In view of the assumptions we have already made, this inconvenience may be removed as follows. *Assuming that $v(x, y)$ may be expanded in a Taylor Series about points on the chord line*, that is on the X-axis, $0 \leq x \leq l$ we write

$$v[x, \eta(x)] = v(x, 0) + \left(\frac{\partial v}{\partial y}\right)_{x,0} \eta(x) + \cdots, \qquad 0 \leq x \leq l$$

Consistent with our previous approximations, one may neglect $(\partial v / \partial y)_{x,0} \eta$ and all the other higher terms. We may therefore set

$$v[x, \eta(x)] = v(x, 0), \qquad 0 \leq x \leq l \tag{17.22}$$

On our making this approximation, (17.21) becomes

$$v(x, 0) = V_\infty \frac{d\eta}{dx} - V_\infty \alpha, \qquad 0 \leq x \leq l \tag{17.23}$$

In view of (17.16) we clarify (17.23) by writing it explicitly as

$$\left. \begin{aligned} v(x, 0_+) &= V_\infty \frac{d\eta_u}{dx} - V_\infty \alpha \\[2mm] v(x, 0_-) &= V_\infty \frac{d\eta_l}{dx} - V_\infty \alpha \end{aligned} \right\} \quad 0 \leq x \leq l \tag{17.24}$$

where 0_+ and 0_-, respectively, denote the top and bottom sides of the X-axis in the strip $0 \leq x \leq l$.

This transfer of the boundary condition as exhibited by (17.24) is possible only if $v(x, y)$ can be expanded in a Taylor Series about the chord line. We have assumed that it is, but our assumption can be checked only a posteriori.

17.4 Frame Work of the Theory of the Thin Airfoil

Introducing the simplified boundary condition and its transfer to the axis of the airfoil, we express the mathematical problem governing the perturbation field as

$$\nabla^2 \phi = \frac{\partial^2 \phi}{\partial x^2} + \frac{\partial^2 \phi}{\partial y^2} = 0 \tag{17.25}$$

$$\left(\frac{\partial \phi}{\partial y}\right)_{x,0} = v(x, 0) = V_\infty \frac{d\eta}{dx} - V_\infty \alpha \qquad 0 \leq x \leq l \tag{17.26*}$$

$$\frac{\partial \phi}{\partial x} \quad \text{and} \quad \frac{\partial \phi}{\partial y} \to 0 \quad \text{at infinity} \tag{17.27}$$

The circulation is such that

$$\frac{\partial \phi}{\partial x} \quad \text{and} \quad \frac{\partial \phi}{\partial y} \quad \text{are finite and continuous at the trailing edge} \tag{17.28}$$

* Recall the explicit meaning of this form which we use for convenience (see 17.24).

Fig. 17.2 Flow past the airfoil at zero angle of attack.

Let $\phi_1(x, y)$ and $\phi_2(x, y)$, respectively, be the solutions of the following two problems:

1. $\nabla^2 \phi_1 = 0$

$$\left(\frac{\partial \phi_1}{\partial y}\right)_{x,0} = V_\infty \frac{d\eta}{dx} \qquad 0 \le x \le l$$

$$\frac{\partial \phi_1}{\partial x}, \frac{\partial \phi_1}{\partial y} \to 0 \quad \text{at infinity}$$

$$\frac{\partial \phi_1}{\partial x}, \frac{\partial \phi_1}{\partial y} \qquad \text{are finite and continuous at the trailing edge}$$

2. $\nabla^2 \phi_2 = 0$

$$\left(\frac{\partial \phi_2}{\partial y}\right)_{x,0} = -V_\infty \alpha, \qquad 0 \le x \le l$$

$$\frac{\partial \phi_2}{\partial x}, \frac{\partial \phi_2}{\partial y} \to 0 \quad \text{at infinity}$$

The circulation being such that

$$\frac{\partial \phi_2}{\partial x}, \frac{\partial \phi_2}{\partial y} \quad \text{are finite and continuous at trailing edge}$$

Then, as may be verified, $\phi_1 + \phi_2$ is a solution of the problem for ϕ. We thus conclude that the solution for ϕ may be obtained as the superposition of the solutions ϕ_1 and ϕ_2 of the respective problems (1) and (2). Problem 1 represents, as shown in Fig. 17.2, the steady flow past the given airfoil at zero angle of attack, the speed of the free stream being V_∞. Problem 2 represents, as shown in Fig. 17.3, the steady flow past a flat plate at an angle of attack equal to α.

Fig. 17.3 Flow past a flat plate at an angle of attack.

Problem 1 may further be represented as the superposition of two other problems. To show this, let us introduce the so-called *camber function* defined by

$$\eta_c = \tfrac{1}{2}(\eta_u + \eta_l) \tag{17.29}$$

and the *thickness function* defined by

$$\eta_t = \tfrac{1}{2}(\eta_u - \eta_l) \tag{17.30}$$

We then have

$$\eta_u = \eta_c + \eta_t$$
$$\eta_l = \eta_c - \eta_t \tag{17.31}$$

The boundary condition for ϕ_1 thus takes the form

$$\left(\frac{\partial \phi_1}{\partial y}\right)_{x,0} = V_\infty\left(\frac{d\eta_c}{dx} + \frac{d\eta_t}{dx}\right) \quad \text{for} \quad y = 0_+$$

$$= V_\infty\left(\frac{d\eta_c}{dx} - \frac{d\eta_t}{dx}\right) \quad \text{for} \quad y = 0_-$$

This we may express in turn as

$$\left(\frac{\partial \phi_1}{\partial y}\right)_{x,0\pm} = V_\infty\frac{d\eta_c}{dx} \pm V_\infty\frac{d\eta_t}{dx} \tag{17.32}$$

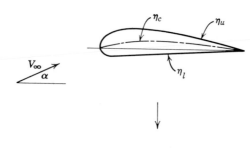

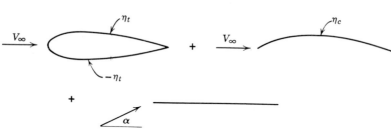

Fig. 17.4 Illustrating the decomposition of the problem of an arbitrary thin airfoil into three simpler problems.

the plus sign corresponding to $y = 0_+$ and the minus sign corresponding to $y = 0_-$. The term $V_\infty(d\eta_c/dx)$ represents the boundary condition for the flow past an airfoil that has only camber and no thickness, the airfoil being at zero angle of attack. The term $\pm V_\infty(d\eta_t/dx)$ represents the boundary condition for the flow past an airfoil that has only thickness and no camber (i.e., past a symmetrical airfoil), the airfoil being at zero angle of atack. We thus conclude that *the solution for the flow past an arbitrary thin airfoil at zero angle of attack may be obtained as the superposition of two solutions, one for a symmetrical airfoil at zero angle of attack and the other for a cambered airfoil at zero angle of attack.* The profiles of these latter airfoils are related to that of the given airfoil through

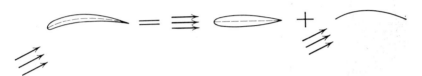

Fig. 17.5 Showing an alternate decomposition of the problem of an arbitrary thin airfoil.

(17.29) and (17.30). It thus follows that the solution for the flow past an arbitrary thin airfoil at a nonzero angle of attack may be obtained as the superposition of the solutions for the three simpler problems represented schematically in Fig. 17.4. Alternatively, as shown in Fig. 17.5, we may represent the problem as the superposition of the problem for a symmetrical airfoil at zero angle of attack and that for a cambered airfoil with zero thickness at an angle of attack.

The solution of these problems may be attempted by the method of complex variables. However, we shall not follow this procedure. In the following we construct the solutions of the three problems by the *method of superposition of singular solutions or singularities.* Such a method finds application also in the problems of flow past a finite wing and a body of revolution.

First, using the approximations of the present simple theory, we obtain a relation for the pressure.

17.5 Pressure Relation in the Simple Theory

The pressure at any point is given by the Bernoulli equation

$$p(x, y) = p_\infty + \frac{\rho}{2} V_\infty{}^2 \left(1 - \frac{V^2}{V_\infty{}^2}\right)$$

In terms of the perturbation $\mathbf{q}$ we have

$$V^2 = (\mathbf{V}_\infty + \mathbf{q}) \cdot (\mathbf{V}_\infty + \mathbf{q}) = V_\infty{}^2 + 2\mathbf{V}_\infty \cdot \mathbf{q} + q^2$$
$$= V_\infty{}^2 + 2(V_\infty u \cos \alpha + V_\infty v \sin \alpha) + u^2 + v^2 \qquad (17.33)$$

The Bernoulli equation then takes the form

$$p = p_\infty - \tfrac{1}{2}\rho V_\infty{}^2 \left(2\,\frac{\mathbf{V}_\infty \cdot \mathbf{q}}{\mathbf{V}_\infty{}^2} + \frac{q^2}{V_\infty{}^2} \right)$$

$$= p_\infty - \frac{\rho}{2} V_\infty{}^2 \left(2\,\frac{u}{V_\infty} \cos \alpha + 2\,\frac{v}{V_\infty} \sin \alpha + \frac{u^2 + v^2}{V_\infty{}^2} \right)$$

Consistent with the approximations made in the simplification of the boundary condition, we set $\cos \alpha = 1$, $\sin \alpha = \alpha$ and neglect those terms in the parenthesis that are smaller than u/V_∞. We then obtain

$$p = p_\infty - \rho V_\infty u = p_\infty - \rho V_\infty \frac{\partial \phi}{\partial x} \qquad (17.34)$$

The corresponding pressure coefficient is given by

$$C_p \equiv \frac{p - p_\infty}{\rho V_\infty^2 / 2} = -2\,\frac{u}{V_\infty} = -\frac{2}{V_\infty}\,\frac{\partial \phi}{\partial x} \qquad (17.35)$$

Equations 17.34 or 17.35 express the pressure as a linear function of the velocity component u. This means that when we solve for u by superposition of separate solutions, we may obtain the pressure by superposition of the pressures corresponding to the separate solutions. Similarly, the forces and moments on the airfoil may be obtained by superposition.

17.6 Symmetrical Airfoil at Zero Angle of Attack: Solution by Source Distribution

We first consider the problem of a thin symmetrical airfoil at zero angle of attack (see Fig. 17.6). The mathematical problem for the disturbance potential ϕ is the following:

$$\nabla^2 \phi = 0 \qquad (17.36)$$

$$v(x, 0_\pm) = \frac{\partial \phi}{\partial y}(x, 0_\pm) = \pm V_\infty \frac{d\eta_t}{dx}, \qquad 0 \le x \le l \qquad (17.37)$$

$$\nabla \phi = 0 \quad \text{at infinity} \qquad (17.38)$$

Since the flow field is symmetrical about the airfoil, there is no circulation around the airfoil. The surface condition, Eq. (17.37), requires

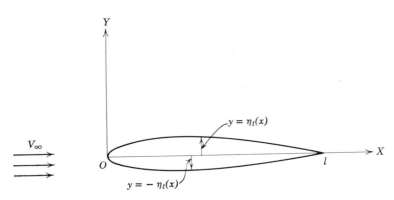

Fig. 17.6 Coordinates for the problem of the symmetric airfoil at zero angle of attack.

that over the strip $0 \le x \le l$ the velocity component v be an odd function of y:

$$v(x, 0_+) = -v(x, 0_-)$$

These considerations suggest that the disturbance field may be represented as that due to a *suitable distribution of sources* along the X-axis in the range $0 \le x \le l$. The velocity field due to such a distribution naturally satisfies Laplace's equation (17.36) and the infinity condition (17.38). The problem of the symmetrical airfoil at zero angle of attack thus reduces to that of the superposition of the uniform stream and a certain source distribution, as represented in Fig. 17.7.

Once the source distribution is determined, all the desired information about the flow field is readily obtained. Let $q = q(x)$ denote the intensity

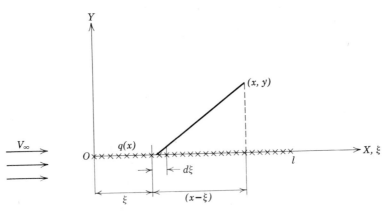

Fig. 17.7 Source distribution for the symmetric airfoil.

of the source distribution;* q is thus the source strength per unit length. Then the perturbation potential and the perturbation velocity components at the field point (x, y) are given by

$$\phi(x, y) = \frac{1}{2\pi} \int_0^l q(\xi) \log \left[(x - \xi)^2 + y^2 \right]^{\frac{1}{2}} d\xi \qquad (17.39)$$

$$u(x, y) = \frac{1}{2\pi} \int_0^l q(\xi) \frac{x - \xi}{(x - \xi)^2 + y^2} d\xi \qquad (17.40)$$

$$v(x, y) = \frac{1}{2\pi} \int_0^l q(\xi) \frac{y}{(x - \xi)^2 + y^2} d\xi \qquad (17.41)$$

To determine the source distribution $q(x)$ we use the boundary condition (17.37). Consequently, we require that

$$\lim_{y \to 0_\pm} \left[\frac{1}{2\pi} \int_0^l q(\xi) \frac{y}{(x - \xi)^2 + y^2} d\xi \right] = v(x, 0_\pm)$$

$$= \pm V_\infty \frac{d\eta_t}{dx} \qquad 0 \le x \le l \quad (17.42)$$

Care is necessary in evaluating the left-hand side of this equation. We observe that for $\xi \ne x$, the integral goes to zero as y goes to zero. For $\xi = x$, the integrand becomes indeterminate. The detailed evaluation of the left-hand side of (17.42) is given later in Appendix E. As shown there, we find that

$$\lim_{y \to 0_\pm} \frac{1}{2\pi} \int_0^l q(\xi) \frac{y}{(x - \xi)^2 + y^2} d\xi = \pm \frac{q(x)}{2} \qquad (17.43)$$

On substituting (17.43) into (17.42) we obtain

$$q(x) = 2v(x, 0_+) = V_\infty \frac{d\eta_t}{dx} \qquad (17.44)$$

This equation determines the source distribution. It shows that the source strength at the point x is proportional to the slope of the airfoil at that point. The result expressed by (17.44) could have been obtained by an elementary argument. Consider an element $q(x)\,dx$ of the source distribution. We know that the outflow of fluid through any circuit enclosing that element and passing through its endpoints is equal to the source

---

* Note that the symbol **q** was used previously to denote the perturbation velocity. In that context q meant the magnitude of **q**. It is hoped that the present use of q to denote source strength will cause no confusion.

strength $q(x)\, dx$. Hence choosing a circuit C, as shown in Fig. 17.8, we have

$$q(x)\, dx = \oint_C \mathbf{V} \cdot \mathbf{n}\, ds$$

$$= \lim_{\substack{+y \to 0_+ \\ \text{and} \\ -y \to 0_-}} \oint_C \mathbf{V} \cdot \mathbf{n}\, ds$$

$$= v(x, 0_+)\, dx - v(x, 0_-)\, dx \qquad (17.45)$$

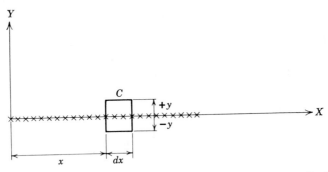

Fig. 17.8 Outflow through a circuit enclosing an element of the source distribution.

Here $\mathbf{n}$ is the outward normal and ds is an element of length along the circuit. Since $v(x, 0_-) = -v(x, 0_+)$, Eq. (17.45) becomes

$$q(x) = \pm 2v(x, 0_\pm)$$

Equations 17.39 to 17.41 giving the perturbation potential and the perturbation velocity components now take the form

$$\phi(x, y) = \frac{V_\infty}{\pi} \int_0^l \frac{d\eta_t}{dx}(\xi) \log \left[(x - \xi)^2 + y^2\right]^{1/2} d\xi \qquad (17.46)$$

$$u(x, y) = \frac{V_\infty}{\pi} \int_0^l \frac{d\eta_t}{dx}(\xi) \frac{x - \xi}{(x - \xi)^2 + y^2}\, d\xi \qquad (17.47)$$

$$v(x, y) = \frac{V_\infty}{\pi} \int_0^l \frac{d\eta_t}{dx}(\xi) \frac{y}{(x - \xi)^2 + y^2}\, d\xi \qquad (17.48)$$

The pressure distribution is given by

$$C_p(x, y) = -2 \frac{u(x, y)}{V_\infty}$$

On the surface of the airfoil, the x-component of the perturbation velocity and the pressure are given by

$$u(x, 0) = \frac{V_\infty}{\pi} \int_0^l \frac{d\eta_t}{dx}(\xi) \frac{i}{x - \xi} \, d\xi \qquad 0 \le x \le l \qquad (17.49)$$

$$C_p(x, 0_\pm) = -2 \frac{u(x, 0)}{V_\infty} \qquad\qquad 0 \le x \le l \qquad (17.50)$$

The y-component of the velocity is, of course, given by the surface condition

$$v(x, 0_\pm) = \pm V_\infty \frac{d\eta_t}{dx}$$

It readily follows that there is neither a resultant force nor a moment on the symmetrical airfoil. We thus conclude that the thickness of a thin airfoil contributes only to the pressure distribution over the airfoil.

To obtain the velocity component u and the pressure distribution over the airfoil we need to evaluate the integral on the right-hand side of (17.49). We observe that this integral is an *improper integral** in the sense that the integrand becomes infinite at the point $\xi = x$, x being in the interval of integration. The value of the integral is given by its so-called *principal* or *Cauchy principal value*. The principal value of an improper integral

$$\int_a^b f(\xi) \, d\xi$$

where $f(\xi)$ becomes infinite at a point $\xi = C$ in the interval (a, b) is defined by the limit

$$\int_a^b f(\xi) \, d\xi = \lim_{\epsilon \to 0} \left[\int_a^{C-\epsilon} f(\xi) \, d\xi + \int_{C+\epsilon}^b f(\xi) \, d\xi \right]$$

For the numerical computation of $u(x, 0)$, from (17.49), it is convenient to express the functions $u(x, 0)$ and $(d\eta_t/dx)(\xi)$ as *conjugate Fourier Series* (see Appendix D). For this purpose we introduce the variable θ such that

$$x = \frac{l}{2}(1 + \cos \theta) \qquad -\pi \le \theta \le \pi \qquad (17.51)$$

(see Fig. 17.9.) Positive values of θ describe the top of the airfoil, whereas negative values of θ describe its bottom surface. The variable ξ may then be expressed as

$$\xi = \frac{l}{2}(1 + \cos \varphi) \qquad\qquad (17.52)$$

---

* For the notion of improper integrals and for certain elementary aspects of their properties, reference may be made to Tom M. Apostol (1961).

Introduce further the notation

$$u(\theta) = u[x(\theta), 0]$$

$$\frac{d\eta_t}{dx}(\theta) = \frac{d\eta_t}{dx}[x(\theta)]$$

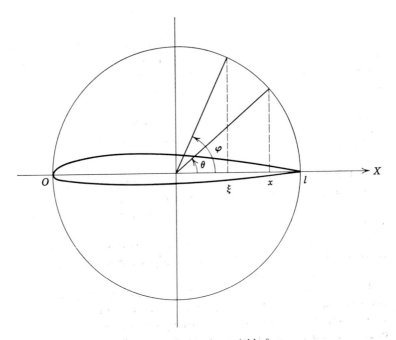

Fig. 17.9 Introducing the variable θ.

Then, in terms of the variables θ and φ, (17.49) takes the form

$$u(\theta) = -\frac{V_\infty}{\pi} \int_0^\pi \frac{d\eta_t}{dx}(\varphi) \frac{\sin \varphi}{\cos \varphi - \cos \theta} \, d\varphi \qquad (17.53)$$

This is known as *Poisson's integral formula*. The integrand in the integral on the right-hand side becomes infinite, as it should, at $\varphi = \theta$. On the basis of the detailed considerations given later in Appendix D, it follows that $[u(\theta)/V_\infty]$ and $(d\eta_t/dx)(\theta)$ are expressible as conjugate Fourier series. We observe that $(d\eta_t/dx)(\theta)$ is an *odd function* of θ. Hence we express it as

$$\frac{d\eta_t}{dx}(\theta) = \sum_{n=1}^\infty A_n \sin n\theta \qquad (17.54)$$

where

$$A_n = \frac{2}{\pi} \int_0^\pi \frac{d\eta_t}{dx}(\theta) \sin n\theta \, d\theta \qquad (17.55)$$

Then $[u(\theta)/V_\infty]$ is given by the conjugate Fourier series

$$\frac{u(\theta)}{V_\infty} = \sum_1^\infty A_n \cos n\theta \qquad (17.56)$$

where the coefficients A_n are obtained from (17.55). As required, $u(\theta)$ is an even function of θ. To treat thin airfoil theory, computational techniques for determining conjugate Fourier series, and also their derivatives and integrals, have been developed, and for these the reader should consult the appropriate references.*

From (17.47) and (17.48) we infer, as shall be shown in detail later (see Appendix B), that the perturbation velocity field exhibits the following features:

1. The velocity components u and v are finite everywhere outside the strip $y = 0$, $0 \leq x \leq l$. They vanish as $r = \sqrt{x^2 + y^2}$ goes to infinity.
2. On the X-axis, for $x < 0$ and $x > l$, the component u is finite while v is zero.

3. Along the strip, *excepting the edges*, the u component is finite. Further, it is continuous across the strip. At the edges, for any nonzero value of the source strength, u becomes infinite.

4. Along the strip the v-component is finite at all points where the source strength is finite. At any point on the strip, the jump in v across the strip is simply equal to the source strength at that point. Since at a round edge the source strength becomes infinite, the v-component becomes infinite at a round edge, that is, at the leading edge of the airfoil.

As expected, we observe that the solution of the linearized problem is not valid at and near the leading and trailing edges of the airfoil.

17.7 Cambered Airfoil of Zero Thickness at Zero Angle of Attack: Solution by Vortex Distribution

Now the mathematical problem for the disturbance potential ϕ is the following (see Fig. 17.10):

$$\nabla^2 \phi = 0 \qquad (17.57)$$

$$v(x, 0_\pm) = \frac{\partial \phi}{\partial y}(x, 0_\pm) = V_\infty \frac{d\eta_c}{dx}, \qquad 0 \leq x \leq l \qquad (17.58)$$

$$\nabla \phi = 0 \qquad \text{at infinity} \qquad (17.59)$$

* See, for instance, Thwaites (1960), where other references may be found.

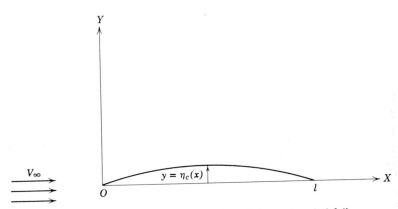

Fig. 17.10 Coordinates for the problem of the cambered airfoil.

Kutta Condition: Now, the circulation around the airfoil is not zero. Its value should be such that the perturbation velocity components $u = \partial\phi/\partial x$ and $v = \partial\phi/\partial y$ should be finite and continuous at the trailing edge.

These requirements, particularly the surface condition (17.58), suggest that the disturbance field may be represented as that due to a suitable distribution of vortices along the X-axis in the interval $0 \leq x \leq l$. We distribute the elementary vortices as shown in Fig. 17.11, the circulation around each of the vortices being designated in the clockwise direction. The velocity field due to such a distribution of vortices of any finite strength automatically satisfies Laplace's equation (17.57) and the infinity

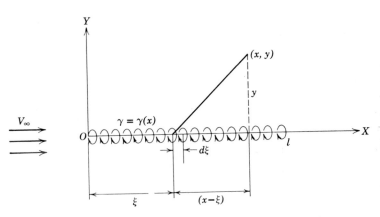

Fig. 17.11 Vortex distribution for the cambered airfoil.

condition (17.59). The strength of the vortices should, however, be distributed in such a way that the surface condition (17.58) and the Kutta condition are satisfied. This, then, is the problem of the cambered airfoil of zero thickness.

Once the vortex distribution is determined, all the desired information about the flow field is readily obtained. Let $\gamma = \gamma(x)$ denote the intensity of the distribution; γ is thus the vortex strength per unit length (see Fig. 17.11). We note that $\gamma \, dx$ is the circulation around the vortices distributed in the element of length dx. The perturbation potential and the perturbation velocity components at the field point (x, y) are then given by

$$\phi(x, y) = -\frac{1}{2\pi} \int_0^l \gamma(\xi) \tan^{-1} \frac{y}{x - \xi} \, d\xi \qquad (17.60)$$

$$u(x, y) = \frac{1}{2\pi} \int_0^l \gamma(\xi) \frac{y}{(x - \xi)^2 + y^2} \, d\xi \qquad (17.61)$$

$$v(x, y) = -\frac{1}{2\pi} \int_0^l \gamma(\xi) \frac{x - \xi}{(x - \xi)^2 + y^2} \, d\xi \qquad (17.62)$$

The pressure at any point is given by

$$C_p(x, y) = -2 \frac{u(x, y)}{V_\infty}$$

where $u(x, y)$ is to be obtained from Eq. (17.61).

We observe that (17.61) and (17.62) are respectively similar to (17.41) and (17.40), which refer to a source distribution along the strip $y = 0$, $0 \leq x \leq l$. We may, therefore, conclude that the u component of the velocity due to the vortex distribution exhibits the same characteristics as the v-component of the velocity due to the source distribution and, that the v-component due to the vortex distribution exhibits* the same characteristics as the u-component due to the source distribution. See the preceding section for detailed description. In particular, we note that the circulation at any point on the distribution is simply equal to the jump at that point in the u-component of the velocity across the vortex sheet. This follows from the result

$$u(x, 0_\pm) = \lim_{y \to 0_\pm} \frac{1}{2\pi} \int_0^l \gamma(\xi) \frac{y}{(x - \xi)^2 + y^2} \, d\xi$$

$$= \pm \frac{\gamma(x)}{2}, \qquad 0 \leq x \leq l$$

* Except for a negative sign.

Compare with (17.43). Therefore we have

$$\gamma(x) = 2u(x, 0_+) = u(x, 0_+) - u(x, 0_-) \qquad (17.63)$$

since $u(x, 0_-) = -u(x, 0_+)$.

Using the relation (17.63) we may immediately express the pressure distribution over the airfoil and the lift and moment on the airfoil in terms of the vortex strength $\gamma(x)$. For the pressure distribution over the airfoil we have

$$C_p(x, 0_\pm) = -2 \frac{u(x, 0_\pm)}{V_\infty} = \mp \frac{\gamma(x)}{V_\infty} \qquad (17.64)$$

The lift on the airfoil is given by

$$L = \int_0^l [p(x, 0_-) - p(x, 0_+)]\, dx$$

$$= \frac{\rho}{2} V_\infty^2 \int_0^l [C_p(x, 0_-) - C_p(x, 0_+)]\, dx$$

$$= \rho V_\infty \int_0^l \gamma(x)\, dx$$

$$= \rho V_\infty \Gamma \qquad (17.65)$$

where $\Gamma = \int_0^l \gamma(x)\, dx$ is the circulation around the airfoil. The moment on the airfoil with respect to the origin (which here is the leading edge) is given by

$$M = \int_0^l [p(x, 0_+) - p(x, 0_-)]x\, dx$$

$$= -\rho V_\infty \int_0^l \gamma(x)x\, dx \qquad (17.66)$$

For the lift and moment coefficients we obtain

$$C_L = \frac{2}{V_\infty} \frac{1}{l} \int_0^l \gamma(x)\, dx \qquad (17.67)$$

$$C_M = -\frac{2}{V_\infty} \frac{1}{l^2} \int_0^l \gamma(x)x\, dx \qquad (17.68)$$

We now consider the question of determining $\gamma(x)$. According to the surface condition (17.58) we require that

$$\lim_{y \to 0_\pm} \left[-\frac{1}{2\pi} \int_0^l \gamma(\xi) \frac{x - \xi}{(x - \xi)^2 + y^2}\, d\xi \right] = v(x, 0_\pm) = V_\infty \frac{d\eta_c}{dx}, \qquad 0 \le x \le l$$

or

$$-\frac{1}{2\pi}\int_0^l \gamma(\xi)\frac{1}{x-\xi}\,d\xi = v(x,0_{\pm}) = V_\infty\frac{d\eta_c}{dx}, \qquad 0 \le x \le l \quad (17.69)$$

In evaluating the left-hand side of this equation, we naturally seek the Cauchy principal value of the integral. Equation (17.69) must, of course, be supplemented by the Kutta condition. We find, as is shown in detail later (see Appendix B), that *for a vortex distribution such as we are considering, the v velocity component becomes infinite at the edges of the distribution if the vortex strength at those edges has a nonzero value* (similar to the behavior of *u* due to a source distribution). The *u*-component is, however, finite at the edges if the vortex strength there is finite. From this it follows that to *satisfy the Kutta condition we should require that the vortex strength at the trailing edge* be zero. We thus have

$$\gamma(x = l) = 0 \qquad\qquad (17.70)$$

The mathematical problem for determining $\gamma(x)$ thus consists of solving the integral equation (17.69) with the condition (17.70).

First we show that the solution of (17.69) is indeterminate to the extent of a certain function. To see this and determine that function, consider the integral

$$I = \int_0^l \gamma(\xi)\frac{d\xi}{x-\xi} \qquad\qquad (17.71)$$

Introducing the variables θ and φ such that

$$x = \frac{l}{2}(1 + \cos\theta)$$

$$\xi = \frac{l}{2}(1 + \cos\varphi)$$

we rewrite the integral as

$$I = -\int_0^\pi \gamma(\varphi)\frac{\sin\varphi}{\cos\varphi - \cos\theta}\,d\varphi \qquad\qquad (17.72)$$

Now, on the basis of the formula

$$\int_0^\pi \frac{\cos n\varphi}{\cos\varphi - \cos\theta}\,d\varphi = \pi\frac{\sin n\theta}{\sin\theta}, \qquad n = 0, 1, 2, \ldots \quad (17.73)$$

which will be proved later (see Appendix E), we have the result

$$\int_0^\pi \frac{d\varphi}{\cos\varphi - \cos\theta} = 0 \qquad\qquad (17.74)$$

Elements of Thin Airfoil Theory

In view of (17.74) we conclude that if we choose

$$\gamma(\theta) = \frac{K}{\sin \theta} \qquad (17.75)$$

where K is a constant, the integral I will become zero. In terms of the original variable x, (17.75) becomes

$$\gamma(x) = \frac{K}{2\sqrt{\frac{x}{l}\left(1 - \frac{x}{l}\right)}} \qquad (17.76)$$

On the basis of these considerations it follows that *the solution of* (17.69) *is indeterminate to the extent of the function represented by* (17.76) *or, equivalently, by* (17.75).

To proceed to the solution of (17.69) we introduce, as before, the variables θ and φ and rewrite the equation as

$$\frac{1}{\pi} \int_0^\pi \frac{\gamma(\varphi)}{2V_\infty} \frac{\sin \varphi}{\cos \varphi - \cos \theta} \, d\varphi = \frac{d\eta_c}{dx}(\theta) \qquad (17.77)$$

where

$$\frac{d\eta_c}{dx}(\theta) = \frac{d\eta_c}{dx}[x(\theta)]$$

Equation (17.77) is again Poisson's integral formula (compare with Eq. (17.53)). On the basis of the detailed considerations given later in Appendix D, it follows that $\gamma(\theta)/2V_\infty$ and $(d\eta_c/dx)(\theta)$ may be expressed as conjugate Fourier series. Since $(d\eta_c/dx)(\theta)$ is an *even function* of θ, let us write

$$\frac{d\eta_c}{dx}(\theta) = \frac{B_0}{2} + \sum_{n=1}^\infty B_n \cos n\theta \qquad (17.78)$$

where

$$B_n = \frac{2}{\pi} \int_0^\pi \frac{d\eta_c}{dx} \cos n\theta \, d\theta \qquad n = 0, 1, 2, \cdots \qquad (17.79)$$

With the use of expression (17.78), (17.77) becomes

$$\frac{1}{\pi} \int_0^\pi \frac{\gamma(\varphi)}{2V_\infty} \frac{\sin \varphi}{\cos \varphi - \cos \theta} \, d\varphi = \frac{B_0}{2} + \sum_1^\infty B_n \cos n\theta \qquad (17.80)$$

Since this equation is linear in γ and its solution is indeterminate to the extent of the function (17.75), we may express the solution as

$$\gamma(\theta) = \gamma_1(\theta) + \gamma_2(\theta) + \frac{K}{\sin \theta} \qquad (17.81)$$

where γ_1 and y_2 are, respectively, the solutions of the equations:

$$\frac{1}{\pi} \int_0^\pi \frac{\gamma_1(\varphi)}{2V_\infty} \frac{\sin \varphi}{\cos \varphi - \cos \theta} \, d\varphi = \sum_1^\infty B_n \cos n\theta \qquad (17.82)$$

$$\frac{1}{\pi} \int_0^\pi \frac{\gamma_2(\varphi)}{2V_\infty} \frac{\sin \varphi}{\cos \varphi - \cos \theta} \, d\varphi = \frac{B_0}{2} \qquad (17.83)$$

The solution of (17.82) is given by the conjugate Fourier series* (see Appendix D)

$$\gamma_1(\theta) = -2V_\infty \sum_{n=0}^\infty B_n \sin n\theta \qquad (17.84)$$

To obtain the solution of (17.83) we first observe that on the basis of the formula (17.73) we have the result

$$\frac{1}{\pi} \int_0^\pi \frac{\cos \phi}{\cos \phi - \cos \theta} \, d\phi = 1 \qquad (17.85)$$

From this it follows that if we set

$$\gamma_2(\theta) = V_\infty B_0 \frac{\cos \theta}{\sin \theta} \qquad (17.86)$$

(17.83) is identically satisfied. Thus the function (17.86) is the required solution for γ_2.

Substituting the expressions (17.84) and (17.86) into (17.81), we obtain

$$\gamma(\theta) = -2V_\infty \sum_1^\infty B_n \sin n\theta + \frac{V_\infty B_0}{\sin \theta}\left(\frac{K}{V_\infty B_0} + \cos \theta\right) \qquad (17.87)$$

To determine the constant K we use the Kutta condition (17.70). Accordingly, we require that

$$\gamma(\theta = 0) = 0 \qquad (17.88)$$

This leads to the condition

$$\frac{1}{\sin \theta}\left(\frac{K}{V_\infty B_0} + \cos \theta\right) = 0 \quad \text{for} \quad \theta = 0 \qquad (17.89)$$

This is satisfied by setting

$$K = -V_\infty B_0 \qquad (17.90)$$

For we have

$$\frac{\cos \theta - 1}{\sin \theta} = \frac{-2 \sin^2 \theta/2}{2 \sin \theta/2 \cos \theta/2} = -\tan \frac{\theta}{2} \qquad (17.91)$$

* This may be verified by substituting the series into (17.28) and noting that

$$\frac{1}{\pi} \int_0^\pi \frac{\sin n\phi \sin \phi}{\cos \phi - \cos \phi} \, d\phi = -\cos n\theta$$

which vanishes for $\theta = 0$. In (17.87) we substitute for K from (17.90) and obtain

$$\gamma(\theta) = -2V_\infty \left(\frac{B_0}{2} \frac{1 - \cos\theta}{\sin\theta} + \sum_1^\infty B_n \sin n\theta \right) \qquad (17.92)$$

Using this relation we may express all the desired aerodynamic properties of the airfoil in terms of the B's, which are the coefficients of the Fourier expansion for $(d\eta_c/dx)$. The pressure distribution over the airfoil is given by

$$C_p(\theta) = C_p[x(\theta), 0_\pm] = -\frac{\gamma(\theta)}{V_\infty}$$

$$= 2\left[\frac{B_0}{2} \frac{1 - \cos\theta}{\sin\theta} + \sum_1^\infty B_n \sin n\theta \right] \qquad (17.93)$$

To obtain the lift and moment coefficients we first rewrite (17.67) and (17.68) in terms of the variable θ:

$$C_L = \frac{1}{V_\infty} \int_0^\pi \gamma(\theta) \sin\theta \, d\theta \qquad (17.94)$$

$$C_M = -\frac{1}{2V_\infty} \int_0^\pi \gamma(\theta)(1 + \cos\theta) \sin\theta \, d\theta \qquad (17.95)$$

Substituting (17.92) into (17.94) and (17.95), and carrying out the integrations, we obtain

$$C_L = -(B_1 + B_1)\pi \qquad (17.96)$$

$$C_M = B_0 \frac{\pi}{4} + B_1 \frac{\pi}{2} + B_2 \frac{\pi}{4}$$

$$= \frac{\pi}{4}(B_0 + B_1) + \frac{\pi}{4}(B_1 + B_2) \qquad (17.97)$$

$$= -\frac{C_L}{4} + \frac{\pi}{4}(B_1 + B_2)$$

where, to recall,

$$B_n = \frac{2}{\pi} \int_0^\pi \frac{d\eta_c}{dx}(\theta) \cos n\theta \, d\theta, \qquad n = 0, 1, 2, \ldots$$

The relations (17.96) and (17.97) show that the force system acting on the airfoil may be represented as consisting of a lift force acting at the *quarter-chord point* from the leading edge and a moment about that point.

We have thus completed, for a cambered airfoil of zero thickness at zero angle of attack, the task of expressing in terms of the airfoil shape all the desired aerodynamic information. We now pass on to consider the problem of a flat plate at an angle of attack.

17.8 Flat Plate Airfoil at an Angle of Attack: Solution by Vortex Distribution

The mathematical problem for the disturbance potential is expressed as follows (see Fig. 17.12):

$$\nabla^2 \phi = 0$$

$$v(x, 0_\pm) = \frac{\partial \phi}{\partial y}(x, 0_\pm) = -V_\infty \alpha \qquad 0 \le x \le l$$

$$\nabla \phi = 0 \text{ at infinity}$$

Kutta Condition: The circulation be such that $\dfrac{\partial \phi}{\partial x}$ and $\dfrac{\partial \phi}{\partial y}$ are finite at the trailing edge.

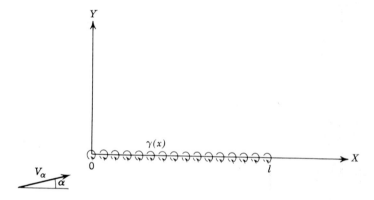

Fig. 17.12 Coordinates and vortex distribution for the problem of the flat plate.

Following similar considerations as those given in the previous section, we immediately conclude that the disturbance field due to the flat plate may be represented as that due to a suitable vortex distribution $\gamma(x)$ along the X-axis in the interval $0 \le x \le l$ and that $\gamma(x)$ is given as a solution of the integral equation

$$\frac{1}{2\pi} \int_0^l \gamma(\xi) \frac{d\xi}{x - \xi} = v(x, 0_\pm) = V_\infty \alpha \qquad 0 \le x \le l \qquad (17.98)$$

with the Kutta condition

$$\gamma(x = l) = 0 \qquad (17.99)$$

Introduce, as before, the variables θ and φ through the relations

$$x = \frac{l}{2}(1 + \cos \theta)$$

$$\xi = \frac{l}{2}(1 + \cos \varphi)$$

Then (17.98) takes the form

$$\frac{1}{\pi} \int_0^\pi \gamma(\varphi) \frac{\sin \varphi}{\cos \varphi - \cos \theta} \, d\varphi = -2V_\infty \alpha \qquad (17.100)$$

In view of the fact that the solution of this equation is indeterminate to the extent of the function $K/\sin \theta$ (see 17.75) and because of the result expressed by (17.85), the solution of (17.100) may be written as

$$\gamma(\theta) = \frac{K}{\sin \theta} - 2V_\infty \alpha \frac{\cos \theta}{\sin \theta}$$

$$= -2V_\infty \alpha \frac{1}{\sin \theta}\left(\frac{-K}{2V_\infty \alpha} + \cos \theta\right) \qquad (17.101)$$

To satisfy the condition (17.99), the constant K should be set equal to $2V_\infty \alpha$. We then obtain

$$\gamma(\theta) = 2V_\infty \alpha \frac{1 - \cos \theta}{\sin \theta} \qquad (17.102)$$

This is the required solution.
    The pressure distribution over the flat plate is then given by

$$C_p(\theta) = C_p[x(\theta), 0_\pm] = -\frac{\gamma(\theta)}{V_\infty}$$

$$= -2\alpha \frac{1 - \cos \theta}{\sin \theta}$$

$$= -2\alpha \tan \frac{\theta}{2} \qquad (17.103)$$

The lift and moment coefficients are given by

$$C_L = \frac{1}{V_\infty} \int_0^\pi \gamma(\theta) \sin \theta \, d\theta$$

$$= 2\pi\alpha \qquad (17.104)$$

$$C_M = -\frac{1}{2V_\infty} \int_0^\pi \gamma(\theta)(1 + \cos \theta) \sin \theta \, d\theta$$

$$= -\frac{\pi}{2}\alpha$$

$$= -\frac{C_L}{4} \qquad (17.105)$$

These relations (17.104) and (17.105) show that the force system on the flat plate may be represented as consisting of only a lift force acting at the quarter-chord point. This point is thus the aerodynamic center. In addition, it is the center of pressure at all angles of attack.

17.9 Aerodynamic Characteristics of a Thin Airfoil

We now superimpose the results of the previous three sections to obtain the characteristics of an arbitrary thin airfoil that has both thickness and camber and is at an angle of attack to the undisturbed stream. Recalling that

$$\frac{d\eta_{\pm}}{dx} = \frac{d\eta_c}{dx} \pm \frac{d\eta_t}{dx}$$

we write in terms of the variable θ [defined by $x = (l/2)(1 + \cos\theta)$]

$$\frac{d\eta}{dx}(\theta) = \frac{d\eta_c}{dx}(\theta) + \frac{d\eta_t}{dx}(\theta) \qquad -\pi \le \theta \le \pi \qquad (17.106)$$

The top of the airfoil corresponds to positive values of θ, whereas the bottom of the airfoil corresponds to negative values of θ. Introducing the Fourier expansions already employed we have

$$\frac{d\eta}{dx}(\theta) = \frac{B_0}{2} + \sum_1^\infty B_n \cos n\theta + \sum_1^\infty A_n \sin n\theta \qquad (17.107)$$

where

$$B_n = \frac{2}{\pi} \int_0^\pi \frac{d\eta_c}{dx}(\theta) \cos n\theta \, d\theta \qquad n = 0, 1, 2, \ldots$$

$$A_n = \frac{2}{\pi} \int_0^\pi \frac{d\eta_t}{dx}(\theta) \sin n\theta \, d\theta \qquad n = 0, 1, 2, \ldots$$

The pressure distribution on the airfoil is given by

$$C_p(\theta) = -2\frac{u(\theta)}{V_\infty}$$

$$= -2\alpha \tan\frac{\theta}{2} + 2\left[\frac{B_0}{2}\tan\frac{\theta}{2} + \sum_1^\infty B_n \sin n\theta\right] - 2\sum_1^\infty A_n \cos n\theta$$

$$= -2\left[\left(\alpha - \frac{B_0}{2}\right)\tan\frac{\theta}{2} - \sum_1^\infty B_n \sin n\theta + \sum_1^\infty A_n \cos n\theta\right] \qquad (17.108)$$

The lift coefficient for the airfoil is given by

$$C_L = 2\pi\alpha - (B_0 + B_1)\pi = 2\pi\left(\alpha - \frac{B_0 + B_1}{2}\right) \qquad (17.109)$$

The slope of the lift curve with respect to the angle of attack is a constant and is equal to 2π:

$$\frac{dC_L}{d\alpha} = 2\pi \tag{17.110}$$

As we know, the thickness contributes nothing to the lift. The camber fixes the lift at zero angle of attack and the zero-lift angle. From Eq. (17.109), the zero-lift angle is obtained as

$$\alpha_0 = \frac{B_0 + B_1}{2} \tag{17.111}$$

The moment coefficient (with respect to the leading edge) is given by

$$C_M = -\frac{\pi}{2}\alpha + (B_0 + B_1)\frac{\pi}{4} + \frac{\pi}{4}(B_1 + B_2)$$

$$= -[2\pi\alpha - (B_0 + B_1)\pi]\frac{1}{4} + \frac{\pi}{4}(B_1 + B_2)$$

$$= -\frac{C_L}{4} + \frac{\pi}{4}(B_1 + B_2) \tag{17.112}$$

As we know, the thickness contributes nothing to the moment on the airfoil. From (17.112) it follows that the force system acting on the airfoil may be represented as consisting of a lift force acting at the quarter-chord point and a moment equal to $(\pi/4)(B_1 + B_2)$ about that point. This moment is independent of the angle of attack and is determined solely by the camber of the airfoil. It follows that the *quarter-chord point is the aerodynamic center*. Thus the moment about the aerodynamic center is given by

$$\tilde{C}_M = \frac{\pi}{4}(B_1 + B_2)$$

$$= \text{the moment coefficient due to camber} \tag{17.113}$$

As we know, it is also the moment coefficient at zero lift.

Chapter 18

Some Features of Flow with Vorticity

Having presented the elements of the theory of lift for the so-called infinite wing, we now wish to take up the elements of the theory of flow past a finite wing. In the formulation of the theory of the finite wing, the disturbance flow field will be represented as that due to a certain spatial distribution of vorticity. Thus, as a preliminary step in the formulation and analysis of the finite-wing problem, we must study some properties of vortex motion and, in particular, must learn how to represent the velocity field. We do so in this chapter. We restrict ourselves only to those features of vortex motion that are pertinent to the study of the flow past a finite wing. For a more detailed study of vortex motion and of the varied problems involving vortex motion, the reader should consult other books cited at the end of this book.

18.1 Recapitulation

We first recall some of the kinematical notions already introduced in Chapter 9. If $\mathbf{V} = \mathbf{V}(\mathbf{r}, t)$ is the velocity field of a fluid in motion, the *angular velocity* $\boldsymbol{\omega} = \boldsymbol{\omega}(\mathbf{r}, t)$ is given by

$$\boldsymbol{\omega} = \tfrac{1}{2}\,\text{curl}\,\mathbf{V} \tag{18.1}$$

The *vorticity* $\boldsymbol{\Omega} = \boldsymbol{\Omega}(\mathbf{r}, t)$ is simply the curl of the velocity

$$\boldsymbol{\Omega} = \text{curl}\,\mathbf{V} \tag{18.2}$$

The *circulation* $\Gamma = \oint_{\mathscr{C}} \mathbf{V} \cdot \mathbf{ds}$ around any closed curve $\mathscr{C}$ is related to the vorticity by the equation

$$\Gamma \equiv \oint_{\mathscr{C}} \mathbf{V} \cdot \mathbf{ds} = \iint_{S} \text{curl}\,\mathbf{V} \cdot \mathbf{n}\, dS$$

$$= \iint_{S} \boldsymbol{\Omega} \cdot \mathbf{n}\, dS \tag{18.3}$$

where S is any (capping) surface whose boundary is the curve $\mathscr{C}$.

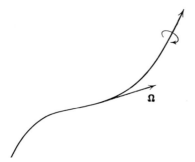

Fig. 18.1   Vortex line.

18.2   Vortex Line, Surface, Tube, and Filament

The field lines of the vorticity field are called *vortex lines*. Analytically they are described by the differential equation

$$\mathbf{\Omega} \times \mathbf{ds} = 0 \tag{18.4}$$

where **ds** is an element of a vortex line. A vortex line is represented as shown in Fig. 18.1. At any point in the flow field, the direction of the vorticity vector (or equivalently of the angular velocity vector) is given by the direction, at that point, of the vortex line passing through that point. In Cartesians, if we write

$$\mathbf{\Omega} = (\Omega_x, \Omega_y, \Omega_z)$$

Eq. (18.4) becomes

$$\frac{dx}{\Omega_x} = \frac{dy}{\Omega_y} = \frac{dz}{\Omega_z}$$

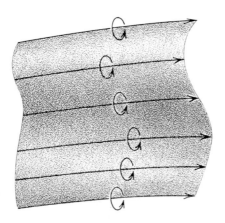

Fig. 18.2   Vortex surface.

If, at any instant of time, we draw an arbitrary line in the flow field and draw the vortex lines passing through that line, a surface is formed. Such a surface is called a *vortex surface* and is represented as shown in Fig. 18.2.

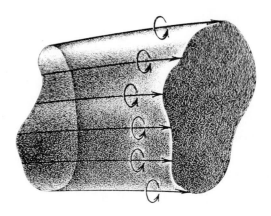

Fig. 18.3   Vortex tube.

If we consider a closed curve and draw all the vortex lines passing through it, a tube is formed. Such a tube is called a *vortex tube* and is represented as shown in Fig. 18.3. A vortex tube of infinitesimal cross-sectional area is known as a *vortex filament*.

18.3   Vorticity Field is a Divergenceless Field

Since the vorticity is the curl of another vector field, we have

$$\operatorname{div} \mathbf{\Omega} = \operatorname{div} (\operatorname{curl} \mathbf{V}) = 0 \qquad (18.5)$$

Thus vorticity is a divergenceless field.

Consider, at any instant, a region of space R enclosed by a closed surface S. We then have

$$\oiint_{S} \mathbf{\Omega} \cdot \mathbf{n} \, dS = \iiint_{R} \operatorname{div} \mathbf{\Omega} \, d\tau = 0 \qquad (18.6)$$

This equation states that *the (net) outflow of vorticity through any closed surface is zero*. This is true at every instant of time.

18.4   Spatial Conservation of Vorticity: Strength of a Vortex Tube

Consider, at any instant, a vortex tube drawn in the flow field. Denote by R the region of space enclosed between the wall of the tube and any two surfaces S_1 and S_2 which cut the tube (see Fig. 18.4). Then, according to (18.6), the outflow of vorticity through the surface S of the region R vanishes. We therefore write

$$\iint_{S_1} \mathbf{\Omega} \cdot \mathbf{n} \, dS + \iint_{S_2} \mathbf{\Omega} \cdot \mathbf{n} \, dS + \iint_{S_w} \mathbf{\Omega} \cdot \mathbf{n} \, dS = \oiint_{S} \mathbf{\Omega} \cdot \mathbf{n} \, dS = 0 \quad (18.7)$$

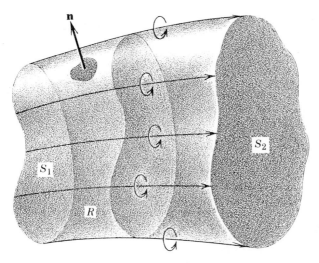

Fig. 18.4 Illustrating the derivation of the spatial conservation of vorticity.

Here S_w denotes the surface of the wall of the tube in the portion under consideration. On the wall of the tube, $\mathbf{\Omega}$ lies in the surface S_w. Hence the integral over S_w vanishes

$$\iint_{S_w} \mathbf{\Omega} \cdot \mathbf{n} \, dS = 0$$

Consequently, we obtain

$$\iint_{S_1} \mathbf{\Omega} \cdot \mathbf{n} \, dS + \iint_{S_2} \mathbf{\Omega} \cdot \mathbf{n} \, dS = 0 \quad (18.8)$$

In this equation $\mathbf{n}$ is an outward normal, outward with reference to the region R. If we *draw the normals on the surfaces S_1 and S_2 in the same sense* and denote them by $\mathbf{n}_1$ and $\mathbf{n}_2$, respectively, Eq. (18.8) may be rewritten as

$$\iint_{S_1} \mathbf{\Omega} \cdot \mathbf{n}_1 \, dS = \iint_{S_2} \mathbf{\Omega} \cdot \mathbf{n}_2 \, dS \quad (18.9)$$

This states that the flow of vorticity through any cross-sectional surface S_1 of a vortex tube is equal to the flow of vorticity through any other cross-sectional surface S_2 of the tube. This is true at every instant of time. If S denotes any cross-sectional surface of the vortex tube (18.9), may be expressed as

$$\iint_S \mathbf{\Omega} \cdot \mathbf{n}\, dS = \text{a constant} \tag{18.10}$$

This states that *the flow of vorticity through any cross-sectional surface of a vortex tube is a constant all along the tube. This is true at every instant of time.*

In view of the intimate relation between circulation and vorticity, the result (18.10) may be expressed equivalently in terms of circulation. Let $\mathscr{C}$ denote any closed curve that embraces the vortex tube ($\mathscr{C}$ encloses the tube and lies on its wall). Then, using Eqs. (18.3) and (18.10), we have

$$\Gamma_{\mathscr{C}} = \iint_S \mathbf{\Omega} \cdot \mathbf{n}\, dS = \text{a constant} \tag{18.11}$$

This states that *the circulation around any closed curve embracing a vortex tube is a constant all along the tube. This is true at every instant of time.*

Equation (18.11) expresses the *spatial conservation of vorticity* in the sense implied by that equation. *For a vortex filament of variable cross-sectional area, this equation takes the form*

$$\Gamma_c = \mathbf{\Omega} \cdot \mathbf{n}\, ds = \text{a constant} \tag{18.12}$$

where $\mathbf{n}\, ds$ *is any cross-sectional area of the filament and c is the boundary curve of* $\mathbf{n}\, ds$. If we take $\mathbf{n}$ in the direction of $\mathbf{\Omega}$, Eq. (18.12) reduces to

$$\Gamma_c = \Omega\, ds = \text{a constant}$$

This shows that the vorticity at any section of a vortex filament is inversely proportional to its cross-sectional area. An important consequence of the spatial conservation of vorticity is that *a vortex tube, and so also a vortex filament or a vortex line, cannot begin or end abruptly in a fluid.* It should either form a closed ring or end at infinity or at a solid or free surface. The circulation around any closed curve embracing a vortex tube, or equivalently the outflow of vorticity through any cross section of the tube, is a characteristic of the tube as a whole and is called the *strength of the vortex tube.*

If we consider a vortex filament of variable cross-sectional area and shrink the area to zero in such a way that the vorticity goes to infinity as the area goes to zero, and the strength of the filament remains constant, we arrive at the conception of *a vortex filament with concentrated vorticity.*

18.5 Consequences of the Theorems of Helmholtz and Kelvin

To the above considerations on the spatial conservation of vorticity we add the theorems of Helmholtz and Kelvin on the permanence of vorticity and circulation. See Sections 9.4 and 9.5. According to these theorems, for an ideal fluid under the action of potential body forces, we have

$$\frac{D}{Dt}\Gamma_{\mathscr{C}} = \frac{D}{Dt}\iint_{S}\mathbf{\Omega}\cdot\mathbf{n}\,dS = 0 \qquad (18.13)$$

where S is any surface bounded by a fluid curve $\mathscr{C}$. For an infinitesimal surface element $\mathbf{n}\,ds$, Eq. (18.13) may be written as

$$\frac{D\Gamma_c}{Dt} = \frac{D}{Dt}(\mathbf{\Omega}\cdot\mathbf{n}\,dS) = 0 \qquad (18.14)$$

where c is the boundary curve of the surface element. The implication of these theorems for irrotational motion has been discussed already (see Section 9.6). We now give some important consequences of (18.13 and 18.14) for rotational motion.

Consider at any instant of time a vortex sheet drawn in the fluid. At that instant

$$\mathbf{\Omega}\cdot\mathbf{n} = 0$$

for every element $\mathbf{n}\,dS$ of the sheet. Choosing an element of the sheet and following that element in its motion, we observe that as a consequence of Eq. (18.14),

$$\mathbf{\Omega}\cdot\mathbf{n}\,dS = 0$$

for all times although $\mathbf{\Omega}$ and $\mathbf{n}\,dS$ may change (see Fig. 18.5). This means $\mathbf{\Omega}\cdot\mathbf{n}$ vanishes at all times for that surface element. We thus conclude that the surface element remains an element of a vortex sheet. From this it follows that *a surface which is a vortex sheet at one instant remains a vortex sheet for all times.* We further state that fluid particles that are part of a vortex sheet at some instant are part of it for all times. Furthermore, it follows that fluid particles that are part of a vortex tube (or of a vortex filament or of a vortex line) at some instant are part of it for all times.

Consider a vortex tube and follow it as it moves along. Let $\mathscr{C}$ be any closed curve embracing the tube. The curve $\mathscr{C}$ moves with the tube and always embraces it. Since $\mathscr{C}$ is a fluid curve, according to Eq. (18.13), the circulation around $\mathscr{C}$ remains constant for all times. This means that *the circulation around a vortex tube, or equivalently the strength of a vortex tube, remains a constant for all times as the tube floats along, regardless of the changes experienced by the vortex tube.*

The spatial conservation of vorticity as expressed by Eq. (18.11) and the consequences, as described above, of the theorem on the permanence of vorticity or circulation, are usually referred to as Helmholtz's *theorems of vortex motion*. The spatial conservation of vorticity is purely a kinematical property, for it directly follows from the fact that the divergence of any

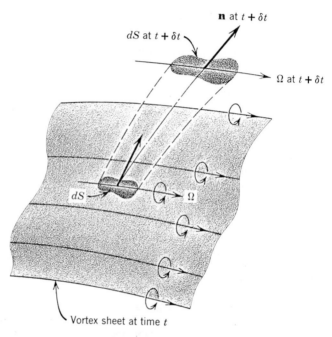

Fig. 18.5 Illustrating that a surface that is a vortex sheet at one instant remains a vortex sheet for all times.

curl vector is zero. The theorem on the permanence of vorticity or circulation is derived with the use of the equation of motion. Consequently, this theorem and the results that follow from it are applicable in the motion of an ideal fluid (or an inviscid compressible fluid for which there is a simple $p - \rho$ relation) under the action of potential body forces (see Section 9.6).

18.6 Velocity Field Due to Vortex Distribution in an Incompressible Fluid

In applications one is concerned with the problem of expressing the velocity field in terms of the vorticity field. To obtain the velocity $\mathbf{V}(\mathbf{r},t)$

in terms of the vorticity $\boldsymbol{\Omega}(\mathbf{r}, t)$ we need to invert the equation

$$\boldsymbol{\Omega} = \text{curl } \mathbf{V} \tag{18.15}$$

We do this as follows.

Considering an *incompressible fluid*, we have

$$\text{div } \mathbf{V} = 0 \tag{18.16}$$

On the basis of this relation, we may express $\mathbf{V}$ as the curl of some other vector field, say of $\mathbf{A}(\mathbf{r}, t)$. Hence we set

$$\mathbf{V} = \text{curl } \mathbf{A} \tag{18.17}$$

Since the curl of any gradient vector is zero, the vector $\mathbf{A}$ is indeterminate to the extent of the gradient of a scalar function of position and time. From (18.17) it follows that

$$\text{curl } \mathbf{V} = \text{curl (curl } \mathbf{A})$$
$$= \text{grad (div } \mathbf{A}) - \nabla^2 \mathbf{A} \tag{18.18}$$

We now stipulate that

$$\text{div } \mathbf{A} = 0 \tag{18.19}$$

This is permissible since $\mathbf{A}$ is indeterminate to the extent of a gradient vector. From (18.18), (18.19), and (18.15) we obtain

$$\nabla^2 \mathbf{A} = -\text{curl } \mathbf{V} = -\boldsymbol{\Omega} \tag{18.20}$$

This is Poisson's equation for $\mathbf{A}$. We call $\mathbf{A}$ a vector potential. Once $\mathbf{A}$ is determined as a solution of (18.20), the velocity field may be deduced from Eq. (18.17). In Cartesian, if we express

$$\mathbf{A} = (A_x, A_y, A_z)$$
$$\boldsymbol{\Omega} = (\Omega_x, \Omega_y, \Omega_z)$$

(18.20) reduces to three scalar equations

$$\nabla^2 A_x = -\Omega_x$$
$$\nabla^2 A_y = -\Omega_y$$
$$\nabla^2 A_z = -\Omega_z$$

The solution of (18.20) is expressed as (see Section 18.10).

$$\mathbf{A}(\mathbf{r}, t) = \frac{1}{4\pi} \iiint_R \frac{\boldsymbol{\Omega}(\mathbf{s}, t)}{|\mathbf{r} - \mathbf{s}|} d\tau \tag{18.21}$$

where $\boldsymbol{\Omega}(\mathbf{s}, t) \, d\tau$ is an element of the vortex distribution situated at the point $\mathbf{s}$ and R·is the region in which the vorticity is distributed (see Fig. 18.6). Note that the integration is with respect to the coordinates of the

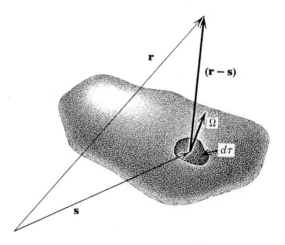

Fig. 18.6 Nomenclature used in the derivation of the velocity resulting from a vortex distribution.

vortex distribution. The velocity field is then given by

$$V = \text{curl } A = \frac{1}{4\pi} \text{curl} \iiint_R \frac{\Omega(s, t)}{|r - s|} \, d\tau \tag{18.22}$$

If we denote by δA the contribution to A at r due to the vortex element $\Omega \, d\tau$ situated at s and similarly by δV the contribution to V at r, we have

$$\delta A(r, t) = \frac{1}{4\pi} \frac{\Omega(s, t)}{|r - s|} \, d\tau \tag{18.23}$$

$$\delta V(r, t) = \frac{1}{4\pi} \text{curl}_r \frac{\Omega(s, t)}{|r - s|} \, d\tau \tag{18.24}$$

We include the subscript r on the curl to emphasize that the curl is to be taken with respect to the coordinates of the point r.

18.7 Velocity Field of a Vortex Filament: Biot-Savart Law

Consider a vortex filament of strength Γ. Choose a volume element $d\tau$ of this filament as the cylinder formed by a cross-sectional surface $n \, dS$ and an element of length dl along the filament (see Fig. 18.7). Then the contribution to the vector potential A at a field point r, from the vortex element at s is given by

$$\delta A(r) = \frac{1}{4\pi} \frac{\Omega(s)}{|r - s|} (n \, dS \cdot dl) \tag{18.25}$$

Since we have

$$\mathbf{dl} = \frac{\boldsymbol{\Omega}}{\Omega}\, dl$$

and

$$\boldsymbol{\Omega} \cdot \mathbf{n}\, dS = \Gamma$$

(18.25) may be rewritten as

$$\delta\mathbf{A}(\mathbf{r}) = \frac{\Gamma}{4\pi} \frac{\mathbf{dl}}{|\mathbf{r} - \mathbf{s}|} \qquad (18.26)$$

The contribution to the velocity at the point $\mathbf{r}$ from the element of the filament is then given by

$$\delta\mathbf{V}(\mathbf{r}) = \mathrm{curl}_{\mathbf{r}} \frac{\Gamma}{4\pi} \frac{\mathbf{dl}}{|\mathbf{r} - \mathbf{s}|} \qquad (18.27)$$

In carrying out the curl operation we keep $\mathbf{dl}$ and $\mathbf{s}$ fixed. Equation (18.27) reduces to

$$\delta\mathbf{V}(\mathbf{r}) = \frac{\Gamma}{4\pi} \frac{\mathbf{dl} \times (\mathbf{r} - \mathbf{s})}{|\mathbf{r} - \mathbf{s}|^3} \qquad (18.28)$$

This is known as the *Biot-Savart law*. The velocity at $\mathbf{r}$ due to the whole

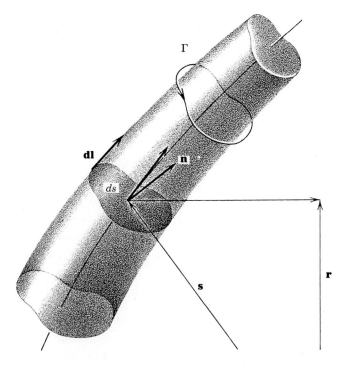

Fig. 18.7 Nomenclature used in the derivation of the Biot-Savart law.

vortex filament is obtained by integration of the expression (18.28) over the length of the filament. We thus have

$$V(r) = \frac{\Gamma}{4\pi} \int \frac{dl \times (r - s)}{|r - s|^3}$$ (18.29)

Since Γ is the strength of the filament, it is a constant and hence appears outside the integral.

18.8 Simple Applications

Two simple examples of the application of Biot-Savart's law are given here and can be used later.

Consider an infinitely long straight vortex filament of strength Γ (Fig. 18.8). To calculate the velocity field choose the origin of coordinates

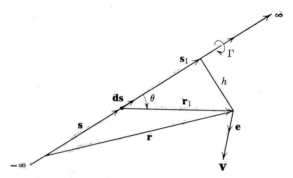

Fig. 18.8 Velocity field due to an infinite straight vortex filament.

at some point on the filament. Then, according to (18.29), the velocity at a field point is given by

$$V(r) = \frac{\Gamma}{4\pi} \int_{-\infty}^{\infty} \frac{ds \times (r - s)}{|r - s|^3}$$

where ds is an element of the filament at s. Denote $r - s$ by r_1 and the direction of $ds \times r_1$ by e (see Fig. 18.8). If θ is the angle measured from ds to r (such that $0 \leq \theta \leq \pi$), we have

$$ds \times r_1 = e r_1 \sin \theta \, ds$$

and the above expression for the velocity becomes

$$V(r) = e \frac{\Gamma}{4\pi} \int_{-\infty}^{\infty} \frac{\sin \theta}{r_1^2} \, ds$$ (18.30)

since e is a constant.

Now, let h denote the normal distance from the field point $\mathbf{r}$ to the filament and let $\mathbf{s_1}$ denote the point of intersection with the filament of the normal to it from the field point (Fig. 18.8). We then have

$$r_1 = h \operatorname{cosec} \theta$$
$$s_1 - s = h \cot \theta$$
$$ds = h \operatorname{cosec}^2 \theta \, d\theta$$

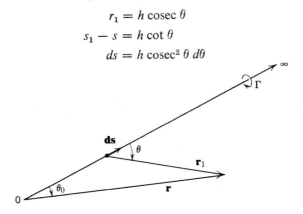

Fig. 18.9 Semi-infinite straight vortex filament.

and

$$\int_{-\infty}^{\infty} \frac{\sin \theta}{r_1^{\,2}} \, ds = \frac{1}{h} \int_0^{\pi} \sin \theta \, d\theta = \frac{2}{h}$$

Hence (18.30) yields

$$\mathbf{V(r)} = \frac{\Gamma}{2\pi h} \, \mathbf{e} \qquad (18.31)$$

We readily conclude that the motion is two-dimensional, the plane of motion being normal to the vortex filament, and that the flow field is that of a two-dimensional or point vortex.

Consider, as a second example, a semi-infinite straight vortex filament extending from 0 to ∞ (Fig. 18.9). The velocity field is then given by

$$\mathbf{V(r)} = \mathbf{e} \, \frac{\Gamma}{4\pi} \int_0^{\infty} \frac{\sin \theta}{r_1^{\,2}} \, ds$$

$$= \mathbf{e} \, \frac{\Gamma}{4\pi h} \int_{\theta_0}^{\pi} \sin \theta \, ds$$

$$= \mathbf{e} \, \frac{\Gamma}{4\pi h} (1 + \cos \theta_0) \qquad (18.32)$$

Thus, if the field point is located in the plane $\theta_0 = \pi/2$, the magnitude of the velocity is $\Gamma/4\pi h$. If the field point is located in a plane situated far away from the origin of the vortex filament, the angle θ_0 tends to zero, and

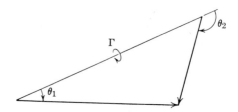

Fig. 18.10 Finite segment of a straight vortex filament.

the magnitude of the velocity tends to $\Gamma/2\pi h$, the value for an infinite filament.

The velocity field due to a finite segment of a straight vortex filament is given by

$$
\mathbf{V}(\mathbf{r}) = \mathbf{e} \frac{\Gamma}{4\pi h} \int_{\theta_1}^{\theta_2} \sin \theta \, ds
$$

$$
= \mathbf{e} \frac{\Gamma}{4\pi h} (\cos \theta_1 - \cos \theta_2) \tag{18.33}
$$

See Fig. 18.10.

18.9 Vortex Sheet

A vortex sheet is defined in a manner similar to that of a vortex filament with concentrated vorticity (see Section 18.4). Consider the narrow region of vorticity enclosed between two neighboring surfaces of vorticity S_1 and S_2 (Fig. 18.11). Let P be a point on an intermediate vortex surface S. Choose at P an element $\mathbf{n} \, dS$ of S and an elemental cylinder with $\mathbf{n} \, dS$ as the cross section through P and with height equal to ϵ the distance at P between S_1 and S_2. If $\mathbf{\Omega}(P)$ denotes the vorticity at P, we can write

$$
\mathbf{\Omega}(P) \, d\tau = \mathbf{\Omega}(P)\varepsilon \, dS
$$

where $d\tau$ is the volume of the infinitesimal cylinder. We now wish to let ϵ go to zero and $\mathbf{\Omega}$ go to infinity in such a way that $\mathbf{\Omega}\epsilon$ remains constant

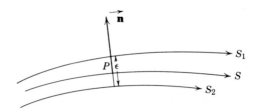

Fig. 18.11 Illustrating the concept of a vortex sheet.

and equal to ζ. Thus we have

$$\lim_{\substack{\varepsilon \to 0 \ \Omega \to \infty \\ \Omega\varepsilon = \zeta}} \Omega(P)\varepsilon \, dS = \zeta \, dS \tag{18.34}$$

In this way we shrink the narrow region of vorticity to a single surface on which the vorticity itself is infinite but ζ defined by (18.34) is finite. We call such a surface a *vortex sheet of concentrated vorticity*, or simply a *vortex sheet*, and ζ the *strength of the vortex sheet*. The strength ζ has the dimensions of *vorticity per unit area*.

We now relate ζ with the velocities on either side of the vortex sheet. According to the definition of a curl vector, we have

$$\mathbf{\Omega} \, d\tau = \operatorname{curl} \mathbf{V} \, d\tau = \iint_{\delta S} \mathbf{n} \times \mathbf{V} \, dS$$

where $\mathbf{V}$ denotes the fluid velocity. We choose $d\tau$ as the infinitesimal cylinder described before and write

$$\zeta \, dS = \lim_{\varepsilon \to 0} \mathbf{\Omega}\varepsilon \, dS = \lim_{\varepsilon \to 0} \iint_{\delta S} \mathbf{n} \times \mathbf{V} \, dS$$

$$= \lim_{\varepsilon \to 0} \left(\iint_{dS_1} \mathbf{n} \times \mathbf{V} \, dS + \iint_{dS_2} \mathbf{n} \times \mathbf{V} \, dS + \iint_{\text{wall}} \mathbf{n} \times \mathbf{V} \, dS \right)$$

where dS_1 and dS_2 are respectively the two faces of the cylinder and "wall" is the wall of the cylinder. The limit of the sum of the integrals over dS_1 and dS_2 is $\mathbf{n} \times (\mathbf{V}_1 - \mathbf{V}_2) \, dS$, and that of the integral over the wall vanishes. Hence we obtain

$$\zeta \, dS = \mathbf{n} \times (\mathbf{V}_1 - \mathbf{V}_2) \, dS$$

or

$$\zeta = \mathbf{n} \times (\mathbf{V}_1 - \mathbf{V}_2) \tag{18.35}$$

We note that ζ, $\mathbf{n}$ and $(\mathbf{V}_1 - \mathbf{V}_2)$ form a right-hand system of orthogonal vectors (Fig. 18.12). It follows that the vector $(\mathbf{V}_1 - \mathbf{V}_2)$ is tangential to the vortex sheet. We see that

$$\mathbf{V}_1 - \mathbf{V}_2 = \zeta \times \mathbf{n} \tag{18.36}$$

$$(\mathbf{V}_1 - \mathbf{V}_2) \cdot \mathbf{n} = 0 \quad \text{or,} \quad \mathbf{V}_1 \cdot \mathbf{n} = \mathbf{V}_2 \cdot \mathbf{n} \tag{18.37}$$

and

$$|\mathbf{V}_1 - \mathbf{V}_2| = \zeta \tag{18.38}$$

We therefore state that *there is a velocity discontinuity across a vortex sheet; the discontinuity occurs only in the tangential component of the velocity, whereas the normal component remains continuous; the magnitude of the discontinuity is equal to the magnitude of the strength of the sheet.* A vortex sheet is thus a surface of tangential discontinuity, and, conversely, a tangential discontinuity is a vortex sheet.

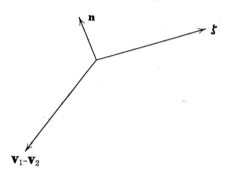

Fig. 18.12 Relation between vortex strength and the velocity discontinuity across a vortex sheet.

Vortex sheets and vortex filaments of concentrated vorticity are not physical possibilities. However, they form suitable analytical approximations when the vorticity is confined to physically narrow regions. The concepts of vortex filaments and sheets are widely applied in wing theory.

18.10   Solution for the Vector Potential

We wish to construct the solution of (18.20)

$$\nabla^2 A = -\mathbf{\Omega}(\mathbf{r})$$

or, in Cartesians, of the following system of equations

$$\nabla^2 A_x = -\Omega_x(\mathbf{r})$$
$$\nabla^2 A_y = -\Omega_y(\mathbf{r}) \qquad\qquad (18.39)$$
$$\nabla^2 A_z = -\Omega_z(\mathbf{r})$$

We consider all space and assume that the field **A** *vanishes sufficiently strongly* at infinity. We shall amplify this vague assumption later.

It is necessary only to consider the solution to one of the equations of the system (18.39). To construct the solution we use Green's theorem in the form given by (2.141)

$$\iiint_R (\phi \, \nabla^2 \psi - \psi \, \nabla^2 \phi) \, d\tau = \oiint_S (\phi \, \text{grad } \psi - \psi \, \text{grad } \phi) \cdot \mathbf{n} \, dS$$

Identify ϕ of this equation with $A_x(\mathbf{r})$ of (18.39) and choose ψ as the function

$$\psi(\mathbf{r}) = (|\mathbf{r} - \mathbf{r}_1|)^{-1} \tag{18.40}$$

We then have

$$\nabla^2 \phi = -\Omega_x(\mathbf{r})$$

$$\operatorname{grad} \phi = \operatorname{grad} A_x$$

$$\operatorname{grad} \psi = -\frac{\mathbf{r} - \mathbf{r}_1}{|\mathbf{r} - \mathbf{r}_1|^3}$$

$$\nabla^2 \psi = 0 \quad \text{everywhere except at } \mathbf{r} = \mathbf{r}_1$$

where it becomes infinite

Because the point $\mathbf{r} = \mathbf{r}_1$ is a singular point, in the sense that ψ, grad ψ, and $\nabla^2 \psi$ become infinite at that point, we surround that point by a small sphere of radius ρ with center at $\mathbf{r}_1$ and apply Green's theorem in the region contained between the sphere and an arbitrarily drawn large surface Σ (Fig. 18.13). To cover all space we remove Σ to infinity and shrink the sphere to the point $\mathbf{r}_1$. In this way we obtain

$$\iiint_R \frac{\Omega_x(\mathbf{r})}{|\mathbf{r} - \mathbf{r}_1|} \, d\tau = -\lim_{\Sigma \to \infty} \oiint_\Sigma \left(A_x \frac{\mathbf{r} - \mathbf{r}_1}{|\mathbf{r} - \mathbf{r}_1|^3} + \frac{\operatorname{grad} A_x}{|\mathbf{r} - \mathbf{r}_1|} \right) \cdot \mathbf{n} \, dS$$

$$-\lim_{\rho \to 0} \oiint_{\text{sphere } \rho} \left(A_x \frac{\mathbf{r} - \mathbf{r}_1}{|\mathbf{r} - \mathbf{r}_1|^3} + \frac{\operatorname{grad} A_x}{|\mathbf{r} - \mathbf{r}_1|} \right) \cdot \mathbf{n} \, dS \tag{18.41}$$

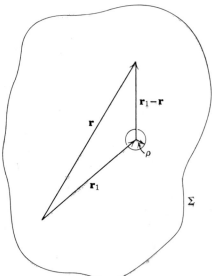

Fig. 18.13 Illustrating the computation of the vector potential.

By our assumption that the field dies out sufficiently strongly at infinity, we require that A_x and grad A_x vanish, as we approach infinity, in such a way that the limit of the surface integral over Σ vanishes as Σ goes to infinity. We thus set equal to zero the limit of the integral over Σ.

Consider the integral over the sphere and introduce spherical coordinates ρ, θ, ϕ with origin at the point $\mathbf{r}_1$. The integral over the sphere is then equal to

$$-\iint \left(A_x + \rho\, \frac{\partial A_x}{\partial \rho} \right) \sin \theta \, d\theta \, d\phi$$

and the limit of the integral as ρ goes to zero is equal to

$$-4\pi\, A_x(\mathbf{r}_1)$$

Therefore

$$\iiint\limits_{R} \frac{\Omega_x(\mathbf{r})}{|\mathbf{r} - \mathbf{r}_1|}\, d\tau = 4\pi A_x(\mathbf{r}_1) \tag{18.42}$$

The variable of integration is $\mathbf{r}$. Switching the roles of $\mathbf{r}$ and $\mathbf{r}_1$, we write

$$A_x(\mathbf{r}) = \frac{1}{4\pi} \iiint\limits_{R} \frac{\Omega_x(\mathbf{r}_1)}{|\mathbf{r} - \mathbf{r}_1|}\, d\tau \tag{18.43}$$

Now the variable of integration is $\mathbf{r}_1$.

The solutions for $A_y(\mathbf{r})$ and $A_z(\mathbf{r})$ are similar to that for $A_x(\mathbf{r})$. From these we conclude that the solution for $\mathbf{A}(\mathbf{r})$ is given by

$$\mathbf{A}(\mathbf{r}) = \frac{1}{4\pi} \iiint\limits_{R} \frac{\boldsymbol{\Omega}(\mathbf{r}_1)}{|\mathbf{r} - \mathbf{r}_1|}\, d\tau \tag{18.44}$$

Compare this with (18.21).

We now amplify the behavior of the field at infinity. Let us choose for the arbitrary surface Σ a large sphere of radius R with its center at $\mathbf{r}_1$. Introduce spherical coordinates R, θ, ϕ with origin at $\mathbf{r}_1$. The limit of the surface integral over Σ is then equal to

$$\lim_{R \to \infty} \iint \left(A_x + R\, \frac{\partial A_x}{\partial R} \right) \sin \theta \, d\theta \, d\phi$$

For this limit to vanish we require that, as R approaches infinity, A_x should approach zero and $\partial A_x / \partial R$ vanish more strongly than $1/R$. Similar behavior is required of A_y and A_z. This is what is meant by the vague assumption that the field vanishes sufficiently strongly at infinity.

Chapter 19

Elements of Finite Wing Theory

We now come to the problem of flow past a finite wing. A finite wing is any three-dimensional body which has the distinctive property that when it is suitably placed in an originally uniform stream the lift on the body is far greater than its drag. By suitably placed we mean that the orientation of the body with respect to the uniform stream is such that any lifting characteristic it may have can come into play. Thus whether a body is a wing or not depends not only on the shape of the body but also on its orientation. In describing the shape of a wing the following terminology is common (see Fig. 19.1).

Wing Section. In a wing there is a fixed direction such that planes normal to that direction cut the wing in cross sections of airfoil shape. Any such cross section of the wing is known as a *wing section.*

Span. The fixed direction itself is known as the *spanwise direction.* The two wing sections that have the greatest distance between them along the spanwise direction are called the wing tips. This spanwise distance between the wing tips is known as the span of the wing and is denoted by b

Section Chord. This is simply the chord of a wing section.

Leading and Trailing Edges. The line joining the leading edges of all the wing sections is called the leading edge of the wing. A trailing edge of the wing is similarly defined.

Midchord Line. This is the line joining the midpoints of the chords of the wing section. It extends from tip to tip.

Straight Wing. If the midchord line is a straight line, the wing is called a straight wing.

Swept Wing. A wing that is not straight is said to be a swept wing.

Rectangular Wing. A straight wing for which the chords of all the wing sections are of the same length is known as the rectangular wing.

Tapered Wing. This is a straight wing for which chord length of the section varies along the span.

Twisted Wing. This is a wing for which the section chords *vary in direction* along the span.

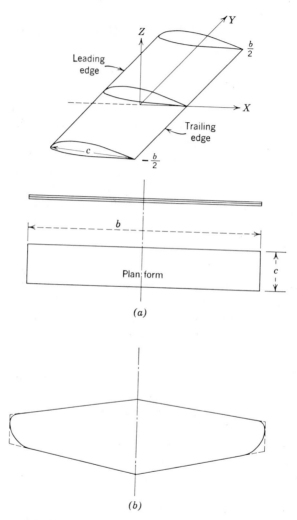

(a)

(b)

Fig. 19.1 Some nomenclature associated with wings: (*a*) rectangular wing; (*b*) plan form of a tapered wing; (*c*) plan form of a swept wing; (*d*) plan forms of tapered and swept wings; (*e*) plan form of a delta wing.

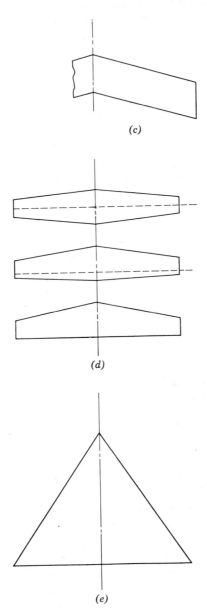

(c)

(d)

(e)

Fig. 19.1   (*Continued*)

Plan Area. This is the projected area of the wing on a plane parallel to the spanwise direction. This area is denoted by S.

Aspect Ratio. This is the ratio of the square of the span to the plan area and is denoted by $\mathcal{R}$. We thus have

$$\mathcal{R} = \frac{b^2}{S}$$

For a rectangular wing the aspect ratio is simply equal to b/c, where c is the chord of a wing section.

Mean Chord. The length defined by the ratio $b/\mathcal{R}$ is defined as the mean chord of the wing and is denoted by $\bar{c}$.

There is a vast variety of possible shapes for a wing (some examples are shown in Fig. 19.1). This being the case, any general discussion of the flow past an arbitrary wing is likely to be complex and extensive. In addition, for arbitrary wing shapes we may not even have a good physical insight into the nature of the flow past the wing. Furthermore, we are naturally not interested in completely arbitrary wing shapes. Therefore the development of the finite wing theory took place with reference to particular wing shapes. The first mathematical formulation of the theory was made by Prandtl (in 1918) for straight wings of large aspect ratios. He was concerned with the problem of extending, in a proper manner, the results of the theory of the infinite wing to develop a theory of the finite wing, a theory that would yield practically applicable results. A presentation of Prandtl's theory is our main concern in this chapter. The ideas underlying Prandtl's theory are important and have served as the basis for further developments of the finite wing theory.

19.1 Prandtl's Theory

The following ideas constitute the formulation of Prandtl's model of the flow past a finite wing of large aspect ratio.

1. Associated with the lift on the wing there is circulation around the wing. At the tips of the wing the lift, and consequently the circulation around the wing, vanishes. Thus the lift and the circulation vary along the span of the wing, their distribution being symmetrical about the midsection of the wing.

2. Since the circulation around any section of the wing is equal to the outflow of vorticity through that section, it follows that the variation of circulation along the wing span must be accompanied by the shedding of vorticity from the wing. Vortex filaments with their vorticity predominantly in the direction of the undisturbed stream originate from the wing and proceed downstream from the trailing edge. If the circulation varies continuously along the wing span, a continuous sheet of *trailing vortices*

must proceed from the wing. Such a sheet is known as a *trailing vortex sheet*. If we assume that the circulation is uniform along the span and drops to zero at the wing tips, then there will be two trailing vortex lines (or vortex filaments with concentrated vorticity) originating at the wing tips. Such a concept of two trailing vortex lines was introduced by Lanchester. Although it is satisfactory for analyzing certain problems, it

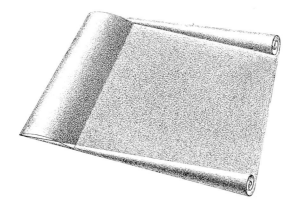

Fig. 19.2   Vortex sheet trailing behind a wing.

is not appropriate for analyzing the flow close to the wing, which is our present concern. For this purpose, the concept of a continuous trailing vortex sheet is more appropriate. This concept was postulated by Prandtl.

In the actual case of a finite wing moving through air one can observe such a trailing vortex sheet. At some distance behind the wing the sheet rolls into discrete vortices. They are eventually dissipated by the action of viscosity.

In the formulation of a theoretical model for the flow past a finite wing, in view of the laws of vortex motion for an ideal fluid, we have to allow the trailing vortices to extend to infinity in the downstream direction. It is known that even in an ideal fluid such a vortex sheet under the mutual influence of its various parts will gradually roll up into a pair of vortices (see Fig. 19.2). For studying the flow close to the wing, this rolling up of the trailing vortex sheet may be ignored. We thus assume that the trailing vortex sheet extends to infinity without any tendency for rolling up.

3. The wing itself may be replaced, for instance as in the case of the infinite wing, by a continuous distribution on the wing surface of vortex lines that are in the spanwise direction. One may use a vortex sheet on the top and a vortex sheet on the bottom of the wing. Such a vortex sheet is called a *sheet of bound vortices* or a *bound vortex sheet*, "bound" meaning

bound to the wing. For analyzing the lift on the wing, we may replace the wing by a single bound vortex sheet. A bound vortex sheet differs in one important respect from the usual vortex sheet, which may be referred to as a *free vortex sheet*. Across the bound vortex sheet, which represents a wing, a pressure difference may exist. Across a free vortex sheet, however, a pressure difference does not exist.

4. The flow past a finite wing may thus be represented as the flow past a certain vortex sheet. Part of this sheet is a bound vortex and the rest is a free vortex sheet.

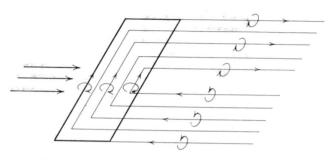

Fig. 19.3 Vortex-sheet model for flow past a wing.

We now assume that the velocity resulting from the vortex sheet alone is small in comparison with the velocity of the undisturbed stream. We then assume that both the free and bound parts of the vortex sheet lie in a plane parallel to the undisturbed stream. The resulting flow model is shown in Fig. 19.3.

For a wing of large aspect ratio (i.e., for which $\bar{c} \ll b$) the bound part of the vortex sheet may be approximated by a single bound vortex line, naturally of varying strength. Such a line is known as the *lifting line*. Prandtl used the lifting line to represent the wing. Therefore his theory is also known as (Prandtl's) *lifting line theory*. The flow model for this theory is as shown in Fig. 19.4. This model is the basis for the rest of our considerations.

5. Consider the velocity at the lifting line (i.e., at the wing). At any point on this line the velocity is the resultant of the velocity V_∞ of the undisturbed stream and the *induced velocity* q resulting from the vortex distribution. We note that q has no component in the direction of V_∞, it has a *spanwise* component (i.e., a component along the lifting line) and a component that is normal to the lifting line and the velocity V_∞ (see Fig. 19.5). We find that this normal component is directed downward (as shown in the figure) all along the lifting line. We refer to it as the *downwash velocity* and denote it by w.

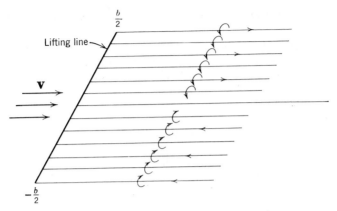

Fig. 19.4 Lifting line model for flow past a wing.

We now assume that the spanwise component of the induced velocity is very small compared to its downward component. As an immediate consequence of this assumption we ignore the spanwise component and take the velocity at any point of the lifting line as equal to the resultant of the downwash w and the velocity $\mathbf{V}_\infty$. We denote this *resultant velocity* by $\mathbf{V}_R$ and note that it is normal to the lifting line. We observe that the present assumption can be valid for a wing of sufficiently large aspect ratio and even then only for those sections of the wing that are sufficiently distant from the wing tips. The assumption cannot be regarded as

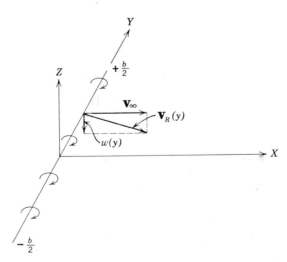

Fig. 19.5 Velocities at the lifting line.

appropriate near the tips of any wing or for any section of a low aspect ratio wing.

In general, the downwash velocity is not uniform along the lifting line. Consequently, the resultant velocity $\mathbf{V}_R$ is also not uniform along the lifting line. Choosing a Cartesian coordinate system as shown in Fig. 19.5 we write

$$w = w(y)$$

$$\mathbf{V}_R = \mathbf{V}_R(y) = \mathbf{V}_\infty - w(y)\mathbf{k} \qquad (19.1)$$

where y is the coordinate along the lifting line or, equivalently, along the span of the wing, and $\mathbf{k}$ is the unit vector in the direction of Z-axis. We

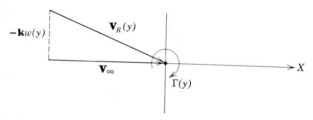

Fig. 19.6 Flow representation at any section of the wing.

note that $\mathbf{V}_R$ is normal to the lifting line all along that line. In this sense the velocity at any section of the lifting line or, equivalently, at any section of the wing, is two dimensional. This result, which is simply the assumption that we have made, enables us to use the results of the theory for an infinite wing to build a quantitative theory for the finite wing.

6. Denote by $\Gamma = \Gamma(y)$ the circulation along the lifting line. Consider the x,z-plane which cuts the lifting line at the distance y. The situation in this plane is the two-dimensional one of a bound point vortex of strength $\Gamma(y)$ in a freestream $\mathbf{V}_R(y)$ at the vortex point (see Fig. 19.6). This, of course, is the vortex representation of the flow past an infinite wing. Thus, according to the present considerations, each section of the finite wing acts exactly as the section of an infinite wing that is placed in an originally undisturbed stream of velocity $\mathbf{V}_R$. This expresses the basic idea of Prandtl's quantitative theory of the finite wing. The velocity $\mathbf{V}_R$, which is to be calculated according to Eq. (19.1), is not the same as the velocity $\mathbf{V}_\infty$ of the undisturbed stream past the finite wing.

7. Consider a slice of the finite wing, the slice being of thickness dy and situated at y. It follows that the force acting on the slice is equal to the force acting on a similar slice of an infinite wing whose section is the same as the section under consideration of the finite wing, and for which the circulation is equal to $\Gamma(y)$ and the velocity of the undisturbed stream is

equal to $\mathbf{V}_R(y)$. Then, according to the Kutta-Joukowski theorem, the force on the slice of the finite wing is given by

$$\delta\mathbf{F}(y) = \rho\mathbf{V}_R(y) \times \mathbf{j}\Gamma(y)\,dy \qquad (19.2)$$

where $\mathbf{j}$ is the direction of the Y-axis. We note that this force is normal to the velocity $\mathbf{V}_R$ and not to the velocity $\mathbf{V}_\infty$. Both the magnitude and direction of the force may vary with y.

8. According to the airfoil theory, the circulation around an infinite wing is proportional to the speed of the undisturbed stream and for small angles of attack to the angle of attack measured from the zero-lift direction

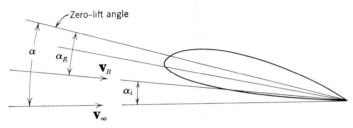

Fig. 19.7   Defining the induced angle of attack.

(see Section 16.3). Denoting by $\alpha_R = \alpha_R(y)$ the angle between $\mathbf{V}_R(y)$ and the zero-lift direction of the section at y of the finite wing we write

$$\Gamma(y) = K(y)\,V_R(y)\,\alpha_R(y) \qquad (19.3)$$

where K is a constant that depends on the form and size of the wing section under consideration. Since the form and size of the wing section may, in general vary with y, we have expressed the constant K as $K(y)$. From the airfoil theory we find that

$$K(y) = \frac{1}{2}\left[\frac{dC_L}{d\alpha_R}(y)\right]c(y)$$

$$= \frac{1}{2}\,a_0(y)c(y) \qquad (19.4)$$

where $c(y)$ is the chord at y of the finite wing and $a_0(y)$ is the slope of the lift versus angle of attack curve for the corresponding airfoil profile (see Section 16.4).

9. Denote by $\alpha(y)$ the angle between the velocity $\mathbf{V}_\infty$ and the zero-lift direction of the wing section at y and by $\alpha_i(y)$ the angle between the velocities $\mathbf{V}_R(y)$ and $\mathbf{V}_\infty$ (see Fig. 19.7). We then have

$$\alpha_R(y) = \alpha(y) - \alpha_i(y) \qquad (19.5)$$

The angle α_i is known as the *induced angle of attack*. It is given by

$$\alpha_i(y) = \tan^{-1} \frac{w(y)}{V_\infty}$$

Assuming w is very small compared to V_∞ we may write approximately

$$\alpha_i(y) = \frac{w(y)}{V_\infty} \tag{19.6}$$

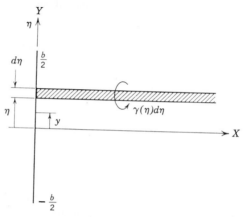

Fig. 19.8 Illustrating the calculation of downwash at the lifting line.

Equation (19.5) then becomes

$$\alpha_R(y) = \alpha(y) - \frac{w(y)}{V_\infty} \tag{19.7}$$

This is a fundamental relation of Prandtl's theory.

10. We now express the downwash $w(y)$ in terms of the circulation $\Gamma(y)$. Let $\gamma = \gamma(y)$ denote the strength of the trailing vortex sheet per unit length along y (see Fig. 19.8). Then the downwash at the section y of the lifting line is given by (see Section 18.8)

$$w(y) = \frac{1}{4\pi} \int_{-b/2}^{b/2} \frac{\gamma(\eta)}{\eta - y} \, d\eta$$

The distributions $\gamma(y)$ and $\Gamma(y)$ are related by the equation (verify this)

$$\gamma(y) = -\frac{d\Gamma}{dy}$$

Hence for the downwash $w(y)$ we obtain

$$w(y) = \frac{1}{4\pi} \int_{-b/2}^{b/2} \frac{d\Gamma}{dy}(\eta) \frac{1}{y - \eta} d\eta \qquad (19.8)$$

Recall that positive w denotes downwash, that is, a velocity directed in the negative z-direction.

We note that the downwash cannot be determined till $\Gamma(y)$ is known. But $\Gamma(y)$ itself is not known until $w(y)$ and consequently $\alpha_R(y)$ are obtained.

11. To proceed, we combine (19.3), (19.7), and (19.8) and obtain

$$\Gamma(y) = K(y)V_R(y)\left[\alpha(y) - \frac{1}{4\pi V_\infty} \int_{-b/2}^{b/2} \frac{d\Gamma}{dy}(\eta) \frac{dy}{y - \eta}\right] \qquad (19.9)$$

Since $w(y)$ is assumed to be much less than V_∞, we approximate $V_R(y)$ by V_∞. Equation (19.9) then becomes

$$\Gamma(y) = K(y)\left[V_\infty\alpha(y) - \frac{1}{4\pi} \int_{-b/2}^{b/2} \frac{d\Gamma}{dy}(\eta) \frac{d\eta}{y - \eta}\right] \qquad (19.10)$$

where, according to Eq. (19.4),

$$K(y) = \tfrac{1}{2}a_0(y)c(y)$$

Equation (19.10) is an *integro-differential equation* for $\Gamma(y)$ and is known as Prandtl's equation of finite wing theory. It is a relation between the circulation around the wing, the geometrical features such as the chord and angle of attack distributions, and the properties of the wing sections. It is thus the basis for obtaining important information about the aerodynamic characteristics of finite wings of sufficiently large aspect ratios. To determine the circulation $\Gamma(y)$, one has to solve (19.10) with the additional condition that the circulation falls to zero at the wing tips:

$$\Gamma\left(-\frac{b}{2}\right) = \Gamma\left(\frac{b}{2}\right) = 0 \qquad (19.11)$$

We shall consider later the problem of solving (19.10).

12. We now give the expressions for calculating the lift and drag of a finite wing and the components of the moment acting on the wing. By definition the lift is the component of the force on the wing in the direction normal to the velocity $\mathbf{V}_\infty$ and the spanwise direction, and the drag is the component in the direction of $\mathbf{V}_\infty$. Accordingly, the lift and drag of a slice dy of the wing are given by

$$\delta L(y) = \delta F(y)\cos\alpha_i \simeq \delta F(y) \qquad (19.12)$$

$$\delta D(y) = \delta F(y)\sin\alpha_i \simeq \delta F(y)\,\alpha_i(y) = \alpha_i(y)\,\delta L(y) \qquad (19.13)$$

where, as before, $\delta F(y)$ is magnitude of the force $\delta\mathbf{F}$ acting on the slice of the wing (see Fig. 19.9). In these relations we substitute $\rho V_R(y)\Gamma(y)\,dy$ for $\delta F(y)$, (see 13.19), approximate $V_R(y)$ by V_∞, and obtain

$$\delta L(y) = \rho V_\infty\,\Gamma(y)\,dy \tag{19.14}$$

$$\delta D(y) = \alpha_i(y)\,\delta L(y)$$

$$= \rho V_\infty\,\alpha_i(y)\Gamma(y)\,dy \tag{19.15}$$

$$= \rho w(y)\,\Gamma(y)\,dy$$

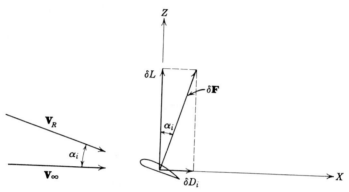

Fig. 19.9 Lift and drag on a slice of the wing.

These equations express the *spanwise distribution* of the lift and drag forces. Integration of these equations yields the total lift L and the total drag D of the wing. We thus obtain

$$L = \rho V_\infty \int_{-b/2}^{b/2} \Gamma(y)\,dy \tag{19.16}$$

$$D = \int_{-b/2}^{b/2} \alpha_i(y)\,\delta L(y)$$

$$= \rho V_\infty \int_{-b/2}^{b/2} \alpha_i(y)\,\Gamma(y)\,dy \tag{19.17}$$

$$= \rho \int_{-b/2}^{b/2} w(y)\,\Gamma(y)\,dy$$

It is thus seen that there is a nonzero drag on a finite wing. It vanishes only if the downwash vanishes all along the span. This happens only if the wing is infinite. For a finite wing experiencing a nonzero lift, the downwash cannot be zero and consequently the drag cannot vanish. Therefore we may say that the drag expressed by Eq. (19.17) is the drag that must be counteracted in order to obtain lift. Such a drag, which originates due to the induced velocity field, is known as the *induced drag*. *To signify this fact, it is denoted by the special symbol D_i.* The work done against this drag force appears as the kinetic energy of the downwash field that must be created to obtain lift. Equivalently, the induced drag may be interpreted as the force that necessitates the expenditure of work in creating the vortices trailing behind a finite wing.

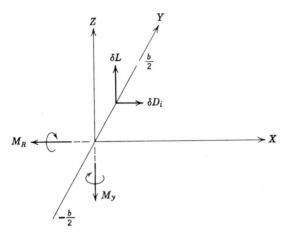

Fig. 19.10 Defining the rolling and yawing moments on the wing.

The moment on the wing with respect to the origin of coordinates is given by

$$\mathbf{M} = \int_{-b/2}^{b/2} y\mathbf{j} \times (\mathbf{i}\delta D_i + \mathbf{k}\delta L)$$

$$= -\mathbf{k}\int_{-b/2}^{b/2} y\,\delta D_i + \mathbf{i}\int_{-b/2}^{b/2} y\,\delta L$$

(see Fig. 19.10). It is customary to call as the *rolling moment* the component of $\mathbf{M}$ in the negative x-direction and as the *yawing moment* the component in the negative z-direction (see figure). Denoting these components by M_R and M_y, respectively, we have

$$M_R = -\int_{-b/2}^{b/2} y\,\delta L = -\rho V_\infty \int_{-b/2}^{b/2} \Gamma(y)y\,dy \qquad (19.18)$$

$$M_y = \int_{-b/2}^{b/2} y\,\delta D_i = \rho \int_{-b/2}^{b/2} w(y)\,\Gamma(y)y\,dy \qquad (19.19)$$

19.2 Problems of Interest

We now consider the type of problems that may be solved by means of Prandtl's theory. The basis for our considerations are the equations (19.8) for the downwash $w(y)$ at the lifting line and the integral equation (19.10) for the circulation $\Gamma(y)$.

A relatively simple problem is to prescribe the distribution of circulation (or equivalently of the lift) $\Gamma(y)$ along the span of the wing and to seek to determine the distributions of the downwash $w(y)$, the angle of attack $\alpha(y)$, and the chord $c(y)$. Recall that the angle of attack is measured from the total zero-lift line that depends on the camber of the particular wing section under consideration. It therefore follows that if the camber varies along the span, the angle of attack will also vary even if the so-called *geometrical angle of attack* is kept constant all along the wing span. The distribution of the chord determines the plan form of the wing. The problem we have posed is known as the *first problem* and may be solved by a straightforward calculation. An important example of this problem is that for an elliptic lift distribution and is given in Section 19.3.

A more interesting and more difficult problem is to determine the aerodynamic characteristics of a wing, such as $\Gamma(y)$, $\delta L(y)$, $\delta D_i(y)$, when its geometry is prescribed, that is, when the distributions of chord $c(y)$, wing section represented by $a_0(y)$ and the zero-lift line, and the angle of attack $\alpha(y)$ along the span are given. This problem is known as the second problem. It requires the solution of the integral equation for the circulation. A method for its solution is given in Section 19.4.

A third problem is the so-called problem of minimum drag. The problem is how to make induced drag as small as possible under given conditions. Some aspects of this question will be discussed in Section 19.6.

19.3 Elliptic Lift Distribution: Elliptic Wing

We now consider the first problem where the spanwise distribution of the circulation $\Gamma(y)$ is given. The distribution of the lift is simply proportional to that of the circulation. The problem becomes particularly simple for the so-called *elliptic lift distribution* expressed by

$$\Gamma(y) = \Gamma_0\sqrt{1 - (2y/b)^2}$$

or

$$\frac{\Gamma^2(y)}{\Gamma_0{}^2} + \frac{y^2}{(b/2)^2} = 1 \tag{19.20}$$

where Γ_0 is a constant. The distribution is represented by the upper half of the ellipse described by this equation (see Fig. 19.11). The constant Γ_0 is the maximum circulation at the midsection of the wing. The solution to the problem of the elliptic lift distribution yields many results of great practical significance.

We first compute the downwash distribution using (19.8) and (19.20). We have

$$w(y) = \frac{1}{4\pi} \int_{-b/2}^{b/2} \frac{d\Gamma}{dy}(\eta) \frac{d\eta}{y - \eta}$$

$$= -\frac{\Gamma_0}{\pi b^2} \int_{-b/2}^{b/2} \left[1 - \left(\frac{2\eta}{b}\right)^2\right]^{-\frac{1}{2}} \frac{\eta}{y - \eta}\, d\eta \qquad (19.21)$$

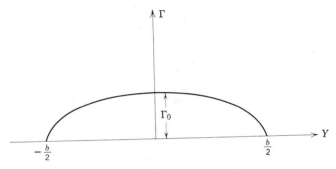

Fig. 19.11 Elliptic lift distribution.

Note that the integral is improper at $\eta = y$, and hence one should obtain the principal value of the integral. Introducing a change of variables through the relation

$$y = \frac{b}{2}\cos\theta \qquad (19.22)$$

we rewrite (19.21) as

$$w(\theta) = -\frac{\Gamma_0}{2\pi b} \int_0^\pi \frac{\cos\varphi}{\cos\theta - \cos\varphi}\, d\varphi \qquad (19.23)$$

The value of the integral is $-\pi$ (see Eq. 17.73). Hence the downwash is given by

$$w(y) = \frac{\Gamma_0}{2b}, \quad \text{a constant} \qquad (19.24)$$

We thus conclude that *the downwash corresponding to an elliptic lift distribution is a constant all along the span.* It follows that in this case the induced angle of attack α_i is also constant along the span.

Since the downwash is a constant, the distribution of the induced drag along the span is also elliptic. We have

$$\delta D_i(y) = \rho w(y)\, \Gamma(y)\, dy$$

$$= \text{constant } \Gamma(y)\, dy$$

The drag of the wing is given by

$$D_i = \rho \int_{-b/2}^{b/2} w(y)\,\Gamma(y)\,dy$$

$$= w\rho \int_{-b/2}^{b/2} \Gamma(y)\,dy = \frac{w}{V_\infty} L \qquad (19.25)$$

$$= \alpha_i L$$

where w and α_i are constant.

The integral $\int_{-b/2}^{b/2} \Gamma(y)\,dy$ is the area of the semiellipse shown in Fig. 19.11. We find

$$\int_{-b/2}^{b/2} \Gamma(y)\,dy = \frac{\pi b \Gamma_0}{4}$$

Therefore the lift and drag are given by

$$L = \frac{\pi}{4}\rho V_\infty \Gamma_0 b \qquad (19.26)$$

$$D_i = \frac{w}{V_\infty} L = \frac{\Gamma_0}{2b}\frac{L}{V_\infty}$$

$$= \frac{\pi}{8}\rho \Gamma_0{}^2 \qquad (19.27)$$

The lift and drag coefficients of the wing are given by

$$C_L \equiv \frac{L}{\frac{1}{2}\rho V_\infty{}^2 S} = \frac{\pi}{2}\frac{b}{S}\frac{\Gamma_0}{V_\infty} \qquad (19.28)$$

$$C_{D_i} \equiv \frac{D_i}{\frac{1}{2}\rho V_\infty{}^2 S} = \frac{\pi}{4S}\left(\frac{\Gamma_0}{V_\infty}\right)^2$$

$$= \frac{1}{\pi}\frac{S}{b^2} C_L{}^2 \qquad (19.29)$$

$$= \frac{1}{\pi}\frac{C_L{}^2}{\mathcal{R}}$$

where S is the plan form area of the wing and $\mathcal{R}$ is the aspect ratio of the wing. A curve that shows the lift coefficient of a wing as a function of the drag coefficient is known as the *polar diagram*. From (19.29) we have

$$C_L = \sqrt{\pi \mathcal{R} C_{D_i}}$$

This shows that the polar diagram of a wing with an elliptic lift distribution is a parabola with its vertex at the origin and its axis along the axis of the drag coefficient (see Fig. 19.12). It is readily verified that the yawing and rolling moments on such a wing are zero no matter how the chord, the angle of attack, and the wing section are arranged.

The spanwise distribution of chord, wing section, and angle of attack are related to the lift distribution by the equation (follows from 19.3)

$$\Gamma(y) = \tfrac{1}{2}a_0(y)c(y)[V_\infty\alpha(y) - w(y)]$$

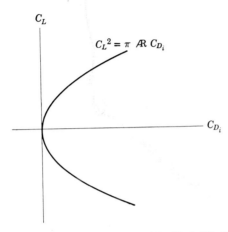

Fig. 19.12   Polar diagram for a wing with elliptic lift distribution.

For a wing with elliptic distribution this becomes

$$\Gamma_0\sqrt{1 - (2y/b)^2} = \Gamma(y) = \frac{1}{2}\,a_0(y)c(y)\left[V_\infty\alpha(y) - \frac{\Gamma_0}{2b}\right] \quad (19.30)$$

We see that this relation may be satisfied by many different combinations of $a_0(y)$, $c(y)$, and $\alpha(y)$. If we choose the same airfoil profile (i.e., same $a_0(y)$ and same camber) all along the span and also keep the same geometrical angle of attack along the span, $a_0(y)$ and $\alpha(y)$ will be constant in (19.30). We may then solve for $c(y)$ and obtain

$$c(y) = \text{constant } \Gamma(y)$$

where the constant is determined by the values of Γ_0, V_∞, α, b, and a_0. *This shows that to obtain an elliptic lift distribution on a (geometrically and aerodynamically) untwisted wing, the spanwise distribution of the chord should be elliptic.* A wing with such a distribution of chord is known as the *elliptic wing.* Examples of the plan form of an elliptic wing are shown in Fig. 19.13.

For an elliptic wing Γ_0 may be expressed readily in terms of the geometrical properties of the wing. To do this we first write

$$c(y) = c_0\sqrt{1 - \left(\frac{2y}{b}\right)^2} \qquad (19.31)$$

where c_0 is the chord of the midsection of the wing. Substituting the relation (19.31) into (19.30) and solving for Γ_0 we obtain*

$$\Gamma_0 = \frac{2bV_\infty\alpha}{1 + (4b/a_0 c_0)} \qquad (19.32)$$

Introduce the aspect ratio defined by

$$\mathcal{R} \equiv \frac{b^2}{S}$$

where S is the plan form area of the wing. For the elliptic wing we have

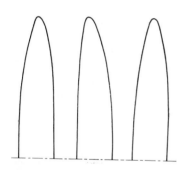

$$S = \frac{\pi}{4} c_0 b$$

and

$$\mathcal{R} = \frac{4b}{\pi c_0}$$

Hence (19.32) may be rewritten as

$$\Gamma_0 = \frac{2bV_\infty\alpha}{1 + (\pi\mathcal{R}/a_0)} \qquad (19.33)$$

Substituting this relation for Γ_0 in (19.28) for the lift coefficient we obtain

Fig. 19.13 Plan forms of elliptic wings.

$$C_L = \frac{\pi b}{2S}\frac{\Gamma_0}{V_\infty} = \frac{a_0}{1 + (a_0/\pi\mathcal{R})}\alpha \qquad (19.34)$$

From this it follows that for a given aspect ratio, the lift varies linearly with the angle of attack. The slope of the C_L versus the α curve is given by

$$a \equiv \frac{dC_L}{d\alpha} = \frac{a_0}{1 + (a_0/\pi\mathcal{R})} \qquad (19.35)$$

Recall that a_0 is the slope of the lift coefficient versus angle of attack curve for an infinite wing that has the same section as the finite wing. This is clear from the (19.35), which shows that as the aspect ratio goes to infinity, the

---

* Note that a_0 and α are constants.

slope of the lift curve becomes a_0. Denoting by $(C_L)_\infty$ the lift coefficient of the infinite wing we write

$$a_0\alpha = (C_L)_\infty \tag{19.36}$$

Then (19.34) and (19.35) may be expressed together by the relation

$$\frac{C_L}{(C_L)_\infty} = \frac{a}{a_0} = \frac{1}{1 + (a_0/\pi \mathcal{R})} \tag{19.37}$$

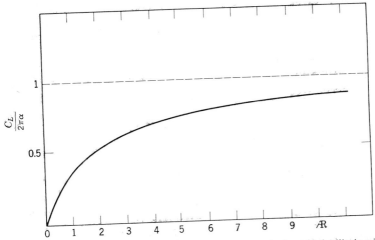

Fig. 19.14 Variation of lift coefficient with aspect ratio for a thin elliptic wing.

This expresses the lift coefficient for a finite (elliptic) wing at some angle of attack in terms of the lift coefficient for the corresponding infinite wing at the same angle of attack.

For a thin wing, the theory of the infinite wing shows that

$$a_0 \simeq 2\pi \tag{19.38}$$

Thus for a thin wing (19.37) takes the approximate form

$$\frac{C_L}{2\pi\alpha} = \frac{C_L}{(C_L)_\infty} = \frac{1}{1 + 2/\mathcal{R}} \tag{19.39}$$

This expresses in a simple way the effect of the aspect ratio on the lift of a thin elliptic wing. The variation of $C_L/(C_L)_\infty$ with $\mathcal{R}$ for such a wing is shown in Fig. 19.14.

The properties of an elliptic wing, as expressed by (19.29) and (19.34), are a fair representation also of the properties of rectangular and trapezoidal wing shapes. Hence the importance of studying the elliptic wing for which simple theoretical formulae may be obtained. It is illuminating to realize that the results expressed by (19.29) and (19.34) were not known, even as empirical facts, before Prandtl's wing theory was developed. Experiments carried out a posteriori have amply confirmed the predictions of the theory. This is really a remarkable feature, particularly in view of the many assumptions made in the theory. But such is the characteristic of a good theory.

19.4 Solution for the Arbitrary Wing: Trefftz's Method

Problem. We now consider the second problem of determining the distributions of the lift, drag, and moments for an arbitrary wing of given shape. For this purpose we must first obtain the distributions of the downwash $w(y)$ and the circulation $\Gamma(y)$, given the distributions $a_0(y)$, $c(y)$, and $\alpha(y)$. We are concerned with the solutions of (19.8) and (19.10), which are

$$w(y) = \frac{1}{4\pi} \int_{-b/2}^{b/2} \frac{\gamma(\eta)}{\eta - y} \, d\eta$$

where

$$\gamma(y) = -\frac{d\Gamma(y)}{dy}$$

and

$$\Gamma(y) = K(y)[V_\infty \alpha(y) - w(y)] \tag{19.10a}$$

$$= K(y)\left[V_\infty \alpha(y) - \frac{1}{4\pi} \int_{-b/2}^{b/2} \frac{d\Gamma}{dy}(\eta) \frac{d\eta}{y - \eta} \right]$$

where

$$K(y) = 1/2 \, a_0(y)c(y)$$

The equation for $\Gamma(y)$ is to be supplemented by the condition

$$\Gamma\left(-\frac{b}{2}\right) = \Gamma\left(\frac{b}{2}\right) = 0$$

Trefftz Plane. Many methods have been developed to obtain $\Gamma(y)$ directly as a solution of the integral equation (19.10). We shall not consider these methods. Instead we describe a method of solving for $w(y)$ and $\Gamma(y)$ that does not use the integral equation directly. Such a method was developed by Trefftz and is known as Trefftz's method. This method depends on the following observations, which are valid under the assumptions of Prandtl's theory:

1. The downwash at any point on the lifting line is half of the downwash at a corresponding point on the trailing vortex sheet far downstream and

2. Far downstream from the wing the induced flow does not depend on the streamwise coordinate and hence may be regarded as two-dimensional in all planes normal to the trailing vortex sheet, that is, to the mainstream direction.

This suggests that the downwash, and thus also the circulation, at the lifting line may be readily obtained if the problem for the two-dimensional flow in the cross planes (i.e., the y,z-planes) far downstream is formulated and solved. The distribution of the downwash on the vortex sheet with the

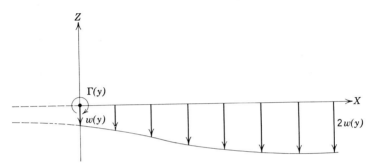

Fig. 19.15 Streamwise distribution of the downwash on the vortex sheet.

streamwise coordinate x is shown in Fig. 19.15. We denote the downwash on the trailing vortex sheet far downstream by $w_\infty(y)$:

$$w_\infty(y) \equiv w(x, y, 0) \quad \text{for} \quad x \to \infty \qquad (19.40)$$

The contribution to the downwash $w_\infty(y)$ from the bound vorticity of the lifting line is insignificant and may be neglected. The contribution from the trailing vortex sheet is that due to rectilinear vortex filaments that extend to infinity in both directions. We thus have

$$w_\infty(y) = \frac{1}{2\pi} \int_{-b/2}^{b/2} \frac{\gamma(\eta)}{\eta - y}\, d\eta = 2w(y) \qquad (19.41)$$

The situation far downstream in a cross plane (i.e., in a plane normal to the vortex sheet) is shown in Fig. 19.16. Such a cross plane is known as the *Trefftz plane*, also sometimes as the *wake plane*. It is seen that the flow field in the Trefftz plane is that due to the vortex distribution $\gamma(y)$ on the strip $z = 0$ and $-b/2 \le y \le b/2$. Thus our problem is essentially reduced to that of finding the flow field resulting from a two-dimensional vortex sheet and satisfying certain boundary conditions.

Relation between the flow in the Trefftz plane and $w(y)$ and $\Gamma(y)$. The flow in the Trefftz plane is irrotational everywhere except on the vortex

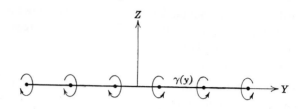

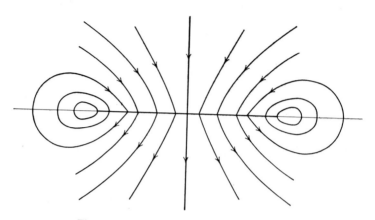

Fig. 19.16 Flow situation in the Trefftz plane.

strip. The velocity field may therefore be represented by a potential

$$\phi = \phi(y, z)$$

Denoting the y- and z-components of the velocity by v' and w' we have

$$v'(y, z) = \frac{\partial \phi}{\partial y} \tag{19.42}$$

$$w'(y, z) = \frac{\partial \phi}{\partial z} \tag{19.43}$$

The downwash at the vortex strip is then given by

$$w_\infty(y) = -w'(y, 0) = -\frac{\partial \phi}{\partial z}(y, 0) \tag{19.44}$$

and for the downwash at the lifting line we have

$$w(y) = -\frac{1}{2} w'(y, 0) = -\frac{1}{2} \frac{\partial \phi}{\partial z}(y, 0) \tag{19.45}$$

The velocity field is continuous everywhere except across the vortex strip. Across the strip the normal component w' is continuous, but the tangential component v' is discontinuous. At any point y on the strip, the jump in v' is simply equal to the vortex strength $\gamma(y)$ at that point. We have

$$v'(y, 0^+) = -\frac{\gamma(y)}{2}$$

$$v'(y, 0^-) = \frac{\gamma(y)}{2}$$

See Fig. 19.16 for sense of $\gamma(y)$. It follows that

$$\gamma(y) = -2v'(y, 0^+) = -2\frac{\partial \phi}{\partial y}(y, 0^+)$$

where 0^+ refers to the upper side of the vortex strip and 0^- refers to its lower side. Since $\gamma(y)$ is equal to $-d\Gamma/dy$, we obtain

$$\frac{d\Gamma}{dy} = -\gamma(y) = 2\frac{\partial \phi}{\partial y}(y, 0^+)$$

or

$$\Gamma(y) = 2\phi(y, 0^+)$$

Similarly, we find

$$\Gamma(y) = -2\phi(y, 0^-)$$

Thus we have

$$\Gamma(y) = 2\phi(y, 0^+) = -2\phi(y, 0^-) \tag{19.46}$$

Problem in the Trefftz Plane. Now we have to determine the potential $\phi(y, z)$. It must have the following properties:

1. ϕ is a harmonic function, that is, it must satisfy Laplace's equation

$$\nabla^2 \phi = 0$$

2. grad ϕ vanishes at infinity.
3. ϕ is an odd function of z:

$$\phi(y, z) = -\phi(y, -z)$$

This follows from Eq. (19.46), which shows that

$$\phi(y, 0^+) = -\phi(y, 0^-) \quad \text{for} \quad -\frac{b}{2} \le y \le \frac{b}{2}$$

and

$$\phi(y, 0) = 0 \quad \text{for} \quad y < -\frac{b}{2} \quad \text{and} \quad y > \frac{b}{2}$$

4. In view of (19.46) we require that

$$\phi\left(-\frac{b}{2}, 0\right) = \phi\left(\frac{b}{2}, 0\right) = 0$$

5. From Eqs. (19.10a), (19.45), and (19.46) it follows that ϕ must satisfy the boundary condition

$$2\phi(y, 0^+) = K(y)\left[V_\infty \alpha(y) + \frac{1}{2}\frac{\partial\phi}{\partial z}(y, 0)\right] \quad \text{for} \quad -\frac{b}{2} \le y \le \frac{b}{2} \quad (19.47)$$

where

$$K(y) = \frac{1}{2}a_0(y)c(y)$$

Solution for the Flow in the Trefftz Plane. To determine ϕ we use the method of the complex variable. Introduce the complex variable ζ defined by

$$\zeta = y + iz \qquad (19.48)$$

and express the complex potential in the ζ-plane as

$$F(\zeta) = \phi(y, z) + i\psi(y, z) \qquad (19.49)$$

where ψ is the harmonic conjugate of ϕ and represents the stream function of the flow. We will not be concerned with it, however. The complex velocity in the ζ-plane is then given by

$$v(y, z) - iw(y, z) \equiv W(\zeta) = \frac{dF}{d\zeta}(\zeta) \qquad (19.50)$$

In terms of this representation the problem reduces to that of determining an analytic $F(\zeta)$ such that its real part has all the properties (1) to (5) outlined above.

To find the function $F(\zeta)$ it is convenient to transform the problem from the ζ-plane to the plane of another complex variable t defined by the mapping

$$\zeta = t + \frac{(b/4)^2}{t} \qquad (19.51)$$

This maps the strip $z = 0$, $-b/2 \le y \le b$ into a circle in the t-plane expressed by

$$t = \frac{b}{4}e^{i\theta} \qquad (19.52)$$

See Fig. 19.17. We are concerned with the regions outside the strip and the circle. Points on the top of the strip defined by $z = 0^+$ and $y = b/2$ to

$y = -b/2$ go into points $r = b/4$ and $\theta = 0$ to $\theta = \pi$, whereas points on the bottom of the strip defined by $z = 0^-$ and $y = b/2$ to $y = -b/2$ go into points $r = b/4$ and $\theta = 0$ to $\theta = -\pi$. For points on the strip we have

$$y = \frac{b}{2} \cos \theta \qquad (19.53)$$

Setting

$$t = re^{i\theta}$$

we express the complex potential in the t-plane by

$$F(t) = \phi_t(r, \theta) + i\psi_t(r, \theta)$$

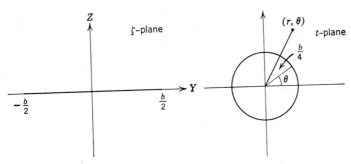

Fig. 19.17 Mapping of the flow in the Trefftz plane.

Then the complex potential in the ζ-plane is obtained as

$$F(\zeta) = F[t(\zeta)]$$

The potential ϕ itself is given by

$$\phi(y, z) = \phi_t[r(y, z), \theta(y, z)] \qquad (19.54)$$

In view of the properties (3) and (4) for $\phi(y, z)$ we require that

$$\phi_t(r, \theta) = -\phi_t(r, -\theta) \quad \text{for} \quad r \geq \frac{b}{4} \qquad (19.55)$$

and

$$\phi_t(r, 0) = \phi_t(r, \pi) = 0 \quad \text{for all} \quad r > \frac{b}{4}$$

In general, the complex potential $F(t)$ may be assumed to be of the form

$$F(t) = \sum_1^\infty \frac{b_n + ia_n}{t^n} = \sum_1^\infty \frac{b_n + ia_n}{r^n} e^{-in\theta}$$

$$= \sum_1^\infty \frac{1}{r^n} [b_n \cos n\theta + a_n \sin n\theta) + i(a_n \cos n\theta - b_n \sin n\theta)]$$

where a_n and b_n are real constants. We then have

$$\phi_t(r, \theta) = \sum_1^\infty \frac{1}{r^n} (b_n \cos n\theta + a_n \sin n\theta)$$

In view of the requirement (19.55) we set the b_n's zero and obtain

$$\phi_t(r, \theta) = \sum_1^\infty \frac{a_n}{r^n} \sin n\theta \tag{19.56}$$

and

$$F(t) = i \sum_1^\infty \frac{a_n}{t^n} \tag{19.57}$$

The potential $\phi(y, z)$, which in terms of r, θ we shall express as $\phi(r, \theta)$, is then given by

$$\phi(y, z) = \phi(r, \theta) = \sum_1^\infty \frac{a_n}{r^n} \sin n\theta \tag{19.58}$$

To determine the coefficients a_n we must use the boundary condition (19.47). However, we must first compute the velocity $w'(y, z) = (\partial\phi/\partial z)$. Now, the complex velocity in the ζ-plane is given by

$$v' - iw' \equiv W(\zeta) = \frac{dF(\zeta)}{d\zeta} = \frac{dF(t)}{dt} \frac{1}{d(\zeta/dt)}$$

Using (19.51) and (19.57) we find

$$v'(y, z) - iw'(y, z) = -\sum_1^\infty n \frac{a_n}{r^{n+1}} \frac{\cos n\theta - i \sin n\theta}{2 \sin \theta}$$

Denoting $w'[y(r, \theta), z(r, \theta)]$ by $w'(r, \theta)$ we have

$$w'(r, \theta) = -\sum_1^\infty n \frac{a_n}{r^{n+1}} \frac{\sin n\theta}{\sin \theta} \tag{19.59}$$

From (19.59) we obtain

$$\frac{\partial\phi}{\partial z} (y, 0) = w'\left(\frac{b}{4}, \theta\right)$$

$$= -\sum_1^\infty n \frac{a_n}{\left(\dfrac{b}{4}\right)^{n+1}} \frac{\sin n\theta}{\sin \theta} \tag{19.60}$$

and from (19.58) we have

$$\phi(y, 0^+) = \phi\left(\frac{b}{4}, 0\right) = \sum_1^\infty \frac{a_n}{\left(\dfrac{b}{4}\right)^n} \sin n\theta, \qquad 0 \le \theta \le \pi \tag{19.61}$$

Substituting the relations (19.60) and (19.61) into the boundary condition (19.47) and rearranging the terms we obtain

$$\sum_1^\infty \frac{1}{bV_\infty} \frac{a_n}{\left(\dfrac{b}{4}\right)^n} \sin n\theta \left[\sin \theta + n \frac{a_0(y)c(y)}{4b} \right] = \frac{a_0(y)c(y)}{4b} \alpha(y) \quad (19.62)$$

We now introduce the notation

$$A_n \equiv \frac{a_n}{V_\infty b \left(\dfrac{b}{4}\right)^n} \quad (19.63)$$

and

$$\mu(\theta) \equiv \frac{a_0[y(\theta)]\, c[y(\theta)]}{4b} \quad (19.64)$$

Then (19.62) takes the form

$$\sum_1^\infty A_n \sin n\theta [n\mu(\theta) + \sin \theta] = \mu(\theta)\alpha(\theta) \quad (19.65)$$

where $\alpha(\theta) \equiv \alpha[y(\theta)]$ and θ ranges from 0 to π. This is the equation that determines the constants A_n or, equivalently, the constants a_n. Later we shall discuss the method of obtaining the solution for the A_n's. This completes the solution of the problem in the Trefftz plane.

Solution for $\Gamma(y)$ and $w(y)$. Introduce the notation

$$\Gamma(\theta) \equiv \Gamma[y(\theta)]$$

and

$$w(\theta) \equiv w[y(\theta)]$$

Then from Eqs. (19.45), (19.46), (19.60), and (19.61), and the notation (19.63) we have

$$\Gamma(\theta) = 2bV_\infty \sum_1^\infty A_n \sin n\theta \quad (19.66)$$

$$w(\theta) = V_\infty \sum_1^\infty nA_n \frac{\sin n\theta}{\sin \theta} \quad (19.67)$$

Once the A_n's are determined from Eq. (19.65), the solutions for the distributions of circulation and downwash for an arbitrary wing of large aspect ratio are complete. The coefficients A_n are functions of the geometrical characteristics of the wing and the free-stream speed. Equations (19.66) and (19.67) thus express the spanwise distributions of the circulation and downwash in terms of the geometry of the wing.

19.5 Forces and Moments on an Arbitrary Wing

Pending the question of determining the coefficients A_n we proceed to obtain expressions for the forces and moments on the wing. The lift and drag of a slice of the wing are given by (19.14) and (19.15). Using the relation

$$y = \frac{b}{2} \cos \theta$$

we rewrite these equations in terms of θ. We thus have

$$\delta L(\theta) = -\rho V_\infty \frac{b}{2} \Gamma(\theta) \sin \theta \, d\theta$$

and

$$\delta D_i(\theta) = -\rho \frac{b}{2} w(\theta) \Gamma(\theta) \sin \theta \, d\theta$$

Substituting into these equations the expressions (19.66) and (19.67) for $\Gamma(\theta)$ and $w(\theta)$ we obtain

$$\delta L(\theta) = -\rho V_\infty{}^2 b^2 \sum_1^\infty A_n \sin n\theta \sin \theta \, d\theta \tag{19.68}$$

$$\delta D_i(\theta) = -\rho V_\infty{}^2 b^2 \sum_1^\infty A_n{}^2 \sin^2 n\theta \, d\theta \tag{19.69}$$

These equations give the *spanwise distribution of lift and drag.*

Integration of these equations from $\theta = \pi$ to $\theta = 0$ yields the total lift and drag of the wing. We obtain

$$L = \int_\pi^0 \delta L(\theta) = \rho V_\infty{}^2 b^2 \frac{\pi}{2} A_1 \tag{19.70}$$

$$D_i = \int_\pi^0 \delta D_i(\theta) = \rho V_\infty{}^2 b^2 \frac{\pi}{2} \sum_1^\infty n A_n{}^2 \tag{19.71}$$

The lift and drag coefficients for the wing are therefore given by

$$C_L \equiv \frac{L}{\frac{1}{2}\rho V_\infty{}^2 S} = \pi \frac{b^2}{S} A_1 = \pi \mathcal{R} A_1 \tag{19.72}$$

$$C_{D_i} \equiv \frac{D_i}{\frac{1}{2}\rho V_\infty{}^2 S} = \pi \mathcal{R} \sum_1^\infty n A_n{}^2 = \mathcal{R} A_1{}^2 \left[1 + \sum_2^\infty n \left(\frac{A_n}{A_1} \right)^2 \right]$$

$$= \pi \mathcal{R} A_1{}^2 (1 + \delta) = \frac{c_L{}^2}{\pi \mathcal{R}} (1 + \delta) \tag{19.73}$$

where

$$\delta \equiv \sum_2^\infty n \left(\frac{A_n}{A_1} \right)^2 \tag{19.74}$$

We observe that while the lift is determined by the coefficient A_1 only, the drag and the distributions of the lift and drag are determined by all the coefficients A_n. The dependence of C_L on the angle of attack is through the coefficient A_1 and that of C_{D_i} on the angle of attack is through all the coefficients A_n. To bear this in mind, for a given wing we write

$$C_L = \pi \mathcal{R} A_1(\bar{\alpha}) \tag{19.75}$$

and

$$C_{D_i} = \frac{C_L^2}{\pi \mathcal{R}} [1 + \delta(\bar{\alpha})] \tag{19.76}$$

where $\bar{\alpha}$ is a representative angle of attack for the wing, say, for example, the angle of attack at the midsection. Recall that the drag coefficient for a wing with an elliptic lift distribution is equal to $C_L^2/\pi \mathcal{R}$. Equation (19.76) describes the polar curve (i.e., the curve of C_L versus C_{D_i}) for an arbitrary wing. Since δ depends on the angle of incidence, it is not, in general, a constant, for C_L and C_{D_i} vary with the angle of incidence. Therefore the polar curve for an arbitrary wing is not, in general, a parabola. This is in contrast to the parabolic polar curve for a wing with an elliptic lift distribution.

The rolling and yawing moments on the wing are expressed by (19.18) and (19.19). In terms of the variable θ they become

$$M_R = -\rho V_\infty \frac{b^2}{8} \int_0^\pi \Gamma(\theta) \sin 2\theta \, d\theta \tag{19.77}$$

$$M_y = \rho \frac{b^2}{8} \int_0^\pi \Gamma(\theta) w(\theta) \sin 2\theta \, d\theta \tag{19.78}$$

Substituting in these for $\Gamma(\theta)$ and $w(\theta)$ from (19.66) and (19.67) and carrying out the integrations we find

$$M_R = -\frac{\pi}{8} \rho V_\infty^2 b^3 A_2 \tag{19.79}$$

$$M_y = \frac{\pi}{8} \rho V_\infty^2 b^3 \sum_1^\infty (2n + 1) A_n A_{n+1} \tag{19.80}$$

The corresponding moment coefficient are given by

$$C_{MR} \equiv \frac{M}{\frac{1}{2}\rho V_\infty^2 S \bar{c}} = -\frac{\pi}{4} (\mathcal{R})^2 A_2 \tag{19.81}$$

$$C_{M_y} \equiv \frac{M_y}{\rho V_\infty^2 S \bar{c}} = \frac{\pi}{4} (\mathcal{R})^2 \sum_1^\infty (2n + 1) A_n A_{n+1} \tag{19.82}$$

Bear in mind that all the coefficients A_n depend on the quantities $a_0(y)$, $c(y)$, $\alpha(y)$, and b.

19.6   Question of the Smallest Drag

We now ask for the conditions under which the smallest value of the induced drag is obtained for a given lift coefficient and aspect ratio of the wing, that is, *for a given value of the coefficient* A_1 (see 19.75). From (19.76) for the drag coefficient it follows that since δ is never negative, in view of (19.74), the drag is smallest when the coefficients A_n vanish for all n greater than one, that is,

$$A_n = 0 \quad \text{for} \quad n > 1 \tag{19.83}$$

When this condition is fulfilled, the distributions of circulation and downwash, according to (19.66) and (19.67), become

$$\Gamma(\theta) = 2bV_\infty A_1 \sin \theta$$

or

$$\Gamma(y) = 2bV_\infty A_1 \sqrt{1 - \frac{2y^2}{b}}$$

and

$$w[y(\theta)] = V_\infty A_1, \text{ a constant}$$

This shows that the distribution of the circulation or of lift is elliptic and that the downwash is uniform along the span. We thus conclude *that the smallest drag occurs for an elliptic lift distribution.*

The geometry of the wing for minimum drag is determined according to Eq. (19.65), which now becomes

$$A_1[\mu(\theta) + \sin \theta] = \mu(\theta)\alpha(\theta)$$

Putting $a_0 c/4b$ for μ and reverting to the variable y we rewrite this equation as

$$\frac{a_0(y)c(y) + 4b\sqrt{1 - (2y^2/b)}}{a_0(y)c(y)\alpha(y)} = \frac{1}{A_1} = \text{a constant}$$

This is identical with (19.30). Note that A_1 is equal to $\Gamma_0/2bV_\infty$. The possible wing shapes that satisfy this equation have been considered previously in Section 19.3. For an untwisted wing this equation demands that the plan form must be elliptic. We thus conclude that *for an untwisted wing, an elliptic distribution of the chord yields the smallest induced drag for a given lift coefficient and aspect ratio.*

19.7   Determination of the Coefficients A_n: Methods of Glauert and Lotz

According to Prandtl's theory, the solution for the problem of the arbitrary wing is complete once the coefficients A_n are determined from

(19.65), which is

$$\sum_{1}^{\infty} A_n \sin n\theta[n\mu(\theta) + \sin\theta] = \mu(\theta)\alpha(\theta)\sin\theta$$

where

$$\mu(\theta) \equiv \frac{a_0(\theta)c(\theta)}{4b}$$

and θ varies from 0 to π. This is an unusual equation. There are an infinite number of unknowns, and they have to satisfy (19.65) for all values of the continuous variable θ from 0 to π. Several methods have been developed to obtain approximate solutions for the problem. We consider here two such methods.

Method of Glauert. Suppose that the first N coefficients $A_1, \ldots, A_N$ are sufficient to describe the distributions of the circulation and downwash to the desired accuracy (see Eqs. 19.66 and 19.67). We then have only N unknowns. To obtain a determinate system of N equations for these N unknowns we content ourselves with the approximation that it is sufficient to satisfy (19.65) only for N suitably chosen values of θ, say $\theta_1, \ldots \theta_N$. In this way we obtain a system of N linear equations for the N unknowns. This system may be expressed by

$$\sum_{k=1}^{N} A_k \sin k\theta_s[k\mu(\theta_s) + \sin\theta_s] = \mu(\theta_s)\alpha(\theta_s), \qquad s = 1, 2, \ldots, N$$

This scheme was developed by Glauert and was used by him for evaluating the properties of different wings.

Method of Irmgard Lotz. The above method* of solving for the A_n's has a serious inconvenience. Each time one attempts to improve the accuracy of the results by taking more terms in the series for the circulation, the whole calculation for the coefficients needs to be repeated. Also, as the number of the coefficients is increased, the construction of the solution becomes increasingly laborious. To avoid these disadvantages Irmgard Lotz devised a method of successive approximations where each new step in the solution is based on the results of the preceding steps. We now describe her method briefly.

Lotz's method is based on the Fourier analysis of the distributions of the chord $c(y)$, the angle of attack $\alpha(y)$, and the sectional lift slope $a_0(y)$. The starting point is again (19.65), which is expressed as

$$\sum_{n=1}^{\infty} nA_n \sin n\theta + \sum_{n=1}^{\infty} A_n \frac{\sin\theta}{\mu(\theta)} \sin n\theta = \alpha(\theta)\sin\theta \qquad (19.84)$$

* Also, some other methods that have been devised.

where θ varies from 0 to π. Recall that

$$\mu(\theta) = \frac{a_0(\theta)c(\theta)}{4b}$$

We now introduce the following expansions:

$$\frac{\sin \theta}{\mu(\theta)} = \sum_{\nu=0}^{\infty} B_\nu \cos \nu\theta, \qquad 0 \leq \theta \leq \pi \qquad (19.85)$$

$$\alpha(\theta) \sin \theta = \sum_{n=1}^{\infty} \alpha_n \sin n\theta, \qquad 0 \leq \theta \leq \pi \qquad (19.86)$$

The coefficients β_ν and α_n may be expressed, following the usual method, in terms of the distributions $\mu(\theta)$ and $\alpha(\theta)$, respectively. On substituting the expressions (19.85) and (19.86) into (19.84) we obtain

$$\sum_{n=1}^{\infty} nA_n \sin n\theta + \frac{1}{2} \sum_{n=1}^{\infty} \sum_{\nu=0}^{\infty} A_n\beta_\nu[\sin (n + \nu)\theta + \sin(n - \nu)\theta] = \sum_{n=1}^{\infty} \alpha_n \sin n\theta$$

$$(19.87)$$

Equating the coefficients of $\sin n\theta$ from both sides of this equation we obtain the following system of equations for the coefficients A_n:

$$\tfrac{1}{2}A_1(\beta_{n-1} - \beta_{n+1}) + \cdots + \tfrac{1}{2}A_{n-1}(\beta_1 - \beta_{2n-1})$$
$$+ A_n(\beta_0 - \tfrac{1}{2}\beta_{2n} + n) + \tfrac{1}{2}A_{n+1}(\beta_1 - \beta_{2n+1})$$
$$+ \tfrac{1}{2}A_{n+2}(\beta_2 - \beta_{2n+2}) + \ldots \text{ ad infinitum}$$
$$= \alpha_n, \qquad n = 1, 2, \ldots, \infty \quad (19.88)$$

An approximate solution for the coefficients is then obtained as follows. It is assumed that in the nth equation of the system the coefficients of the terms A_{n+1}, A_{n+2}, and so forth, decrease rather rapidly. On the basis of such an assumption (which may be verified *a posteriori*), the system (19.88) is approximated by the following determinate system:

$$\tfrac{1}{2}A_1(\beta_{n-1} - \beta_{n+1}) + \cdots + \tfrac{1}{2}A_{n-1}(\beta_1 - \beta_{2n-1})$$
$$+ A_n(\beta_0 - \tfrac{1}{2}\beta_{2n} + n) = \alpha_n, \qquad n = 1, 2, 3, \ldots, \infty \quad (19.89)$$

This system is readily solved by iteration. The equation for A_1 contains only A_1 and hence may be solved immediately. The equation for A_2 contains only A_1 and A_2 and, since A_1 is already known, A_2 may be readily obtained. In this manner the coefficients A_1, A_2, and so forth may be determined successively.

The accuracy of the coefficients A_1, A_2, etc., may be improved by the process of successive approximation. Denote by $A_n^{(1)}$ and $A_n^{(2)}$ the first and second approximations to the coefficients A_n. Then $A_n^{(1)}$ are the solutions of the system (19.89). The second approximation $A_n^{(2)}$ is then the solution of the system:

$$\tfrac{1}{2}A_1^{(2)}(\beta_{n-1} - \beta_{n+1}) + \cdots + \tfrac{1}{2}A_{n-1}^{(2)}(\beta_1 - \beta_{2n-1})$$
$$+ A_n^{(2)}(\beta_0 - \tfrac{1}{2}\beta_{2n} + n) = \alpha_n - \tfrac{1}{2}A_{n+1}^{(1)}(\beta_1 - \beta_{2n+1})$$
$$- \tfrac{1}{2}A_{n+2}^{(1)}(\beta_2 - \beta_{2n+2})$$
$$- \cdots \text{ ad infinitum}$$

$$(19.90)$$

This system can again be solved by iteration. Higher approximations for the A_n's may be constructed by repeating the same procedure.

It is seen that the method of Lotz has the great advantage of improving successively the accuracy of the solution for the A_n's to any desired degree. The ideas of Lotz served as the basis for the developments of further practical schemes to obtain the spanwise distribution of lift over an arbitrary wing. One well-known scheme is the numerical method of Multhopp. For applications of the methods of Glauert, Lotz, and Multhopp, reference should be made to the literature.*

* In particular, consult the following where further references may be found: Glauert (1947), Mises (1959), Kármán and Burgers (1935), Thwaites (1960).

Chapter 20

Elements of the Theory for the Flow Past a Slender Body of Revolution

The problem of solving for the flow of an ideal fluid past a body of revolution, which may be at an angle of attack, was discussed in Chapter 11 (see Section 11.10). It was then pointed out that it is difficult to obtain a solution for the exact problem in closed form and that an approximate closed form solution may be obtained, however, if the body is a so-called slender body of revolution and the angle of attack is small. In this chapter we consider the development of such a solution. The corresponding theory is known as the slender body theory for the flow past a body of revolution.

The results of the simplified theory are of practical significance only for certain flow conditions. The results, however, give rise to significant concepts that form the basis of the so-called *slender body theory*. This theory is an approximate theory for the flow past a sufficiently slender body of any cross-sectional shape.

It is well to bear in mind the difficulty with the choice of a proper theoretical model for the actual flow of a real fluid past a body of revolution at an angle of attack.

In the actual case, the flow separates from the body and extends from it in the form of vortex sheets and vortices. There is no clear indication as to how such a flow field may be represented uniquely by an ideal-fluid flow field using distribution of vortex sheets or vortices.

20.1 Formulation of the Problem in Terms of the Perturbation Field

We consider the steady flow past a body of revolution that is at an angle of attack to the direction of the undisturbed stream (see Fig. 20.1). We choose the origin of the coordinate system at the nose (or leading edge) of the body and the X-axis along the axis of the body, the positive direction of the X-axis being that from the leading edge to the trailing edge. The direction of the undisturbed velocity (which we denote as usual by $\mathbf{V}_\infty$) and the axis of the body define a plane. We choose this plane as the x,z-plane and dispose of the Y- and Z-axes as shown in the figure. In analyzing the flow past a body of revolution it is convenient to use cylindrical coordinates r, θ, x as shown in the figure. Then the surface of the body is described by an equation of the form

$$r = R(x) \tag{20.1}$$

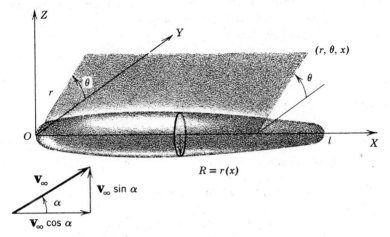

Fig. 20.1 Coordinates for flow past a body of revolution.

This may also be expressed as

$$F(r, \theta, x) = r - R(x) = 0 \tag{20.2}$$

The mathematical problem, in terms of the velocity potential $\Phi = \Phi(r, \theta, x)$, for the flow past the body is

$$\nabla^2 \Phi = 0 \tag{20.3}$$

$$\operatorname{grad} \Phi \cdot \operatorname{grad} F = \mathbf{V} \cdot \operatorname{grad} F = 0 \text{ on } F(r, \theta, x) = 0 \tag{20.4}$$

$$\operatorname{grad} \Phi = \mathbf{V}_\infty \text{ at infinity} \tag{20.5}$$

We now rewrite the problem in terms of the perturbation field due to the body. Let us denote by $\mathbf{q} = \mathbf{q}(r, \theta, x)$ the perturbation velocity and by $\phi = \phi(r, \theta, x)$ the perturbation potential. We then have

$$\operatorname{grad} \Phi = \mathbf{V}_\infty + \mathbf{q} = \mathbf{V}_\infty + \operatorname{grad} \phi \tag{20.6}$$

The differential equation (20.3) then takes the form

$$\nabla^2 \phi = 0 \tag{20.7}$$

The surface condition (20.4) may be rewritten as

$$(\mathbf{V}_\infty + \mathbf{q}) \cdot \text{grad } F = 0 \text{ on } F(r, \theta, x) = 0 \tag{20.8}$$

The infinity condition (20.5) becomes

$$\text{grad } \phi = \mathbf{q} = 0 \text{ at infinity} \tag{20.9}$$

To express the equations (20.7) to (20.9) explicitly in terms of the chosen coordinate system we denote the components of $\mathbf{q}$ by u_r, u_θ, u_x in the directions $\mathbf{e}_r$, $\mathbf{e}_\theta$, $\mathbf{e}_x$. We then have

$$\frac{\partial \phi}{\partial r} = u_r, \quad \frac{1}{r}\frac{\partial \phi}{\partial \theta} = u_\theta \quad \frac{\partial \phi}{\partial x} = u_x \tag{20.10}$$

The components of V_∞ are expressed by

$$\mathbf{V}_\infty = (V_\infty \sin \alpha \sin \theta, \; V_\infty \sin \alpha \cos \theta, \; V_\infty \cos \alpha) \tag{20.11}$$

In cylindrical coordinates we have

$$\nabla^2 \phi \equiv \text{div}(\text{grad } \phi) = \text{div } \mathbf{q}$$

$$= \frac{1}{r}\left[\frac{\partial}{\partial r}(ru_r) + \frac{\partial}{\partial \theta}(u_\theta) + \frac{\partial}{\partial x}(ru_x)\right]$$

$$= \frac{\partial u_r}{\partial r} + \frac{u_r}{r} + \frac{1}{r}\frac{\partial u_\theta}{\partial \theta} + \frac{\partial u_x}{\partial x}$$

$$= \frac{\partial^2 \phi}{\partial r^2} + \frac{1}{r}\frac{\partial \phi}{\partial r} + \frac{1}{r^2}\frac{\partial^2 \phi}{\partial \theta^2} + \frac{\partial^2 \phi}{\partial x^2} \tag{20.12}$$

Equations 20.7 to 20.9 then take the following form:

$$\frac{\partial^2 \phi}{\partial r^2} + \frac{1}{r}\frac{\partial \phi}{\partial r} + \frac{1}{r^2}\frac{\partial^2 \phi}{\partial \theta^2} + \frac{\partial^2 \phi}{\partial x^2} = 0 \tag{20.13}$$

$$(V_\infty \sin \alpha \sin \theta + u_r)\frac{\partial F}{\partial r} + (V_\infty \cos \alpha + u_x)\frac{\partial F}{\partial x} = 0$$

$$\text{on } F(r, \theta, x) = 0 \tag{20.14}$$

$$u_r, u_\theta, u_x \text{ go to zero at infinity} \tag{20.15}$$

Here u_r, u_θ, u_x are to be expressed in terms of ϕ according to (20.10).

The solution of the mathematical problem represented by (20.13) to (20.15) may be expressed as the superposition of the solutions of two separate problems. To see this we proceed as follows. From (20.2) we obtain

$$\frac{\partial F}{\partial r} = 1$$

$$\frac{\partial F}{\partial x} = -\frac{dR}{dx} \tag{20.16}$$

The surface condition (20.14) may then be rewritten as

$$\frac{\partial \phi}{\partial r} [r = R(x), \theta, x] = u_r(R, \theta, x)$$

$$= (V_\infty \cos \alpha + u_x) \frac{dR}{dx} - V_\infty \sin \alpha \sin \theta$$

$$= (U + u_x) \frac{dR}{dx} - W \sin \theta, \qquad 0 \le x \le l \tag{20.17}$$

where we have set

$$V_\infty \cos \alpha \equiv U \tag{20.18}$$

$$V_\infty \sin \alpha \equiv W \tag{20.19}$$

Writing

$$u_x = u_{x_1} + u_{x_2} \tag{20.20}$$

we may express (20.17) as

$$\frac{\partial \phi}{\partial r}(R, \theta, x) = u_r(R, \theta, x)$$

$$= (U + u_{x_1}) \frac{dR}{dx} + \left(u_{x_2} \frac{dR}{dx} - W \sin \theta \right) \tag{20.21}$$

Let us write

$$\phi = \phi_1 + \phi_2$$

and set

$$\frac{\partial \phi_1}{\partial r}(R, \theta, x) = (U + u_{x_1}) \frac{dR}{dx} \tag{20.22}$$

$$\frac{\partial \phi_2}{\partial r}(R, \theta, x) = u_{x_2} \frac{dR}{dx} - W \sin \theta \tag{20.23}$$

We then readily see that (20.22) represents the surface condition for axisymmetric flow past the body of revolution (see Fig. 20.2), where the speed of the undisturbed stream is U, and that (20.23) represents the surface condition for lateral flow past the same body of revolution (see Fig. 20.3), where the speed of the undisturbed stream is W. In view of this, the nature of the infinity condition and the linearity of the governing differential equation, the *mathematical problem represented by the equations (20.13) to (20.15) may be expressed as the superposition of two problems:*

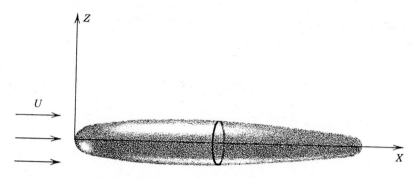

Fig. 20.2 Illustrating axisymmetric flow past the body of revolution.

(1) *the problem representing the axisymmetric flow past the body of revolution, and* (2) *the problem representing the lateral flow past the same body of revolution.* The speed of the undisturbed stream for the axisymmetric flow is given by (20.18), whereas that for the lateral flow is given by (20.19).

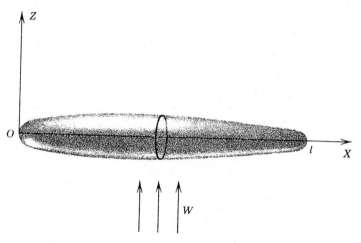

Fig. 20.3 Illustrating lateral flow past the body of revolution.

20.2 Boundary Condition for a Slender Body of Revolution

Consider the flow past a *slender body of revolution*. By such a body we mean an elongated body of revolution whose cross-sectional radius changes slowly with distance along the axis of revolution. We assume that the magnitude of the perturbation velocity components is small compared to the speed of the undisturbed stream. On the basis of these considerations we may neglect the term $u_x(\partial F/\partial x)$ from (20.14) or, equivalently, the term $u_x(dR/dx)$ from (20.17). The surface condition then takes the approximate form

$$\frac{\partial \phi}{\partial r}(R, \theta, x) = u_r(R, \theta, x)$$

$$= V_\infty \cos \alpha \frac{dR}{dx} - V_\infty \sin \alpha \sin \theta \qquad 0 \le x \le l \qquad (20.24)$$

This is known as the linearized form of the boundary condition for flow past a slender body of revolution. It is linear in the sense that $(\partial \phi / \partial x)$ (i.e., u_x) does not appear in the equation expressing it.

We immediately observe that the assumption of small perturbations is not valid at and near the stagnation points on the body. Furthermore, the assumption that the body is slender implies that $\partial F / \partial x$ is much less than $\partial F / \partial r$ or that dR/dx is much less than unity. This, however, is violated at and near the ends of the body if those ends are rounded. We should, therefore, expect that any solution obtained on the basis of the linearized surface condition (20.24) cannot be valid near stagnation points or rounded ends. These regions where the solution is not valid are, however, small.

In view of Eq. (20.24), the solution for the flow past a slender body of revolution at an angle of attack may be obtained as the superposition of the solutions of the following two mathematical problems:

Axisymmetric Flow Past a Slender Body of Revolution

$$\phi = \phi(r, x) \qquad (20.25)$$

$$\nabla^2 \phi = \frac{\partial^2 \phi}{\partial r^2} + \frac{1}{r}\frac{\partial \phi}{\partial r} + \frac{\partial^2 \phi}{\partial x^2} = 0 \qquad (20.26)$$

$$\frac{\partial \phi}{\partial r}(R, x) = u_r(R, x) = U\frac{dR}{dx} \qquad 0 \le x \le l \qquad (20.27)$$

where

$$U = V_\infty \cos \alpha$$
$$\nabla \phi = 0 \text{ at infinity} \qquad (20.28)$$

Cross Flow or Lateral Flow Past a Slender Body of Revolution

$$\phi = \phi(r, \theta, x) \qquad (20.29)$$

$$\nabla^2 \phi = \frac{\partial^2 \phi}{\partial r^2} + \frac{1}{r}\frac{\partial \phi}{\partial r} + \frac{1}{r^2}\frac{\partial^2 \phi}{\partial \theta^2} + \frac{\partial^2 \phi}{\partial x^2} = 0 \qquad (20.30)$$

$$\frac{\partial \phi}{\partial r}(R, \theta, x) = u_r(R, \theta, x) = -W\sin\theta \qquad 0 \le x \le l \qquad (20.31)$$

where

$$W = V_\infty \sin \alpha$$

$$\nabla \phi = 0 \text{ at infinity} \qquad (20.32)$$

We note that (20.27) and (20.31) are the linearized boundary conditions for their respective problems.

Recalling that in the problem of the thin airfoil we were able to transfer the boundary condition from the surface of the airfoil to its axis, we may inquire whether such a procedure is also feasible in the problem of the slender body of revolution. As we shall see later, the perturbation velocities in the problem of the body of revolution become infinite on its axis. This means that they cannot be expanded in Taylor series about the axis and that one cannot approximate their values on the surface of the body by their values on the axis. We thus conclude that the boundary condition (20.24) (or, equivalently, the conditions (20.27) and (20.31)) cannot be imposed on the axis of the body instead of on its surface. An alternative procedure for transferring the boundary condition to the axis in a modified form, however, becomes possible. We show this in its proper context.

20.3 Axisymmetric Flow Past a Slender Body of Revolution: Solution by Source Distribution

The mathematical problem is represented by (20.25) to (20.28). The angle of attack α is zero. Hence $V_\infty \equiv U$. On the basis of our past experience we know that the disturbance field in such a flow may be represented as that due to a suitable distribution of point sources along the axis of the body in the interval $0 \leq x \leq l$. Let $q = q(x)$ denote the source strength per unit length of such a distribution (Fig. 20.4). The disturbance potential and the disturbance velocity components at any field point r, θ, x are then given by

$$\phi(r, x) = -\frac{1}{4\pi} \int_0^l q(\xi) \frac{d\xi}{\sqrt{(x - \xi)^2 + r^2}} \qquad (20.33)$$

$$u_r(r, x) = \frac{\partial \phi}{\partial r} = \frac{1}{4\pi} \int_0^l q(\xi) \frac{r}{[(x - \xi)^2 + r^2]^{3/2}} d\xi \qquad (20.34)$$

$$u_x(r, x) = \frac{\partial \phi}{\partial x} = \frac{1}{4\pi} \int_0^l q(\xi) \frac{x - \xi}{[(x - \xi)^2 + r^2]^{3/2}} d\xi \qquad (20.35)$$

According to the boundary condition (20.27), the source distribution $q = q(x)$ must satisfy the integral equation

$$\left\{ \frac{1}{4\pi} \int_0^l q(\xi) \frac{r}{[(x - \xi)^2 + r^2]^{3/2}} d\xi \right\}_{r=R(x)} = u_r(r = R(x)) = U \frac{dR}{dx}$$

$$0 \leq x \leq l \quad (20.36)$$

To this we add the condition that since the body is closed, the total source strength should be zero:

$$\int_0^l q(x)\, dx = 0 \qquad (20.37)$$

Explicit determination of $q(x)$ from (20.36) and (20.37) is not possible. Numerical or approximate methods are to be employed to obtain $q(x)$. For

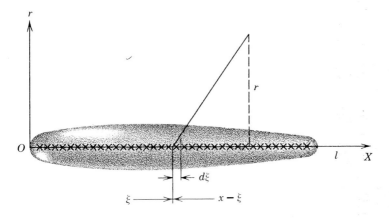

Fig. 20.4 Illustrating the determination of the disturbance field for the axisymmetric flow by source distribution.

a very slender body of revolution we may obtain a good approximation to the source distribution by employing the so-called *slender-body approximation* for the boundary condition.

To develop this approximation we first seek the value of the integral

$$u_r = \frac{\partial \phi}{\partial r} = \frac{1}{4\pi} \int_0^l q(\xi)\, \frac{r}{[(x - \xi)^2 + r^2]^{3/2}}\, d\xi$$

for small values of r, that is, as $r \to 0$. It is seen that care is necessary in obtaining this value of the integral, since for $r = 0$ at $\xi = x$ the integrand becomes indeterminate. As shown in Appendix E

$$(r u_r)_{r \to 0} = \frac{q(x)}{2\pi} \qquad (20.38)$$

This result may be obtained by an elementary argument. Consider, as shown in Fig. 20.5, a small cylinder of radius r and length dx, surrounding an element of the source distribution. Then as $r \to 0$, the volume outflow of fluid through the surface of the cylinder is equal to the volume of the fluid put out by the source element:

$$(2\pi r u_r\, dx)_{r\to 0} = q(x)\, dx$$

or

$$(r u_r)_{r\to 0} = \frac{q(x)}{2\pi}$$

This shows that for *small r* we have*

$$u_r \sim \frac{1}{r}, \qquad 0 \le x \le l$$

and hence u_r is infinite on the axis of the body in the interval $0 \le x \le l$. It appears, therefore, that in determining $q(x)$ from the surface condition

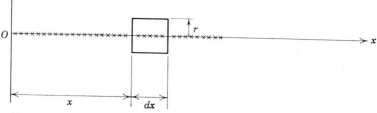

Fig. 20.5 Illustrating the determination of the source strength by considering the outflow through a cylinder enclosing an element of the source distribution.

(20.36), we should not attempt to approximate u_r at $r = R(x)$ by u_r at $r = 0$ in analogy with the transfer of the boundary condition from the surface to the axis of a thin airfoil. It is observed, however, that in the present case $r u_r$ is finite as $r \to 0$ provided $q(x)$ is finite. This being the case, we may express $r u_r$ at $r = R(x)$ as a Taylor series about the axis:

$$(r u_r)_{r=R(x)} = (r u_r)_{r=0} + \left[\frac{\partial}{\partial r}(r u_r)\right]_{r=0} R(x) + \left[\frac{\partial^2}{\partial r^2}(r u_r)\right]_{r=0} \frac{R^2(x)}{2} + \cdots$$

Neglecting the terms involving the derivatives and the powers of $R(x)$ we write

$$(r u_r)_{r=R(x)} = (r u_r)_{r=0} \tag{20.39}$$

In view of this equation, we express the boundary condition (20.27)

$$u_r(r = R(x)) = U\frac{dR}{dx} \qquad 0 \le x \le l$$

in the modified approximate form

$$(r u_r)_{r=0} = (r u_r)_{r=R(x)} = UR\frac{dR}{dx} \qquad 0 \le x \le l \tag{20.40}$$

* Note that this is unlike the behavior of v for $y \to 0$ in the case of uniform flow past a thin symmetrical airfoil at zero angle of attack.

In this form the boundary condition is imposed on the axis of the body instead of on its surface. Equation (20.40) is known as the *slender body approximation* for the boundary condition for the axisymmetric flow past a slender body of revolution.

From (20.38) and (20.40) we obtain

$$q(x) = 2\pi U R \frac{dR}{dx}$$

$$= U \frac{dS(x)}{dx} \equiv US'(x)$$

(20.41)

where $S(x) = \pi R^2(x)$ is the cross-sectional area of the body at x.

With the source strength thus determined, the perturbation potential and the perturbation velocity components may be obtained from Eqs. (20.33) to (20.35). We are particularly interested in their values near and on the body. They are obtained by constructing the values of the integrals in (20.33) to (20.35) for small values of r, that is, for $r \ll (l - x)$. We find that *for small r*

$$\phi(r, x) = 2f(x) \log \frac{r}{2} - \left[f(0) \log x + f(l) \log (l - x) \right.$$

$$\left. + \int_0^l f'(\xi) \log (x - \xi) \, d\xi \right]$$

(20.42)

$$u_x(r, x) = 2f'(x) \log \frac{r}{2} - \left[\frac{f(0)}{x} - \frac{f(l)}{l - x} \right.$$

$$+ f'(0) \log x + f'(l) \log (l - x)$$

$$\left. + \int_0^l f''(\xi) \log (x - \xi) \, d\xi \right]$$

(20.43)

and, as we already know,

$$u_r(r, x) = 2 \frac{f(x)}{r}$$

(20.44)

Here we used the following notation

$$f(x) \equiv \frac{q(x)}{4\pi} = \frac{U}{4\pi} S'(x) = \frac{U}{2} R \frac{dR}{dx}$$

(20.45)

$$f'(x) \equiv \frac{df}{dx}, \quad \text{etc}$$

We observe that the results (20.42) to (20.44) are not valid near the ends of the body. If the body is rounded at the ends, $f(0)$ and $f(l)$ become infinite. If we choose pointed ends, then $R(0)$ and $R(l)$ are zero, and consequently $f(0)$ and $f(l)$ also vanish. Usually a pointed body is assumed.

20.4   Cross Flow Past a Slender Body of Revolution: Solution by Doublet Distribution

Now the flow field is no longer axially symmetric. The mathematical problem for the disturbance potential is expressed by Eqs. (20.29) to (20.31)

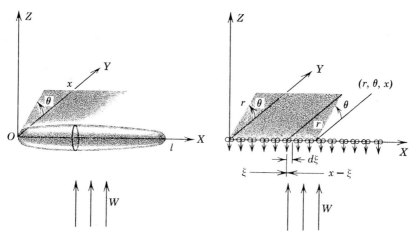

Fig. 20.6   Illustrating the determination of the cross flow by doublet distribution.

where we simply set $\alpha = \pi/2$. We shall retain W to denote the undisturbed stream.

We represent the disturbance field as that due to a suitable distribution of doublets along the axis of the body, with the axes of the doublets opposing the undisturbed stream (see Fig. 20.6). See Section 11.9. Let $\mu = \mu(x)$ denote the doublet strength per unit length of the distribution. As we know, the flow field due to such a doublet distribution satisfies Laplace's equation (20.30), and the infinity condition (20.32). It follows that the distribution $\mu(x)$ should be chosen so as to satisfy the surface condition (20.31). Once $\mu(x)$ is determined, all the desired information about the flow field may be found.

To express the perturbation potential and the perturbation velocity components in terms of $\mu(x)$ we first recall the pertinent results for a doublet **M** situated at the point **s** from the origin of coordinates (see Section 11.8). If **R** denotes the position vector from the origin to a field point, the potential and the velocity respectively, are, given by

$$\phi(\mathbf{R}) = -\frac{1}{4\pi} \frac{\mathbf{M} \cdot (\mathbf{R} - \mathbf{s})}{|\mathbf{R} - \mathbf{s}|^3} \qquad (20.46)$$

$$\mathbf{q}(\mathbf{r}) = -\frac{1}{4\pi} \text{grad}\left[\frac{\mathbf{M} \cdot (\mathbf{R} - \mathbf{s})}{|\mathbf{R} - \mathbf{s}|^3}\right] \qquad (20.47)$$

Equation (20.46) is (11.65). Now, consider the element $-\mathbf{e}_z \mu(\xi) \, d\xi$ of the doublet distribution in the present problem. We have $\mathbf{s} = \xi \mathbf{e}_x$. The potential at the field point $R = (r, \theta, x)$ due to this element is then given by

$$\delta\phi(r, \theta, x) = \frac{1}{4\pi} \mu(\xi) \frac{r \sin \theta}{[(x - \xi)^2 + r^2]^{3/2}} \, d\xi$$

Therefore the potential at the field point due to the whole doublet distribution is given by

$$\phi(r, \theta, x) = \frac{1}{4\pi} \int_0^l \mu(\xi) \frac{r \sin \theta}{[(x - \xi)^2 + r^2]^{3/2}} \, d\xi \qquad (20.48)$$

For the velocity components we obtain

$$u_r(r, \theta, x) = \frac{\partial \phi}{\partial r} = \frac{1}{4\pi} \int_0^l \mu(\xi) \frac{\sin \theta}{[(x - \xi)^2 + r^2]^{3/2}} \, d\xi$$

$$- \frac{3}{4\pi} \int_0^l \mu(\xi) \frac{r^2 \sin \theta}{[(x - \xi)^2 + r^2]^{5/2}} \, d\xi \qquad (20.49)$$

$$u_\theta(r, \theta, x) = \frac{1}{r} \frac{\partial \phi}{\partial \theta} = \frac{1}{4\pi} \int_0^l \mu(\xi) \frac{\cos \theta}{[(x - \xi)^2 + r^2]^{3/2}} \, d\xi \qquad (20.50)$$

$$u_x(r, \theta, x) = \frac{\partial \phi}{\partial x} = -\frac{3}{4\pi} \int_0^l \mu(\xi) \frac{(x - \xi) r \sin \theta}{[(x - \xi)^2 + r^2]^{5/2}} \, d\xi \qquad (20.51)$$

According to the boundary condition (20.31), the doublet distribution $\mu(x)$ must satisfy the integral equation

$$\left\{ \sin \theta \frac{\partial}{\partial r} \int_0^l \mu(\xi) \frac{r}{[(x - \xi)^2 + r^2]^{3/2}} \, d\xi \right\}_{r=R(x)} = u_r[R(x), \theta, x]$$

$$= -W \sin \theta \qquad 0 \leq x \leq l$$

$$(20.52)$$

Explicit determination of $\mu(x)$ from this equation is not possible. Numerical or approximate methods are to be employed to obtain $\mu(x)$. For a very slender body of revolution we may obtain a good approximation to the doublet strength by employing a so-called *slender body approximation* for the boundary condition.

To develop this approximation we first seek the value of u_r for small values of r, that is, as $r \to 0$. We find that

$$(r^2 u_r)_{r \to 0} = -\frac{\mu(x)}{2\pi} \sin \theta \qquad 0 \le x \le l$$

or
(20.53)

$$(u_r)_{r \to 0} = -\frac{\mu(x)}{2\pi} \frac{\sin \theta}{r^2} \qquad 0 \le x \le l$$

It is seen that although u_r becomes infinite on the axis of the body, $r^2 u_r$ is finite on that axis. Hence we may express $r^2 u_r$ on the surface of the body as Taylor series about the axis:

$$(r^2 u_r)_{r=R(x)} = (r^2 u_r)_{r=0} + \left[\frac{\partial}{\partial r}(r^2 u_r)\right]_{r=0} R(x) + \left[\frac{\partial^2}{\partial r^2}(r^2 u_r)\right]_{r=0} \frac{R^2(x)}{2} + \cdots$$

Neglecting the terms involving the derivatives, and the powers of $R(x)$ we may write approximately

$$(r^2 u_r)_{r=R(x)} = (r^2 u_r)_{r=0} \qquad (20.54)$$

In view of this equation, the boundary condition (20.52) may be expressed in the modified approximate form

$$(r^2 u_r)_{r=0} = (r u_r)_{r=R(x)} = -WR^2 \sin \theta \qquad 0 \le x \le l \qquad (20.55)$$

In this form, known as the slender body approximation, the boundary condition is imposed on the axis of the body instead of on its surface. From Eqs. (20.53) and (20.55) we obtain

$$\mu(x) = 2\pi WR^2(x)$$
$$= 2WS(x) \qquad (20.56)$$

where, as before, $S(x)$ is the cross-sectional area of the body of revolution.

With the strength of the doublet distribution thus determined, the perturbation potential and the perturbation velocity components may be obtained from Eqs. (20.48) to (20.51). Near and on the body they are given by the values for small r of the integrals in these equations. We find that for *small r*

$$\phi(r, \theta, x) = \frac{\mu(x)}{2\pi} \frac{\sin \theta}{r} \qquad (20.57)$$

$$u_\theta(r, \theta, x) = \frac{\mu(x)}{2\pi} \frac{\cos \theta}{r^2} \qquad (20.58)$$

$$u_x(r, \theta, x) = \frac{1}{2\pi} \frac{\sin \theta}{r} \frac{d\mu(x)}{dx} \qquad (20.59)$$

and, as we have already seen,

$$u_r(r, \theta, x) = -\frac{\mu(x)}{2\pi}\frac{\sin\theta}{r^2}$$

In these relations the expression given by (20.56) must be substituted for $\mu(x)$.

By combining the results of this section and the previous section, we may obtain the perturbation potential and the perturbation velocity components in the flow past a slender body of revolution at a small angle of yaw or attack. In combining these relations we replace U by $V_\infty \cos \alpha$ and W by $V_\infty \sin \alpha$.

20.5 Pressure Distribution

The pressure at any point is given by the Bernoulli equation

$$p = p_\infty + \frac{1}{2}\rho V_\infty{}^2\left[1 - \frac{V^2}{V_\infty{}^2}\right]$$

In terms of the perturbation field we have

$$p = p_\infty - \frac{1}{2}\rho V_\infty{}^2\left(2\frac{\mathbf{V}_\infty \cdot \mathbf{q}}{V_\infty{}^2} + \frac{q^2}{V_\infty{}^2}\right) \tag{20.60}$$

The pressure coefficient is given by

$$C_p \equiv \frac{p - p_\infty}{\frac{1}{2}\rho V_\infty{}^2} = -2\frac{\mathbf{V}_\infty \cdot \mathbf{q}}{V_\infty{}^2} - \frac{q^2}{V_\infty{}^2} \tag{20.61}$$

Axial Flow Past a Slender Body of Revolution. For axial flow we have

$$\mathbf{V}_\infty = U\mathbf{e}_x$$

$$\mathbf{V}_\infty \cdot \mathbf{q} = Uu_x$$

In terms of cylindrical coordinates r, θ, x introduced before, we obtain

$$C_p(r, \theta, x) = -2\frac{u_x}{U} - \frac{u_r{}^2 + u_x{}^2}{U^2} \tag{20.62}$$

To obtain the pressure distribution over the body we evaluate the right-hand side of this equation for $r = R(x)$ in the interval $0 \leq x \leq l$. From Eqs. (20.43) and (20.44) we find that near the body $(u_r/U)^2$ is of a magnitude comparable with that of u_x/U. This means that in any attempt at simplifying the pressure relation (20.62) we cannot neglect the term $(u_r/U)^2$, arguing that it involves the square of a perturbation velocity. We may, however, neglect the term $(u_x/U)^2$ in comparison with the rest. Thus the pressure on the body is given approximately by

$$C_p(R, \theta, x) = -2\left(\frac{u_x}{U}\right)_{r=R(x)} - \left(\frac{u_r^2}{U^2}\right)_{r=R(x)} \qquad 0 \le x \le l \quad (20.63)$$

Now, from the surface condition we have

$$\left(\frac{u_r}{U}\right)_{r=R(x)} = \frac{dR}{dx}$$

Hence (20.63) may be rewritten as

$$C_p(R, \theta, x) = -2\left(\frac{u_x}{U}\right)_{r=R(x)} - \left(\frac{dR}{dx}\right)^2 \qquad 0 \le x \le l \quad (20.64)$$

Lateral Flow Past a Slender Body of Revolution. For lateral flow we have

$$\mathbf{V}_\infty = W\mathbf{e}_z$$

$$\mathbf{V}_\infty \cdot \mathbf{q} = Wu_r \sin\theta + Wu_\theta \cos\theta$$

Equation (20.61) for the pressure distribution then takes the form

$$C_p(r, \theta, x) = -\frac{2}{W}(u_r \sin\theta + u_\theta \cos\theta) - \frac{u_r^2 + u_\theta^2 + u_x^2}{W^2} \quad (20.65)$$

To obtain the pressure distribution over the body, we first observe that, on the basis of Eqs. (20.58), (20.59), (20.53), and (20.56), we may neglect the term $(u_x/W)^2$ in (20.65) while retaining the rest. Thus the pressure distribution over the body is given approximately by

$$C_p(R, \theta, x) \doteq -\frac{2}{W}(u_r \sin\theta + u_\theta \cos\theta) - \frac{u_r^2 + u_\theta^2}{W^2} \qquad 0 \le x \le l \quad (20.66)$$

Flow Past a Slender Body of Revolution at an Angle of Yaw. Now we have

$$\mathbf{V}_\infty = V_\infty (\sin\alpha \sin\theta \mathbf{e}_r + \sin\alpha \cos\theta \mathbf{e}_\theta + \cos\alpha \mathbf{e}_x)$$

Equation (20.61) for the pressure at any point, therefore, takes the form

$$C_p(r, \theta, x) = -2\frac{u_x \cos\alpha}{V_\infty} - 2\frac{\sin\alpha}{V_\infty}(u_r \sin\theta + u_\theta \cos\theta)$$

$$- \frac{u_r^2 + u_\theta^2 + u_x^2}{V_\infty^2} \quad (20.67)$$

Any component of the perturbation velocity is now given by the sum of the corresponding components in the axial and lateral flows.

To obtain the pressure distribution over the body we note that the term $(u_x/V_\infty)^2$ in (20.67) may be neglected while retaining the rest. Thus the pressure distribution over the body is given approximately by

$$C_p(R, \theta, x) = -2 \frac{u_x \cos \alpha}{V_\infty} - 2 \frac{\sin \alpha}{V_\infty} (u_r \sin \theta + u_\theta \cos \theta)$$
$$- \frac{u_r^2 + u_\theta^2}{V_\infty^2} \qquad 0 \le x \le l \qquad (20.68)$$

We see that *this pressure distribution cannot be obtained by simply combining the pressure distributions of the axial and lateral flows.* This is so because now (unlike in the case of the thin airfoil) the pressure relation involves the squares of the velocity components.

Equation 20.68 may be rewritten in a form that will express the pressure on the body as made of two parts, one part which is the same as the pressure distribution in the axial flow and another part which involves the angle of yaw and the cross-flow velocity components only. To do this let us first rewrite (20.68) setting $\cos \alpha \simeq 1$ and $\sin \alpha \simeq \alpha$. We then have

$$C_p(R, \theta, x) = -2 \frac{u_x}{V_\infty} - 2 \frac{\alpha}{V_\infty} (u_r \sin \theta + u_\theta \cos \theta) - \frac{u_r^2 + u_\theta^2}{V_\infty^2}$$
$$0 \le x \le l \quad (20.69)$$

On the body according to the approximate boundary condition (20.24), we have

$$u_r(R, \theta, x) = V_\infty \frac{dR}{dx} - V_\infty \alpha \sin \theta, \qquad 0 \le x \le l \qquad (20.70)$$

Using this relation in (20.69), we obtain

$$C_p(R, \theta, x) = -2 \frac{u_x}{V_\infty} - \left(\frac{dR}{dx}\right)^2 + \alpha^2 \sin^2 \theta - 2\alpha \frac{u_\theta}{V_\infty} \cos \theta - \left(\frac{u_\theta}{V_\infty}\right)^2$$
$$0 \le x \le l \quad (20.71)$$

Now, express the u_x-component as

$$u_x = u_{x_1} + u_{x_2} \qquad (20.72)$$

where u_{x_1} is the component in the axial flow and u_{x_2} is the component in the lateral flow. Using (20.72) we rewrite (20.71) as

$$C_p(R, \theta, x) = C_{p_1}(R, \theta, x) + C_{p_2}(R, \theta, x) \qquad (20.73)$$

where

$$C_{p_1}(R, \theta, x) = -2 \frac{u_{x_1}}{V_\infty} - \left(\frac{dR}{dx}\right)^2 \qquad 0 \le x \le l \qquad (20.74)$$

$$C_{p_2}(R, \theta, x) = -2 \frac{u_{x_2}}{V_\infty} + \alpha^2 \sin^2 \theta - 2\alpha \frac{u_\theta}{V_\infty} \cos \theta - \left(\frac{u_\theta}{V_\infty}\right)^2$$
$$0 \le x \le l \quad (20.75)$$

We see that C_{p_1} is the pressure distribution in the axial flow (see 20.64). The component C_{p_2} represents the effects of the angle of yaw and involves α and the velocity components of the cross flow only.

The expression (20.75) for C_{p_2} may be put into a simpler form. Using (20.56), (20.58), (20.59) and setting $W \simeq V_\infty \alpha$, we reduce (20.75) to the form

$$C_{p_2}(R, \theta, x) = -4\alpha \frac{dR}{dx} \sin \theta + \alpha^2 (1 - 4\cos^2 \theta), \qquad 0 \le x \le l \quad (20.76)$$

20.6 Forces on the Body of Revolution

On the basis of the general considerations that lead to d'Alembert's paradox, we assert that the preceding results for the flow past a body of revolution, whether it is at an angle of attack or not, must lead to a zero force on the body. Although the total force on the body is zero, it is of interest to know the distribution of the force with distance along the axis of the body. We shall, therefore, set up the relevant expressions for calculating such a distribution.

Consider an element $\mathbf{n}\, dA$ of the surface of the body (see Fig. 20.7). The element is situated at the point R, θ, x. We have

$$dA = R\, d\theta\, ds$$

where ds is an element of the curve of intersection between the body surface and the plane $\theta = \text{constant}$. The force on the body is given by

$$\mathbf{F} = -\oiint_A p\mathbf{n}\, dA = -\iint_A p\mathbf{n}R\, d\theta\, ds \qquad (20.77)$$

where A denotes the surface of the body. We are usually interested in the components of this force with respect to the x, y, z-directions. Denoting these components respectively by F_x, F_y, F_z, we have

$$F_x = -\iint_A R p (\mathbf{n} \cdot \mathbf{e}_x)\, d\theta\, ds \qquad (20.78)$$

$$F_y = -\iint_A R p (\mathbf{n} \cdot \mathbf{e}_y)\, d\theta\, ds \qquad (20.79)$$

$$F_z = -\iint_A R p (\mathbf{n} \cdot \mathbf{e}_z)\, d\theta\, ds \qquad (20.80)$$

Drawing the outward normal[*] $\mathbf{n}$ as shown in Figure 20.7b, we denote by

[*] Note that in terms of the system r, θ, x we have

$$\mathbf{n} = \pm \frac{1}{|1 - (dR/dx)^2|^{1/2}} \left(\mathbf{e}_r - \frac{dR}{dx} \mathbf{e}_x \right)$$

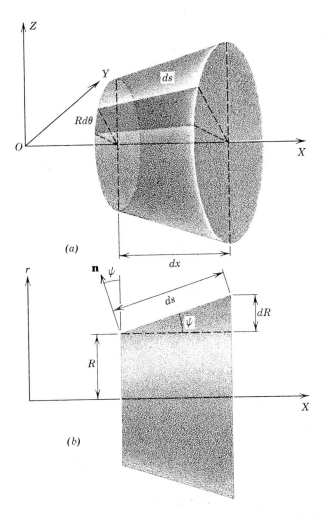

Fig. 20.7   Elemental area on the surface of the body of revolution.

ψ the angle between $\mathbf{n}$ and the direction of the coordinate r. We then have

$$\mathbf{n} = -\sin \psi \mathbf{e}_x + \cos \psi \cos \theta \mathbf{e}_y + \cos \psi \sin \theta \mathbf{e}_z$$

Also, we note that

$$ds \cos \psi = dx$$
$$ds \sin \psi = dR = \frac{dR}{dx} dx$$

On using these relations, Eqs. (20.78) to (20.80) become

$$F_x = \int_0^l \int_0^{2\pi} R \frac{dR}{dx} p \, d\theta \, dx$$

$$F_y = -\int_0^l \int_0^{2\pi} Rp \cos\theta \, d\theta \, dx$$

$$F_z = -\int_0^l \int_0^{2\pi} Rp \sin\theta \, d\theta \, dx$$

The distributions of the force components F_x, F_y, F_z are then given by

$$\frac{dF_x}{dx} = R \frac{dR}{dx} \int_0^{2\pi} p \, d\theta \tag{20.81}$$

$$\frac{dF_y}{dx} = -R \int_0^{2\pi} p \cos\theta \, d\theta \tag{20.82}$$

$$\frac{dF_z}{dx} = -R \int_0^{2\pi} p \sin\theta \, d\theta \tag{20.83}$$

We now rewrite these relations in terms of the pressure coefficient C_p. We have

$$p = q_\infty C_p + p_\infty \tag{20.84}$$

where

$$q_\infty = \tfrac{1}{2}\rho V_\infty^2$$

Using this relation, we rewrite (20.81) to (20.83) as

$$\frac{dF_x}{dx} = q_\infty R \frac{dR}{dx} \int_0^{2\pi} C_p \, d\theta + p_\infty \frac{dS}{dx} \tag{20.85}$$

$$\frac{dF_y}{dx} = -q_\infty R \int_0^{2\pi} C_p \cos\theta \, d\theta \tag{20.86}$$

$$\frac{dF_z}{dx} = -q_\infty R \int_0^{2\pi} C_p \sin\theta \, d\theta \tag{20.87}$$

where $C_p = C_p(R, \theta, x)$, $0 \leq x \leq l$. Note that S is πR^2.

For the flow past a body of revolution at a small angle of yaw, the pressure coefficient is given by (20.73):

$$C_p(R, \theta, x) = C_{p_1}(R, \theta, x) + C_{p_2}(R, \theta, x)$$

where C_{p_1} given by (20.74) represents the pressure for the axial part of the flow and C_{p_2} given by (20.76) represents the effect of the angle of yaw on the pressure. From (20.74) and (20.76), and (20.85) to (20.87) we obtain

$$\frac{dF_y}{dx} = 0 \qquad (20.88)$$

$$\frac{dF_x}{dx} = \left\{ p_\infty - q_\infty \left[\alpha^2 + \frac{u_{x_1}}{V_\infty} (R, x) + \left(\frac{dR}{dx} \right)^2 \right] \right\} \frac{dS}{dx} \qquad (20.89)$$

$$\frac{dF_z}{dx} = 2q_\infty \alpha \frac{dS}{dx} \qquad (20.90)$$

where $u_{x_1}(R, x)$ is known from Eq. (20.43). We see that both the axial and cross-flow parts contribute nothing to the distribution dF_y/dx. They both contribute to the distribution of the axial force, the contribution from the cross-flow part being $-q_\infty \alpha^2 (dS/dx)$. Only the cross-flow part contributes to the distribution of the normal force.

20.7   Moment on the Body of Revolution

Although there is no (resultant) force on the body of revolution, there is, in general, a moment on it. The moment about the origin of coordinates is given by

$$\mathbf{M} = -\oiint_A \mathbf{R} \times p\mathbf{n} \, dA$$

$$= -\oiint_A (\mathbf{R} \times \mathbf{n}) pR \, d\theta \, ds \qquad (20.91)$$

where $\mathbf{R}$ is the position vector to the surface element $\mathbf{n} \, dA$ at the point (R, θ, x). On the basis of this equation we find, as we expect in view of the symmetry of pressure distribution about the x,z-plane, that there is no (rolling) moment about the X-axis and there is no (yawing) moment about Z-axis:

$$M_x = 0, \qquad M_z = 0 \qquad (20.92)$$

The pitching moment M_y is given by

$$M_y = \int_0^l \int_0^{2\pi} R\left(x + R \frac{dR}{dx} \right) p \sin \theta \, d\theta \, dx$$

$$= q_\infty \int_0^l \int_0^{2\pi} R\left(x + R \frac{dR}{dx} \right) C_p \sin \theta \, d\theta \, dx \qquad (20.93)$$

Using (20.87) and (20.90) we obtain

$$
\begin{aligned}
M_y &= -\int_0^l \left(x + R\,\frac{dR}{dx} \right)\frac{dF_z}{dx}\,dx \\
&= -2q_\infty \alpha \int_0^l \left(x + R\,\frac{dR}{dx} \right)\frac{dS}{dx}\,dx \\
&= -2q_\infty \alpha \int_0^l \left[x\,\frac{dS}{dx} + \frac{1}{2\pi}\left(\frac{dS}{dx}\right)^2 \right] dx
\end{aligned}
\tag{20.94}
$$

Neglecting the term $(dS/dx)^2$ we have *approximately*

$$
\begin{aligned}
M_y &= -2q_\infty \alpha \int_0^l x\,\frac{dS}{dx}\,dx \\
&= -2q_\infty \alpha \left[xS(x)\Big|_0^l - \int_0^l S(x)\,dx \right]
\end{aligned}
\tag{20.95}
$$

Since $S(l)$ is zero and the integral $\int_0^l S(x)\,dx$ is the volume of the body, the pitching moment on the body is given approximately by

$$
M_y = 2q_\infty \alpha(\text{volume of the body})
\tag{20.96}
$$

It is customary to define a moment coefficient by the relation

$$
C_M \equiv \frac{M_y}{q_\infty(\text{volume of the body})}
\tag{20.97}
$$

Hence the moment coefficient for a slender body of revolution at a small angle of attack is given approximately by

$$
C_M = 2\alpha
\tag{20.98}
$$

Selected Problems

CHAPTER 2

2.1 Find the position vector of a point P that divides a line AB in the ratio $k:1$.

2.2 Find the vector equation of the straightline through two points A and B.

2.3 Find the vector equation of a plane in space.

2.4 Find the vector equation of the sphere with center at a point A and radius C.

2.5 Express vectorially the projection of a plane area $\mathbf{S}$ on some given plane.

2.6 Show that the vectorial sum of the areas of a tetrahedron is zero.

2.7 Using the vector equation for a line in space, find the condition for two lines to intersect.

2.8 Suppose we are given a line and a plane in space. If we visualize this situation we see that (a) the line and the plane meet in a single point, (b) the line lies parallel to, but not in the plane, or (c) the line lies entirely in the plane. Derive these results by using the vector equations of a line and a plane.

2.9 Let $\mathbf{r}_A$ and $\mathbf{r}_B$ be the radii vectors of two fixed points, A and B. Find the region of space for which the radius vector $\mathbf{r}_i$ of *any* point i belonging to this region satisfies the relation:

$$|\mathbf{r}_i - \mathbf{r}_A| = |\mathbf{r}_i - \mathbf{r}_B|$$

2.10 Consider the three free vectors $\mathbf{A}$, $\mathbf{B}$, and $\mathbf{C}$. Given that they satisfy the relation

$$\mathbf{A} + \mathbf{B} + \mathbf{C} = 0$$

determine how the vectors $\mathbf{A} \times \mathbf{B}, \mathbf{B} \times \mathbf{C}, \mathbf{C} \times \mathbf{A}$ are related and interpret geometrically their magnitude.

2.11 Find the free vector $\mathbf{R}$ for which

$$\mathbf{R} \cdot \mathbf{A} = C = \text{constant}$$

and

$$\mathbf{R} \times \mathbf{A} = \mathbf{k} = \text{constant vector}$$

where $\mathbf{A}$ and $\mathbf{k}$ are given free vectors and C is a given scalar. (Note. Your answer should be of the form $\mathbf{R} = $ function of $\mathbf{A}$, $\mathbf{k}$, and C.)

2.12 Let A_x, A_y, A_z be the components of $\mathbf{A}$ with respect to an orthogonal right-handed system $\mathbf{e}_x$, $\mathbf{e}_y$, $\mathbf{e}_z$. If this system is given a rotation, keeping the origin fixed, a new orthogonal system is obtained. Denote the new system by $\mathbf{e}_{x_1}$, $\mathbf{e}_{y_1}$, $\mathbf{e}_{z_1}$. If A_{x_1}, A_{y_1}, A_{z_1} are the components of $\mathbf{A}$ with respect to this new system, how are the components $A_{x_1}, A_{y_1}, A_{z_1}$ related to A_x, A_y, A_z and vice versa? (Note. The relations thus obtained define rules for transforming a set of three numbers i.e., the components, defining a vector under a rotation of the reference system of unit vectors. Such transformation rules can then be made the basis of the definition of a vector.)

2.13 Consider a rectangular Cartesian coordinate system (primed) that is a reflection of another Cartesian system (unprimed). Find the transformation relations.

2.14 Obtain the component form of the equation for the field lines in Cartesian, cylindrical, and spherical coordinates.

2.15 Consider a mass particle moving in space. Obtain the component form of the velocity and acceleration of the mass particle in (a) Cartesian; (b) cylindrical; (c) spherical; and, (d) "natural" or "intrinsic" coordinates. The natural coordinates are set up so that at each point of the space curve (traced by the moving particle) one of the reference unit vectors is in the direction of the tangent to the curve at the point considered; in other words, the space curve is one of the coordinate curves.

2.16 Consider a mass particle moving in space.

1. Obtain the component form of the velocity and acceleration of the particle in orthogonal curvilinear coordinates

$$q_1, q_2, q_3$$

2. Specialize the results in (1) for spherical coordinates.

2.17 A rigid body is rotating with a constant angular velocity $\boldsymbol{\omega}$. What is the acceleration of a point of the body? What is the acceleration if $\boldsymbol{\omega} = \boldsymbol{\omega}(t)$?

2.18 Express Newton's laws of motion in vector form.

2.19 Consider the motion of two interacting masses m_1 and m_2. The only forces that act on the masses are those of mutual interaction. The magnitude of the interaction force is a function of the distance between the masses and the direction of the force is always along the line that joins the masses. Using Newton's second law of motion, show that the motion of the masses takes place in a certain plane.

2.20 1. Show that the vector field of position $\mathbf{A} = r^n \mathbf{r}$ is irrotational for any value of the scalar n.

2. Find the value of n for which $\mathbf{A}$ is also solenoidal (divergenceless).

2.21 Using Cartesians prove the vector identity:

$$\nabla \times (\nabla \times \mathbf{A}) = \nabla(\nabla \cdot \mathbf{A}) - \nabla^2 \mathbf{A}$$
$$\nabla \cdot \mathbf{r} = 3$$
$$\nabla \times \mathbf{r} = 0$$
$$\nabla \cdot (\mathbf{U} \times \mathbf{V}) = \mathbf{V} \cdot \nabla \times \mathbf{U} - \mathbf{U} \cdot \nabla \times \mathbf{V}$$

2.22 Consider the functions (a) $\nabla(\mathbf{a} \cdot \mathbf{r})$, (b) $\nabla \cdot (\mathbf{a} \times \mathbf{r})$, (c) $\nabla \times (\mathbf{a} \times \mathbf{r})$, where $\mathbf{a}$ is a constant vector and $\mathbf{r}$ is the position vector of any point in space. Show that these functions are (at the most) only functions of $\mathbf{a}$ and find their value in terms of $\mathbf{a}$ (use Cartesian components).

2.23 Compute the values of the integrals: (a) $\oint d\mathbf{r}$; (b) $\oint \mathbf{r} \cdot d\mathbf{r}$.

2.24 Express in vector form Fourier's law of heat conduction (the rate of heat flow through a surface element is proportional to the spatial derivative of temperature in the direction of the normal to the surface element). Consider an element of volume fixed in a heat-conducting medium at rest. The time rate of change of heat energy in the volume element is equal to the inflow of heat into the element. Express this statement in vector form to give the equation of heat conduction.

2.25 According to the Newtonian law of gravitation, there is a force of attraction between any two mass particles, the magnitude of the force being proportional to the masses and inversely proportional to the square of the distance between the particles. Write symbolically the expression for the force. Take the proportionality constant to be unity; this can be done by choosing the unit of mass appropriately. Show that the force is irrotational. Consider now a particle of mass m that is stationary. The force exerted by m on any other mass particle of unit mass is said to be the gravitational force field of the mass m. Write symbolically the expression for the gravitational field. Since the field, denoted by $\mathbf{F}$, is irrotational, it may be represented as the gradient of a scalar field, say ϕ:

$$\mathbf{F} = \text{grad } \phi = \nabla\phi$$

Show that $\phi(\mathbf{r})$, where $\mathbf{r}$ is the distance from mass m, is the work done in bringing a unit mass from infinity by any path whatsoever to the point r. We call ϕ the potential at $\mathbf{r}$ due to a particle of mass m situated at $\mathbf{r} = 0$. Show further that ϕ obeys the equation

$$\nabla^2\phi = 0$$

except at $\mathbf{r} = 0$. Consider now a continuous distribution of matter in a region R bounded by a closed surface S. What is the potential at a field point P (a) when P is outside R and (b) when P is inside R? (In the second case, if you see any difficulty, make a meaningful definition.) What is the corresponding expression for the force $\mathbf{F}$? Show that

$$\nabla^2\phi\,(\mathbf{r}) = \text{div } \mathbf{F}(\mathbf{r}) = 0$$

if $\mathbf{r} \neq \mathbf{s}$, where $\mathbf{r}$ is a field point and $\mathbf{s}$ the location of a mass element. Using the integral definition of the divergence of a vector field (in this case $\mathbf{F}$) show that at points $\mathbf{r} = \mathbf{s}$

$$\nabla^2\phi = \text{div } \mathbf{F} = -4\pi\rho$$

where ρ is the density of the mass distribution. Hence the potential obeys Poisson's equation. Comment on the results obtained.

2.26 An electromagnetic medium at rest (i.e., not moving) is characterized by the following fields:*

$\mathbf{E} = \mathbf{E}(\mathbf{r}, t)$	known as electric field
$\mathbf{D} = \mathbf{D}(\mathbf{r}, t)$	known as electric excitation
$\mathbf{J} = \mathbf{J}(\mathbf{r}, t)$	known as conduction current
$\mathbf{C} = \dfrac{\partial \mathbf{D}}{\partial t} + \mathbf{J}$	known as total current
$\rho = \rho(\mathbf{r}, t)$	known as charge density (charge distribution per unit volume)
$\mathbf{B} = \mathbf{B}(\mathbf{r}, t)$	known as magnetic field
$\mathbf{H} = \mathbf{H}(\mathbf{r}, t)$	known as magnetic excitation

We define the following quantities:
Electromotive force. The circulation of $\mathbf{E}$ around a closed curve.

* In solving the problem no importance need be attached to the actual physical meaning of the various quantities.

Magnetomotive force. The circulation of **H** around a closed curve. The basic laws governing the behavior of such a medium are the following: (a) *Faraday's law of induction*: The time rate of change of the outflow of the magnetic field through any fixed open surface *S*, is equal and opposite to the electromotive force around the curve bounding the surface *S*. (b) *Ampere's law*: The outflow of the current through any fixed open surface *S* is equal to the magnetomotive force around the curve bounding the surface *S*. (c) *Gauss's law*: The outflow of the electric excitation through any closed surface is equal to the total charge enclosed by that surface.

1. Express the above three laws in vector (integral) form.
2. Reduce the resulting equations to differential form.
3. Using the differential form show that (a) The divergence of the magnetic field is divergenceless at all times. Hence div **B** = 0. (b) The divergence of the conduction current is equal and opposite to the time rate of change of the charge density.
4. Assume that

$$\mathbf{D} = \varepsilon\mathbf{E}, \ \mathbf{J} = \sigma\mathbf{E}, \ \mathbf{B} = \mu\mathbf{H}$$

where ε, σ, μ, respectively, are known as the dielectric constant, the conductivity and permeability of the medium. Using these relations and assuming that ε, σ, μ are constants (both in time and space), reduce the basic equations of part (2) to forms that involve only **E**, **H**, and ρ. (These are the well-known Maxwell's equations that describe the time and space behavior of electromagnetic fields in nonmoving media.)
5. From the equations in part (4) obtain a single equation governing the electric field. (*Hint.* Use the vector identity of Problem 2.21.)
6. Assuming further that there are no charges present and that the conductivity is zero, simplify the equation in part (5). (This is the equation of propagation of electric waves.)
7. Finally, supposing that the electric field is irrotational, show that the equation in (6) can be reduced to a single scalar equation of the same form.

CHAPTER 3

3.1 Obtain an expression for the resultant of the surfaces forces acting on an element of an elastic medium or of a viscous fluid in motion.
3.2 Write the equations of equilibrium of an element of an elastic medium.
3.3 Determine the distribution of hydrostatic pressure in a fluid of constant density under the action of a constant gravity field. Discuss the implications.
3.4 Determine the distribution of hydrostatic pressure in a compressible fluid subject to constant gravity. Assume a pressure-density relation of the form $p = $ constant ρ^n. Discuss the result. If the fluid is a perfect gas with the equation of state given by $p = \rho RT$, determine the temperature distribution.
3.5 A cylindrical drum containing a fluid of constant density rotates with a constant angular velocity about its axis, which is vertical. Determine the

pressure distribution and the shape of the free surface. The fluid surface at the top of the drum is open to the atmosphere.

3.6 Derive the equation of static equilibrium for a large mass of fluid whose separate parts are held together by gravitational attraction.

CHAPTER 4

4.1 Determine the streamlines and describe the flow field in each of the following cases:

(a)
$$V = U, \text{ a constant}$$

(b)
$$V = A \frac{e_r}{r^2} = A \frac{r}{r^3}$$

where A is a constant, r and e_r, respectively, are the magnitude and direction of position vector $\mathbf{r}$ from a fixed reference point to a field point.

(c)
$$V = U + A \frac{e_r}{r^2}$$

(d)
$$V = -\frac{1}{4\pi} \text{grad} \frac{\mathbf{B} \cdot \mathbf{r}}{r^3}$$

where $\mathbf{r}$ is the position vector as before and $\mathbf{B}$ is a constant.

(e)
$$V = U - \frac{1}{4\pi} \text{grad} \frac{\mathbf{B} \cdot \mathbf{r}}{r^3}$$

where U is also a constant and the direction of $\mathbf{B}$ is opposite to that of U.

4.2 Consider two-dimensional motion such that

$$\frac{\partial}{\partial z}(\ ) = 0 \quad \text{and} \quad w = 0$$

For such a motion determine the streamlines in each of the following cases; r, θ are polar coordinates in the plane of motion.

(a)
$$V = \frac{q}{2\pi} \frac{e_r}{r}$$

(b)
$$V = \frac{B}{2\pi} \frac{1}{r^2} (\cos \theta e_r + \sin \theta e_\theta)$$

(c)
$$V = Ui - \frac{B}{2\pi} \frac{1}{r^2} (\cos \theta e_r + \sin \theta e_\theta)$$

where $\mathbf{i}$ is the direction of the axis from which θ is measured.

(d)
$$V = K \frac{e_\theta}{r} \quad \text{where } K \text{ is a constant}$$

CHAPTER 5

5.1 Write the Lagrangian equations for one-dimensional motion of an ideal fluid.

5.2 Consider the motion of a compressible, viscous, heat-conducting fluid. Write the equations of constancy of mass, motion, and energy. Comment on the need for additional relations.

5.3 Consider the motion of an incompressible, viscous, heat-conducting fluid. Deduce the equations of mass, motion, and energy from the corresponding equations of a compressible fluid. Comment on the status of solving for the motion.

5.4 Consider the motion of an incompressible, viscous, heat-conducting fluid. Assume that the internal stresses are given by

$$\sigma_{ij} = -p\delta_{ij} + \tau_{ij}$$

where p is pressure, τ_{ij} is the viscous stress, and δ_{ij} is such that it is 0 if $i \neq j$ and equal to unity if $i = j$. Also, assume that

$$\tau_{ij} = 2\mu\varepsilon_{ij}$$

where μ is the viscosity coefficient and ε_{ij} is the rate-of-strain tensor given by

$$\varepsilon_{ij} = \frac{1}{2}\left(\frac{\partial u_i}{\partial x_j} + \frac{\partial u_j}{\partial x_i}\right)$$

where i or j may be 1, 2, or 3 and refer to the directions of a Cartesian coordinate system x_1, x_2, x_3. The velocity components are u_1, u_2, u_3. Assume further that the heat-flow vector $\mathbf{q}$ is related to the gradient of temperature T as follows:

$$\mathbf{q} = -k \text{ grad } T$$

where k is the coefficient of thermal conduction. Write the equation of motion and energy and comment on the status of solving for the motion.

5.5 Consider steady two-dimensional motion of a viscous incompressible fluid. Express the viscous stresses in terms of the velocity derivatives. Write the equation of motion in terms of the stream function.

5.6 Express the equations for an ideal fluid in terms of *streamline coordinates*.

5.7 Obtain the equations for an ideal fluid in terms of a reference frame which itself is translating and rotating.

CHAPTER 6

6.1 Write the integral form of the equations of change for mass, momentum, and energy for a viscous, heat-conducting, compressible fluid.

CHAPTER 8

8.1 Consider the motion of an inviscid compressible fluid. Discuss the possibility of integrating the equation of motion and obtain an integral when the motion is steady and rotational and when the motion is unsteady but irrotational.

CHAPTER 9

9.1 Show that acyclic irrotational motion is impossible if the fluid is bounded entirely by fixed rigid walls.

9.2 Show that acyclic irrotational motion is impossible if the fluid is at rest at infinity and bounded internally by fixed rigid walls.

9.3 Show that irrotational motion in a simply connected region has less kinetic energy than any other motion consistent with the same boundary conditions.

9.4 Show that for a compressible fluid in irrotational motion the velocity potential is given by $\int d\bar{\omega}/\rho$, where $\bar{\omega}$ is the impulsive pressure.

CHAPTER 10

10.1 Consider the acyclic irrotational motion generated in an ideal fluid by a translating and rotating rigid body. Extend to such a case the considerations given in Sections 10.5 through to 10.9.

10.2 Develop results corresponding to those given in Sections 10.5 through 10.11 for two-dimensional acyclic motion.

·10.3 The surface of a sphere immersed in an infinitely extending ideal fluid pulsates harmonically. Determine the velocity and pressure fields. Assume, if necessary, that the amplitude of the pulsation is small and that consequently the boundary condition may be applied at the mean position of the sphere surface.

10.4 A rigid sphere executes an oscillatory motion in an ideal fluid. Determine the virtual mass.

10.5 Determine the motion of an ideal fluid contained between a solid sphere and a fixed, concentric spherical boundary when the sphere moves with a given velocity. What is the apparent additional mass?

10.6 A spherical hole of radius a is suddenly formed in an ideal fluid that extends through all space. Determine the resulting motion.

10.7 An infinitely long circular cylinder moves perpendicular to its axis through an infinitely extending ideal fluid. Determine the flow field and the force on the cylinder. What is the additional apparent mass?

CHAPTER 11

11.1 Find the exact solution for steady irrotational flow past an ellipsoid of revolution. The direction of the undisturbed velocity is inclined to the axis of symmetry of the ellipsoid. Use semielliptic coordinates μ, ζ, θ, which are defined in terms of the cylindrical coordinates r, θ, x by the relations

$$x = l\mu\zeta, \qquad r = l(1 - \mu^2)^{1/2}(\zeta^2 - 1)^{1/2}, \qquad \theta = \theta$$

First show that the surfaces $\zeta = $ constant define a family of confocal ellipsoids of revolution; the surfaces $\mu = $ constant define hyperbolas in the $x - r$ plane. For $\zeta \to 1$ the ellipsoids become very elongated and for $\zeta \to \infty$ they approach a spherical shape.

11.2 Calculate the apparent additional mass for axial and transverse motion of an ellipsoid of revolution.

11.3 Find the steady flow past a sphere which is due to a source placed at a point exterior to the sphere. Find the force on the sphere.

11.4 Find the steady flow past a sphere which is due to a doublet placed at a point exterior to the sphere. The axis of the doublet is directed radially toward the sphere. Find the force on the sphere.

11.5 Point sources are distributed continuously along a circle, thus generating a *source ring*. Give integral expressions for the velocity components. Show that for a source distribution of constant strength the integrals can be reduced to complete elliptic integrals.

11.6 Consider the acyclic motion generated by a body of arbitrary shape. Express the kinetic energy of the fluid in terms of the singularities used to represent the body.

CHAPTER 12

12.1 Find the steady flow past a circle which is due to a source placed at a point exterior to the circle. Find the force on the corresponding circular cylinder.

12.2 Find the steady flow past a circle which is due to a doublet placed at a point exterior to the circle. The doublet axis is directed radially toward the circle. Find the force on the corresponding cylinder.

12.3 Find the steady two-dimensional flow past an elliptic cylinder by choosing appropriate coordinates, the undisturbed stream being along the major axis.

12.4 Find the two-dimensional flow generated by an expanding circular cylinder. Find the pressure distribution and the force on the cylinder.

12.5 Find the two-dimensional flow generated by a pulsating circular cylinder.

12.6 Find the two-dimensional flow generated by an oscillating circular cylinder.

12.7 Express the kinetic energy of the fluid for two-dimensional motion in terms of the stream function.

CHAPTER 14

14.1 Find the real and imaginary parts of the following

$$(1 + i)^3 \qquad \frac{1 + i}{1 - i} \qquad e^{i\pi/2} \qquad e^{2 + i\pi/4} \qquad \sin\left(\frac{\pi}{4} + 2i\right) \qquad \cosh\left(2 + \frac{i\pi}{4}\right)$$

14.2 Describe geometrically the regions given by the following

$$|Z| = 1 \qquad |\operatorname{Re}(Z)| \leq |Z| \qquad |\operatorname{Im}(Z)| \leq |Z|$$

$|Z - a| = p$, where a is a given complex number and p is a real positive number.

$$\left|\frac{Z - 1}{Z + 1}\right| = 1 \qquad \arg\left(\frac{Z - 1}{Z + 1}\right) = \text{a constant}$$

$\operatorname{Re}(Z) \geq 0 \qquad \operatorname{Re}(Z) < 0 \qquad \operatorname{Im}(Z) \leq 0 \qquad \operatorname{Im}(Z) > 0.$

14.3 Do the following satisfy the Cauchy-Riemann conditions? $Z, \bar{Z}, Z^2, Z^n,$ $e^Z, \sin Z, \cos Z, \tan Z.$

14.4 Discuss the multivaluedness of the function $A \log (Z - a)$.

14.5 Show that

$$\sin Z = \frac{e^{iZ} - e^{-iZ}}{2i}$$

$$\sinh Z = \frac{e^{Z} - e^{-Z}}{2}$$

$$\sin iZ = i \sinh Z$$

$$\sin iZ = i \sin Z$$

$$\tan^{-1} Z = \frac{i}{2} \log \frac{1 - iZ}{1 + iZ}$$

$$\tanh^{-1} Z = \frac{1}{2} \log \frac{1 + Z}{1 - Z}$$

14.6 Discuss the properties of the transformation $\zeta = Z^2$.

14.7 Discuss the properties of the transformation $\zeta = \log Z$.

14.8 Find the singularities of

$$\frac{Z}{(Z - a)(Z + b)^2}$$

and find the residues.

14.9 Evaluate the following integrals by considering the corresponding complex functions

$$\int_0^\infty \frac{\cos x}{x^2 + a^2} \, dx$$

$$\int_0^{2\pi} \frac{\cos n\theta}{\cos \theta - \cos \phi} \, d\theta$$

CHAPTER 15

15.1 Describe the flow given by $F(Z) = e^{Z}$.

15.2 Using complex variables, rework the problems (12.1) and (12.2). Give the expressions for the complex potential and the velocity in each case.

15.3 Find the acyclic flow past a flat plate when the undisturbed velocity is normal to the plane of plate. Sketch the streamlines and the pressure distribution.

15.4 Find the acyclic flow past a flat plate at an angle of attack. Sketch the streamlines and the pressure distribution.

15.5 Find the acyclic flow past an elliptic cylinder at an angle of attack. Sketch the streamlines and the pressure distribution.

15.6 Find the apparent masses of an elliptic cylinder.

15.7 Find the apparent mass of a flat plate moving normal to its plane.

15.8 A vortex is placed outside a circle. Find the flow field and the force.

15.9 A vortex is placed inside a circle. Find the flow field and the force.

15.10 A vortex is placed at a certain distance from the edge of a flat plate along the chord line. Find the flow field and the force.

15.11 Consider a two-dimensional channel formed by two parallel flat plates placed at a finite distance. Otherwise the plates extend to infinity. A vortex is placed midway between the plates. Find the flow field and the pressure distribution on the walls. Discuss the limiting case when the distance between the plates becomes very large.

15.12 Show that the complex velocity is itself an analytic function. Consequently the problem of flow past an arbitrary cylinder may be formulated directly in terms of the complex velocity. Give the details of such a formulation.

CHAPTER 16

16.1 Choose an arbitrary Joukowski airfoil at an angle of attack and compute the acyclic and cyclic flows. For the latter determine the circulation according to the Kutta condition. Plot the pressure distribution over the airfoil.

16.2 Choose an arbitrary airfoil at an angle of attack. Find the pressure distribution, using Theodorsen's method.

CHAPTER 17

17.1 Obtain the aerodynamic characteristics of a symmetric biconvex airfoil (composed of circular arcs) at zero angle of attack. Plot the distribution of the velocity components and the pressure over the airfoil surface.

17.2 Obtain the aerodynamic characteristics of a parabolic-arc airfoil placed at an angle of attack. Plot the distribution of the velocity components and the pressure over the airfoil.

17.3 Obtain the aerodynamic properties of a cubic-arc airfoil described by

$$n(x) = -gc\,\frac{2x}{c}\left(1 - \frac{2x^2}{c}\right)$$

Assume it is at zero angle of attack. Plot the velocity components and the pressure over the surface.

17.4 Consider a flapped flat plate airfoil, that is, an airfoil composed of two straight segments at an angle. The point of contact of the two segments is referred to as the hinge. Obtain the aerodynamic characteristics of such an airfoil when placed at an angle of attack. Compute the hinge moment coefficient. Obtain the derivatives of the hinge moment with respect to the angle of attack and the flap angle (i.e., the angle between the two segments). Exhibit the results graphically.

17.5 Obtain the camber line that gives a uniform lift distribution across the chord of the airfoil.

17.6 Consider the flat plate airfoil at an angle of attack. The general Blasius relations give only a lift for the resultant force on the plate. If we attempt to obtain the force by integration of the pressures on it, we are led to a normal force on the plate; thus there is an apparent drag in addition to lift. Discuss the origin of this contradiction and show by calculation how the drag component is canceled.

17.7 Formulate the thin airfoil theory using the method of conformal mapping.

17.8 Using complex variables, formulate the thin airfoil theory directly in terms of the complex velocity which itself is an analytic function.

CHAPTER 18

18.1 Describe the motion of a vortex pair (i.e., two infinitely long straight-vortex filaments parallel to each other) when its circulations are equal and in the same direction and when equal and in opposite directions.

18.2 Consider a vortex pair of equal and opposite circulations. Bring them together in such a way that as their separation goes to zero their strength increases indefinitely, thus keeping the product of the strength and separation distance constant. Such a limiting vortex pair is called a vortex doublet. Find the flow field of a vortex doublet and compare it with that of a doublet.

18.3 Generate the flow past a circular cylinder by suitable superposition of a uniform stream and a vortex doublet.

18.4 Find the path of a vortex bounded by two walls perpendicular to each other.

18.5 Find the flow field due to an infinite row of point vortices of equal strength distributed along a straight line at equal intervals.

18.6 Consider two parallel infinite rows of vortices. The magnitude of the circulations of all the vortices is the same. The sense of circulation of the vortices in one row is opposite that of the vortices in the other row. The vortices are placed at the same constant interval in both the rows. They are, however, arranged in such a way that a vortex in one row is midway between two corresponding vortices in the other row. This arrangement is known as Karman's vortex street. Determine the flow field.

18.7 A vortex filament is in the form of a circular ring (vortex ring). Determine the flow field.

18.8 A horseshoe vortex filament is formed by two parallel semi-infinite straight segments joined by another straight segment normal to them. Determine the flow field.

18.9 Consider a plane vortex sheet formed by a distribution of horseshoe vortex filaments. Determine the flow field.

CHAPTER 19

19.1 Choose a suitable plan form for a finite wing (other than those for which explicit results are given in the text) and compute the spanwise distributions of lift, drag, and moments.

19.2 A flat monoplane wing of elliptic plan form has flaps over the center half of the wing span. The flaps are deflected so that the angle from zero lift of this region is increased by δ. Thus, if α_0 is the angle of attack from zero lift of the basic wing,

$$\alpha = \alpha_0 + \delta \quad \text{over the center half}$$
$$= \alpha_0 \quad \text{over the tip quarters}$$

Assume that α_0 is constant across the span. Compute the ratio α_0/δ which gives zero lift for the whole wing.

19.3 Consider a thin finite wing, not necessarily of large aspect ratio at a small angle of attack. In steady flow past the wing assume that the plane vortex sheet with its normal perpendicular to the undisturbed stream follows behind the trailing edge. Formulate the problem for a linearized theory of the flow past the wing on lines similar to that of the thin airfoil.

CHAPTER 20

20.1 Find the force and moment distributions over a slender ellipsoid of revolution which is at a small angle of attack.

20.2 Find the approximate surface pressure distribution for an ellipsoid of revolution of fineness ratio 5 at zero angle of yaw. Compare it with the exact distribution.

References

The following list consists only of those works to which reference is made in the text. For a detailed list of references on incompressible aerodynamics see Thwaites (1960).

Apostle, Tom M., 1961, *Calculus*, Vol. I, Blaisdell, New York.

Birnbaum, W., 1923, Die Tragende Wirbelfläche als Hilfsmittel zur Behandlung des Ebenen Problems der Tragflügeltheorie. *Z. angew. Math. Mech.* **3,** 290.

Courant, R., 1934, *Calculus*, Vol. I, Interscience-Wiley, New York.

Dhawan, S., 1953, Direct Measurements of Skin Friction. *Rep. N.A.C.A.* 1121.
Dryden, H. E., and A. M., Kuethe, 1929, Effects of Turbulence in Wind Tunnel Measurements. *Rep. N.A.C.A.* 342.
Durand, W. F., 1934, *Aerodynamic Theory*, Vol. I, Springer-Verlag, New York.

Furhrmann, G. 1911, Widerstands-und Druckmessungen an Ballonmodellen. *Z. Flugtech.* **11,** 165.

Glauert, H., 1926, *The Elements of Aerofoil and Airscrew Theory*, 1st Edition, Cambridge University Press, Cambridge.
Glauert, H., 1947, *The Elements of Aerofoil and Airscrew Theory*, 2nd Edition, Cambridge University Press, Cambridge.
Goldstein, S., 1938, *Modern Developments in Fluid Dynamics*, Oxford University Press, New York.
Goursat, E., 1959, *Course in Mathematical Analysis*, Dover, New York.

Helmholtz, H. von. 1858, Über Integrale der Hydrodynamischen Gleichungen, Welche den Wirbelbewegungen Entsprechen. *J. reine angew. Math.* **55,** 25.
Hess, J. L., 1962, Calculation of Potential Flow about Bodies of Revolution, Having Axes Perpendicular to the Free-Stream Direction. *J. Aero. Sci.* **29,** 726.
Hess, J. L., and A. M. O. Smith, 1962, Calculation of Nonlifting Potential Flow about Arbitrary Three Dimensional Bodies. *Douglas Aircraft Co.*, Report No. ES 40622; also *J. Ship Research* **8,** No. 2, September 1964.
Hildebrand, F. B., 1949, *Advanced Calculus for Engineers*, Prentice-Hall, Englewood Cliffs, N.J.
Homann, F., 1936, Einfluss Grosser Zähigkeit bei Strömung um Zylinder. *Forsch. Arb. IngWes.* **7,** 1–10.

Joukowski, N. E., 1906, Sur les Tourbillons Adjoints. *Trans. Phys. Sec., Imp. Soc. Friends Natl. Sci. Moscow* **23.**

601

Kane, T., 1961, *Analytical Elements of Mechanics*, Vol. 2, Academic Press, New York.

Kármán, Th. von, 1927, Calculation of Pressure Distribution on Airship Hulls. *Tech. Memor, N.A.C.A.*, 574.

Kármán, Th. von, 1954, Aerodynamics, McGraw-Hill, New York.

Kármán, Th. von, 1958, The first Lanchester Memorial Lecture. *J. R. Aero. Soc.* **62,** 79.

Kármán, Th. von, and M. A. Biot, 1940, *Mathematical Methods in Engineering*, McGraw-Hill, New York.

Kármán, Th. von, and J. M. Burgers, 1935, General Aerodynamics Theory—Perfect Fluids. Section E in W. F. Durand (1935), *Aerodynamic Theory*, Vol. II, Springer-Verlag, New York.

Kelvin, 1869, *see* Thomson (1869).

Knopp, Konrad, 1952, *Elements of the Theory of Functions* and *Theory of Functions*, Part One, Dover, New York.

Kochin, N. E., I. A. Kibel, and N. V. Roze, 1964, *Theoretical Hydrodynamics*, Wiley, New York.

Kuethe, A. M., and J. D. Schetzer, 1959, *Foundations of Aerodynamics*, 2nd Edition, Wiley, New York.

Kutta, M. W., 1902, Auftriebskräfte in Strömenden Flüssigkeiten. *Ill. Aero. Mitt.* **6,** 133.

Lamb, H., 1932, *Hydrodynamics*, Cambridge University Press, Cambridge.

Liepmann, H. W., and S. Dhawan, 1951, 1953, Direct Measurement of Local Skin Friction in Low-speed and High-speed flow. *Proc. First. U.S. Natl. Congr. Appl. Mech.* 1951, *also see* S. Dhawan, (1953).

Lighthill, M. J., 1963, See Chapter I and II of *Laminar Boundary Layers*, edited by L. Rosenhead, Oxford University Press, New York.

Lotz, I. F., 1931, The Calculation of Potential Flow Past Airship Bodies in Yaw. *Tech. Memor, N.A.C.A.* 675.

Love, A. E. H., 1944, *A Treatise on the Mathematical Theory of Elasticity*, Dover, New York.

Margenau, H., and G. M. Murphy, 1956, *Mathematicas of Physics and Chemistry*, Vol. I, D. Von Nostrand, Princeton, N.J.

Mises, R. von, 1959, *Theory of Flight*, Dover, New York.

Munk, M. M., 1922, General Theory of Thin Wing Sections. *Rep. N.A.C.A.* 142

Munk, M. M., 1934, *Fluid Mechanics*, Part II. Section C in W. F. Durand (1934), *Aerodynamic Theory*, I. Springer-Verlag, New York.

Nikuradse, J., 1942, Laminare Reibungsschichten an der Längsangeströmten Platte. *Monograph, Zentrale f. Wiss. Berichtswesen*, Berlin.

Pankhurst, R. C., and Holder, D. W., 1952, *Windtunnel Techniques*, Pitman, New York.

Prandtl, L., 1904, Über Flussigkeitsbewegung Sehr Kleiner Reibung. *Verh. 3. Int. Math. Kongr.*, Heidelberg, Germany, **484.**

Prandtl, L., 1918, Tragflügeltheorie. *Nachr. Ges. Wiss. Göttingen*, 107 and 451.

Prandtl, L., 1925, Applications of Modern Hydrodynamics to Aeronautics. *Rep. N.A.C.A.* 116.

Prandtl, L., 1935, The Mechanics of Viscous Fluid. Section G in W. F. Durand (1935), *Aerodynamic Theory*, Vol. III. Springer-Verlag, New York.

Prandtl, L., and O. G. Tietjens, 1934, *Fundamentals of Hydro- and Aero-mechanics*, Vol. 1; *Applied Hydro- and Aero-mechanics*, Vol. 2, McGraw-Hill, New York.
Pritchard, J. L., 1957, The Dawn of Aerodynamics. *J. R. Aero. Soc.* **61,** 152.

Reynolds, O., 1883, An Experimental Investigation of the Circumstances Which Determine Whether the Motion of Water Shall be Direct or Sinuous, and the Laws of Resistance in Parallel Channels. *Phil. Trans. Roy. Soc. London,* **174.**

Schlichting, H., 1955, *Boundary Layer Theory*, McGraw-Hill, New York.
Smith, A. M. O., and J. Pierce, 1958, Exact Solutions of the Neumann Problem; Calculation of Noncirculatory Plane and Axially Symmetric Flows About or Within Arbitrary Boundaries. *Proc. 3rd. U.S. Natl. Congr. Appl. Mechanics.* Providence, R.I.
Sneddon, I., 1957, *Elements of Partial Differential Equations*, McGraw-Hill, New York.
Sommerfeld, A., 1950, *Mechanics of Deformable Bodies*, Academic Press, New York.

Theodorsen, T., 1932, Theory of Wing Sections of Arbitrary Shape, *Rep. N.C.A.A.* 411.
Thawites, B., Editor 1960, *Fluid Motion Memoirs Volume Incompressible Aerodynamics* Oxford University Press, New York.
Thomson, W. (Lord Kelvin), 1869, On Vortex Motion. *Transactions R. Soc. Edinburgh,* **25.**

Some Books

The following books, besides those cited in the list of references, are of interest.

Fluid Mechanics

Abbott, I. H., and A. E. von Doenhoff, 1949, *Theory of Wing Sections*, McGraw-Hill, New York (also Dover, New York).

Betz, A., 1935, Applied Airfoil Theory. Section J in W. F. Durand (1935), *Aerodynamic Theory*, Vol. IV. Springer-Verlag, New York.

Goldstein, S., 1960, *Lectures on Fluid Mechanics*, Interscience-Wiley, New York.

Landau, L. D., and E. M. Lifshitz, 1959, *Fluid Mechanics* Addison-Wesley, Reading, Mass.

Liepmann, H. W., and A. Roshko, 1957, *Elements of Gasdynamics*, Wiley, New York.

Milne-Thomson, L. M., 1962, *Theoretical Hydrodynamics*, 4th Edition, Macmillan, New York.

Prandtl, L., 1952, *Essentials of Fluid Dynamics*, Blackie, London.

Robertson, J. M., 1965, *Hydrodynamics in Theory and Application*, Prentice-Hall, Englewood Cliffs, N.J.

Robinson, A., and J. A. Laurmann, 1956, *Wing Theory*, Cambridge University Press, Cambridge.

Sedov, L. I., 1965, *Two-dimensional Problems of Hydrodynamics and Aerodynamics*, Wiley, New York.

Streeter, V. L., 1948, *Fluid Dynamics*, McGraw-Hill, New York.

Tietjens, O. G., 1960, *Strömungslehre*, Vol. 1, Springer-Verlag, New York.

Mathematics

Courant, R., 1948, *Theory of Functions of a Complex Variable*, Notes by A. A. Blank, New York University.

Hildebrand, F. B., 1962, *Advanced Calculus for Applications*, Prentice-Hall, Englewood Cliffs, N.J.

Karamcheti, K. (in publication), *Vector Analysis and Cartesian Tensors With Selected Applications*, Holden-Day, San Francisco.

Dimensional Analysis

Bridgman, P. W., *Dimensional Analysis*, Yale University Press, New Haven.

Sedov, L. I., 1959, *Similarity and Dimensional Methods in Mechanics*, Academic Press, New York.

Appendix A

Theorems of Linear
and Angular Momentum

Consider steady irrotational motion of an ideal fluid and assume that the effects of any body forces acting on the fluid are taken into account separately in terms of the hydrostatic pressure distribution that would exist under the action of the body forces. Let R denote a finite region of space enclosed by a fixed surface Σ. The equation of change for the momentum in R is

$$\rho \oiint_{\Sigma} \mathbf{V}(\mathbf{V} \cdot \mathbf{n}) \, dS + \oiint_{\Sigma} p\mathbf{n} \, dS = 0 \qquad (A.1)$$

where the pressure distribution is given by the Bernoulli equation

$$p = H - \frac{\rho}{2} V^2 \qquad (A.2)$$

H being a constant. Using (A.2), we express (A.1) in terms of the velocity as

$$\rho \oiint_{\Sigma} \mathbf{V}(\mathbf{V} \cdot \mathbf{n}) \, dS - \frac{\rho}{2} \oiint_{\Sigma} V^2 \mathbf{n} \, dS = 0 \qquad (A.3)$$

We refer to (A.1) or equivalently (A.3) as the theorem of linear momentum for steady irrotational motion.

An expression similar to the equation of change for linear momentum is that for the equation of change for angular momentum; that is,

the total time rate of increase of the angular momentum with respect to a point 0 of the fluid within R

= the net inflow per unit time of angular momentum through the surface enclosing R

+ the resultant of the moments about 0 of the pressure forces acting on the fluid in R across its surface, where the effects of any potential body forces are taken into account separately in terms of the hydrostatic pressure distribution. For steady motion we have

$$\rho \oiint_{\Sigma} \mathbf{r} \times \mathbf{V}(\mathbf{V} \cdot \mathbf{n}) \, dS + \oiint_{\Sigma} \mathbf{r} \times p\mathbf{n} \, dS = 0 \qquad (A.4)$$

Using (A.2) we express (A.4) in terms of the velocity as

$$\rho \oiint_{\Sigma} \mathbf{r} \times \mathbf{V}(\mathbf{V} \cdot \mathbf{n}) \, dS - \frac{\rho}{2} \oiint_{\Sigma} \mathbf{r} \times V^2 \mathbf{n} \, dS = 0 \qquad (A.5)$$

We refer to (A.4) or equivalently (A.5) as the theorem of angular momentum for steady irrotational motion.

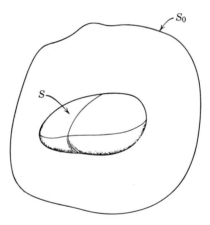

Fig. A.1

Consider now steady flow past a fixed rigid body. Let S denote the surface of the body and S_0 an arbitrary closed surface fixed in space and enclosing the surface S (Fig. A.1). It is usual to refer to S_0 as a *control surface*. Applying (A.1) in the region included between S and S_0, we obtain

$$\rho \oiint_{S} \mathbf{V}(\mathbf{V} \cdot \mathbf{n}) \, dS + \oiint_{S} p\mathbf{n} \, dS = -\rho \oiint_{S_0} \mathbf{V}(\mathbf{V} \cdot \mathbf{n}) \, dS - \oiint_{S_0} p\mathbf{n} \, dS \qquad (A.6)$$

Similarly, (A.4) applied in the region between S and S_0 yields

$$\rho \oiint_S \mathbf{r} \times \mathbf{V}(\mathbf{V} \cdot \mathbf{n}) \, dS + \oiint_S \mathbf{r} \times p\mathbf{n} \, dS$$

$$= -\rho \oiint_{S_0} \mathbf{r} \times \mathbf{V}(\mathbf{V} \cdot \mathbf{n}) \, dS - \oiint_{S_0} \mathbf{r} \times p\mathbf{n} \, dS \quad (A.7)$$

Since S is the surface of a fixed rigid body, $\mathbf{V} \cdot \mathbf{n}$ is zero on S. Consequently, the surface integrals over S involving $\mathbf{V} \cdot \mathbf{n}$ in (A.6) and (A.7) vanish. Furthermore, the force on the body is equal to

$$\mathbf{F} = \oiint p\mathbf{n} \, dS$$

and the moment on the body is equal to

$$\mathbf{M} = \oiint \mathbf{r} \times p\mathbf{n} \, dS$$

Note that $\mathbf{n}$ is the outward normal with respect to the region between S_0 and S.

It follows therefore that

$$\mathbf{F} = -\rho \oiint_{S_0} \mathbf{V}(\mathbf{V} \cdot \mathbf{n}) \, dS - \oiint_{S_0} p\mathbf{n} \, dS \quad (A.8)$$

and

$$\mathbf{M} = -\rho \oiint_{S_0} \mathbf{r} \times \mathbf{V}(\mathbf{V} \cdot \mathbf{n}) \, dS - \oiint_{S_0} \mathbf{r} \times p\mathbf{n} \, dS \quad (A.9)$$

Using (A.2), we express (A.8) and (A.9) in terms of the velocity as

$$\mathbf{F} = \frac{\rho}{2} \oiint_{S_0} V^2 \mathbf{n} \, dS - \rho \oiint_{S_0} \mathbf{V}(\mathbf{V} \cdot \mathbf{n}) \, dS \quad (A.10)$$

$$\mathbf{M} = \frac{\rho}{2} \oiint_{S_0} \mathbf{r} \times V^2 \mathbf{n} \, dS - \rho \oiint_{S_0} \mathbf{r} \times \mathbf{V}(\mathbf{V} \cdot \mathbf{n}) \, dS \quad (A.11)$$

Equations (A.8) and (A.10), and (A.9) and (A.11), respectively, relate the force and moment on the body to integrals over an arbitrarily chosen surface enclosing the body. They thus enable us to calculate the force and moment without a detailed knowledge of the flow field, in particular, of the flow in the immediate neighborhood of the body.

For steady two-dimensional motion the above force and moment relations take the following form

$$\mathbf{F} = -\rho \oint_{\mathscr{C}_0} \mathbf{V}(\mathbf{V} \cdot \mathbf{n})\, dl - \oint_{\mathscr{C}_0} p\mathbf{n}\, dl$$

$$= \frac{\rho}{2} \oint_{\mathscr{C}_0} V^2 \mathbf{n}\, dl - \rho \oint_{\mathscr{C}_0} \mathbf{V}(\mathbf{V} \cdot \mathbf{n})\, dl \tag{A.12}$$

and

$$\mathbf{M} = -\rho \oint_{\mathscr{C}_0} \mathbf{r} \times \mathbf{V}(\mathbf{V} \cdot \mathbf{n})\, dl - \oint_{\mathscr{C}_0} \mathbf{r} \times p\mathbf{n}\, dl$$

$$= \frac{\rho}{2} \oint_{\mathscr{C}_0} \mathbf{r} \times V^2 \mathbf{n}\, dl - \oint_{\mathscr{C}_0} \mathbf{r} \times \mathbf{V}(\mathbf{V} \cdot \mathbf{n})\, dl \tag{A.13}$$

where $\mathbf{F}$ and $\mathbf{M}$ are measured per unit length of the cylinder, $\mathscr{C}_0$ is an arbitrary curve enclosing the cylinder, and dl is an element of $\mathscr{C}_0$ (see Fig. A.2).

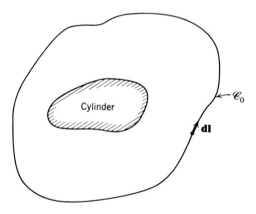

Fig. A.2

In Cartesians we write

$$\mathbf{F} = (F_x, F_y, 0)$$
$$\mathbf{r} = (x, y, 0)$$
$$\mathbf{V} = (u, v, 0)$$

Then we have

$$\mathbf{n} = \left(\frac{dy}{dl}, -\frac{dx}{dl}, 0 \right)$$

$$\mathbf{V} \cdot \mathbf{n}\, dl = u\, dy - v\, dx$$

(See Section 12.2, Eqs. 12.6 and 12.7.) The force and moment relations then become

$$F_x = -\rho \oint_{\mathscr{C}_0} u(u\, dy - v\, dx) - \oint_{\mathscr{C}_0} p\, dy$$

$$= \frac{\rho}{2} \oint_{\mathscr{C}_0} V^2\, dy - \rho \oint_{\mathscr{C}_0} u(u\, dy - v\, dx) \tag{A.14}$$

$$F_y = -\rho \oint_{\mathscr{C}_0} v(u\, dy - v\, dx) + \oint_{\mathscr{C}_0} p\, dx$$

$$= -\frac{\rho}{2} \oint_{\mathscr{C}_0} v^2\, dx - \rho \oint_{\mathscr{C}_0} v(u\, dy - v\, dx) \tag{A.15}$$

$$\mathbf{M} = -\mathbf{k}\left[\rho \oint_{\mathscr{C}_0} (vx - uy)(u\, dy - v\, dx) - \oint_{\mathscr{C}_0} p(x\, dx + y\, dy)\right]$$

$$= -\mathbf{k}\left[\frac{\rho}{2} \oint_{\mathscr{C}_0} V^2(x\, dx + y\, dy) + \rho \oint_{\mathscr{C}_0} (vx - uy)(u\, dy - v\, dx)\right] \tag{A.16}$$

Characteristics of the
Flow Fields of Two-dimensional
Source and Vortex Distributions

B.1 Source Distribution

Consider first a source distribution of uniform strength q along the X-axis from $x = l_1$ to $x = l_2$ (Fig. B.1). The velocity components at any

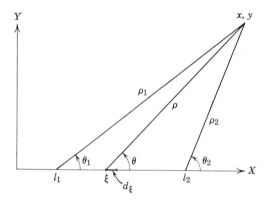

Fig. B.1

field point x, y are then given by

$$u(x, y) = \frac{q}{2\pi} \int_{l_1}^{l_2} \frac{x - \xi}{(x - \xi)^2 + y^2}\, dl = \frac{q}{2} \int_{l_1}^{l_2} \frac{\cos \theta}{\rho}\, d\xi \tag{B.1}$$

$$v(x, y) = \frac{q}{2\pi} \int_{l_1}^{l_2} \frac{y}{(x - \xi)^2 + y^2}\, dl = \frac{q}{2\pi} \int_{l_1}^{l_2} \frac{\sin \theta}{\rho}\, d\xi \tag{B.2}$$

where ρ and θ are local polar coordinates with origin at the source element (see Section 17.6).

612

As we can readily verify, we have

$$(x - \xi)^2 + y^2 = \rho^2$$

$$d\xi = - \frac{d\rho}{\cos \theta}$$

$$\frac{\cos \theta}{\rho} d\xi = - \frac{d\rho}{\rho}$$

$$\frac{\sin \theta}{\rho} d\xi = d\theta$$

Hence (B.1) and (B.2) yield

$$u(x, y) = - \frac{q}{2\pi} \int_{\rho_1}^{\rho_2} \frac{d\rho}{\rho} = \frac{q}{2\pi} \log \frac{\rho_1}{\rho_2} \tag{B.3}$$

$$v(x, y) = - \frac{q}{2\pi} \int_{\theta_1}^{\theta_2} d\theta = \frac{q}{2\pi} (\theta_2 - \theta_1) \tag{B.4}$$

where ρ_1, ρ_2, θ_1, θ_2 are as shown in Fig. B.1.

These relations completely determine the velocity field. Both velocity components vanish at infinity (we know this to begin with). The u-component is finite everywhere except at the edges of the distribution where ρ_1 or ρ_2 is zero. At the edges it becomes infinite. The v-component is finite everywhere.

Consider the situation along the Y-axis outside the distribution, that is, for $x < l_1$, and for $x > l_2$ with $y = 0$. The u-component has a nonzero value, the v-component is zero, for

$$\theta_1 = \theta_2 = \pi \quad \text{if} \quad x < l_1$$

and

$$\theta_1 = \theta_2 = 0 \quad \text{if} \quad x > l_2$$

Now consider a field point, such as P in Fig. B.2, which is just above the X-axis with x between l_1 and l_2. At such a point we have

$$u(x, 0_+) = \frac{q}{2\pi} \log \frac{\rho_1}{\rho_2}$$

and

$$v(x, 0_+) = \frac{q}{2\pi} (\pi - 0) = \frac{q}{2}$$

For a corresponding point Q just below the X-axis, we have

$$u(x, 0_-) = \frac{q}{2\pi} \log \frac{\rho_1}{\rho_2}$$

$$v(x, 0_-) = \frac{q}{2\pi}(-\pi - 0) = -\frac{q}{2}$$

Hence it follows that

$$u(x, 0_+) = u(x, 0_-) \tag{B.5}$$

$$v(x, 0_+) = -v(x, 0_-) = \frac{q}{2} \tag{B.6}$$

and

$$v(x, 0_+) - v(x, 0_-) = q \tag{B.7}$$

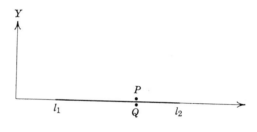

Fig. B.2

We conclude that across the source distribution the tangential velocity component, namely u, is continuous, whereas the normal component, namely v, is discontinuous. The magnitude of the discontinuity is simply equal to the source strength.

Now consider the case in which the source distribution is not of uniform strength but in which $q = q(x)$. The field of the nonuniform distribution is the superposition of the fields of individual source elements, each of which is of uniform strength. It follows, therefore, that the characteristics of the field of a nonuniform source distribution are identical to those of the field of a uniform distribution,* which were summarized in Section 17.6. The same considerations apply even if the source distribution is along a curved line.

---

* In superposition the infinite values for the u-components of the individual source elements will cancel one another all along the distribution except at the edges.

Appendix B

B.2   Vortex Distribution

Consider a vortex distribution of uniform strength γ along the X-axis from $x = l_1$ to $x = l_2$ (Fig. B.3). The velocity components at a field point x, y are given by

$$u(x, y) = \frac{\gamma}{2\pi} \int_{l_1}^{l_2} \frac{y}{(x - \xi)^2 + y^2} \, d\xi = \frac{q}{2\pi} \int_{l_1}^{l_2} \frac{\sin \theta}{\rho} \, d\xi \qquad (B.8)$$

$$v(x, y) = -\frac{\gamma}{2\pi} \int_{l_1}^{l_2} \frac{x - \xi}{(x - \xi)^2 + y^2} \, d\xi = -\frac{q}{2\pi} \int_{l_1}^{l_2} \frac{\cos \theta}{\rho} \, d\xi \qquad (B.9)$$

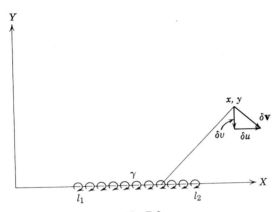

Fig. B.3

Comparing relations (B.8) and (B.9) with (B.1) and (B.2), we see that if the magnitudes of the source and vortex strengths are equal

$$u_\gamma = v_q$$
$$v_\gamma = -u_q \qquad (B.10)$$

where the subscripts γ and q, respectively, denote reference to the vortex and source distribution. On the basis of this correspondence we can describe the characteristics of the field of a nonuniform vortex distribution along the X-axis. Specifically the features are as follows:

1. Both velocity components vanish at infinity. They are finite everywhere outside the distribution.
2. The v-component is finite everywhere except at the edges of the distribution, where it becomes infinite. The u-component is finite everywhere.
3. Along the X-axis, outside the distribution, the v-component has a nonzero value, whereas the u-component is zero.

4. Across the distribution the u-component is discontinuous, whereas the v-component is continuous.

5. The magnitude of the discontinuity in u is equal to the local strength of the distribution. We have

$$u(x, 0_+) = -u(x, 0_-) = \frac{\gamma(x)}{2} \tag{B.11}$$

hence

$$u(x, 0_+) - u(x, 0_-) = \gamma(x) \tag{B.12}$$

Appendix C

Poisson's Integral Formulas

To introduce Poisson's formulas and conjugate Fourier series it is convenient and instructive to consider the problem of determining conjugate harmonic functions (see Section 14.8) or, equivalently, an analytic function of a complex variable in a domain given the boundary values of the function. Such a problem is the essence of airfoil theory.

We start with Cauchy's integral formula (Eq. 14.59). If $F(Z)$ is an analytic function in the domain bounded by a curve C, the value of the function at any point Z in the domain is given by*

$$2\pi i\, F(Z) = \int_C \frac{F(\alpha)}{\alpha - Z}\, d\alpha \qquad (C.1)$$

We may consider, without loss of generality, the boundary curve C as simply a circle of radius a with center at the origin (see Fig. C.1). A point on the circle is given by the complex number

$$\alpha = ae^{i\theta} \qquad (C.2)$$

We then have

$$d\alpha = i\alpha\, d\theta$$

Equation C.1 therefore becomes

$$2\pi\, F(Z) = \int_0^{2\pi} F(\alpha) \frac{\alpha}{\alpha - Z}\, d\theta \qquad (C.3)$$

Separating the real and imaginary parts of this formula yields the solution to our problem; that is, we will have succeeded in expressing the real and imaginary parts of $F(Z)$ in terms of their values on the boundary.

---

* In airfoil theory we are concerned with an exterior problem, but by suitable mapping it can be reduced to an interior problem. The following considerations are therefore equally valid in our case. The forms of Poisson's formulas and conjugate Fourier series naturally do not depend on whether an interior or exterior problem is considered.

We do this as follows:
  We observe that

$$\int_C \frac{F(\alpha)}{\alpha - a^2/\overline{Z}}\, d\alpha = 0 \qquad\qquad (C.4)$$

for the integrand is analytic within C; $\overline{Z}$ is the complex conjugate of Z. From (C.3) and (C.4) we obtain

$$2\pi F(Z) = \int_0^{2\pi} F(\alpha)\left(\frac{\alpha}{\alpha - Z} \pm \frac{\alpha}{\alpha - a^2/\overline{Z}}\right) d\theta \qquad\qquad (C.5)$$

This is the basic relation for further considerations.

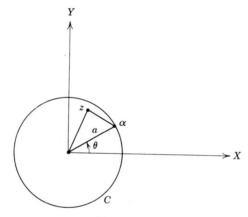

Fig. C.1

As may be verified we have the following relations:

$$\frac{\alpha}{\alpha - a^2/\overline{Z}} = -\frac{\overline{Z}}{\overline{\alpha} - \overline{Z}}$$

$$\frac{\alpha}{\alpha - Z} \pm \frac{\alpha}{\alpha - a^2/\overline{Z}} = 1 + 2i\frac{\text{Im}\,(\overline{\alpha}Z)}{|\alpha - Z|^2} \quad \text{for the plus sign in (C.5)}$$

$$= \frac{\alpha\overline{\alpha} - Z\overline{Z}}{|\alpha - Z|^2} \quad \text{for the negative sign in (C.5).}$$

Represent Z by

$$Z = re^{i\phi} \qquad\qquad (C.6)$$

We then obtain the following relations:

$$|\alpha - Z|^2 = a^2 + r^2 - 2ar\cos(\phi - \theta)$$
$$\text{Im}\,(\overline{\alpha}Z) = ar\sin(\phi - \theta)$$
$$\alpha\overline{\alpha} - Z\overline{Z} = a^2 - r^2$$

Using these relations and setting

$$F(Z) = f(r, \phi) + ig(r, \phi) \qquad (C.7)$$

and

$$F(\alpha) = f(a, \theta) + ig(a, \theta)$$
$$= f(\theta) + ig(\theta) \qquad (C.8)$$

we can separate (C.5) into its real and imaginary parts. In this way we obtain

$$2\pi f(r, \phi) = \int_0^{2\pi} f(\theta) \, d\theta - 2\int_0^{2\pi} g(\theta) \frac{ar \sin(\phi - \theta)}{a^2 + r^2 - 2ar \cos(\phi - \theta)} \, d\theta$$

$$= \int_0^{2\pi} f(\theta) \frac{a^2 - r^2}{a^2 + r^2 - 2ar \cos(\phi - \theta)} \, d\theta \qquad (C.9)$$

$$2\pi g(r, \phi) = \int_0^{2\pi} g(\theta) \, d\theta + 2\int_0^{2\pi} f(\theta) \frac{ar \sin(\phi - \theta)}{a^2 + r^2 - 2ar \cos(\phi - \theta)} \, d\theta$$

$$= \int_0^{2\pi} g(\theta) \frac{a^2 - r^2}{a^2 + r^2 - 2ar \cos(\phi - \theta)} \, d\theta \qquad (C.10)$$

These equations, known as *Poisson's integral formulas*, relate conjugate harmonic functions to their boundary values. The integrals $\int_0^{2\pi} f(\theta) \, d\theta$ and $\int_0^{2\pi} g(\theta) \, d\theta$ are constants and are denoted by f_0 and g_0, respectively. They are actually, as may be verified, proportional to the values of f and g at the center of the circle.

For further considerations we choose the following form of the formulas:

$$2\pi f(r, \phi) = \int_0^{2\pi} f(\theta) \frac{a^2 - r^2}{a^2 + r^2 - 2ar \cos(\phi - \theta)} \, d\theta \qquad (C.11)$$

$$2\pi g(r, \phi) = g_0 + 2\int_0^{2\pi} f(\theta) \frac{ar \sin(\phi - \theta)}{a^2 + r^2 - 2ar \cos(\phi - \theta)} \, d\theta \qquad (C.12)$$

Note that $f(\theta) \equiv f(a, \theta)$.

Equation C.12 may be used to obtain the value of g on the circle itself by putting $r = a$. Thus we have

$$2\pi g(\phi) \equiv 2\pi g(a, \phi) = g_0 + 2\int_0^{2\pi} f(\theta) \frac{\sin(\phi - \theta)}{1 - \cos(\phi - \theta)} \, d\theta$$

$$= g_0 + \int_0^{2\pi} f(\theta) \cot \frac{\phi - \theta}{2} \, d\theta \qquad (C.13)$$

This is known as *Poisson's integral*.

If $f(\theta)$ is an *even function*, that is, if

$$f(\theta) = f(-\theta) = f(2\pi - \theta)$$

we obtain

$$g(\phi) - g_0' = \frac{\sin \phi}{\pi} \int_0^\pi \frac{f(\theta)}{\cos \theta - \cos \phi} \, d\theta \tag{C.14}$$

where $g_0' = g_0/2\pi$. If $f(\theta)$ is an *odd function*, that is, if

$$f(\theta) = -f(-\theta) = -f(2\pi - \theta)$$

we obtain

$$g(\phi) - g_0' = \frac{1}{\pi} \int_0^\pi f(\theta) \frac{\sin \theta}{\cos \theta - \cos \phi} \, d\theta \tag{C.15}$$

The formulas (C.11) and (C.12) may be expressed in complex notation as

$$2\pi f(r, \phi) = \int_0^{2\pi} f(\theta) \frac{\alpha\bar{\alpha} - Z\bar{Z}}{|\alpha - Z|^2} \, d\theta$$

$$2\pi g(r, \phi) = g_0 + \int_0^{2\pi} f(\theta) \frac{2 \operatorname{Im}(\bar{\alpha}Z)}{|\alpha - Z|^2} \, d\theta$$

where it is recalled that

$$Z = re^{i\phi}$$

$$\alpha = ae^{i\phi}$$

Combining these relations into an analytic function $F(Z)$, we have

$$2 F(Z) \equiv 2\pi[f(r, \phi) + ig(r, \phi)]$$

$$= ig_0 + \int_0^{2\pi} f(\theta) \frac{\alpha\bar{\alpha} - Z\bar{Z} + 2i \operatorname{Im}(\bar{\alpha}Z)}{|\alpha - Z|^2} \, d\theta$$

$$= ig_0 + \int_0^{2\pi} f(\theta) \frac{\alpha + Z}{\alpha - Z} \, d\theta \tag{C.16}$$

Note that

$$2i \operatorname{Im}(\bar{\alpha}Z) = \bar{\alpha}Z - \overline{\bar{\alpha}Z}$$

$$= \bar{\alpha}Z - \alpha\bar{Z}$$

Equation C.16 not only puts Poisson's formulas into a simple form but represents at the same time the solution for an analytic function in terms of its boundary values.

In conclusion we point out that *if two functions are related by Poisson's integral formulas they form conjugate harmonic functions.*

Appendix D

Conjugate Fourier Series

In Appendix C we expressed conjugate harmonic functions in terms of integrals taken over their boundary values. Instead, it is possible to express these functions in terms of Fourier series whose coefficients depend only on the boundary values.

We start with Poisson's formula expressed by (C.16) and note that

$$\frac{\alpha + Z}{\alpha - Z} = 1 + 2\frac{Z}{\alpha}\left(1 - \frac{Z}{\alpha}\right)^{-1}$$

$$= 1 + 2\sum_{n=1}^{\infty}\left(\frac{Z}{\alpha}\right)^n$$

$$= 1 + 2\sum_{n=1}^{\infty}\left(\frac{r}{a}\right)^n e^{in(\phi-\theta)}$$

Note that Z/α is less than unity; hence (C.16) takes the form

$$2\pi F(Z) = ig_0 + \int_0^{2\pi} f(\theta)\frac{\alpha + Z}{\alpha - Z}\,d\theta$$

$$= ig_0 + 2\int_0^{2\pi} f(\theta)\left[\sum_1^{\infty}\left(\frac{r}{a}\right)^n e^{in(\phi-\theta)}\right]d\theta \qquad (D.1)$$

Expressing $e^{in(\phi-\theta)}$ in its trigonometric form and setting $F(Z)$ equal to $f(r, \phi) + ig(r, \phi)$, we find from (D.1) the following expressions:

$$f(r, \phi) = \frac{a_0}{2} + \sum_{n=1}^{\infty}\left(\frac{r}{a}\right)^n (a_n \cos n\phi + b_n \sin n\phi) \qquad (D.2)$$

$$g(r, \phi) = g_0' + \sum_{n=1}^{\infty}\left(\frac{r}{a}\right)^n (a_n \sin n\phi - b_n \cos n\phi) \qquad (D.3)$$

where $g_0' = g_0/2\pi$, and the coefficients are defined by

$$a_0 = \frac{1}{\pi} \int_0^{2\pi} f(\theta) \, d\theta \tag{D.4}$$

$$a_n = \frac{1}{\pi} \int_0^{2\pi} f(\theta) \cos n\theta \, d\theta \qquad n > 1 \tag{D.5}$$

$$b_n = \frac{1}{\pi} \int_0^{2\pi} f(\theta) \sin n\theta \, d\theta \tag{D.6}$$

Recalling that $f(\theta) \equiv f(a, \theta)$, we have thus expressed conjugate harmonic functions in terms of Fourier series whose coefficients depend on the boundary values.

A pair of Fourier series, such as those in (D.2) and (D.3), in which the coefficients of the cosine and sine terms are interchanged with appropriate changes of sign, is known as a *conjugate Fourier series*. Thus *any two conjugate harmonic functions, or, equivalently, any two functions related by Poisson's formulas, can be represented by conjugate Fourier series.*

For $r = a$, the expansions (D.2) and (D.3) reduce to

$$f(\phi) \equiv f(a, \phi) = \frac{a_0}{2} + \sum_1^{\infty} (a_n \cos n\phi + b_n \sin n\phi) \tag{D.7}$$

$$g(\phi) \equiv g(a, \phi) = g_0' + \sum_1^{\infty} (a_n \sin n\phi - b_n \cos n\phi) \tag{D.8}$$

When the *function* $f(\theta)$ *is an even* function, Poisson's integral for the conjugate function $g(\phi)$ is given by (C.14):

$$g(\phi) = g_0' + \frac{\sin \phi}{\pi} \int_0^{\pi} \frac{f(\theta)}{\cos \theta - \cos \phi} \, d\theta \tag{D.9}$$

The corresponding conjugate series are

$$f(\phi) = \frac{a_0}{2} + \sum_1^{\infty} a_n \cos n\phi \tag{D.10}$$

$$g(\phi) = g_0' + \sum_1^{\infty} a_n \sin n\phi \tag{D.11}$$

where the a_n's are given by (D.4) and (D.5); the b_n's are now zero. We note that the conjugate function g is an odd function.

When the *function* $f(\theta)$ *is an odd* function, Poisson's integral for $g(\phi)$ is given by (C.15):

$$g(\phi) = g_0' + \frac{1}{\pi} \int_0^{\pi} f(\theta) \frac{\sin \theta}{\cos \theta - \cos \phi} \, d\theta \tag{D.12}$$

The corresponding series are

$$f(\phi) = \sum_1^\infty b_n \sin n\phi \qquad\qquad \text{(D.13)}$$

$$g(\phi) = g_0' - \sum_1^\infty b_n \cos n\phi \qquad\qquad \text{(D.14)}$$

where the b_n's are given by (D.6). The a_n's are now zero. We note that the conjugate function g is an even function.

Some Integrals

1. We wish to show that

$$\int_0^\pi \frac{\cos n\phi}{\cos \phi - \cos \theta}\, d\phi = \pi \frac{\sin n\theta}{\sin \theta}, \qquad n = 0, 1, 2, \ldots \qquad \text{(E.1)}$$

Consider the integral

$$I = \int_0^\pi \frac{\cos n\phi}{\cos \phi - \cos \phi}\, \sin \theta\, d\phi$$

With the use of the formula

$$\frac{2 \sin \theta}{\cos \phi - \cos \theta} = \cot \frac{\theta - \phi}{2} + \cot \frac{\theta + \phi}{2}$$

the integral I becomes

$$I = \frac{1}{2} \int_0^\pi \left(\cot \frac{\theta - \phi}{2} + \cot \frac{\theta + \phi}{2} \right) \cos n\phi\, d\phi$$

$$= \frac{1}{2} \int_{-\pi}^\pi \cot \frac{\theta + \phi}{2} \cos n\phi\, d\phi$$

$$= \frac{\cos n\theta}{2} \int_{\theta-\pi}^{\theta+\pi} \cos nx \cot \frac{x}{2}\, dx + \frac{\sin n\theta}{2} \int_{\theta-\pi}^{\theta+\pi} \sin nx \cot \frac{x}{2}\, dx$$

With the result

$$\int_{\theta-\pi}^{\theta+\pi} \cos nx \cot \frac{x}{2}\, dx = 0, \qquad \text{for } n = 1, 2, 3, \ldots$$

(note that the integrand is an odd function) the integral I reduces to

$$I = \frac{\sin n\theta}{2} \int_{\theta-\pi}^{\theta+\pi} \sin nx \cot \frac{x}{2}\, dx$$

$$= \frac{\sin n\theta}{2} J_n$$

Appendix E

where J_n denotes the integral on the right. To evaluate this we note that for $n > 1$

$$J_n - J_{n-1} = \int_{\theta - \pi}^{\theta + \pi} [\sin nx - \sin (n-1)x] \cot \frac{x}{2} \, dx$$

$$= \int_{\theta - \pi}^{\theta + \pi} \cos nx \, dx + \int_{-\pi}^{\pi} \cos (n-1)x \, dx$$

$$= 0$$

Hence it follows that

$$J_n = J_{n-1} = \cdots = J_1 \quad \text{for} \quad n = 1$$

where

$$J_1 = \int_{\theta - \pi}^{\theta + \pi} \sin x \cot \frac{x}{2} \, dx$$

$$= 2 \int_{\theta - \pi}^{\theta + \pi} (1 + \cos x) \, dx$$

$$= 2\pi$$

Finally we obtain

$$I = \pi \sin n\theta$$

and

$$\int_0^\pi \frac{\cos n\phi}{\cos \phi - \cos \theta} \, d\phi = \frac{I}{\sin \theta} = \pi \frac{\sin n\theta}{\sin \theta}, \qquad n = 0, 1, 2, \ldots$$

2. We wish to show that

$$I \equiv \int_0^\pi \frac{\sin n\phi \sin \phi}{\cos \phi - \cos \theta} \, d\phi = -\pi \cos n\theta \qquad \text{(E.2)}$$

Using the formula

$$\sin n\phi \sin \phi = \tfrac{1}{2} \cos (n-1)\phi - \tfrac{1}{2} \cos (n+1)\phi$$

we express the integral I as

$$I = \frac{1}{2} \int_0^\pi \frac{\cos (n-1)\phi}{\cos \phi - \cos \theta} \, d\phi - \frac{1}{2} \int_0^\pi \frac{\cos (n+1)\phi}{\cos \phi - \cos \theta} \, d\phi$$

The values of the integrals on the right are readily found by using (E.1). Thus we obtain

$$I = \frac{\pi}{2 \sin \theta} [\sin (n-1)\theta - \sin (n+1)\theta]$$

$$= -\pi \cos n\theta$$

3. We wish to find

$$I(x, 0_\pm) \equiv \lim_{y \to 0_+ \text{ or } 0_-} \int_0^l f(\xi) \frac{y}{(x - \xi)^2 + y^2} \, d\xi \quad \text{for} \quad 0 \le x \le l$$

The evaluation of the integral on the right, as pointed out previously, offers no difficulty as long as the field point x, y is outside the strip $0 \leq x \leq l$, $y = 0$. Care is necessary to evaluate it for points at the strip. We write

$$I(x, 0_{\pm}) = \lim_{\substack{y \to 0_{\pm} \\ \varepsilon \to 0}} \left(\int_0^{x-\varepsilon} + \int_{x-\varepsilon}^{x+\varepsilon} + \int_{x+\varepsilon}^l \right)$$

For $y = 0$ the integrals from 0 to $x - \varepsilon$ and from $x + \varepsilon$ to l vanish. They are evaluated, since $\xi \neq x$, by setting $y = 0$.

Therefore we have

$$I(x, 0_{\pm}) = \lim_{\substack{y \to 0_{\pm} \\ \varepsilon \to 0}} \int_{x-\varepsilon}^{x+\varepsilon} \frac{f(\xi)}{y[1 + (x - \xi)^2/y^2]} \, d\xi$$

In view of the limit, $f(\xi)$ may be assumed to be constant in the interval $x - \varepsilon$ to $x + \varepsilon$ and may be set equal to $f(x)$. Making the substitution

$$\frac{x - \xi}{y} = \lambda$$

we may rewrite the above relation as

$$I(x, 0_{\pm}) = f(x) \lim_{\substack{y \to 0_{\pm} \\ \varepsilon \to 0}} \int_{-\varepsilon/y}^{\varepsilon/y} \frac{d\lambda}{1 + \lambda^2}$$

$$= 2f(x) \lim_{\substack{y \to 0_{\pm} \\ \varepsilon \to 0}} \tan^{-1} \frac{\varepsilon}{y}$$

In taking the limits we let y and ε go to zero in such a way that $\varepsilon/y \to \infty$. Also, we distinguish the limits $y \to 0_+$ and $y \to 0_-$. In this way we obtain

$$I(x, 0_{\pm}) = \pm \pi f(x) \qquad \qquad \text{(E.3)}$$

4. We wish to find the value of

$$I(x, r) \equiv \int_0^l f(\xi) \frac{r}{[(x - \xi)^2 + r^2]^{3/2}} \, d\xi$$

for small values of r, that is, for $r \to 0$ when x lies in the interval 0 to l. To do this we first note that

$$\frac{\partial}{\partial \xi} \frac{x - \xi}{[(x - \xi)^2 + r^2]^{1/2}} = - \frac{r^2}{[(x - \xi)^2 + r^2]^{3/2}}$$

Therefore we write

$$rI(x, r) = - \int_0^l f(\xi) \frac{\partial}{\partial \xi} \frac{x - \xi}{[(x - \xi)^2 + r^2]^{1/2}} \, d\xi$$

Appendix E

Integration by parts leads to

$$rI(x, r) = -f(\xi)\frac{x - \xi}{[(x - \xi)^2 + r^2]^{1/2}}\bigg|_0^l + \int_0^l f'(\xi)\frac{x - \xi}{[(x - \xi)^2 + r^2]^{1/2}}\, d\xi$$

where

$$f'(\xi) = \frac{df(\xi)}{d\xi}$$

Note that $[(x - \xi)^2 + r^2]^{1/2}$ implies positive square root. We then have

$$rI(x, r) = -f(l)\frac{x - l}{[(x - l)^2 + r^2]^{1/2}} + f(0)\frac{x}{(x^2 + r^2)^{1/2}}$$
$$+ \int_0^l f'(\xi)\frac{x - \xi}{[(x - \xi)^2 + r^2]^{1/2}}\, d\xi$$

We are obliged to speak about the limit of rI, and not of I alone, as $r \to 0$; for $I \to \infty$ as $r \to 0$, if the right hand member is nonzero as $r \to 0$.

For small r with x between 0 and l, that is for $r < |x - l|$ and for $r < x$, we obtain, noting $(x - l)/|x - l| = -1$,

$$(rI)_{r\to 0} \simeq f(l) + f(0) + \int_0^l f'(\xi)\frac{x - \xi}{|x - \xi|}\, d\xi$$

The integral in this relation is evaluated as follows

$$\int_0^l f'(\xi)\frac{x - \xi}{|x - \xi|}\, d\xi = \lim_{\varepsilon \to 0}\left[\int_0^{x-\varepsilon} f'(\xi)\, d\xi - \int_{x+\varepsilon}^l f'(\xi)\, d\xi\right]$$
$$= \lim_{\varepsilon \to 0}[f(\xi)\,|_0^{x-\varepsilon} - f(\xi)\,|_{x+\varepsilon}^l]$$
$$= 2f(x) - f(l) - f(0)$$

Note that for $0 \le \xi < x$, $(x - \xi)/|x - \xi| = 1$, and for $x < \xi \le l$, $(x - \xi)/|x - \xi| = -1$.

Finally, we obtain

$$(rI)_{r\to 0} \simeq 2f(x) \qquad\qquad (E.4)$$

Index

Acceleration potential, 229, 243
Acyclic motion, 259
  steady, 312
    two-dimensional, 359
  unsteady, 278
Aerodynamics, 1, 54, 55
  center of, 472
Airfoil, 33, 35, 390
  characteristics of, 516
  circular-arc, 480
  flat-plate, 480, 514
  lift coefficient for, 516
  moment on, 471
  pressure distribution on, 474, 516
  problem of the, 466
  profiles of, 390
  symmetrical, 481, 500
  velocity distribution on, 474
Airplane, wings of, 1
Airshiplike body, 1, 26
Analytic function, 410, 414
  unlimited differentiability of, 437
Angle, of attack, 33, 35, 51, 467
  induced, 544
  of no-lift, 485
Angular momentum, rate of change of, 304
Apparent mass, additional, 291, 301
  tensor, 297, 301
Arbitrary body, flow past, 344
  force on, 354
  of revolution, flow past, 343
Aspect ratio, 538
  effect on lift, 553
Axisymmetric flow, over closed bodies of revolution, 327
  over slender bodies of revolution, 331

Barriers, 257, 274
Bernoulli equation, the, 226, 230
  along a streamline, 224
  unsteady, 226

Biot-Savart law, 526
Blasius, H., 46
  distribution, 49
  relations, 453
Bluff body, 27, 30, 33, 55
Body forces, 148, 149
Body of revolution, 35, 55
  at an angle of yaw, 582
  axisymmetric flow past, 342, 573
  cross flow past, 342, 573, 578
  forces on, 584
  lateral flow past, 342, 573, 578
  moment on, 587
  slender, moment coefficient for, 588
  theory for, 568
Boundary conditions, 249
  for slender body approximation, 575, 577, 579
  for slender body of revolution, 572
  linearized form of, 495
  transfer of, 495
Boundary layer, 40, 41, 43, 46, 49, 53
  concept of Prandtl, 40, 41, 48, 52, 54, 55
  edge of, 45, 46, 47, 53
  flow of, characteristics of, 54
  laminar, 46, 49, 51
    characteristics of, 41
  separation of, 50
  theory for, 54, 55
  thickness of, 41, 42, 43, 46, 54
    turbulent, 49
  turbulent, 47, 49
Boundary value problem, 249
Branches, 416
Branch point, 416
Bulk modulus, 7
Bulk properties, 2
Bulk viscosity, coefficient of, 7

Cambered airfoil, 506
Camber function, 498
Camber line, 467

Cauchy, integral formula of, 435, 437
  integral theorem of, 426, 428
  principal value of, 504
Cauchy-Riemann, conditions, the, 411,
  412, 427
  equations, the, consequences of, 413
Center of pressure, 474
Chord, 33, 467
  mean, 538
  section, 535
Circuits, irreconcilable, 255
  irreducible, 254
  reconcilable, 255
  reducible, 254
Circulation, 235, 355, 450, 468, 518
  generation of, 393
  line integrals, 111
  of **A**, 113, 133
  rate of change of, 239
  theory of lift for, mathematical prob-
    lem of, 395
  problem of, 389
Circulatory flow, with concentrated
    vorticity, 380
  with constant vorticity, 376, 378
  without rotation, 379
Circulatory motion, 30
Coefficients, local friction, 49
Complex conjugate, 404
Complex function, geometrical significance of, 416
Complex integrals, 425
Complex numbers, 404
  algebra of, 404
  exponential form of, 407
  imaginary part of, 405
  nomenclature of, 404
  polar form of, 407
  real part of, 405
Complex plane, 405
Complex potential, 444, 445
Complex variable, 404
  function of, 404, 409
  introduction of, 402
Complex velocity, 444, 445
Component, of a curl as the circula-
    tion, 125
  of grad ϕ, in Cartesians, 102
  in cylindrical coordinates, 102
  in spherical coordinates, 102

Compressibility, 3, 4, 8
  effects of, 35
Conduction, heat, 4, 10
Connectivity, 252
Conservation, equations of, 199
  of energy, 201
    equation of, 185, 202
  of mass, 199
    equation of, 181, 183
  of momentum, 199, 201
Conservative field, 229
Continuous medium, 2
Continuum, 2
Convection, 10
  currents, 4
  forced, 10
  free, 10
  terms, 222
Convective derivative, 175, 177
Convention rule, right-handed screw,
  67
Coordinates, curvilinear, 46, 78
  general orthogonal, 79, 100, 105,
    140
  cylindrical, 79
    changes in unit vectors of, 94
  spherical, 81
Cross product; see Outer product
Curl **A**, 107, 109, 120, 145
  alternative definition for, 127
Cyclic motion, 260

D'Alembert's paradox, 296, 354
Del, in Cartesians, 108
  in cylindrical coordinates, 108
  in spherical coordinates, 108
Del operator, 108, 145
De Moivre's theorem, 408
Density, 2
  material derivative of, 181
Derivative, directional, 98
Dhawan, S., 46
Dilatation, 183
Dimensional analysis, 21
Dipole, 335
Dirichlet's problem, 364
Discontinuity, moving, 213
  normal, 220
  tangential, 213, 219
Disturbance, potential of, 284

Disturbance, velocity of, 284
Divergence, of **A**, 106, 107, 119, 120, 145
  theorem of, 131, 135
Dot product; *see* Scalar product
Doublets, 335, 369, 448
  axis of, 335
  distribution of, 578
    in a uniform stream, 342
  moment of, 335
  potential, 333, 335
  strength of, 335
Doubly-connected region, 256, 273
Drag, 1, 545
  induced, 547
  spanwise distribution of, 562
  viscous, 43
Dryden, H., 13
Dyadic product, 200

Eddy, 380
Edge, leading, 33, 35, 466, 535
  trailing, 33, 35, 390, 466, 535
Element, differential volume, 143
  surface, 143
  volume, 143
Energy flux vector, 202
Enthalpy, 3
Entropy, 3
Equations, characteristic, 3
  of an ideal fluid, 190
  of an inviscid compressible fluid, 187
  of change, 198
  of continuity, 183
  of discontinuous motion, 210
  of energy, 18, 184, 185, 189
    mechanical, 185, 230
    thermodynamic form of, 186
    total, 185
  of irrotational motion, 245
  of mass, 183
  of motion of a fluid, 179
  of state, 3, 4, 186
Equipotential line, 363
Eulerian equations, 160, 175
Eulerian method, 158, 159
Euler's equation, 178, 181, 221, 223, 226
Extension, 183

Field, 87
  irrotational, 136, 137
  nonstationary, 87
  point-of-view, 158
  stationary, 87
  steady, 87
  unsteady, 87
Flight, science of, 1
Flow, in a boundary layer, 47
  inner, 53, 54
  laminar, 10
  mapping of, 456
  outer, 54
  over a cylinder, 372
  past a circular cylinder with circulation, 382
  patterns, 35, 55
  reverse, 49
  Reynolds number, 52
  separation of, 30, 50
  transition to turbulent, 47
  turbulent, 10
  with circulation past arbitrary cylinder, 389
  with vorticity, some features of, 518
Fluctuations, turbulent, 48
Fluid, 1
  curve, 239
  definition of, 154
  element, 2, 158
    acceleration of, 179
    general motion of, 231
  incompressible homogeneous, 188
  inviscid, 41, 54
  mechanics, science of, 1, 55
  motion, description of, 158
  Newtonian, 7
  particles, 2, 158
  region, 175, 203, 205
Fluid-fluid boundary, 190
Force, 21
  field, irrotational, 227
  frictional, 54
  gravitational, 14
  inertial, 14, 41, 42
  internal, 149
  on a translating body of arbitrary shape, 291
  potential of, 229
  surface, 148

Force, viscous, 15, 41, 42
Form drag, 353
Fourier series, conjugate, 504, 511
Free boundary, 190
Free surface, 190, 219
    condition at, 193, 249
Friction, 5
    coefficient of, local skin, 46
        skin, 46
    internal, 5
    local skin, 46
Frictional force, 54
Frictional stresses, 6, 154
Froude number, 14, 16, 23, 26
Functions, harmonic, 413
    conjugate, 413

Gasdynamics, 1
Gauss, theorem of, 131
Glauert, method of, 565
Gradient, integral definition of, 116
    of **A**, 98, 99, 101, 144
    theorem, 131
Gravity, effect of, 14, 26
Green's theorem, 134
    in the first form, 135
    in the second form, 136

Heat, conduction of, 4, 10
    flux, 4
    transfer of, 10, 21, 55
Helmholtz's theorem, 239
    consequences of, 523
Hydraulics, 55
Hydrodynamics, 1, 54, 55
Hydrostatic pressure, 153, 154
Hypersonic range, 40

Ideal fluid, 55
    boundary conditions for, 190
    theory for, 55
Imaginary unit, 403
Impulse, 247, 297, 301
    potential of, 246
Impulsive pressure, 297
Incompressibility, condition of, 187
    consequences of, 188
Incompressible flow, 40
    steady, in a laminar boundary layer, 46
Incompressible fluid, 3, 25

Incompressible inviscid fluid, 55
Inertial force, 14
Infinitesimal distance, 142
Infinite wing, 390
Infinity, conditions of, 194, 263
    velocity components at, 268
Initial conditions, 190
Inner product, 62, 83
Integral, surface, 113
    volume, 115, 116
Integral form of conservation equations for inviscid fluid, 203
Internal energy, 3, 184
Internal forces, 149
    state of, 151
Inviscid flow theory, 54, 55
Inviscid motion, 55
Irrotationality, condition of, 244
Irrotational motion, in a doubly-connected region, 259
    in a simply-connected region, 258
    properties of, 269

Joukowski, airfoils, the, 477, 479
    properties of, 484
    transformation, the, 477

Kármán-Trefftz airfoils, 488
Kelvin, circulation theorem of, 243
    consequences of, 523
Kinematical relation, 183
Kinetic energy, 184, 301
Kuethe, A. M., 13
Kutta condition, 390, 393, 395, 468
Kutta-Joukowski theorem, 359, 389, 453

Lagrangian, description, the, 159
    equations, the, 159
    method, the, 158
    variables, the, 160
Lamb, H., 159
Laminar motion, 10, 48
Laminar sublayer, 49
Laplace operator, 133, 134, 145
    in Cartesians, 134
    in cylindrical coordinates, 134
    in spherical coordinates, 134
Laplace's equation, 138
    general solution of, in two dimensions, 402

Laplacian; *see* Laplace operator
Laurent series, 440
Laws of conservation, of energy, 178, 230
  of mass, 178
Legendre, differential equation of, 286
  functions of, 286
Liepmann, H. W., 46
Lift, 1, 388, 545
  coefficient of, 470
  distribution of elliptic, 548, 551
  force, 358
  spanwise distribution of, 562
Lifting line, 540
Lines, coordinate, 78
  field, 87
Local derivative, 175, 177
Lotz, method of, 565

Mach number, 14, 17, 23, 35
  critical, 35, 40
  local, 40
Magnus effect, 388
Mapping, notion of, 416
Mass flux vector, 199
Material derivative, 175, 177
Method of separation of variables, 313
Mid-chord line, 535
Mises airfoils, 487
Moment, about aerodynamic center, 472
  at no-lift, 472
  coefficient of, 474
  impulse of, 307
  on a translating body, 304
  rolling, 547
  yawing, 547
Momentum flux tensor, 201
Motion, acyclic, 259
  steady, 312
  unsteady, 278
  circulatory, 30
  cyclic, 260
  impulsively generated, 246
  inviscid, 55
  irrotational, 231, 244
  laminar; *see* Laminar motion
  potential, 244
  steady rotational, 223
  turbulent; *see* Turbulent motion

Motion, unsteady, 162
Multiply-connected region, 256

Neumann exterior problem, 252, 263, 275
  solution of, 272
Newton's second law of motion, 14, 157, 178
Nikuradse, J., 46
No-lift angle, 470
Normal stress, 150
No-slip condition, 53

Outer problem, 54
Outer product, 64
Outflow of **A**, 114

Parallelogram law, 59
Path, irreconcilable, 253
  reconcilable, 253
Path line, 162
Particle derivative, 177
Particle point of view, 158
Pascal's law, 153
Pfaffian differential equations, the, 165
Plan area, 538
Point, field, 87
  general coordinates of, 77
Point vortex, 380
Poisson, equation of, 139, 525
  integral formula of, 505, 511
Polar diagram, 550
Polynomial solutions, 313
Position, scalar function of, 85
  vector function of, 85, 86
Potential, 229
  single-valued, 249
  total, 284
Potential fields, 229
Prandtl, L., 40, 46, 50, 162
  boundary layer concept of; *see* Boundary layer concept
  equation of, 545
  lifting line theory of, 540
  number, the 20, 21
  theory of, 538
    fundamental relation, 544
Pressure, 2, 3, 4, 154
  coefficient, 349
  drag, 353
  force to estimate the, 14

Pressure, gradient adverse, 50, 51, 55
    relation for, in simple theory, 499

Quarter-chord point, 483

Rankine bodies, 330
Rankine ovals, 372
Rate of strain tensor, 233
Reference frame, 96, 165
Residue theorem, 441
Retardation layer, 41
Reynolds, O., 10
    number, the, 14, 16, 23, 26, 30, 35,
        42, 47
        critical, 47
        large, 40, 41, 51
Rotation, 233
    of **A**, 107

Scalars, 56
    components of, 73
    differentiation of a vector function
        of, 89
    fields of, 87
    function of, Laplacian of, 145
    integration of a vector function of,
        111
    potential of, 137
    product of; see Inner product
    triple product of, 70, 84
Scale factors, 143
Schlichting, H., 50
Separation, 35, 49, 55
    bubble, 30
    lines, 30
    point, 30, 50
    possible location of, 55
    region, 30
Shadowgraph pictures, 40
Shearing stress, 151
Shear stress, local, 49
Shear viscosity, coefficient of, 6
Shock wave, 35, 40, 220
    detached, 35
Simply-connected region, 256
Singularity method, 313
Sink, 319
Skew product; see Outer product
Skin friction, 55
Skin-friction drag, 54
Slender body theory, 568

Slip condition, 54
Solenoidal field, 137, 138
Solid-fluid boundary, 190
    condition at, 191, 249
Sommerfeld, A., 159
Sound, speed of, 2, 16
Source, 319, 366, 448
    distribution of, 574
    in a uniform flow, 322
    potential of, 317, 319
    strength of, 320, 450
    two-dimensional, 366
Span, 535
Spherical coordinates, changes in unit
        vectors of, 94
Stagnation point, 172, 315
Stagnation streamline, 315
Starting vortex, 393
    development of, 393
Static equilibrium, 4, 6, 13
Stationary discontinuity in a steady
        flow, 210
Steady motions, 162
Stokes, stream function of, 172
    theorem of, 131, 133
Strain, of **A**, 106
    rates of, in a fluid, 7, 154
Stream function, 165, 167, 276, 362
    for axisymmetric motion, 170
    for incompressible flow, 194
    for two-dimensional flow, 168
Streamline, 26, 87, 162
    differential equation for, 162
    dividing, 50, 326
    separation, 50
Streamlined body, 30, 51
    stationary, 52
Stream surface, 164
Stream tube, 164
Stress, 3
    at a point, 149
    components of, 152
    in a fluid, 148
        at rest, 153
        in motion, 154
    normal, 3
    state of, 151
        in an inviscid fluid in motion, 155
    tangential, 3
    tensor, 152

Index

Stress, vector, 149
Substantial derivative; *see* Material
    derivative
Superposition, principle of, 223
Supersonic airfoil, 35
Supersonic flow, 40
Surfaces, curvilinear coordinate, 78
   level, 87, 103
   of discontinuity, 194, 210
   of tangential discontinuity, 194
Surface forces, 148
Surface integral, 113
System of orthogonal vectors, 76

Tangential stress, 151
Taylor series, the, 438
Tietjens, O. G., 50, 162
Temperature, 2, 4
Tensor, 56
   gradient of **A**, 106
   second-order, 104
Thermal conduction, 4
Thermal conductivity, 4
   coefficient of, 4
Thermodynamic equilibrium, 4
Thermodynamics, first law of, 186
   properties of, 2
Thermodynamic system, simple, 186
Thickness function, 498
Thin airfoil theory, 492
Topological notions, 252
Trailing edge, mapping at, 467
Trailing-edge angle, 468, 479
Transformations, 417
   by analytic functions, 420
   conformal, 422
   critical points of, 422
   fixed points of, 418
   local scale factors of, 421
Translation, permanent, 311
Transonic flow, 40
Transonic range, 40
Trefftz, method of, 554
   plane, the, 554
      problem in, 557
Turbulent motion, 10, 41, 47, 48

Unsteady motions, 162

V, curl of, 124
   rotation of, 124

Vectors, 56
   addition of, 58
   analysis of, 84
   axial, 69
   bound, 58
   changes in the reference unit, 145
   collinear, 58
   components of, 72
   coplanar, 58
   definition of, 60
   field, 87
   free, 58
   localized, 58
   orthogonal, 76
   plane area as, 64
   polar, 69
   potential of, 138, 229, 246, 276, 525
      for incompressible flow, 196
   position, 57
   product of, 64, 83
   representation of, 57
   specification of, 74
   subtraction of, 58
   triple product of, 70, 84
   unit, 61, 140
   vortex, 124
   zero, 61
Velocity, angular, 235
   distribution of, across boundary layer,
      46, 49
   downwash, 540
   field of, induced, 547
   total, 284
Virtual mass, 291, 301
   tensor, 301
Viscosity, 5
   coefficient of, 7, 154
   flow theory of, 54
   force of, 41
      to estimate magnitude of, 15
   kinematic, 8, 43
   stresses of, 6, 43, 154
Vortex, 30, 448
   distribution, velocity field due to,
      524
   filament, 381, 519
   flow, 378
   free, 380
   line, 381, 519
   motion of, theorems of, 524

Vortex, sheet, 219, 508, 523
  bound, 539
  free, 540
  Kármán's, 30
  trailing, 539
  strength of, 381
  surface, 219, 519
  trailing, 538
  tube, 519
    strength of, 520, 522
  vector, 235
Vorticity, 30, 124, 133, 180, 233, 235, 518
  field of, 520
  material rate of change of, 238
  rate of change of, 236

Vorticity, spatial conservation of, 520, 522

Wings, arbitrary solution for, 554
  components of the moment acting on, 545
  elliptic, 548, 551
  finite, 535
  forces on, 562
  moments on, 562
  rectangular, 535
  straight, 535
  swept, 535
  tapered, 535
  theory of, 535
  twisted, 535
  section of, 535